THE LEGO® MINDSTORMS® NXT 2.0 DISCOVERY BOOK

THE LEGO® MINDSTORMS® NXT 2.0 DISCOVERY BOOK

a beginner's guide to building and programming robots

laurens **valk**

14 13 12 11   3 4 5 6 7 8 9

ISBN-10: 1-59327-211-1
ISBN-13: 978-1-59327-211-1

Publisher: William Pollock
Production Editor: Riley Hoffman
Cover and Interior Design: Octopod Studios
Cover Photograph: Jochem de Klerk
Technical Reviewers: Damien Kee, Martijn Boogaarts, and Richard Li
Copyeditor: Kim Wimpsett
Compositor: Riley Hoffman
Proofreaders: Nancy Sixsmith and Xander Soldaat
Indexer: Valerie Haynes Perry

For information on book distributors or translations, please contact No Starch Press, Inc. directly:
No Starch Press, Inc.
38 Ringold Street, San Francisco, CA 94103
phone: 415.863.9900; fax: 415.863.9950; info@nostarch.com; www.nostarch.com

Library of Congress Cataloging-in-Publication Data
Valk, Laurens.
The LEGO Mindstorms NXT 2.0 discovery book : a beginner's guide to building and programming robots / Laurens Valk.
p. cm.
Includes index.
ISBN-13: 978-1-59327-211-1
ISBN-10: 1-59327-211-1
1. Robots--Design and construction--Popular works. 2. Robots--Programming--Popular works. 3. LEGO toys. I. Title.
TJ211.15.V353 2010
629.8'92--dc22
2010011157

about the author

Laurens Valk is a member of the MINDSTORMS Community Partners (MCP), a group of MINDSTORMS enthusiasts who help test and develop new NXT products. He has been inventing robots with the MINDSTORMS NXT sets since their introduction. Laurens enjoys designing robots that can be built with just one NXT set, making it easy for MINDSTORMS fans worldwide to follow his building instructions. One of his robot designs, Manty, appears on the back of the NXT 2.0 set's box as a bonus robot. He is a coauthor of *LEGO MINDSTORMS NXT One-Kit Wonders* (No Starch Press) and a contributor to *The Unofficial LEGO MINDSTORMS NXT 2.0 Inventor's Guide* (No Starch Press). He's also a contributor to the popular NXT STEP blog (*http://thenxtstep.blogspot.com/*). Laurens lives in the Netherlands, where he studies Mechanical Engineering at Delft University of Technology, and he maintains his website about robotics at *http://www.laurensvalk.com/*.

about the technical reviewer

Dr. Damien Kee holds a PhD in robotics and a bachelor's degree in electrical engineering, both from the University of Queensland, Australia. He has built a wide variety of robots, from maze-solving mice to humanoids to robots that dispense traffic cones. Damien has been heavily involved with the RoboCup Junior competition since 2001, and in 2009 he was elected chairman of RoboCup Junior Australia and technical chair of the RoboCup Junior International Rescue Committee. Since 2003, Damien has been conducting robotics workshops for teachers, educators, and students worldwide, and he has written several teacher resource books. He is a member of the MINDSTORMS Community Partners; a contributor to the NXT STEP blog; and editor in chief of The NXT Classroom (*http://theNXTclassroom.com/*), a website providing resources and support for teachers.

brief contents

contents in detail

PART II BUILDING AND PROGRAMMING ROBOTS WITH SENSORS

7
using the touch, color, and rotation sensors

8
shot-roller: a robotic defense system

PART IV ADVANCED ROBOT PROJECTS

acknowledgments

The book you hold in your hands is the result of more than a year of hard work, and I would never have completed writing it without the help of many others. First, I'd like to thank Fay Rhodes and Jim Kelly for introducing me to book writing. If I hadn't had the opportunity to coauthor *LEGO MINDSTORMS NXT One-Kit Wonders* with them in 2008, I might never have considered writing my own book, especially in a language that is not my native tongue.

Next, I'd like to thank the LDraw community for developing many tools required to make clear building instructions, especially Philippe Hurbain for making detailed 3D drawings of the NXT set components; Travis Cobbs for developing LDView to visualize the building steps; and Kevin Clague for developing LPUB4, which allowed me to organize my building instructions in a way that would make them easy to follow. Thanks too to John Hansen for letting me use his tool to take screenshots of the NXT display.

Thanks also to all who have been directly involved in this book: Richard Li and Martijn Boogaarts for carefully testing my robots and their building instructions; Jochem de Klerk for taking the cover photograph; Micah Edelblut for taking the photographs used in Figure 9-4, Figure 10-1, and Figure 10-16; my parents for their advice on how to manage working on this book; and Xander Soldaat for proofreading the book for technical accuracy. Special thanks go to technical editor Damien Kee for reading all of the chapters early on to make sure they were understandable and clear.

Further, many thanks go to the people at No Starch Press, who have been wonderful colleagues in the past year. I especially thank William Pollock for converting the text in this book into clear and smooth English and for his constructive criticism to make every section in this book useful and worth reading. I also thank Riley Hoffman for managing this project and for making all the pages in this book look as good as they do.

Thanks to everyone else who supported me while I worked on this book. My friends and the NXT builders eagerly awaiting the publication of this book encouraged me to complete this book and kept me going until it was finished.

Finally, thanks to the LEGO company for LEGO MINDSTORMS NXT, the wonderful product on which this book is based. Not only is the NXT fun to play with, but it brings people all over the world together who otherwise might never have met.

introduction

You probably already know that LEGO MINDSTORMS NXT 2.0 is a robotics kit that lets you build and program your own robots. I first became involved with MINDSTORMS in 2005 when I was 13 years old. Back then I had just enough money to buy a MINDSTORMS Robotics Invention System, the version available at that time. Thus began my new hobby, and as time passed, I became more and more involved in the world of MINDSTORMS. The result is the book you're holding (published in 2010). Its purpose is to help you explore the possibilities with MINDSTORMS in hopes that you'll have just as much fun with this robotics kit as I have.

why this book?

The LEGO MINDSTORMS NXT 2.0 robotics kit includes numerous parts and plans for four robots that you program using a computer. I think you'll find that it's a lot of fun to build and program the robots, but the going can be a bit rough when you're just getting started. The kit provides you with the tools you need to make the robots work, but the kit's user guide covers only a fraction of what you need to know to build and program your own robots.

This book is designed as a guidebook to help you discover the power of the LEGO MINDSTORMS NXT 2.0 robotics kit; you'll learn the skills you'll need to make your robots really do what you want them to do. In other words, with this book, your robotics kit won't end up gathering dust on the shelf!

is this book for you?

This book assumes no previous experience with either building or programming LEGO MINDSTORMS. As you read, you'll move from basic to advanced programming and build increasingly sophisticated robots. New users should begin in Chapter 1 and then follow the step-by-step instructions in Chapter 2 to build and program a basic robot, while more experienced MINDSTORMS users might simply start with a chapter they find challenging and move on from there. The advanced programming section in Part III and the robot designs in Part IV will be especially interesting for more advanced readers.

how does this book work?

Although you could use it as such, this book isn't intended as a reference manual; it's more like a workbook. I've mixed together building, programming, and robotic challenges to avoid long, theoretical chapters that can be hard to wade through. For example, you'll learn basic programming techniques at the same time that you learn to make your first robot move, but you'll learn about sensors as you build a new robot. The reason for this approach is that I think that *doing* is the best way to learn how to build and program MINDSTORMS robots.

the discoveries

As you learn to program your robots and as you build the robots in this book, you'll notice many "Discovery" sections throughout the chapters. Each new programming technique is supported with examples and these discoveries. I encourage you to try the discoveries and not just take what's written for granted.

This book includes 72 programming discoveries to help you build your programming skills and 15 building discoveries, showing you how to further expand on the book's robot designs and get you started creating your own original robots. If you have trouble with any of the discoveries, visit the book's companion website (*http://www.discovery.laurensvalk.com/*) to ask questions or to share your experiences.

what to expect in each chapter

The book is split into four parts. Here's a brief overview of each part. Some of the terms used here may be new to you, but you'll learn them as you read the book.

part I: getting started

Part I begins by taking you through the pieces in the NXT 2.0 robotics kit in Chapter 1. In Chapter 2, you'll build your first robot and learn about the NXT (often referred to as the *NXT brick*). In Chapter 3, you'll meet NXT-G, the software included in the NXT 2.0 robotics kit, which you'll use to program robots. In Chapter 4, you'll use NXT-G to make your robot move as you create your first programs with basic programming blocks. Finally, in Chapter 5, you'll learn programming techniques such as repeating and how to make your robot do more than one thing at the same time.

part II: building and programming robots with sensors

This part teaches you all about sensors, an essential aspect of MINDSTORMS robots. In Chapter 6, you'll begin by adding a sensor to the robot that you built in Part I in order to expand its capabilities while you learn the programming techniques required to use sensors. In Chapter 7, you'll learn about the other sensors in the NXT 2.0 robotics kit, and in Chapter 8, you'll build the Shot-Roller, a robotic defense system, while learning about several new programming blocks. Chapter 9 will dig deeper into sensors as you build and program Strider, a six-legged creature that walks around and interacts with its environment.

part III: creating advanced programs

Part III is devoted to advanced programming concepts. Here is where you'll learn how to make your robots perform complex actions. In Chapter 10, you'll learn about data hubs and data wires while you build SmartBot, a platform for testing advanced programs. Then, in Chapter 11, you'll use data wires to control data programming blocks while you learn some other advanced programming tricks. Finally, Chapter 12 will teach you how to use variables and constants and how to combine all of the programming techniques you've learned up to this point to play a game on the NXT.

part IV: advanced robot projects

Having learned about the NXT 2.0 robotics kit, motors, and sensors, as well as how to program your robots, this last part will have you combine all of your newly learned skills as you create three new robots. In Chapter 13, you'll build and program the Snatcher, an autonomous robotic arm that can find, grab, and lift objects. In Chapter 14, you'll build the Hybrid Brick Sorter, a machine that sorts LEGO bricks by color and size. Finally, in Chapter 15, you'll build CCC, a vertical climber that balances as it moves.

getting help: the companion website

The instructions and explanations in this book have been tested and reviewed extensively, but you may still have questions. On the companion website (*http://www.discovery.laurensvalk.com/*), you'll find links to other helpful websites, downloadable versions of all of the programs in this book, and solutions to some of the discoveries presented in this book. (These solutions will get you started, but you'll need to be creative in order to solve many of the challenges that you won't find available for download!)

If you need more help or if you can't find the answer to your questions at the website, try posting your question on the book's user forum, where you'll find solutions to this book's discoveries from other readers and where you can share your own solutions.

conclusion

MINDSTORMS can spark the imagination and creativity of anyone who uses it, whether child or adult. Now grab your NXT 2.0 robotics kit, start reading Chapter 1, and enter the creative world of LEGO MINDSTORMS. I hope that my offering to you will help spark your imagination!

getting started

collecting the equipment for your robot

In Chapters 2 through 5, you'll create a working robot that can move around a room by itself. But before you start building this robot, you need to determine what you need. There have been various versions of LEGO MINDSTORMS NXT sets in the past, but all the robot models in this book require only one LEGO MINDSTORMS NXT 2.0 kit (LEGO catalog #8547). If you have this kit, shown in Figure 1-1, you're ready to go. If you have a different version of the NXT system and you would still like to use this book, visit the companion website (*http://www.discovery.laurensvalk.com/*) for suggestions on how to find the pieces that you'll need in order to complete the projects in this book.

Figure 1-1: The LEGO MINDSTORMS NXT 2.0 set

what's in the box

The LEGO MINDSTORMS NXT 2.0 kit comes with a lot of TECHNIC building pieces, as well as some electronic ones (including motors, sensors, the NXT brick, and cables). Figure 1-2 shows each type of NXT 2.0 piece. You'll learn how to use each type of part as you read this book.

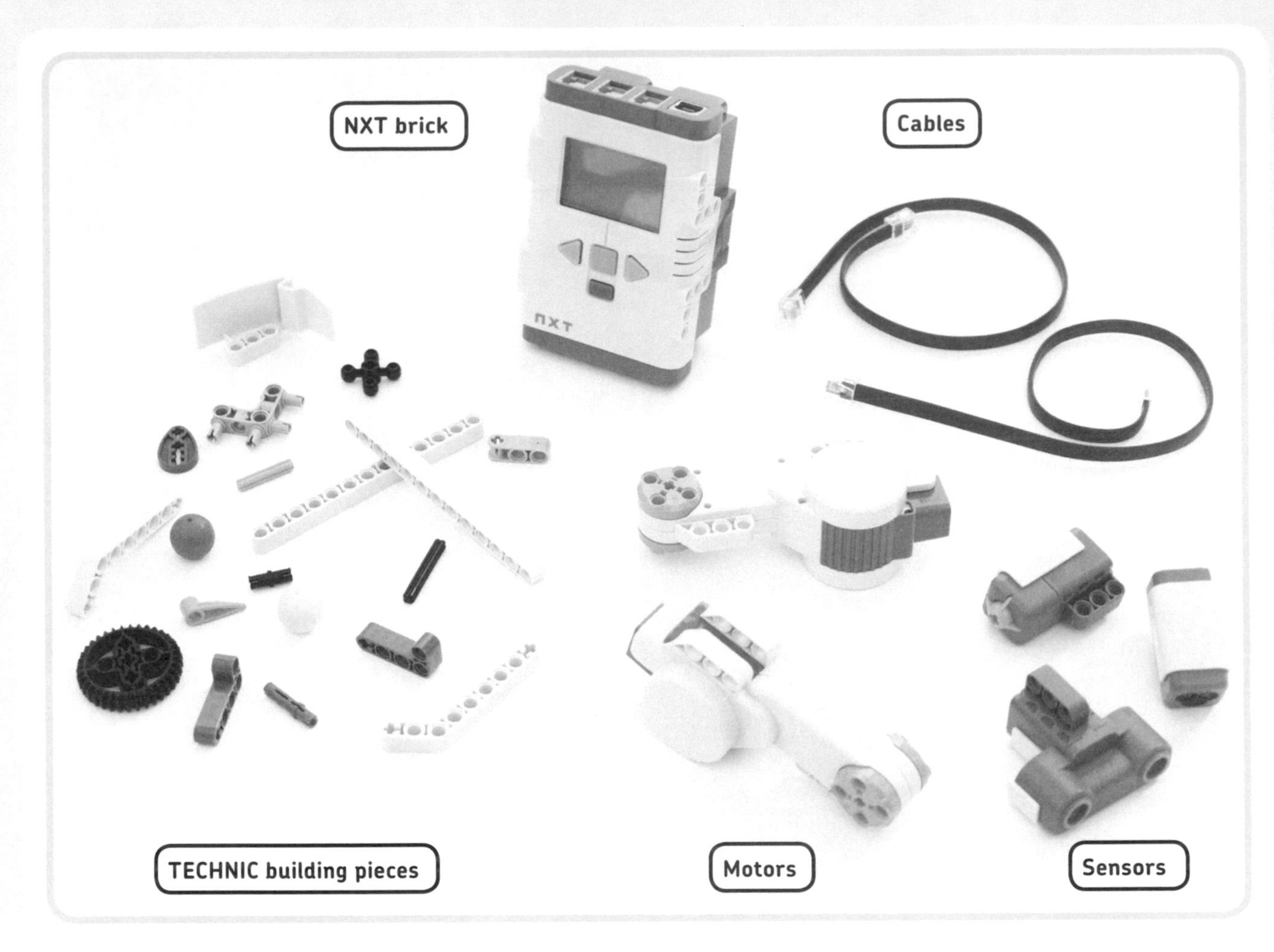

Figure 1-2: The NXT kit contains various types of parts.

NXT robots use *motors* for movements such as grabbing an object or driving. They use *sensors* to take input from their surroundings, such as the intensity of a lightbulb, a color, or the distance to an object.

The *cables* connect the motors and sensors to the *NXT brick*.

the NXT brick

The NXT brick, or simply the NXT, is a small computer that controls the motors and sensors so that your robot can perform actions by itself. For instance, you could build a robot that automatically turns on a light switch when it is getting dark outside. A sensor (which can measure light) tells the NXT that it is getting dark, and the NXT brick then triggers a motor to press the light switch.

But your robot won't do anything without a program.

the NXT-G programming software

You'll find the NXT-G programming software on the CD included in the LEGO MINDSTORMS NXT 2.0 kit (Figure 1-3). You create an NXT program using a computer and the NXT-G programming software. An NXT *program* contains instructions for all the actions that your robot should perform. Once you've finished creating your program, you transfer it to the NXT with a USB cable (supplied in the kit). The combination of motors, sensors, the NXT, and your program form your LEGO MINDSTORMS *robot*.

NOTE If you've mixed up the parts in your LEGO MINDSTORMS NXT 2.0 kit with your other LEGO pieces, you can sort them out using the inventory sheet on the back inside cover.

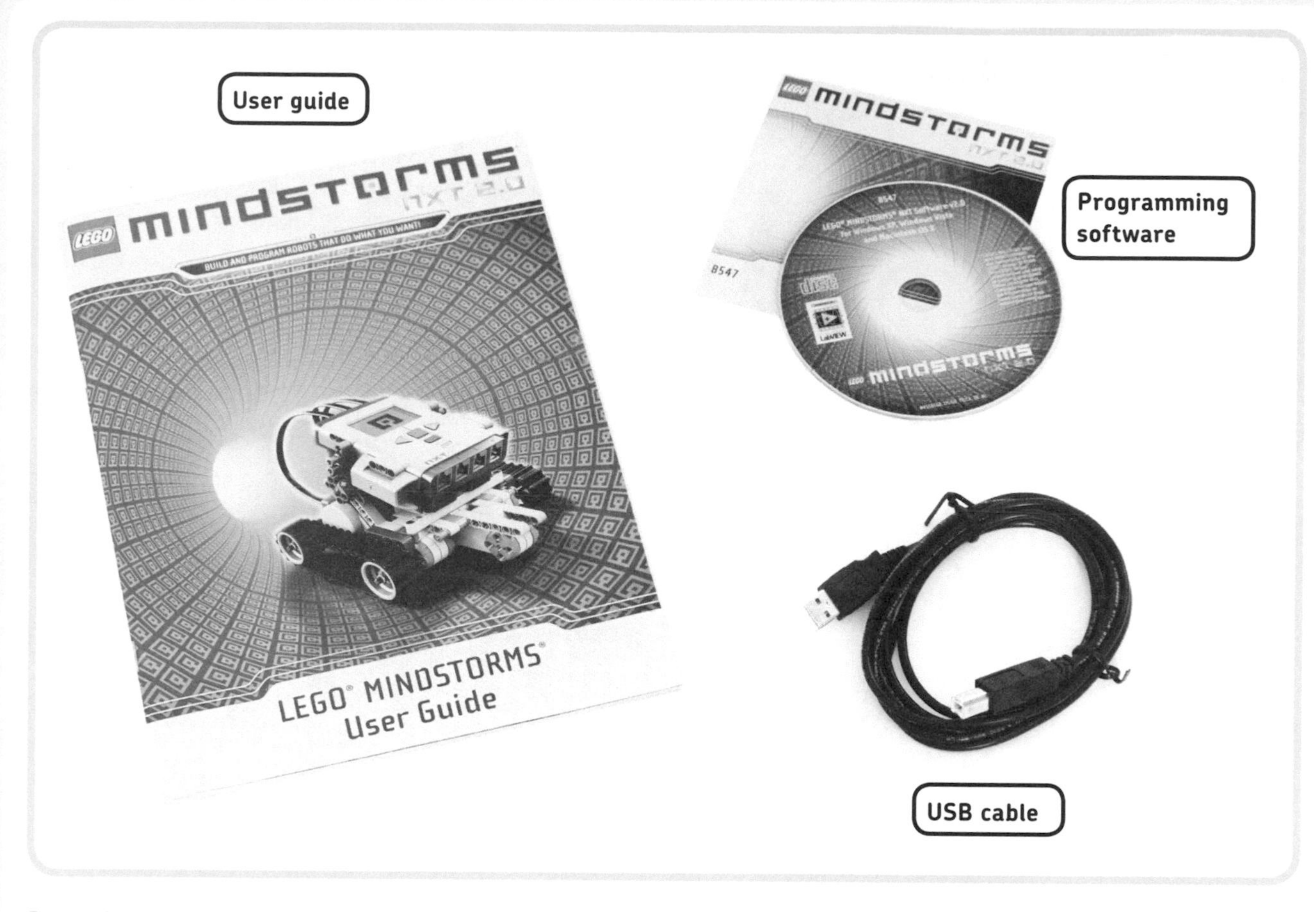

Figure 1-3: The user guide, the NXT-G programming software, and the USB cable. The user guide contains additional information about the NXT set, but I'll be covering much of that information in this book.

installing the software

Before you can create programs to control your robots, you need to install the LEGO MINDSTORMS NXT 2.0 software. Insert the CD included with your kit into your CD drive, and you'll see an installation screen. Just follow the instructions that appear on the computer screen to install the software. You can install the software in any language you like, but this book is based on the English version. If this book asks you to enter a decimal number such as **2.5**, but you typically enter numbers with a comma in place of the period, just enter **2,5** the way you are used to doing.

the test pad

You'll use the Test Pad (Figure 1-4) for some of your robot activities. For example, you'll create a robot that follows the thick black line on this pad later in this book.

inserting batteries

To power the robot, insert six AA batteries into the NXT, as shown in Figure 1-5. You can use standard rechargeable or nonrechargeable batteries or the LEGO rechargeable battery (LEGO catalog #9798) with the transformer (#9833). You may also use the new LEGO rechargeable battery (#9693) with its transformer (#8887).

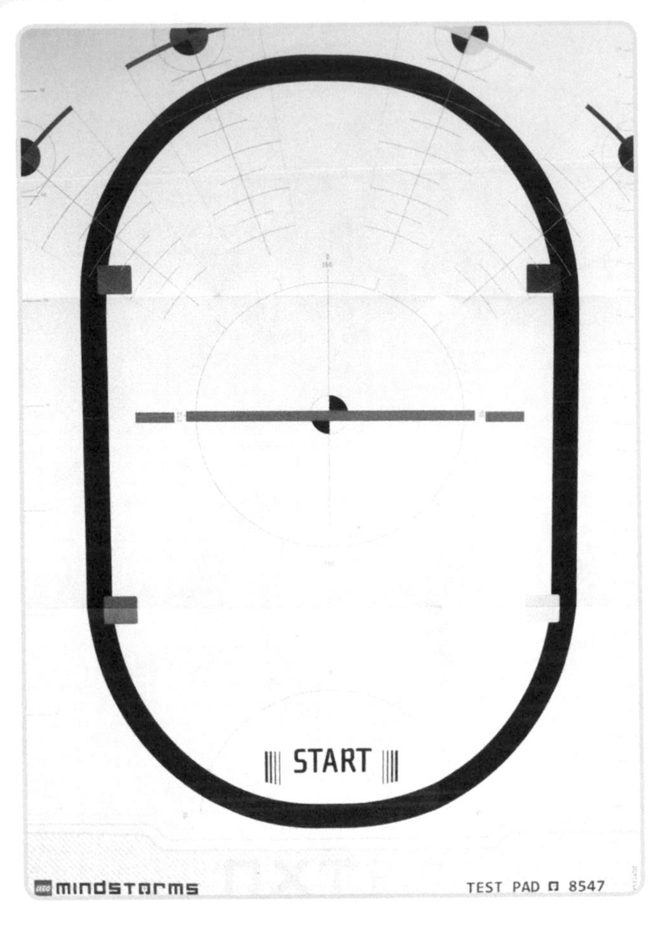

Figure 1-4: The Test Pad

conclusion

Now that you've collected everything you need to build and program a working robot, you're ready to start building one. In Chapter 2 you'll learn more about the NXT, motors, and cables as you build your first robot.

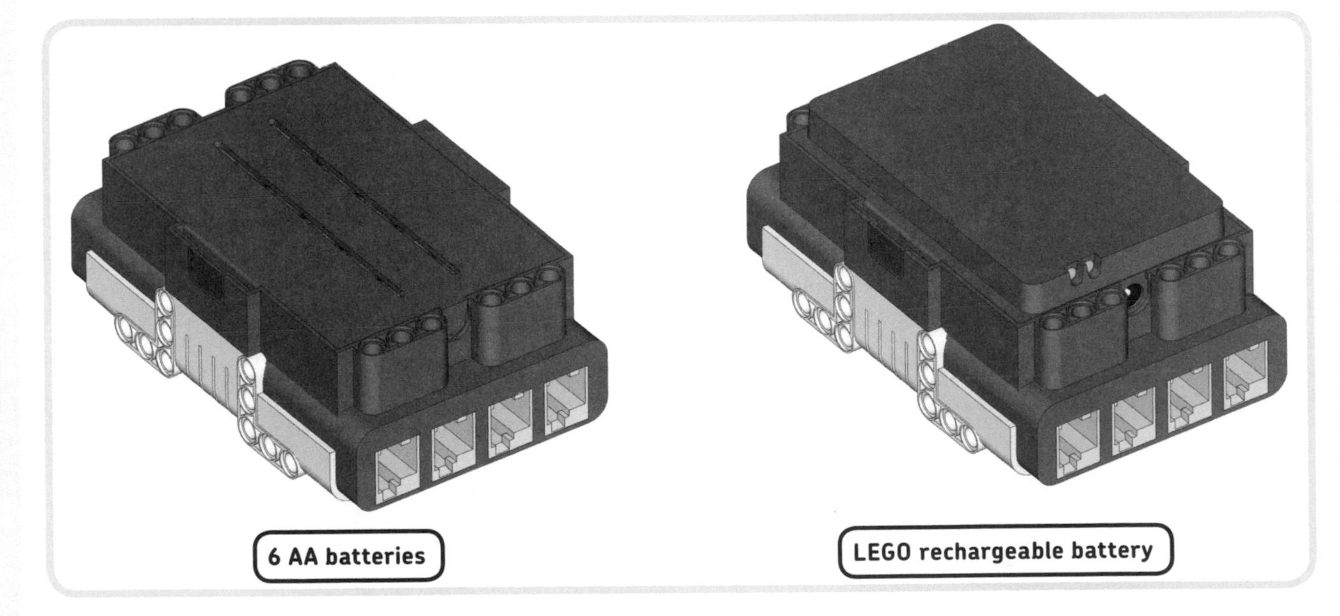

Figure 1-5: You can power the NXT brick with six AA batteries or the LEGO rechargeable battery.

building your first robot

In Chapter 1, you learned that a robot consists of several important components. To make it easier for you to understand how each of these works, you'll begin by working with only some of them. Specifically, you'll first learn to work with the NXT motors and the brick as you build a wheeled vehicle (the Explorer, shown in Figure 2-1) that can drive around your room. Once you've built the Explorer, you'll do a quick test to see whether you've correctly assembled the robot and to make it move!

Figure 2-1: The Explorer

building the explorer

To begin, select the pieces you'll need for the Explorer by referencing the *bill of materials* shown in Figure 2-2. Then, assemble the robot as shown in the step-by-step building images on the next pages.

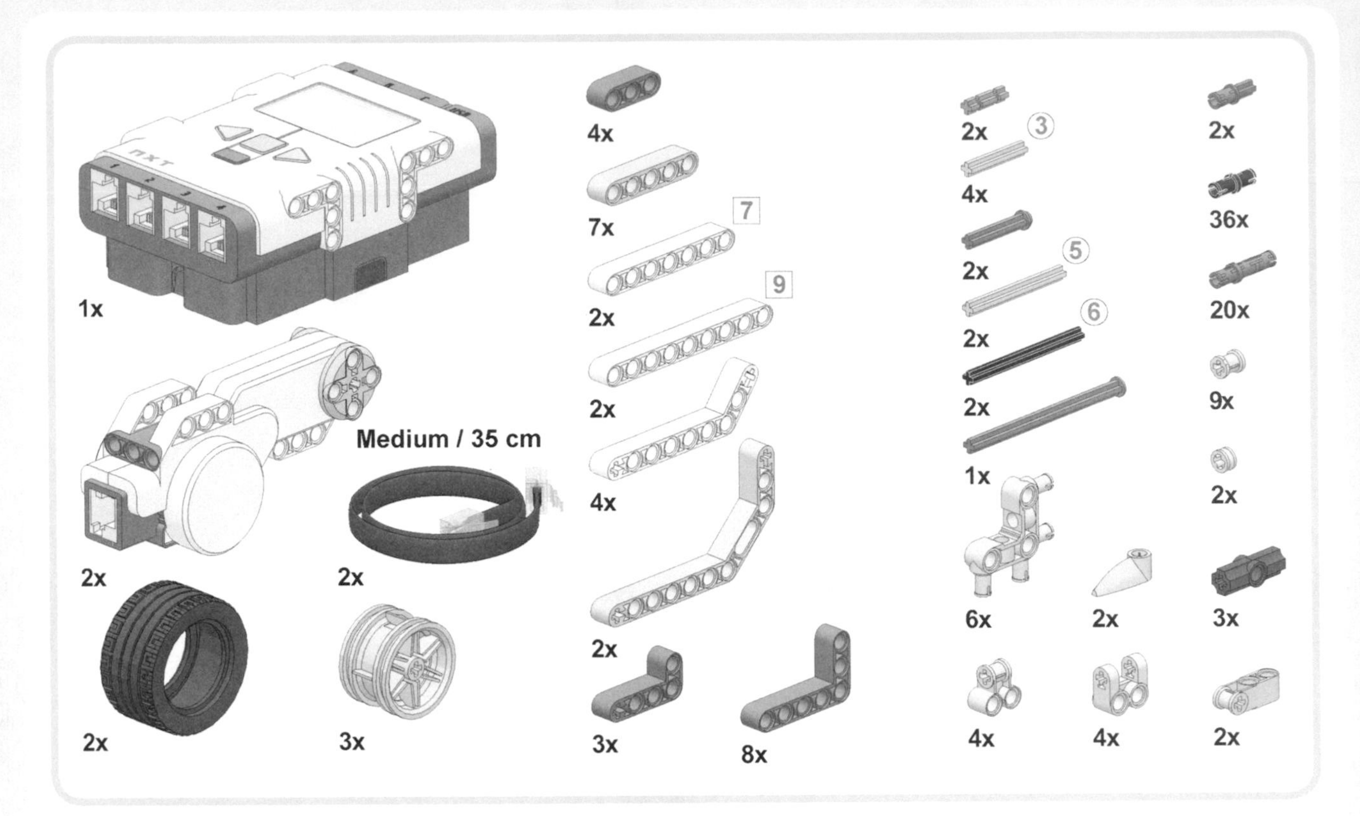

Figure 2-2: Bill of materials

building tip: beams and axles

The LEGO MINDSTORMS NXT 2.0 robotics kit contains a lot of *beams* and *axles*. Because these parts come in a variety of lengths, it's sometimes hard to figure out which one you'll need. To help you sort out which one you'll need, I'll indicate the length as shown in the labels on the diagram in Figure 2-3. The numbers in boxes refer to beams; those in circles refer to axles.

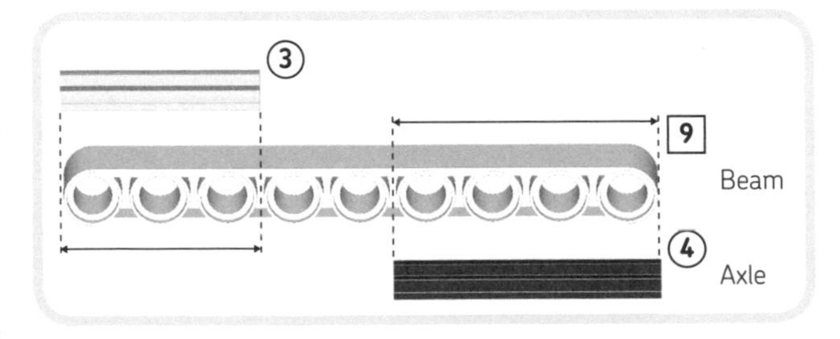

Figure 2-3: Beams and axles come in different lengths, so make sure you pick the correct ones while going through the building instructions. Instead of determining the length of beams and axles yourself, you can also use the 1:1 reference chart on the front inside cover.

To find the length of a beam, simply count the number of holes. For instance, the beam shown in Figure 2-3 has nine holes in it, as indicated by the "9" in the box next to it. The numbers in circles tell you the length of axles. To determine the length of an axle, place it next to a beam, and count how many holes it covers, as shown in Figure 2-3. The black axle shown here covers four holes, so it shows "4" in the circle.

building tip: friction and nonfriction pins

The NXT set also includes *pins*, which you'll use to connect two or more parts. The set contains *friction pins*, which won't turn easily if you connect them to a beam, and *nonfriction pins*, which do rotate easily when connected to a beam.

Friction and nonfriction pins may be the same shape, but the pin's color tells you what type of pin it is, as shown on the front inside cover. When you're following the building instructions, be sure to pick the correct type of pin as you build. The front inside cover also shows how the pin colors look in the black-and-white images in the book.

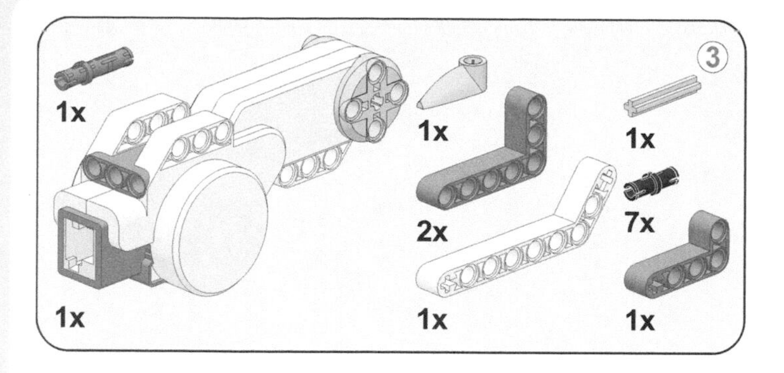

3
1x
1x
1x
2x
7x
1x
1x
1x

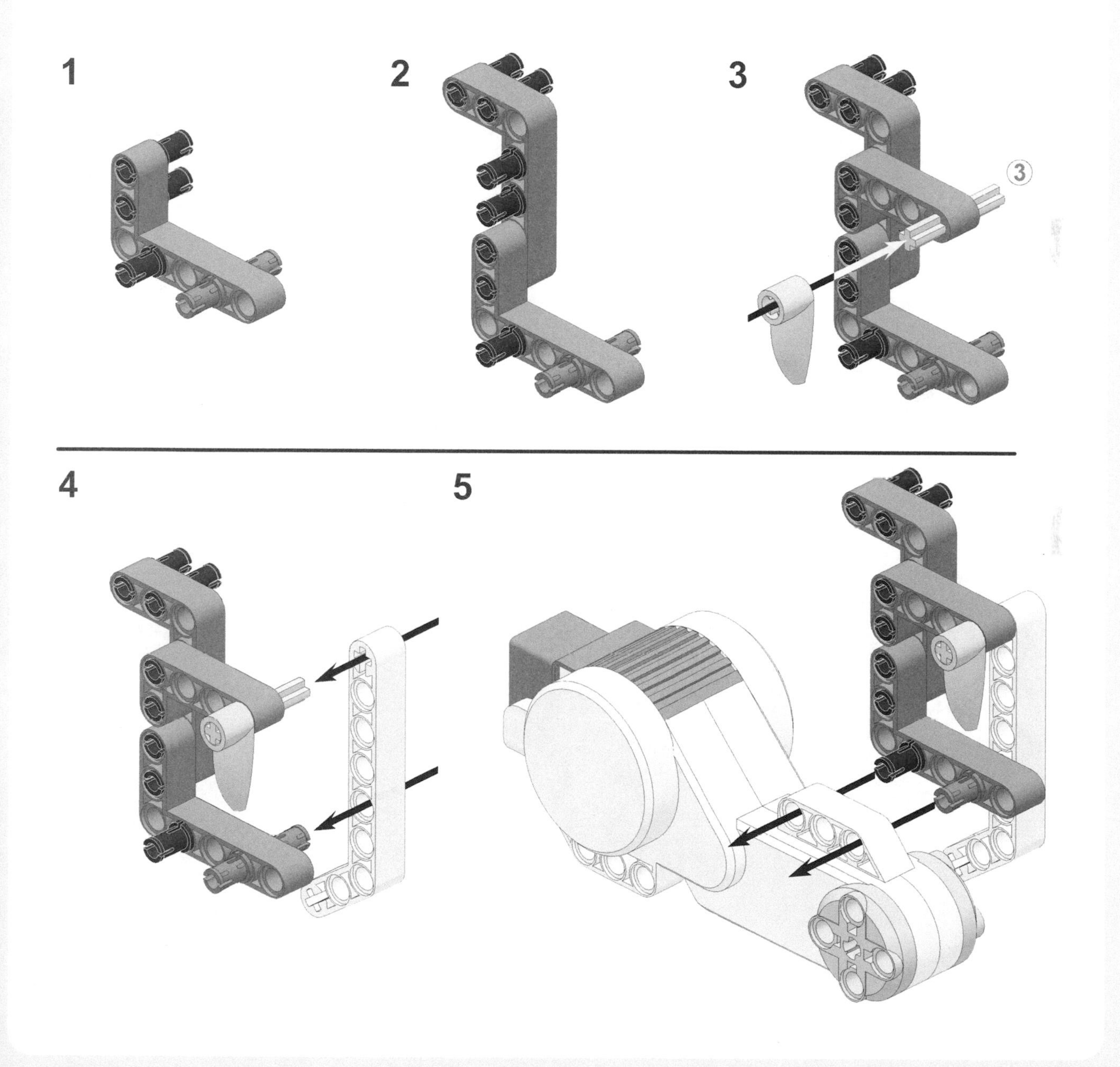

1
2
3
3
4
5

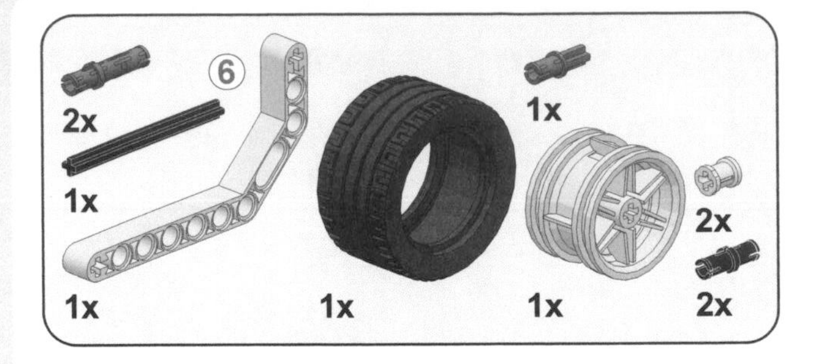
2x
1x
1x
1x
1x
1x
2x
2x

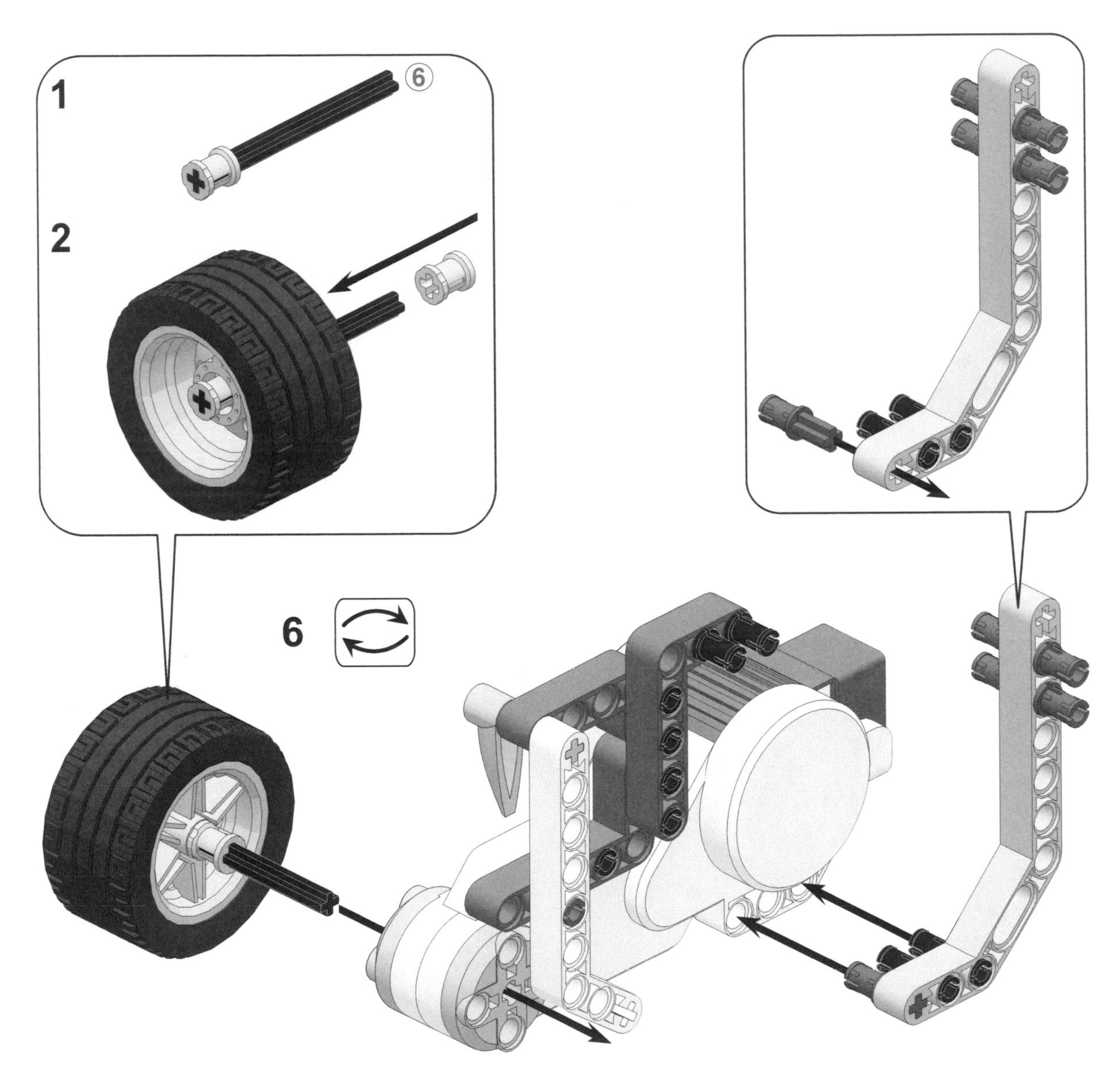
1
2
6

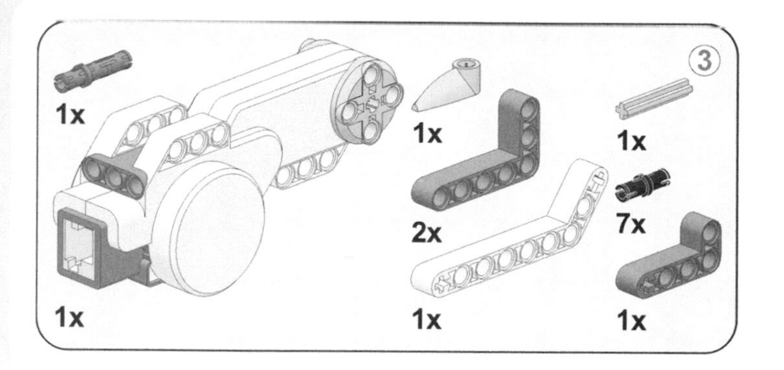
1x
1x
1x
2x
7x
1x
1x
1x
3

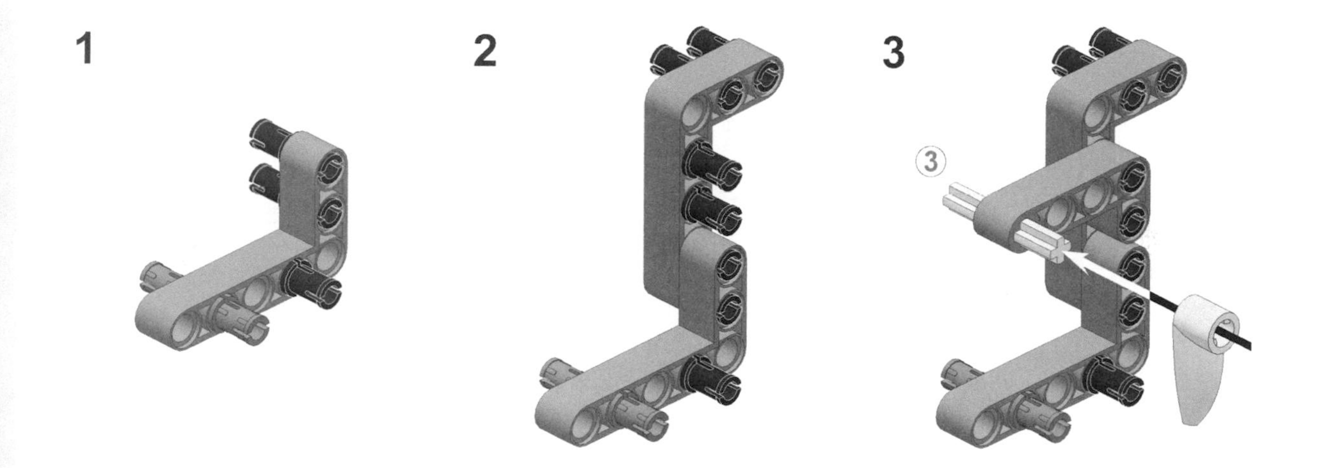
1
2
3
3

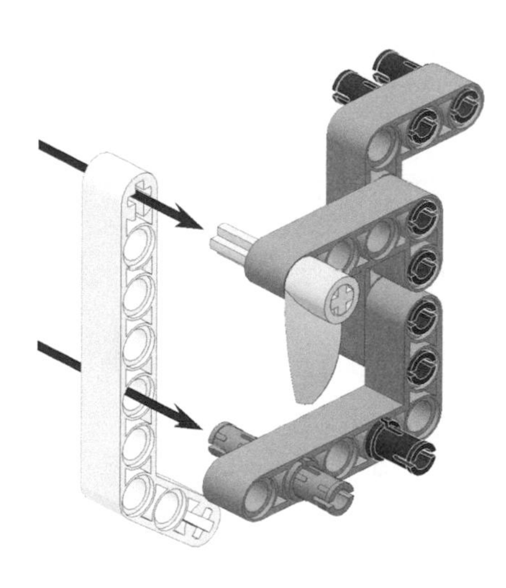
4

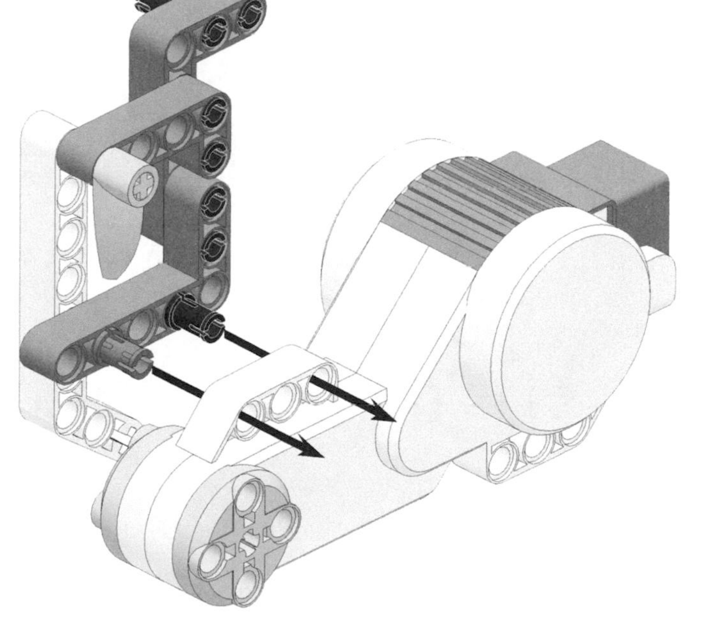
5

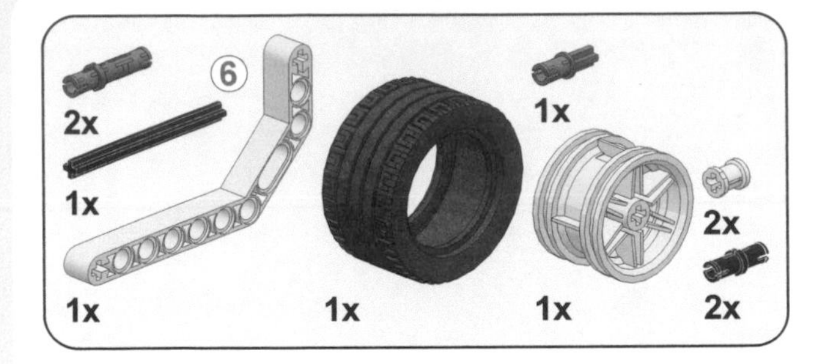

2x
6
1x
1x
1x
1x
1x
2x
2x

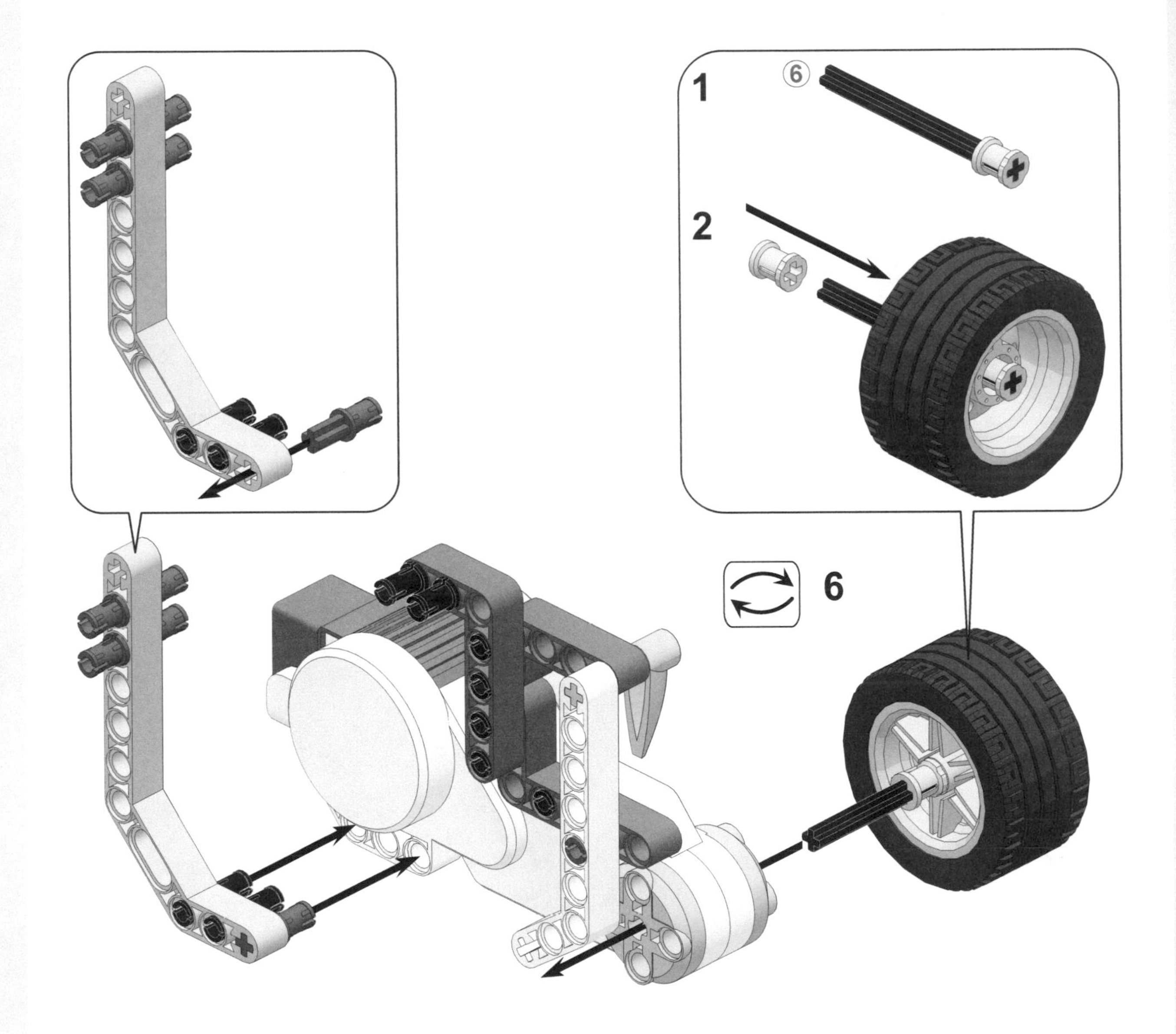

1
6
2
6

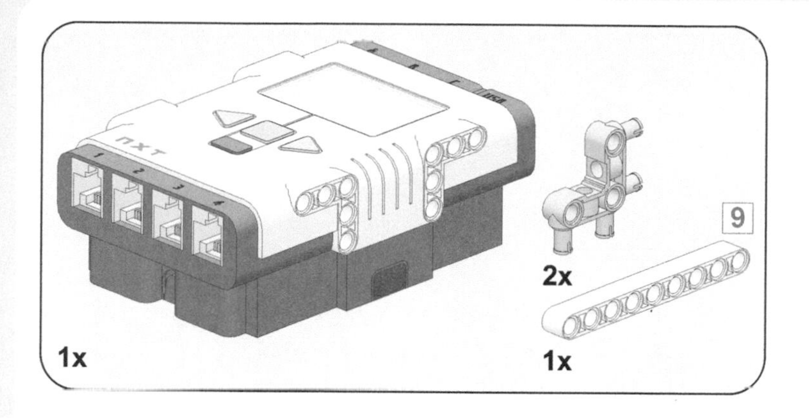
1x
2x
9
1x

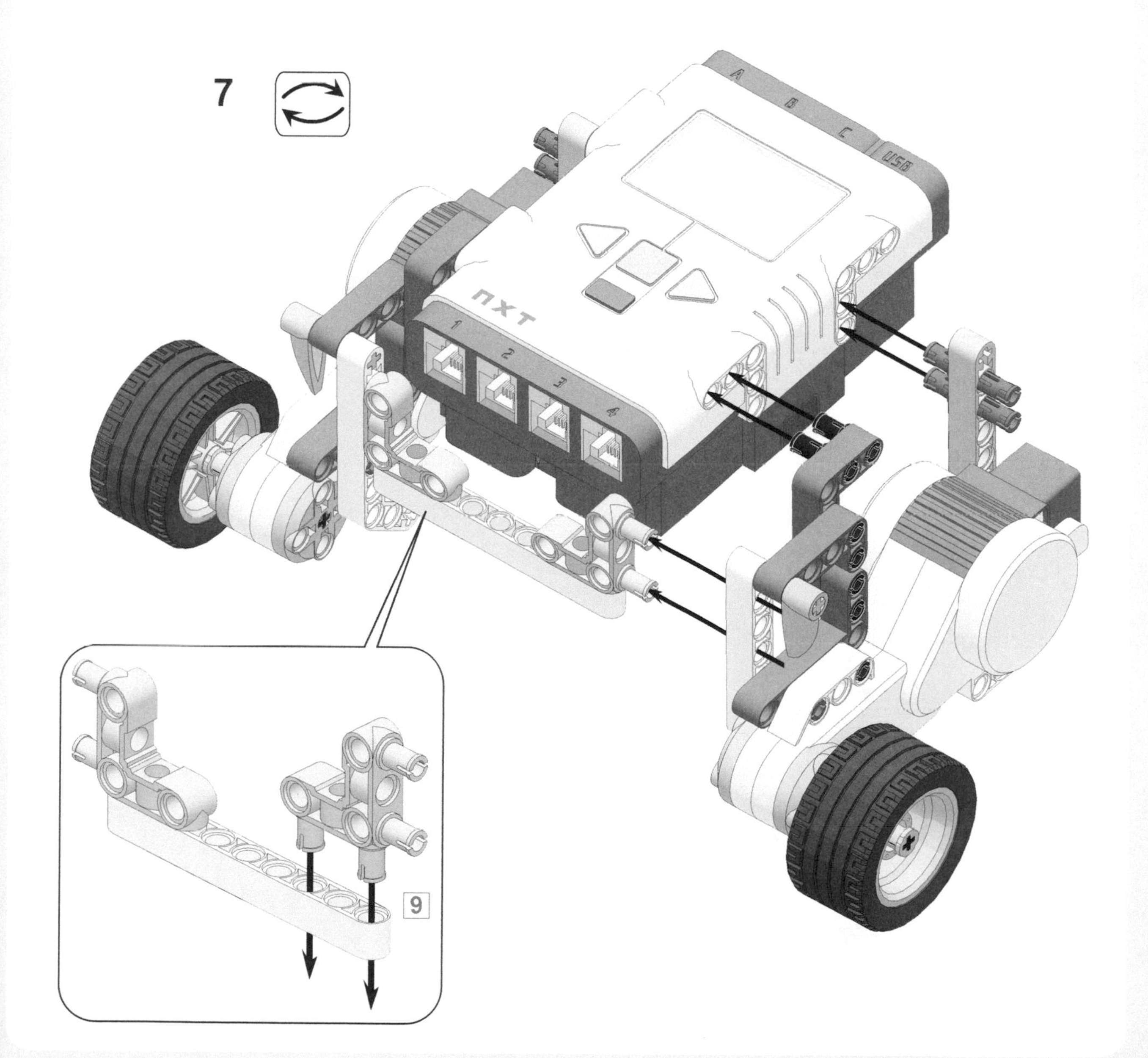
7
9

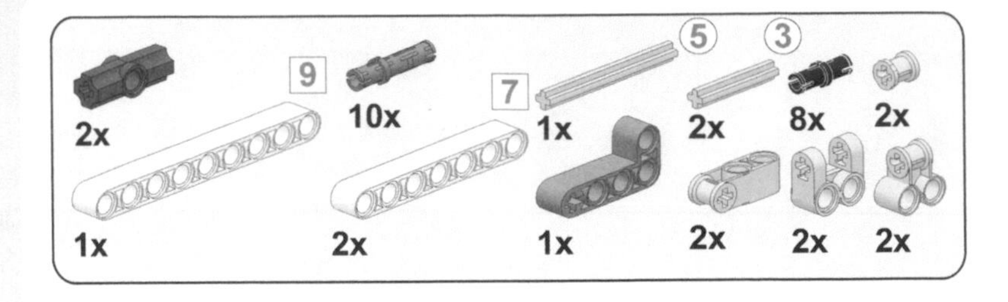

1

7

7

2

3

9

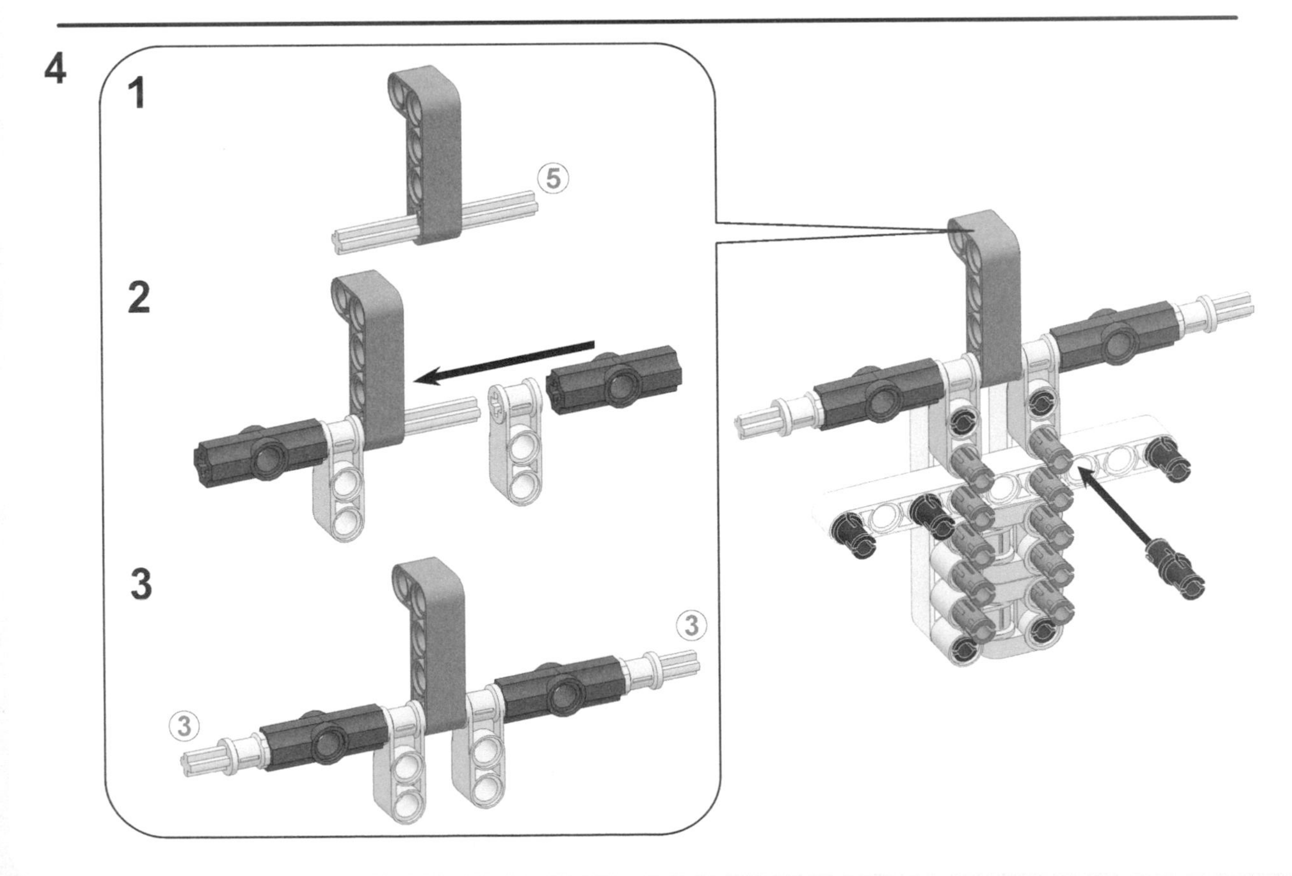

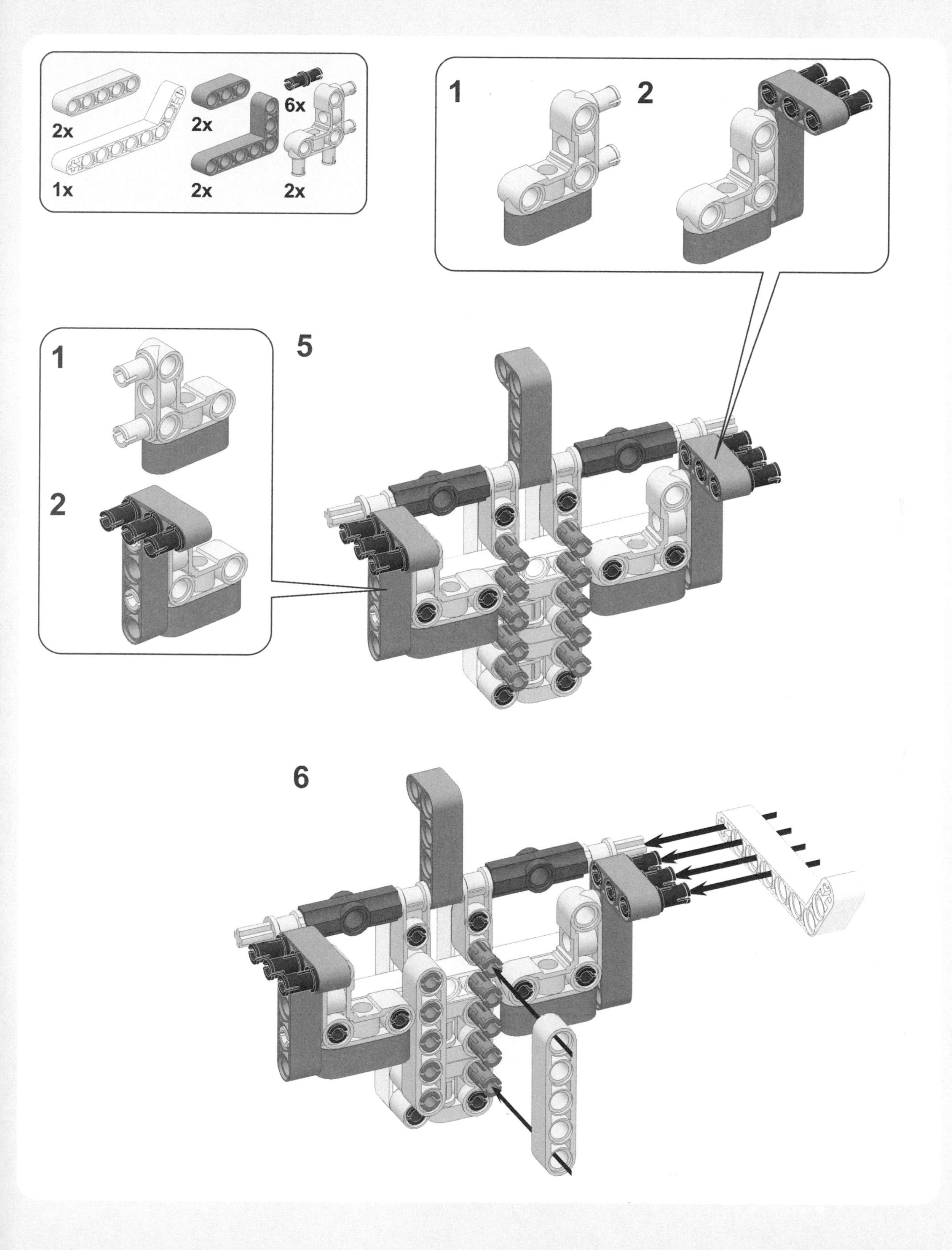
2x
2x
6x
1x
2x
2x
1
2
1
2
5
6

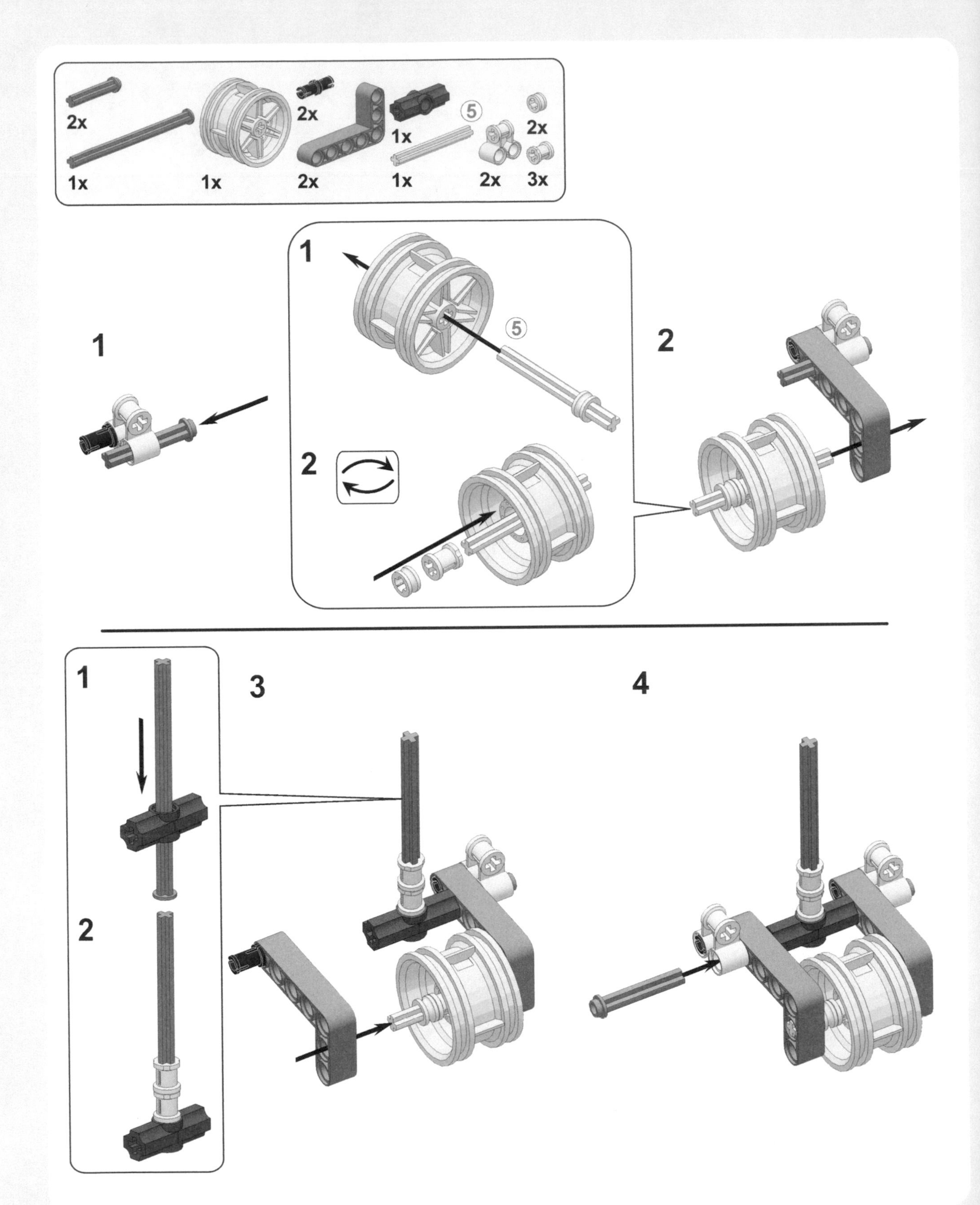
2x
1x
1x
2x
2x
1x
5
1x
2x
2x
3x
1
5
2
1
2
2
1
3
4
2

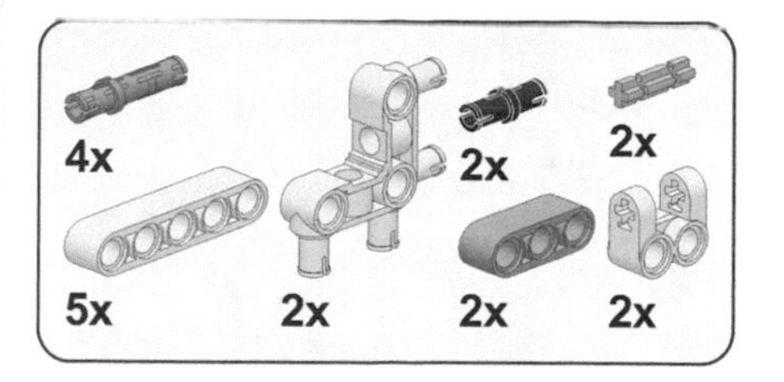
4x
2x
2x
5x
2x
2x
2x

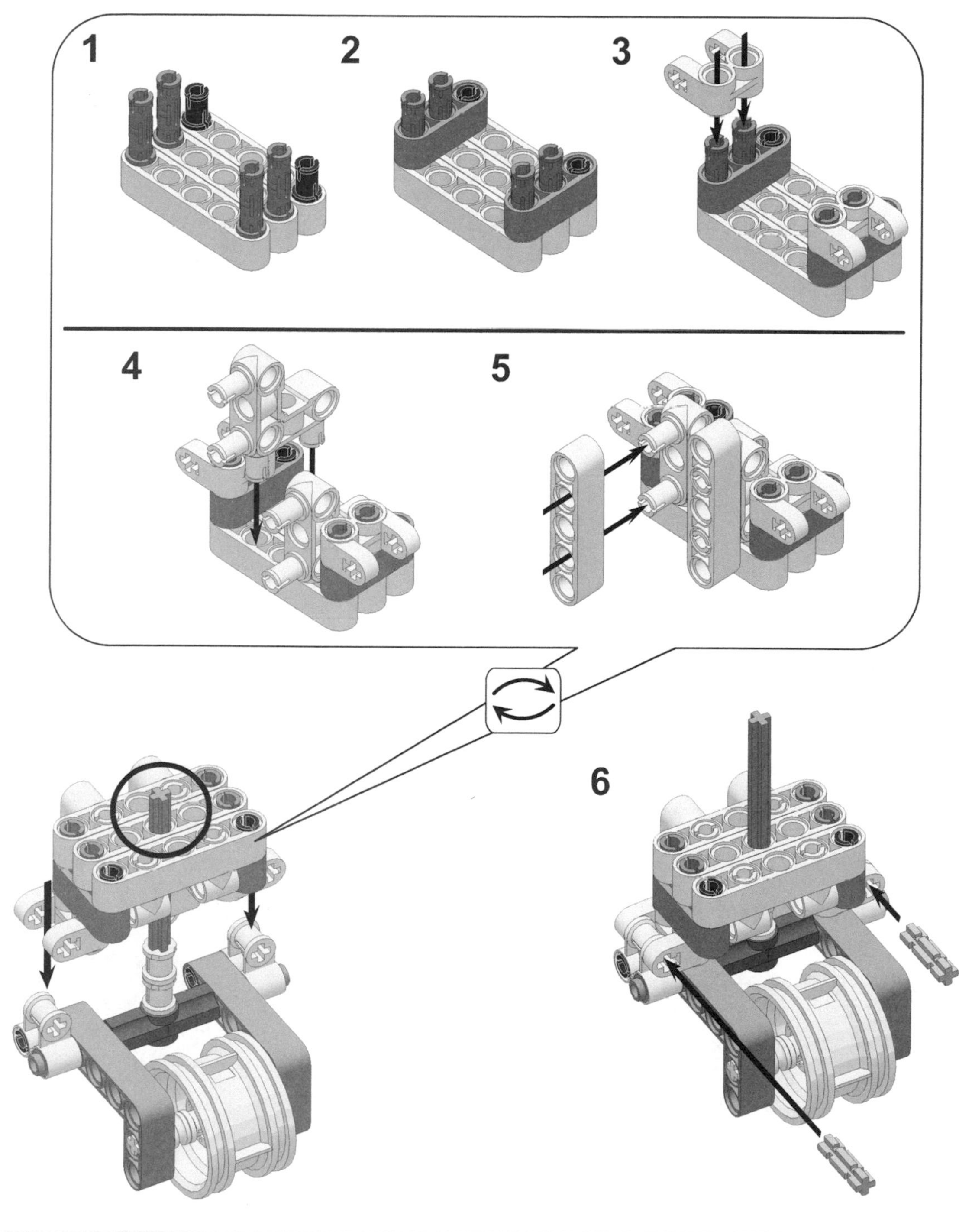
1
2
3
4
5
5
6

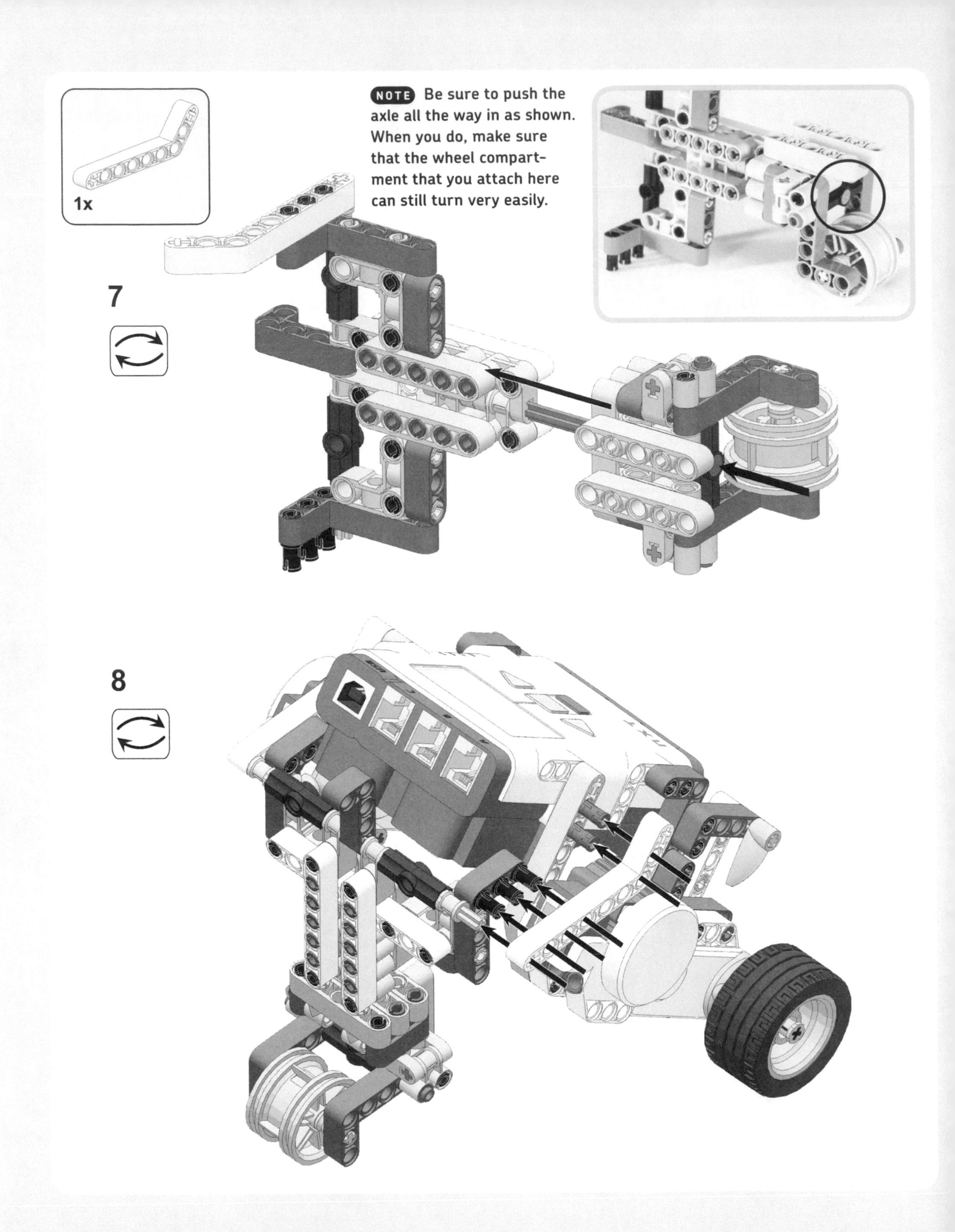
1x
NOTE Be sure to push the axle all the way in as shown. When you do, make sure that the wheel compartment that you attach here can still turn very easily.
7
8

connecting the cables

To use the NXT motors, you need to connect them to the NXT brick using cables. Motors are connected to *output ports* called A, B, and C, as shown in Figure 2-4.

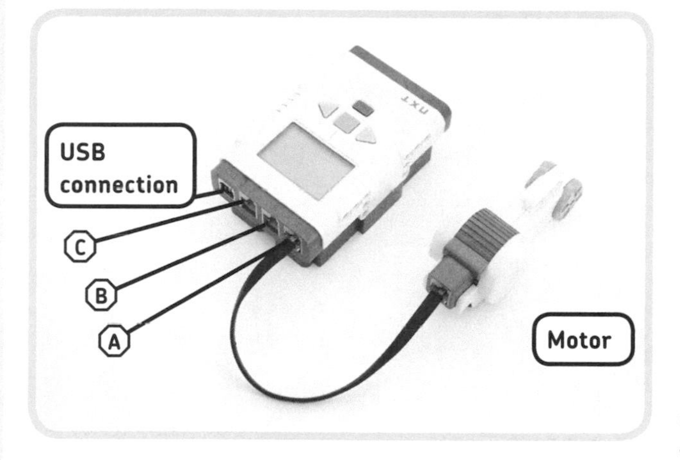

Figure 2-4: Output ports A, B, and C are used to connect motors to the NXT. In this example, the motor is connected to output port A.

Your LEGO MINDSTORMS NXT 2.0 robotics kit has three types of cables: a short cable (20 cm, or 8 inches), four medium-sized ones (35 cm, or 15 inches), and two long cables (50 cm, or 20 inches). For the Explorer robot, you'll use medium-sized cables to connect the motors to ports B and C, as shown in Figure 2-5.

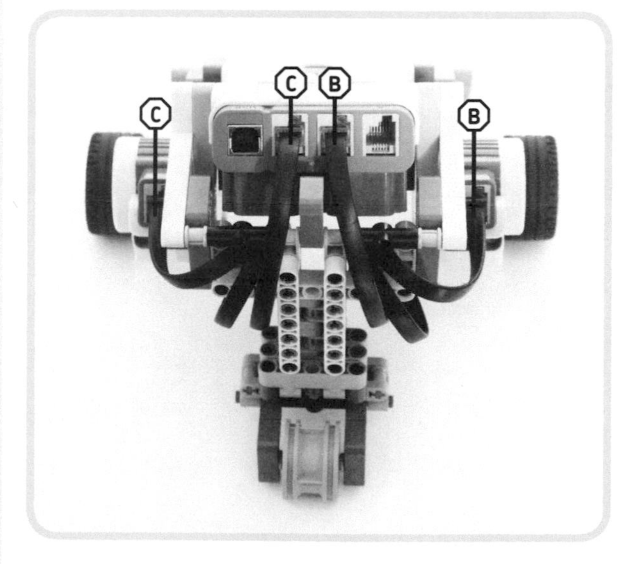

Figure 2-5: When connecting the medium-sized cables to the Explorer, plug one end of a cable into the NXT brick, wrap the cable around the LEGO pieces a few times, and then plug it into the appropriate motor.

When connecting your cables, make sure they do not interfere with the movement of the front wheels or the supporting wheel in the back of the model. To move the cables out of the way, tie them around some of the LEGO pieces in the robot so that the wheels and the caster wheel can spin freely. For example, you can wire the cables as shown in Figure 2-5.

using the NXT buttons to navigate on the NXT brick

Congratulations, you've finished building your training vehicle! Now, before you move on to programming in Chapter 3, you'll learn how to use the buttons on the NXT brick (shown in Figure 2-6) to navigate around the NXT brick's menus and to run programs stored on the brick.

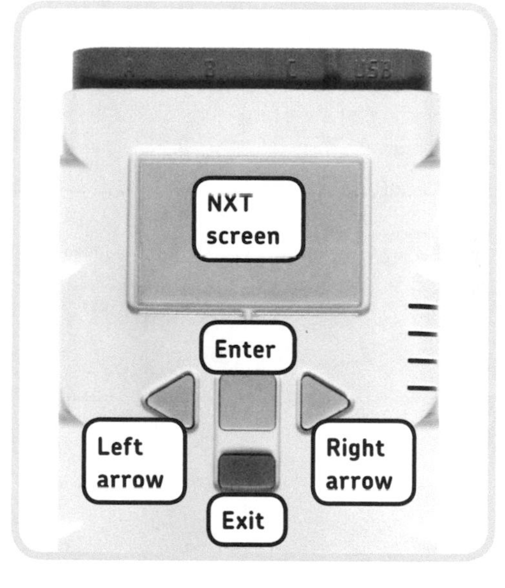

Figure 2-6: The NXT screen and the NXT buttons

turning on the brick

To turn on the brick, press the (orange) **Enter** button, as shown in Figure 2-7. After you hear the startup sound, you should see the main menu on the brick's screen, displaying a couple of different icons, as shown in the figure.

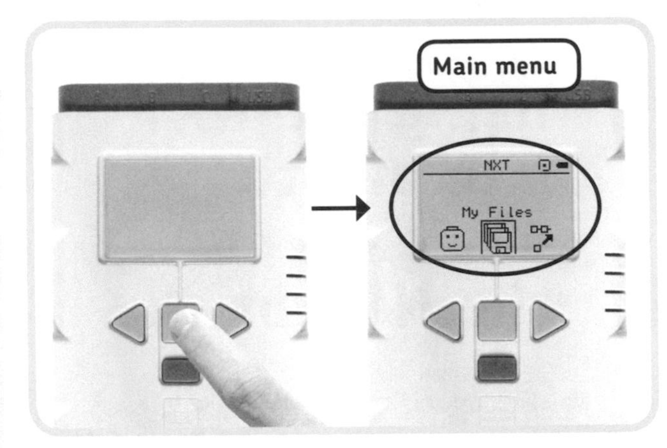

Figure 2-7: Turning on the NXT Brick with the Enter button will open the main menu.

selecting and choosing items

The selected item is always in the middle of the screen, as shown in Figure 2-8. You can switch to icons on the left and the right with the light gray arrow buttons. To select an item (Figure 2-8), press the **Enter** button.

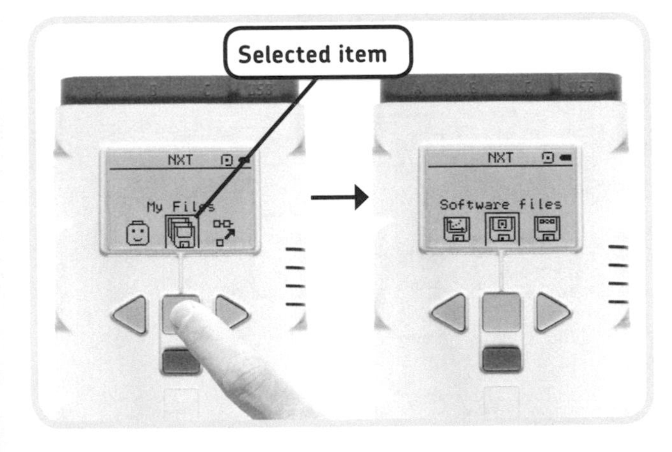

Figure 2-8: You press the Enter button to choose a selected item.

To return to the previous menu (Figure 2-9), press the dark gray **Exit** button.

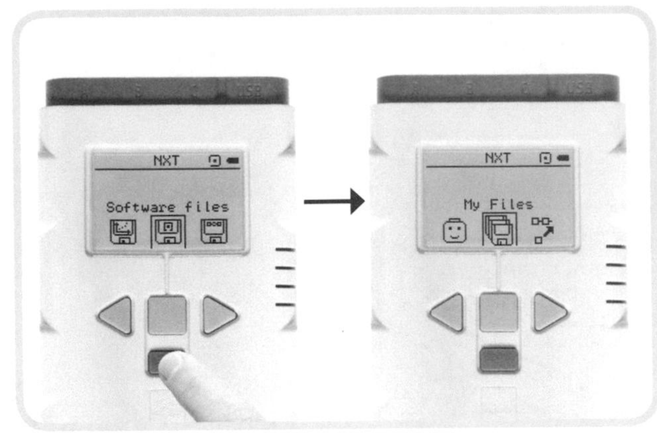

Figure 2-9: You can return to the previous menu by pressing the Exit button.

turning off the brick

To turn off the brick, return to the main menu, and press the **Exit** button. When you see the option for turning off the brick, either select the check mark to turn it off or select the X icon to cancel (Figure 2-10).

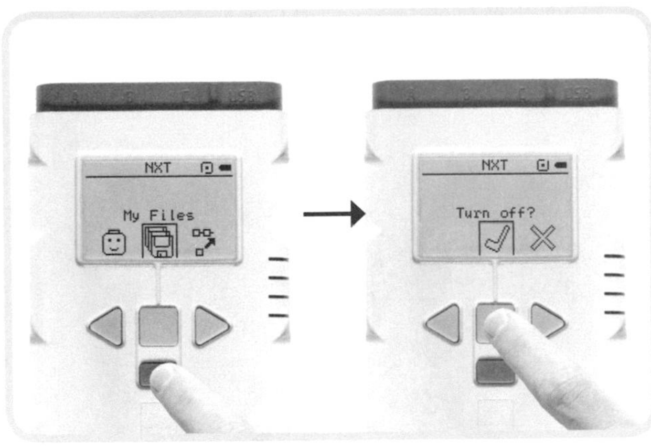

Figure 2-10: Turning off the NXT brick

running a program

NXT robots begin performing their actions when you select and run a program that has been transferred to the brick. You haven't transferred a program to the brick yet, but you can try a sample program that is already on the brick, called DemoV2. To test your Explorer, run this DemoV2 program by navigating through the menu on the NXT brick, as shown in Figure 2-11.

NOTE If for some reason you can't find the DemoV2 program on your NXT brick, just skip this step.

If you've put together everything properly, your robot should move around and make some sounds. To abort the running program, press the **Exit** button.

Now that you know how to start and stop programs, you're ready to create your own programs!

conclusion

You have just learned to work with two essential robot components: the NXT brick and motors. When you ran the DemoV2 program, the program turned the motors, which made the robot move. In Chapters 3 and 4, you'll learn how these programs work, as well as how you can make your own programs.

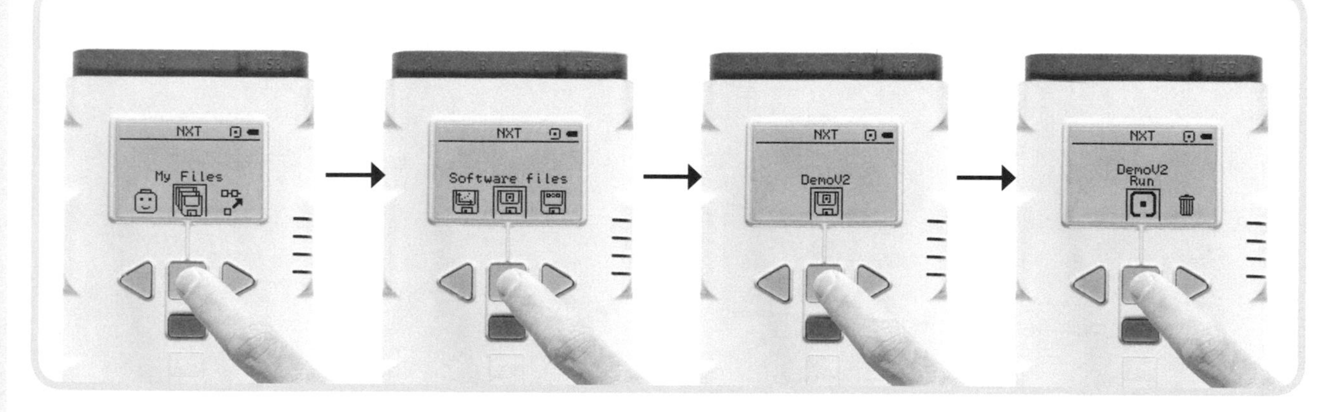

Figure 2-11: Run the DemoV2 program by navigating to My Files ▸ Software files ▸ DemoV2 ▸ Run.

creating and modifying programs

Now that you've built your robot, the next thing the robot needs in order to function is a *program*. Programs tell your robot what to do. For example, a program can make the Explorer drive forward and then steer. This chapter will teach you how to create and edit programs.

a quick first program

You'll first take a quick look at the NXT-G programming software by creating and downloading a small program to the robot. To create the program, follow these steps:

Figure 3-1: The robot connected to the computer

1. Connect the robot to the computer using the USB cable that came with the kit (Figure 3-1), and then turn on the NXT brick by pressing the orange **Enter** button. (You have to connect the robot to the computer each time you want to download a program to the robot.) If you want to use Bluetooth to transfer programs to the NXT, see the appendix.
2. Launch the LEGO MINDSTORMS NXT 2.0 software (also called NXT-G). Once it finishes loading, you should see a screen like the one shown in Figure 3-2. This is the main menu of the programming software. You'll use it to create new programs or to open programs you've made.
3. Create a new program by entering **Explorer-1** in the Create new program box (located in the middle of the screen), and then click the **Go >>** button (Figure 3-3). You can choose any name for your program, but it's a good idea to give it a descriptive name that will be easy for you to recognize later. For example, in this case, you can call your program Explorer-1 because it is the first program for the Explorer robot.

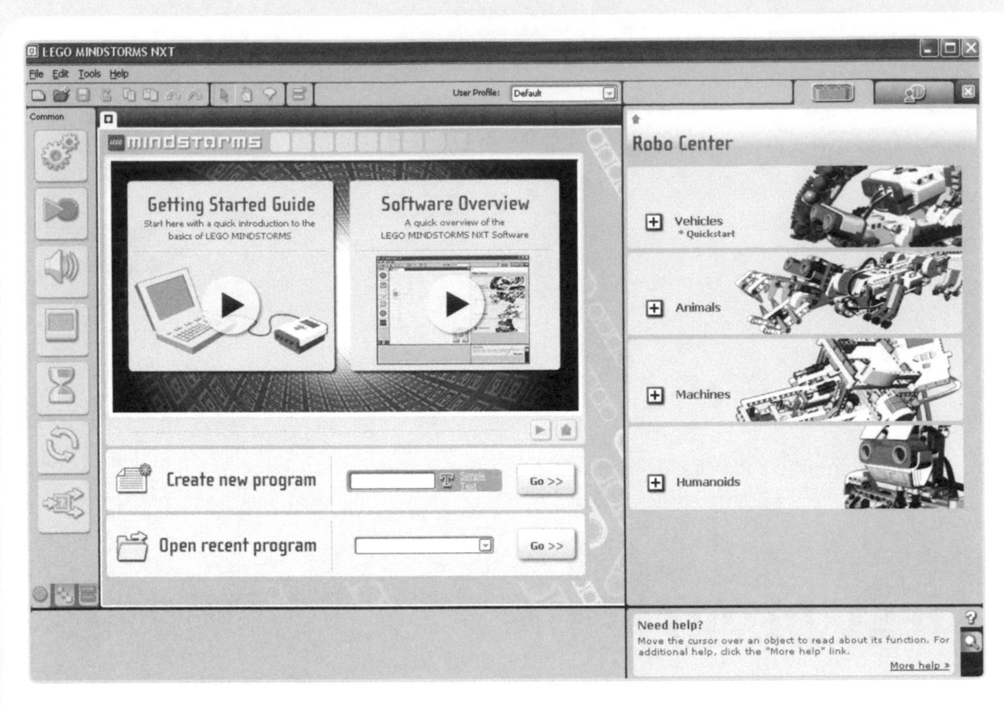

Figure 3-2: The NXT-G startup screen

Figure 3-3: Creating a new program

4. Pick a block and place it on the location indicated by Figure 3-4. Remember that a program is basically a list of instructions for actions that the robot should perform. The single block you place here is just such an instruction—one that makes the robot move forward for a short while.

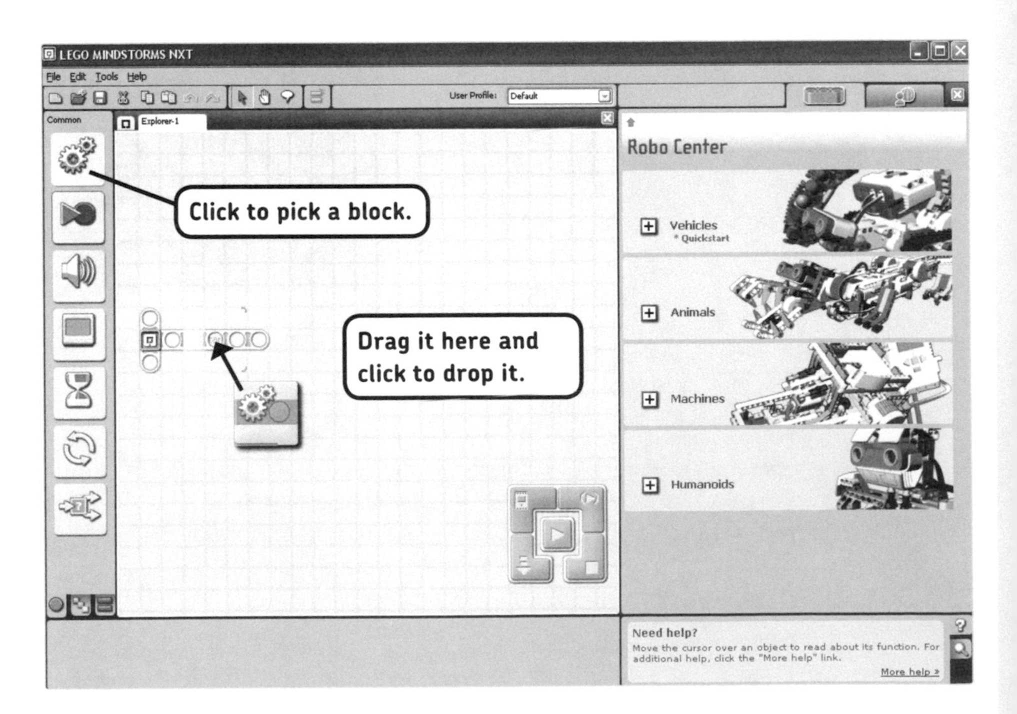

Figure 3-4: Picking and placing a block

5. Once you've finished steps 1 through 4, click the **Download and Run** button, and wait for your robot to start moving. (Figure 3-5). Each time you want to download the program again, simply click this button.

Congratulations! If your robot moves forward a short distance, you've successfully created your first program.

Figure 3-5: Downloading the program to the robot

NOTE If for some reason your robot did not move as described or if you got an error message, something may be wrong with the USB connection. Try turning the NXT brick off and then back on again. If that doesn't help, see the appendix for more instructions.

creating a basic program

It was cool to see that you could make the robot move, but what exactly did you do to accomplish this? You'll now look at several sections of the NXT-G software to better understand how to create and edit basic programs before you move on to creating more complicated ones.

Figure 3-6 shows what your screen should look like just after you start working on a new program. I'll discuss each of the marked sections in turn.

1. programming palette

A program for your robot consists of one or more *programming blocks*. Each block instructs the robot to do something, such as move forward or make a sound. You can select any of these blocks from the *Programming Palette* (Figure 3-7), and each of them will make the robot do something different.

There are multiple Programming Palettes, but for now you need only one. If your Programming Palette doesn't look like the one shown in Figure 3-7, click the tab at the bottom of the palette to open the correct one.

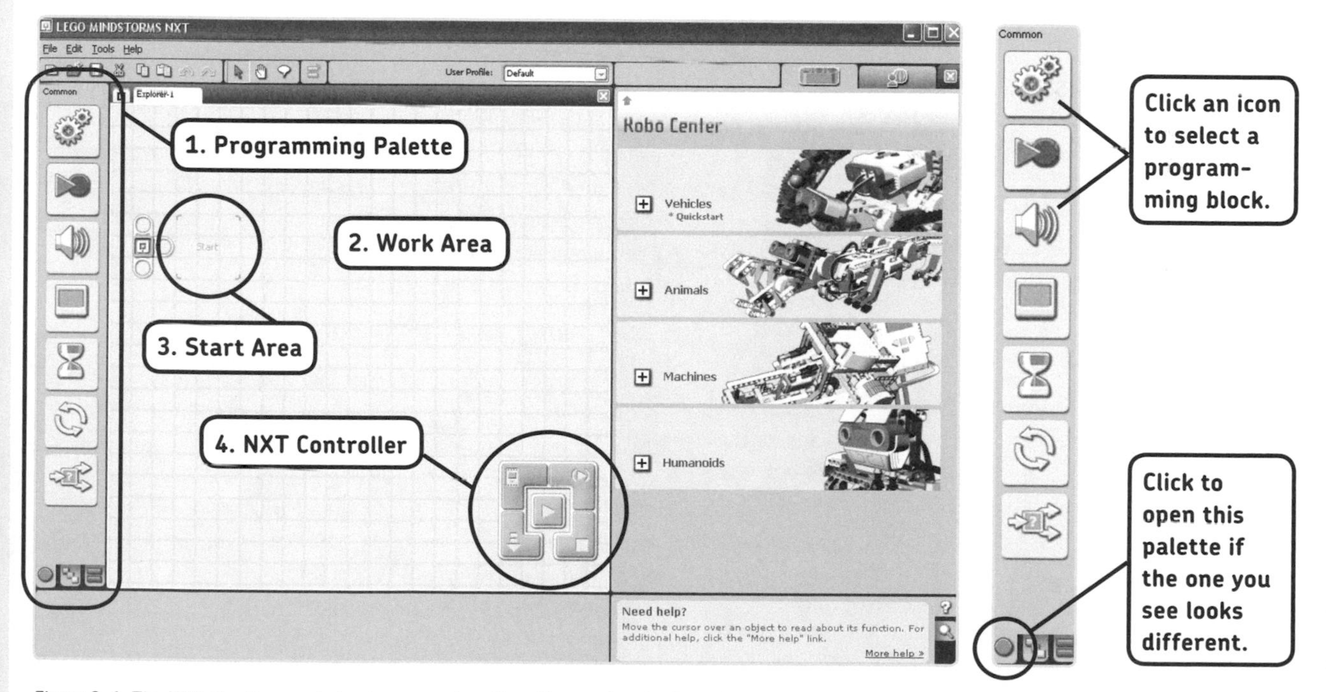

Figure 3-6: The NXT-G software window has several sections. You used each of the labeled ones when you created your first program.

Figure 3-7: The Programming Palette (a close-up of section 1 in Figure 3-6)

2. work area

Once you've selected a block to use in your program, you place that block on the *Work Area*, as shown in Figure 3-8. The Work Area is where you create your program. Programs often consist of more than one block.

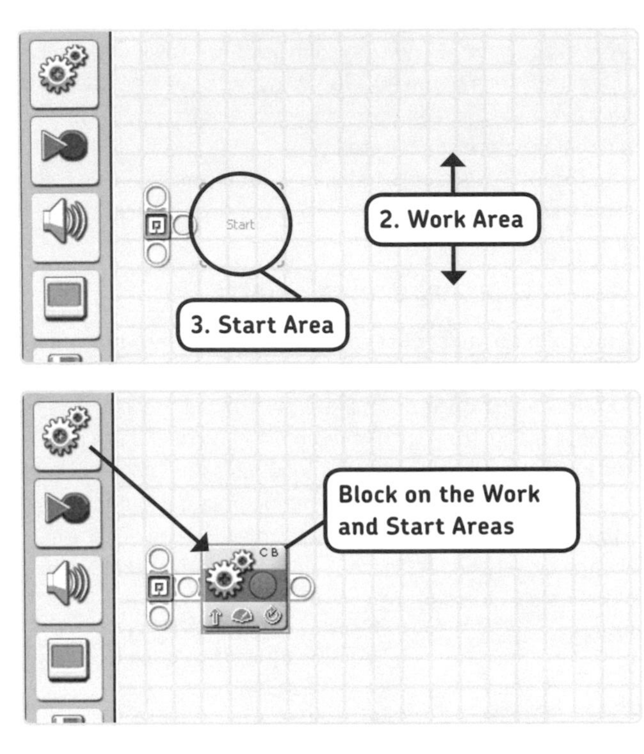

Figure 3-8: Once you've picked a block by clicking an icon on the Programming Palette, you place it on the Work Area. If it's the first block in a program, you place it on the Start Area.

To move a block to the Work Area, simply click an icon on the Programming Palette, and then click the Work Area where you want to place the block.

moving and deleting blocks

Once you've placed a block on the Work Area, you can *move* it by clicking it with the left mouse button and keeping the button pressed as you move the block around. To *delete* a block from the Work Area, click to select it, and press the DELETE key.

3. start area

The *Start Area*, shown in Figure 3-8, is where you'll place your program's first programming block. You'll place all subsequent programming blocks to the right, next to the previous block. (You'll learn about programs with more than one block in Chapter 4.)

4. NXT controller

You use the *NXT controller* (shown in Figure 3-9) to download (transfer) your programs to the NXT brick.

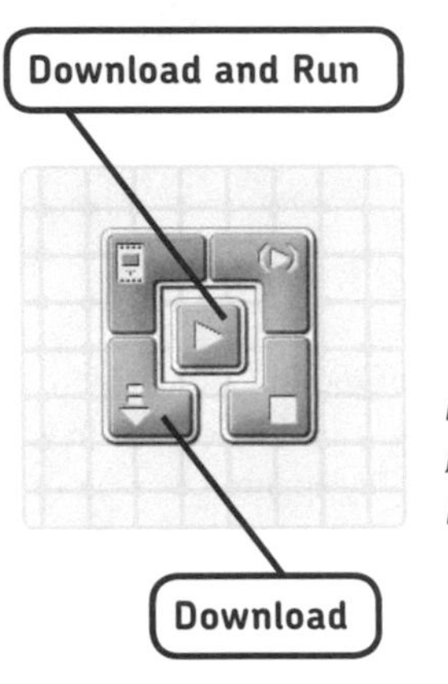

Figure 3-9: The NXT controller has five buttons, but for now you'll use only the Download and Download and Run buttons.

downloading and running a program

To transfer a program to the NXT brick, make sure that the brick is connected to your computer (either via USB or via Bluetooth), and then click the **Download and Run** button on the NXT controller. The robot should beep to indicate that the program has been transferred successfully, and the program should begin running automatically. The program will stop when it finishes running each block of the program.

Once a program has been sent to the NXT brick, the robot can run the program independent of the computer. Even when you unplug the robot's USB cable, the program should continue to run.

manually running a program

When a program ends or when you abort a program by pressing the Exit button on the NXT brick, you can restart the program manually using the NXT buttons, as discussed in Chapter 2. You can find all the programs downloaded to the brick by navigating to **My files ▸ Software Files**. Your programs remain in the brick's memory even when you turn off the NXT, allowing you to run them whenever you want, including when you're not near a computer.

downloading a program without running it

Sometimes it is not very helpful to have a program run automatically once you've finished downloading it to your robot. For example, if you program your robot to drive around the room and you still have it to connected to your computer by a USB cable, your robot might get stuck because of that connected USB cable.

To transfer a program to the brick without having the program run automatically, click the **Download** button on the NXT controller. Once the program finishes downloading (as indicated by the beeping sound), you can disconnect the USB cable and then start the program yourself using the buttons on the brick.

using Bluetooth to download programs to the NXT

You can transfer programs to the brick using Bluetooth instead of a USB cable. If you'd like to use Bluetooth, please see the appendix.

working with the NXT-G software

You've just seen what creating programs is all about, but you'll have to look at some more options of the NXT-G software before you can create more complex programs. Figure 3-10 shows you a few more important sections of the software.

5. configuration panel

If you click a programming block on the Work Area, the *Configuration Panel* shows up at the bottom of the software window (Figure 3-11).

You learned that each block makes the robot perform a certain action. The block you see in Figure 3-10 makes the Explorer move, but the exact movement is specified with the

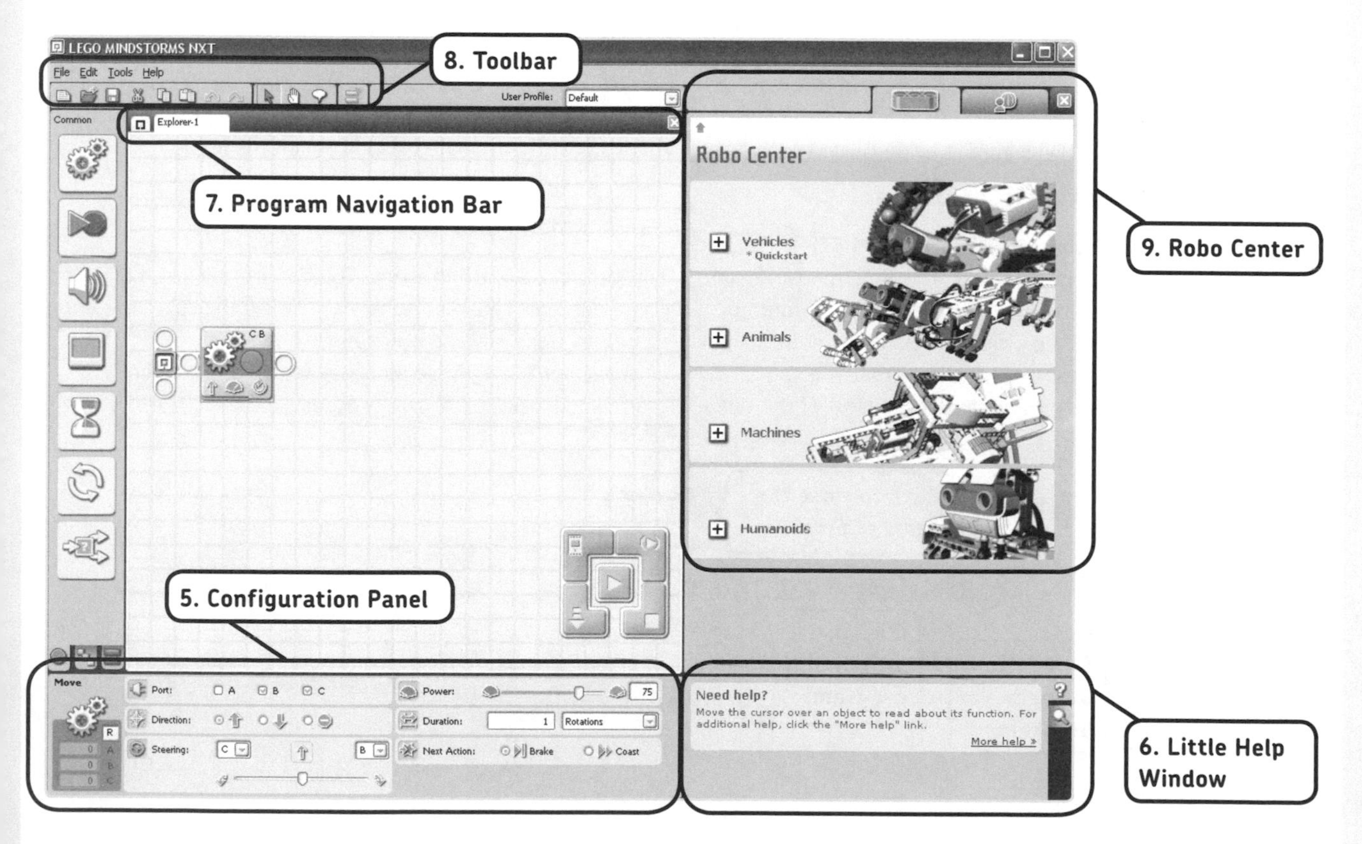

Figure 3-10: The section "Working with the NXT-G Software" discusses these remaining sections.

Figure 3-11: The Configuration Panel

Figure 3-12: The Little Help Window gives you brief information about a selected programming block.

Configuration Panel. For instance, you can use the panel to *configure* the robot to move backward instead of forward.

Many different programming blocks exist, each with its own Configuration Panel. Subsequent chapters will teach you about many of these blocks and their Configuration Panels.

6. little help window

If you keep your mouse on a programming block for a few seconds, the *Little Help Window* shows up at the bottom right of the screen (Figure 3-12). This window tells you what kind of block you selected and gives you brief information about its function.

To learn more about the block in question, click **More help**.

We'll discuss many programming blocks in this book, but if you are looking for specific information, such as the settings of a certain block, try using the Little Help Window.

7. program navigation bar

You can work on more than one program at the time. Each program has a corresponding tab on the *Program Navigation Bar* at the top of the Work Area. To navigate to a program, simply click the tab with the program's name on it, as shown in Figure 3-13. You'll also find a tab here that you can click to return to the main menu and, at the top right of the Work Area, a button to close the program on which you are currently working.

8. toolbar

At the top left of the screen is the *Toolbar* (see Figure 3-14). The buttons on this bar are used to manage and modify your programs. (The Toolbar also has a few drop-down menus, but you won't be using them right now.)

managing programs

Use the three buttons on the left of the Toolbar to create new programs (New Program), open ones you've made (Open Program), or save programs you are currently working on (Save Program).

When you save a newly created program for the first time, you'll be asked to enter a name for the program. (Note that you can also create or open programs in the main menu, as you saw when creating your first program.)

modifying programs

The next three buttons on the Toolbar are the Cut, Copy, and Paste buttons. These buttons are just like the cut, copy, and paste functions in word processors. Instead of copying text, however, you'll use these buttons to copy one or more programming blocks, as shown in Figure 3-15.

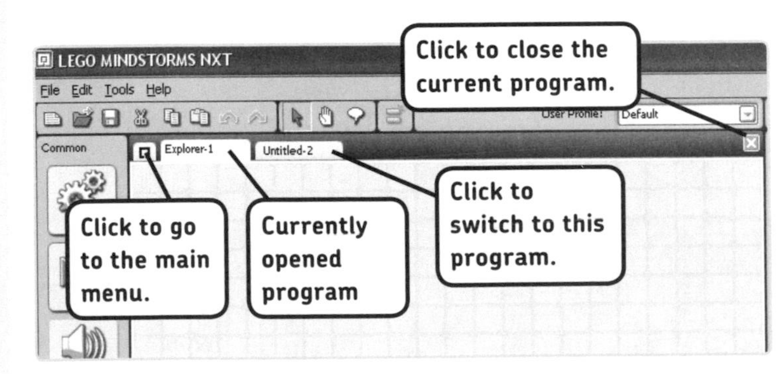

Figure 3-13: You can navigate between different programs and the main menu by clicking the corresponding tab. To close a program you're currently working on, click the red button at the top right of the Work Area.

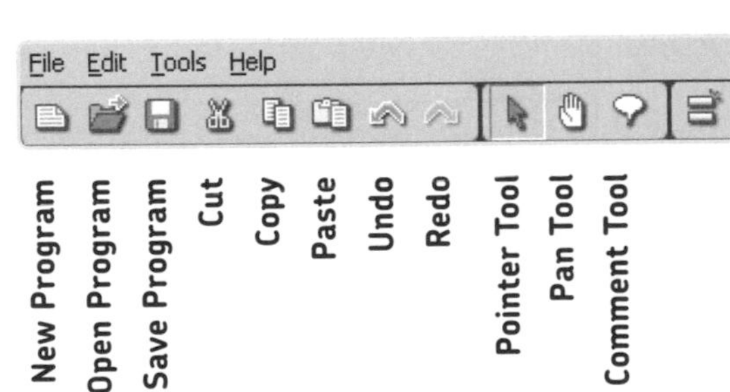

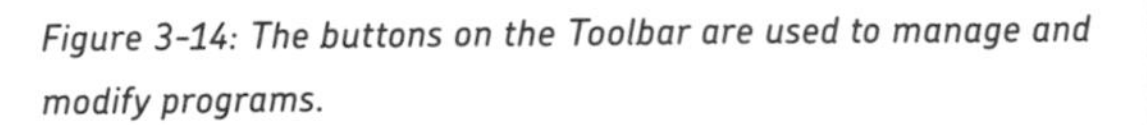

Figure 3-14: The buttons on the Toolbar are used to manage and modify programs.

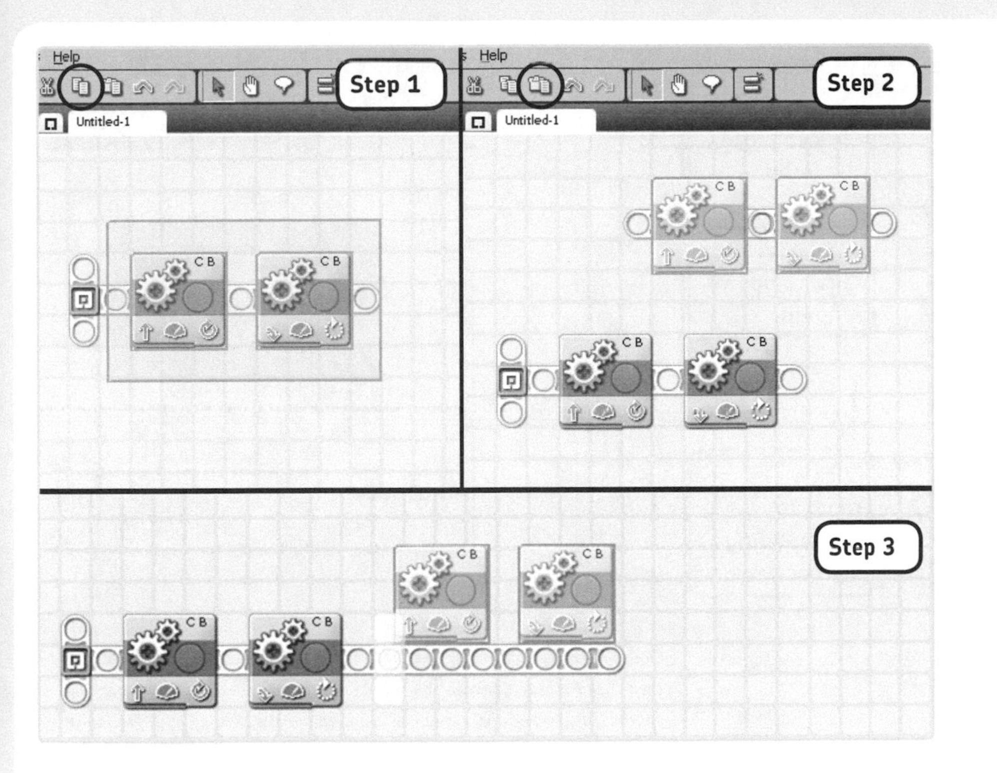

*Figure 3-15: Copying a series of blocks. Step 1: Keep the left mouse button pressed while you drag a selection around the blocks you want to copy; then press the **Copy** button. Step 2: Click the **Paste** button. Step 3: Use the left mouse button to drag the new blocks next to the blocks that were already there.*

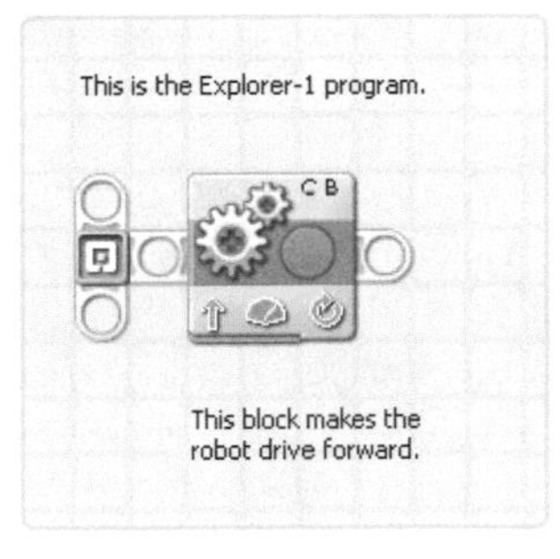

Figure 3-16: Comments in a program

The Undo and Redo buttons are used to undo or redo changes you make to a program. For instance, you can undo the event in which you accidentally deleted a block.

using the pointer, pan, and comment tools

When creating programs, you can use three tools to edit or navigate around programs. The Pointer Tool is most commonly used. When the Pointer Tool is selected on the Toolbar (Figure 3-14), you can use the mouse to place, move, and configure programming blocks on the Work Area.

If you select the Pan Tool, the mouse is used to move the Work Area. This is especially useful when you make large programs that don't fit on your computer screen. To navigate to a specific part in your program, select the **Pan Tool**, click somewhere in the Work Area, and drag this area around by moving the mouse while you keep the left mouse button pressed.

Next on the Toolbar is the Comment Tool. Use this tool to place your own *comments* on the Work Area. These comments do not affect the actions of your program, but they will help you remember what each part in a program does when you review the program later. To add a comment, select the **Comment Tool** on the Toolbar, click where you want to place the comment on the Work Area, and enter the comment. You can see an example of a program with comments in Figure 3-16.

9. robo center

On the right of the software window is the *Robo Center*, as shown in Figure 3-17. This section contains building and programming plans for the four robots featured on the LEGO MINDSTORMS NXT 2.0 box. To view these instructions, click one of the robots.

When you are creating programs for the robots in this book, you won't be using the Robo Center, so go ahead and close it (by clicking the red X button) to free up space on your computer screen to work on your program. You can open it at any time by clicking the orange beam, as shown in Figure 3-17.

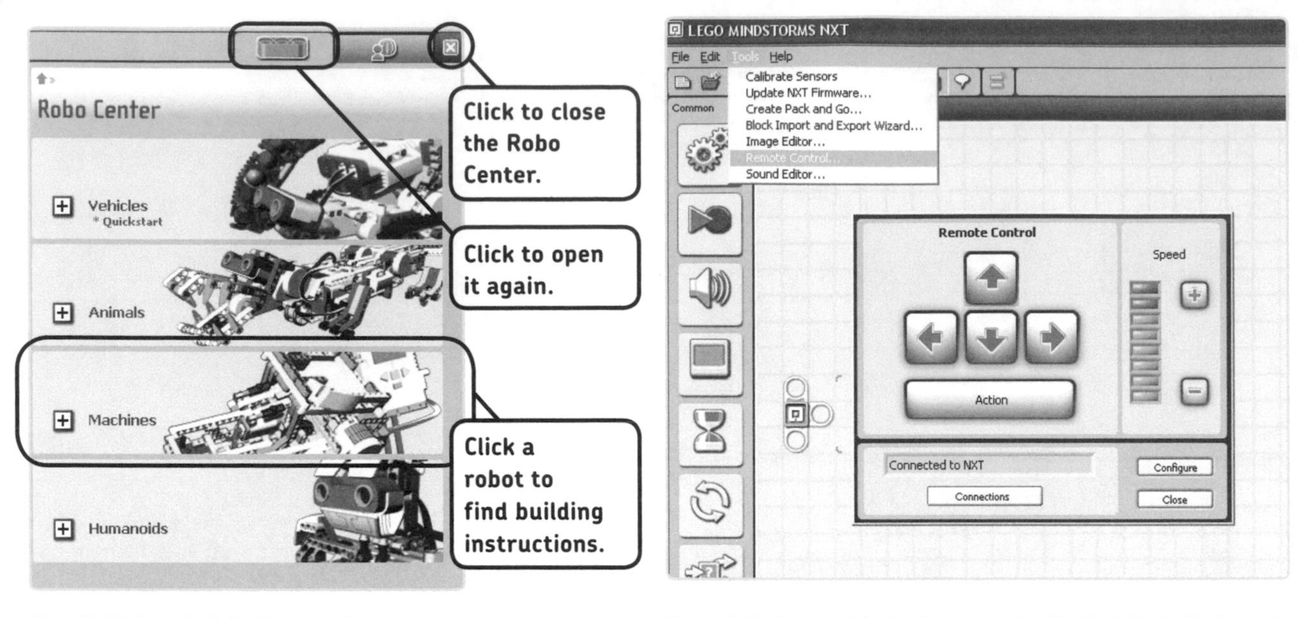

Figure 3-17: Close the Robot Center to free up space on the programming field. To reopen it, click the orange LEGO beam.

*Figure 3-18: To control the Explorer remotely, click **Tools** in the Toolbar and then click **Remote Control**. You should now be able to control the robot by clicking the arrows onscreen or by using the arrow keys on your keyboard. To adjust the robot's speed, click the + and – buttons. To return to programming, click the **Close** button.*

controlling the robot remotely

Programming a robot to move autonomously (as you'll learn in Chapter 4) is certainly fun, but you can also control it remotely from your computer keyboard using the arrow keys. To do so, make sure that the robot is connected to the computer (either via the USB cable or via Bluetooth), and follow the instructions in Figure 3-18.

conclusion

In this chapter, you've learned a lot about working with the NXT-G programming software. You should now know how to create, edit, and save programs, as well as how to transfer programs to the NXT brick.

Now that you've gained this essential bit of information, you're ready to take on some serious programming challenges in Chapter 4.

working with programming blocks: move, sound, and display

In Chapter 3, you learned the basics of how to create a new program and transfer it to the Explorer robot. You also learned that the programs you make are collections of *programming blocks*, which are instructions that tell the robot what to do. In this chapter, you'll learn more about these programming blocks and how to use them to make working programs.

You'll begin by learning more about how to make the Explorer move as you explore NXT programming in more detail. You'll also learn how to make the robot "talk" and display text or images on the NXT screen. As you practice with the sample programs in this chapter, you'll be challenged to solve some programming puzzles by yourself!

what do programming blocks do?

NXT programs consist of a series of programming blocks, each of which is used to make the robot do something specific, such as move forward for one second. All the blocks are put on the *Sequence Beam*, as shown in Figure 4-1.

The NXT program runs the blocks on the Sequence Beam one by one, beginning with the first block. Once the first block finishes running, the program continues with the second block, and so on. Once the last block finishes running, the program ends.

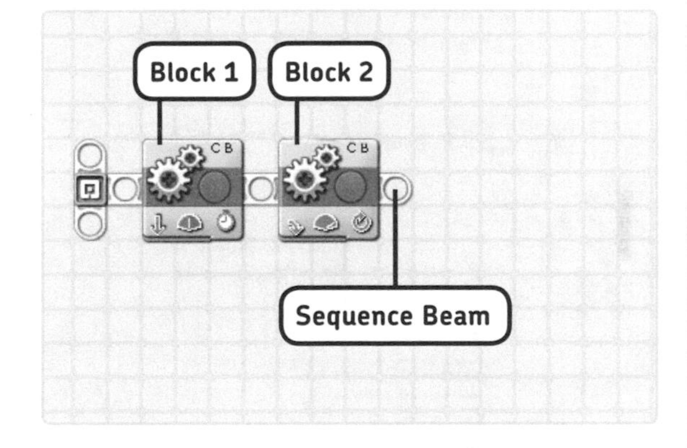

Figure 4-1: Two programming blocks on the Sequence Beam

using blocks to create programs

As you learned in Chapter 3, you add a block to a program by selecting the block from the Programming Palette and placing it on the Work Area. Once you've done that, you can modify the block's actions in the Configuration Panel. For example, you can configure a block to make the robot move backward instead of forward.

When you've finished creating your program, you download it to the NXT brick and run it.

using different programming blocks

Many different programming blocks exist, including ones that make the robot move or make sounds. Each block has its own name and unique look so you can easily tell the difference between the blocks that you place on the Work Area. Different combinations of blocks and settings will make your robot behave differently, and this chapter will teach you how some of the essential programming blocks work.

the move block

The first block you'll learn to use is the *Move block*, which controls the movement of a robot's motors. By using this block in your program, you can make the Explorer move forward and backward and steer left or right.

For example, you used a Move block in Chapter 3 to make your Explorer move forward for a short while.

seeing the move block in action

Before you learn how the Move block works, you'll make a small program to see its functionality in action. This program will make the Explorer drive backward for three seconds and then quickly spin to the right. Because these are two different actions, you'll use two Move blocks.

1. Create a new program called **Explorer-Move**, and then pick two Move blocks from the Programming Palette and place them on the Work Area, as shown in Figure 4-2.

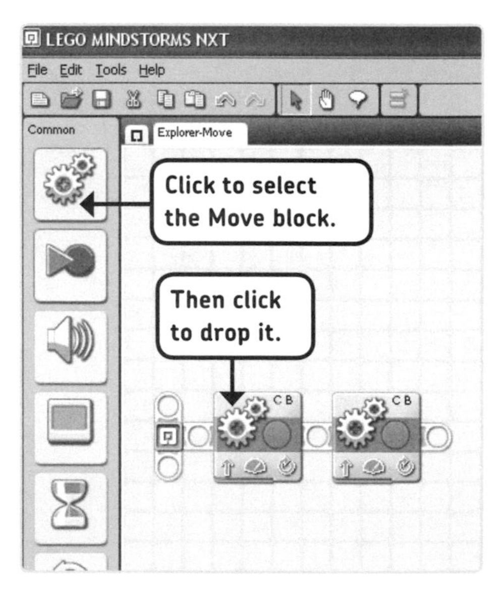

Figure 4-2: Choose a Move block from the Programming Palette, and place it on the Start Area. To place the second block, choose another block, and put it next to the first one.

2. By default, the blocks you've just placed are configured to make the robot go forward for a little while. However, you want the first Move block to make the robot drive backward and the other block to steer the robot. To accomplish this, you'll change the settings in the Configuration Panel of each block, as shown in Figure 4-3.

3. Next, you'll modify the configurations of the second block, as shown in Figure 4-4. This block will make the Explorer spin quickly to the right. When both wheels have made two complete rotations, the motors will stop as specified in the block. (You'll learn more about each of the settings in the next section.)

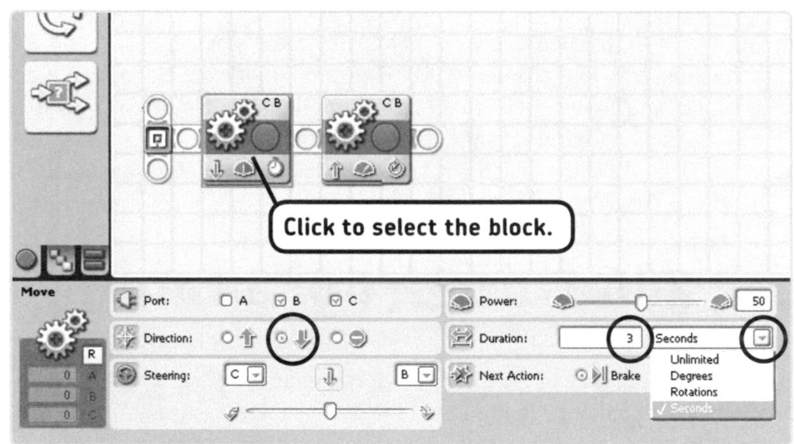

Figure 4-3: The configuration of the first block in the Explorer-Move program. Click the block to select it, and configure your block settings to look like those in this figure. The selected block makes the Explorer drive backward slowly for three seconds.

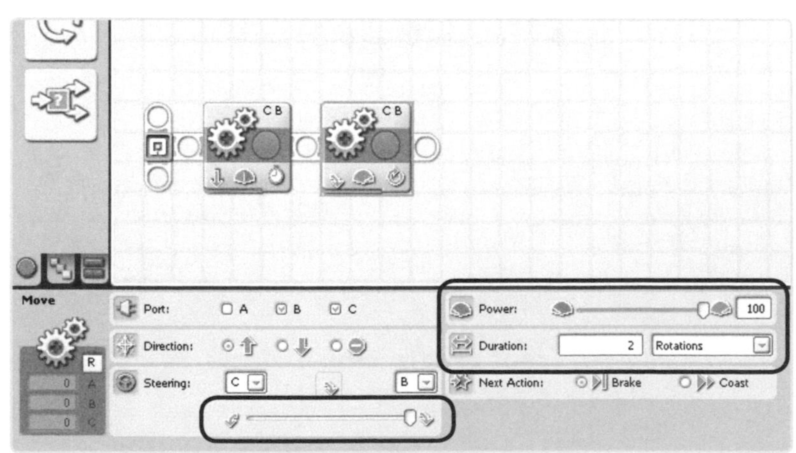

Figure 4-4: The configuration of the second block in the Explorer-Move program. Before applying these settings, click the second Move block to open its Configuration Panel.

4. Once you have configured both Move blocks, you can download the program to your robot and run it. The Explorer should go backward for exactly three seconds and then turn around quickly.

understanding the configuration panel

You'll now take a closer look at the settings on the Configuration Panel for each block to better understand how the sample program really worked. The combination of all the settings determines what the block will do.

Figure 4-5 shows the Move block's Configuration Panel. At the top left of the panel you see the word *Move* and an image of two gears indicating that this is a Move block. Each block has its own name and image. (You'll also see a small section with several letters and numbers just below this image, but you won't use these values just yet. When configuring blocks, you can ignore these values.)

The rest of the Configuration Panel is divided into light gray boxes, each of which configures one setting. For example, you use the Power setting to set the robot's speed. You'll now look at the functions of each of the settings.

port

Because the Explorer is driven by two motors connected to output ports B and C with cables, ports B and C are selected in this block's panel. If you build a robot with different connections (such as ports A and B), select the appropriate ports in the Ports box.

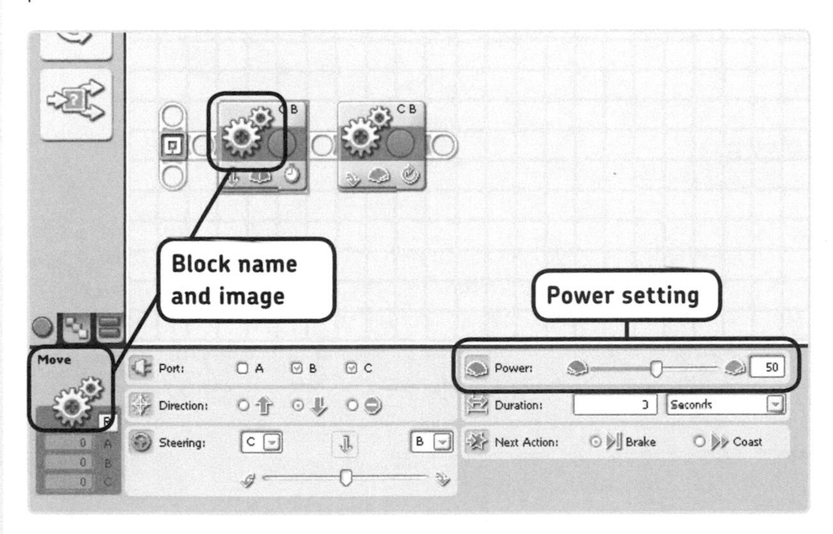

Figure 4-5: The Move block Configuration Panel contains several subsections to change the way the robot moves.

direction

In the Direction box, you select whether you want the robot to move forward or backward. In other words, you select the direction in which the motors must turn. If you select the upward-pointing arrow, the robot will go forward; select the downward-pointing arrow, and it will move backward.

You can also use the Move block to make the motors stop moving by selecting the stop sign in this Direction box. (This is useful only when the robot is already moving because of another block, as you'll see in "More on the Move Block: Moving Unlimited" on page 44.)

steering

As you saw with the Explorer-Move program, you can also use a Move block to make the robot steer. To adjust your robot's steering, drag the Steering slider to the left (to make the robot steer to the left) or right.

But how can a vehicle turn without a steering wheel? Figure 4-6 shows the effect of different combinations of Direction and Steering configurations. Notice that the Explorer can turn by altering the speed and direction of both wheels.

power

Use the Move block's Power box to control the speed of the motors. Zero power means that the wheels do not move at all, while 100 sets the motors to maximum speed.

duration

In the Explorer-Move program you saw that you can control how long a certain movement will run. For instance, setting Duration to **3 Seconds** made the Explorer move for three seconds. Other options here include Rotations and Degrees:

* Rotations controls the number of times that the wheels go through a complete rotation. For example, when you set Duration to 2 Rotations in the sample program, both wheels made two complete rotations.
* Degrees works like Rotation, except that it controls the number of degrees that the wheels turn. For example, you could set Duration to 180 Degrees to make the wheels turn one-half rotation.

(You'll learn about the Unlimited option in Chapter 5.)

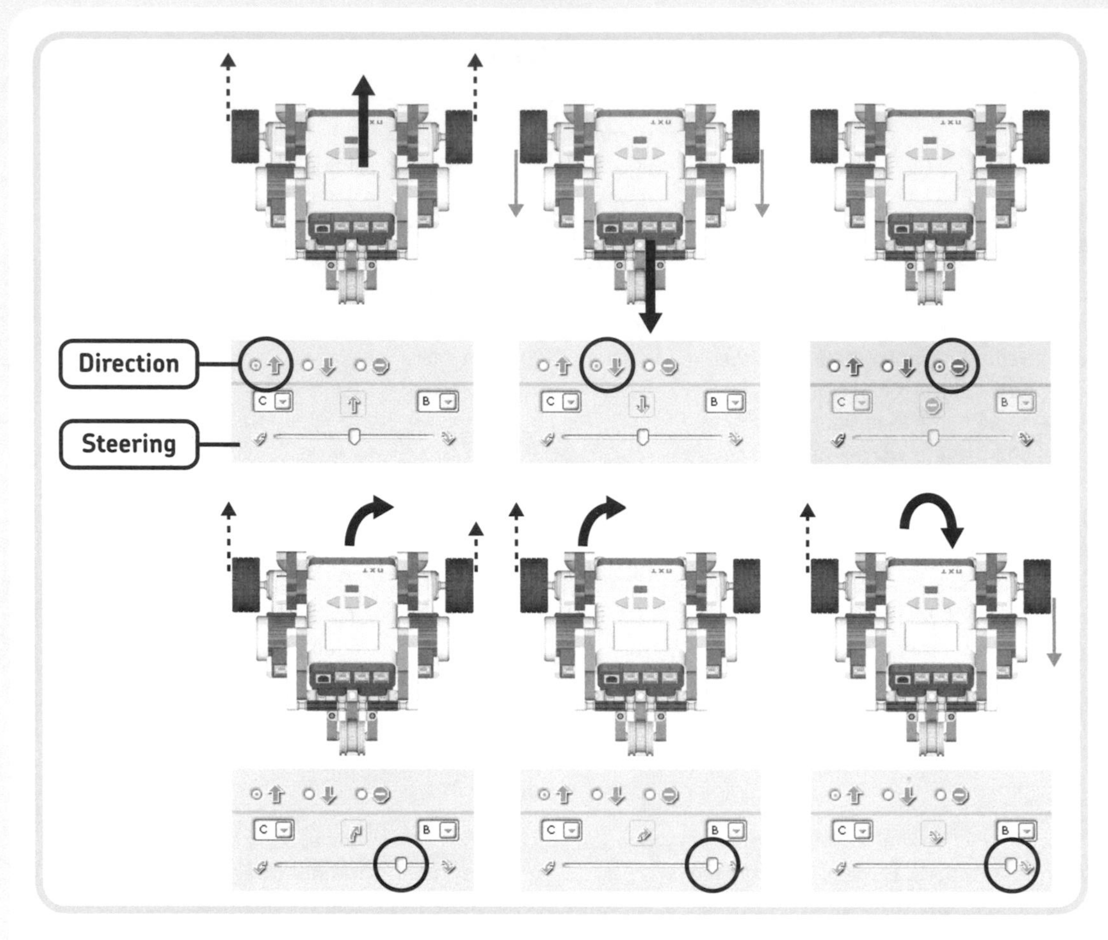

Figure 4-6: To make the robot turn, adjust the Steering setting in the Move block's Configuration Panel. When you do so, the NXT should control the motor's speed and direction to make the robot turn. The dashed arrows indicate that a wheel rolls forward, and the gray arrows mean that the wheel rolls backward. The big black arrow shows the direction in which the robot eventually moves.

DISCOVERY #1: ACCELERATE!

Difficulty: Easy

Now that you've learned some important information about the Move block, you're ready to experiment with it. The goal in this discovery will be to create a program that makes the robot move slowly at first but accelerate as the program progresses. To begin, place 10 Move blocks on the Work Area, and configure the first one as shown in Figure 4-7. Configure the second one in the same way, but set the motor's Power setting to 20. In subsequent blocks, set the Power setting to 30, and so on, incrementing by 10 with each block. What happens when you run this program?

Figure 4-7: The Configuration Panel of the first block of the program for Discovery #1

NOTE **One complete motor rotation is equal to a 360-degree turn. To find the number of rotations, divide the number of degrees by 360. To find the number of degrees, multiply the number of rotations by 360. For example, instead of entering 180 degrees in the Duration box, you could also use 0.5 rotations.**

next action

The Next Action box controls what happens after the Move block completes its movement:

* Brake stops the motors immediately.
* Float stops the motors gently.

understanding the configuration icons

As you change a block's settings in the Configuration Panel, the *configuration icons* on the block change as well, as shown in Figure 4-8. By looking at these icons, you can roughly determine what a block does. This is a useful way to get a general overview of how a program works.

For instance, as you can see in Figure 4-8, the second Move block is configured to turn right. You can also see that the robot will turn at the maximum speed and that the Duration setting is configured as Rotations. To determine what an icon on a block means, you can either check how it changes as you change the Configuration Panel or click **More help** in the Little Help Window.

NOTE **When creating the programs in this book, you don't need to remember what each specific configuration icon stands for because the figures show you the Configuration Panel of each block, making it easy for you to re-create the program.**

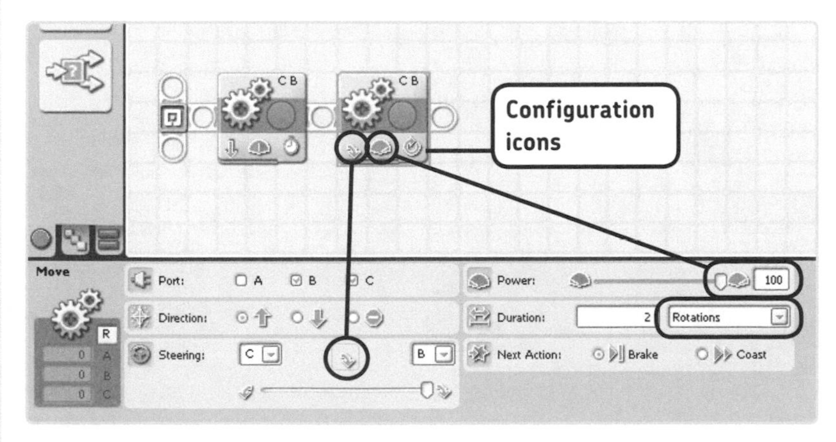

Figure 4-8: Configuration icons on the blocks in the Explorer-Move program

making accurate turns

When you use a Move block to have the robot make a 90-degree turn, you might think you need to set the Duration setting to 90 degrees, but this is *not* correct. The Duration setting specifies only how many degrees the *motors* (and therefore the wheels) turn. The actual number of degrees that the motors should turn in order to enable the robot to make a 90-degree turn is different for every robot. Discovery #2 gets you started finding the appropriate number of degrees for your robot.

DISCOVERY #2: LOOKING BACK!

Difficulty: Easy
Can you get your robot to make a 180-degree turn in place? Create a new program with one Move block, and make sure that the Steering slider in the Configuration Panel is shifted all the way to the right. Now set the Duration box to **Degrees**. What number do you need to fill in for Degrees in order to make an accurate 180-degree turn? Begin by setting Degrees to **500**. If this is not enough, try **550**, **600**, or maybe even something higher.

DISCOVERY #3: MOVE THAT 'BOT!

Difficulty: Medium
Make a program that uses three Move blocks to move the Explorer forward for three seconds at 50 percent power, make a 180-degree turn, and then return to its starting position. When configuring the Move block that lets the robot turn around (the second block), use the Duration value that you found in Discovery #2.

the sound block

It's fun to make programs that allow the Explorer to drive around, but things become even more fun when you can program the NXT with sounds using the *Sound block*. Your robot can play two types of sounds: a simple *tone* (like a beep) or a *sound file*, such as applause or a spoken word such as "Yes." When you use a Sound block in your programs, the robot will be appear to be more interactive and lifelike because it can "talk."

understanding the sound block configurations

Even though every programming block allows the robot to do something different, all blocks are used in the same way. In other words, you can simply pick a Sound block from the Programming Palette and place it on the Work Area as you did with the Move block. Once it's in place, you adjust the block's settings in the Configuration Panel.

Before you start creating bigger programs with Sound blocks, you'll take a quick look at the Sound block's Configuration Panel. Create a new program called **Explorer-Sound**, and place one Sound block on the Work Area, as shown in Figure 4-9.

action

Use the Action setting on the block's Configuration Panel to set whether the block should play a sound file or a tone.

DISCOVERY #4: ROBOSPELLER!

Difficulty: Medium
Use Move blocks to make a program that enables the Explorer to drive as if it were drawing the first letter of your name. How many blocks do you need for your letter?

HINT For curved turns, use the Steering slider to adjust the tightness of turns.

Depending on the option you select, the Configuration Panel changes slightly, as explained next.

control

Normally, the Control box is specified as Play. To abort a sound that is currently playing, select **Stop**.

volume

Use the Volume setting to make the sound softer or louder.

function

In the Function box, you can specify whether to repeat a sound by checking the **Repeat** checkbox. To stop the sound from repeating, use another Sound block with the Control setting configured to stop the sound.

file

If you've configured the Sound block's Action setting to play a sound file, you can select one from a list in the File box. You can choose from various sounds like words (such as "Hello"), numbers (such as "Two") or short phrases (such as "You're good"). If you use multiple Sound blocks in a row, with each configured to play a sound file, you can make your robot appear to talk.

You can also create your own sound files by recording your voice with a microphone or by using an existing music file. To do so, select **Tools ▸ Sound Editor** on the Toolbar.

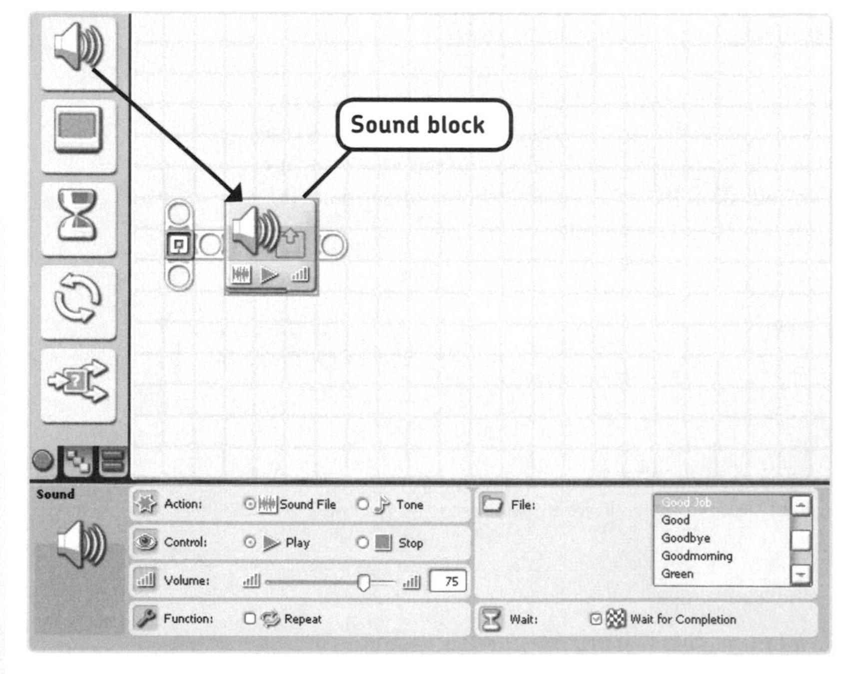

Figure 4-9: The Sound block and its Configuration Panel

(To learn more, see page 56 in the NXT's user guide.) Once you've made your own sound file, you can use it in your programs with the Sound block, just like a regular sound file.

NOTE If you use sounds files in your program, they will be downloaded to the robot as you transfer your program. If you use a lot of different sounds, you might get the message that "the NXT device is out of memory." To solve this problem, see the appendix.

note

If you configure the Sound block's Action setting to play a tone, a different box appears in the Configuration Panel, as shown in Figure 4-10. Instead of selecting a sound file, you can now select a note from a keyboard, as well as the length of time that the tone should sound.

wait

You'll find the last setting in the Wait box. The Wait for Completion checkbox should be checked if you want the program to wait for the sound to finish playing before continuing with the rest of the program, as explained in the next section.

seeing the sound block in action

You'll now learn how to make a program that uses Sound blocks so that you can see how they work. The program will allow the robot to move around while making sounds.

To begin, open the Explorer-Sound program (or create it if you haven't already done so), and place two Sound blocks and two Move blocks in it, as shown in Figure 4-11. Configure the blocks as shown in the Configuration Panels. When you've finished creating your program, download it to your robot, and run it.

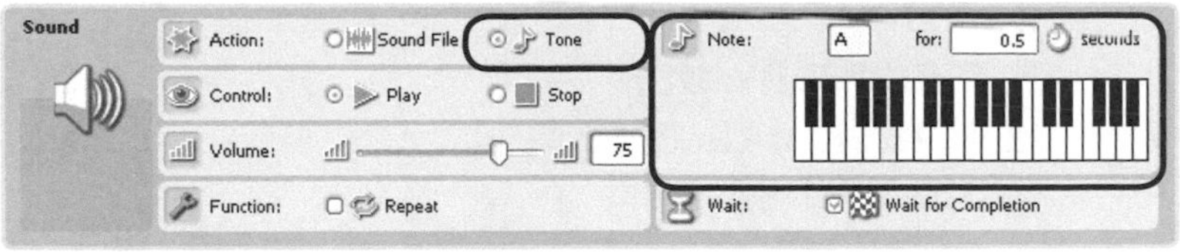

Figure 4-10: The Configuration Panel of a Sound block that is configured to play a tone

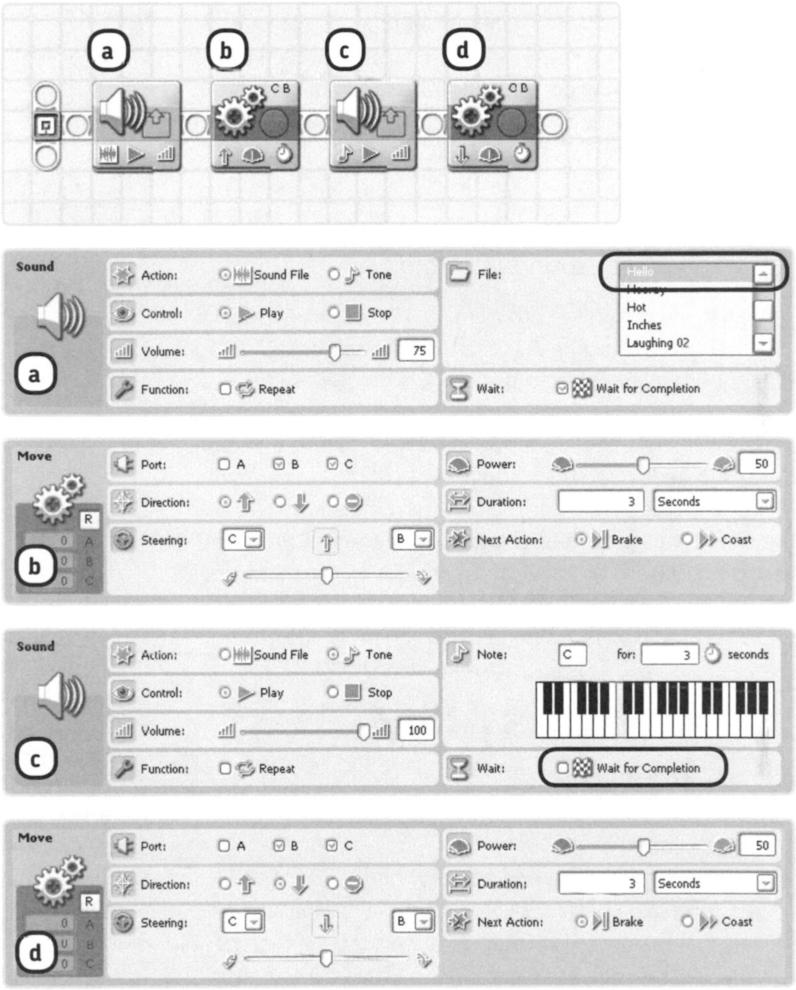

Figure 4-11: The four blocks of the Explorer-Sound program. Use these images of the Configuration Panels to configure each of the programming blocks. Panel "a" shows the settings of block "a."

NOTE The letters *a*, *b*, *c*, and *d* are there to help you understand which Configuration Panel belongs to which block. They're not part of the actual program.

DISCOVERY #5: WHICH DIRECTION DID YOU SAY?!

Difficulty: Easy
Create a program like the Explorer-Sound program that has the robot announce the direction in which it's moving as it moves. While going forward, it should say "Forward," and while going backward, it should say "Backward." How do you configure the Wait for Completion settings in the Sound blocks?

DISCOVERY #6: BE THE DJ!

Difficulty: Medium
By making a program with a series of Sound blocks configured to play notes, you could play your own musical compositions. Can you play a well-known tune on the NXT or create your own cool tune? Post your program to the companion website (*http://www.discovery.laurensvalk.com/*) to see what others think!

understanding the explorer-sound program

Now that you've run the program, you'll learn how it works. The first Sound block makes the Explorer say "Hello." The Wait for Completion box in this block is checked, so the robot waits until the robot finishes saying "Hello." Once it has, a Move block makes the robot drive forward for three seconds. Next, another Sound block causes the NXT to play a tone. This block doesn't wait for the tone to complete, so while the sound is playing, the second Move block gets the robot to drive backward for three seconds. Finally, the robot stops moving. Because the tone was also configured to play for three seconds, the program ends.

the display block

In addition to moving around and playing sounds, an NXT program can also control what is displayed on the NXT's LCD screen. For example, you could create a program that makes the NXT display look like Figure 4-12. (The LCD screen is 100 pixels wide and 64 pixels tall. Pixels are like small dots.)

You use the *Display block* to display an image (like a LEGO minifigure head), text (a word like "Hello"), or drawings (such as a line) on the NXT screen.

NOTE The Display block can't put multiple images or text lines on the screen at once, so you'll need to use a series of Display blocks in your program in order to create the display shown in Figure 4-12.

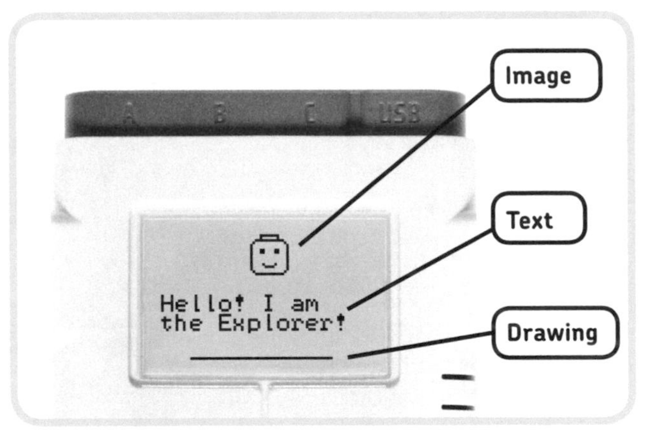

Figure 4-12: Using Display blocks, you can show images, text, and drawings on the screen of the NXT brick.

Once the block has put something on the NXT screen, the program moves on to the next block, say, a Move block. The NXT screen keeps showing the image until another Display block is used to display something else. So, in this example, the image remains on the screen while the robot moves.

When a program ends, the NXT automatically displays the NXT's menu. Therefore, when the last block in your program is a Display block, you won't have time to see what's on the display because the program has already ended. To see it, you'd need to add another block, such as a Move block, to keep the program from ending instantly.

understanding the display block configurations

Before you can use Display blocks in your programs, you need to understand how they work. In this section, I'll discuss the settings of the block's Configuration Panel.

Create a new program (call it **Explorer-Display**), and place one Display block on the Work Area, as shown in Figure 4-13.

action

Use the Action setting to choose whether the screen should display an image, some text, or a drawing. The Configuration Panel will change slightly, depending on the option you choose. Reset will make the screen look like the display of a program that doesn't use any Display blocks at all.

display

When the Clear checkbox in the Display box is checked, the NXT removes everything from the screen before displaying something new. This is useful if you don't want different images to overlap each other. Uncheck this option when the screen should *not* be cleared before displaying something, such as when you want to display more than one text line on the NXT screen. Because the screen isn't cleared, the new text line will simply be added below the previous text line on the screen. (You'll practice with this option in the sample program.)

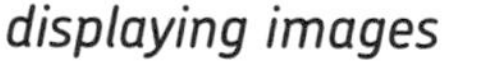

displaying images

The Configuration Panel that you see displayed will depend on which option you select in the action box. For example, if you select Image, the panel should look like the one in Figure 4-13.

* Use the File box to select an image from a list of images. When you do this, you should see a preview of this image on the right side of the panel.
* Use the Position box to define where to place the image. Drag the image around with your mouse, or enter values in the X box (0 is the most left and 63 is the most right) and the Y box (0 is the lowest and 99 is the highest) to set the image's position. (The X and Y coordinates would set the position of the bottom-left corner of the image.)

Just as you can create your own sound files, you can also create your own image files using drawings of squares, lines, and circles, or you can simply use an existing picture. To do so, select **Tools ▸ Image Editor** on the Toolbar. (For more information, see page 57 of the NXT user guide.) When you're done creating your image, you can display it with the Display block as you would any other image.

displaying text

If you select **Text** in the Action box, you'll see a Configuration Panel as shown in Figure 4-14.

Enter the text you want to display in the Text box. Use the settings in the Position box to define where the text should be displayed either by dragging the text with your mouse or by specifying the X and Y values as with displaying images.

You can also position lines of text by setting the Line number. Selecting 1 for Line will display text at the top of the screen, while selecting 8 for Line puts text on the bottom of the screen. This configuration is useful when you want to display a longer sentence that doesn't fit on one line: You can split the sentence into multiple parts and display each part on one text line to keep your text nicely aligned.

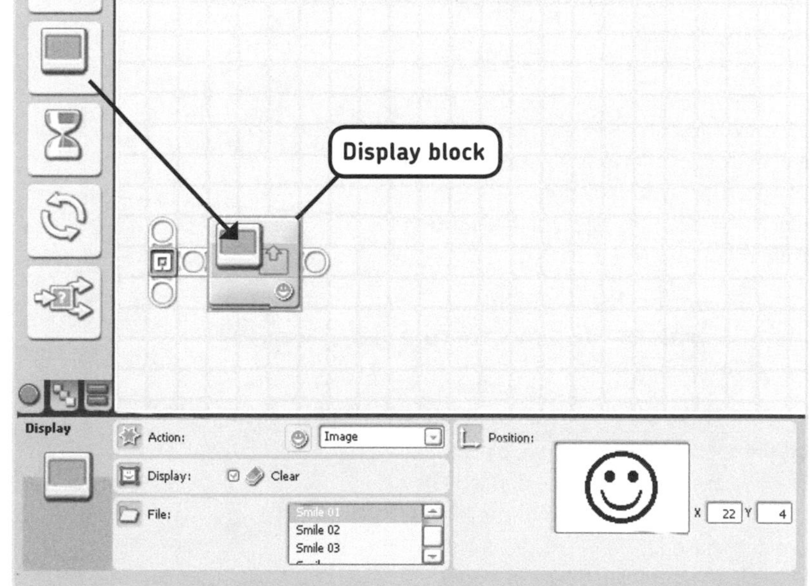

Figure 4-13: The Display block and its Configuration Panel

Figure 4-14: The Configuration Panel of a Display block configured to display text

displaying drawings

To display lines, points (dots), or circles, select **Drawing** in the action box. Your Configuration Panel should now look like Figure 4-15.

In the Type box, select whether to display a point, a line, or a circle. You can adjust the position of each by setting the appropriate values in the X and Y boxes or by dragging the object with your mouse in the Position area. When displaying a line, you can enter the coordinates of the line's starting and ending points. When displaying a circle, use the Radius box to set the circle's size.

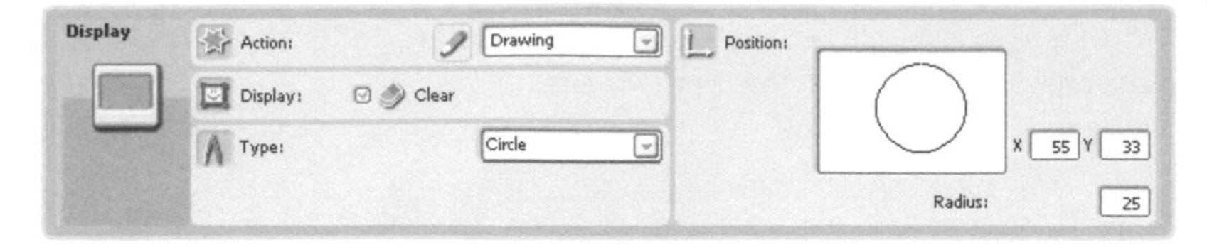

Figure 4-15: The Configuration Panel of a Display block configured to display a drawing

seeing the display block in action

Now that you understand most of the Display block's features, you'll test its functionality by creating a program that puts things on the NXT screen while the robot moves.

To do so, place three Display blocks and two Move blocks on the Work Area, as shown in Figure 4-16; then configure each block as shown in the Configuration Panels. Once you've configured all the blocks, you can transfer the program to your robot and run it.

understanding the explorer-display program

In the Explorer-Display program, the Display blocks simply put several things on the screen, and the Move blocks allow the Explorer to drive around. (Notice how the Clear checkboxes in the blocks are configured.)

The first Display block (block *a*) clears the screen before it displays an image. The robot then starts to move, and the image stays on the screen while the Move block (*b*) is running. The next Display block (*c*) is also configured to clear the screen, so it removes the image before it writes a line of text on the screen. The program then moves on to another Display block (*d*), which puts

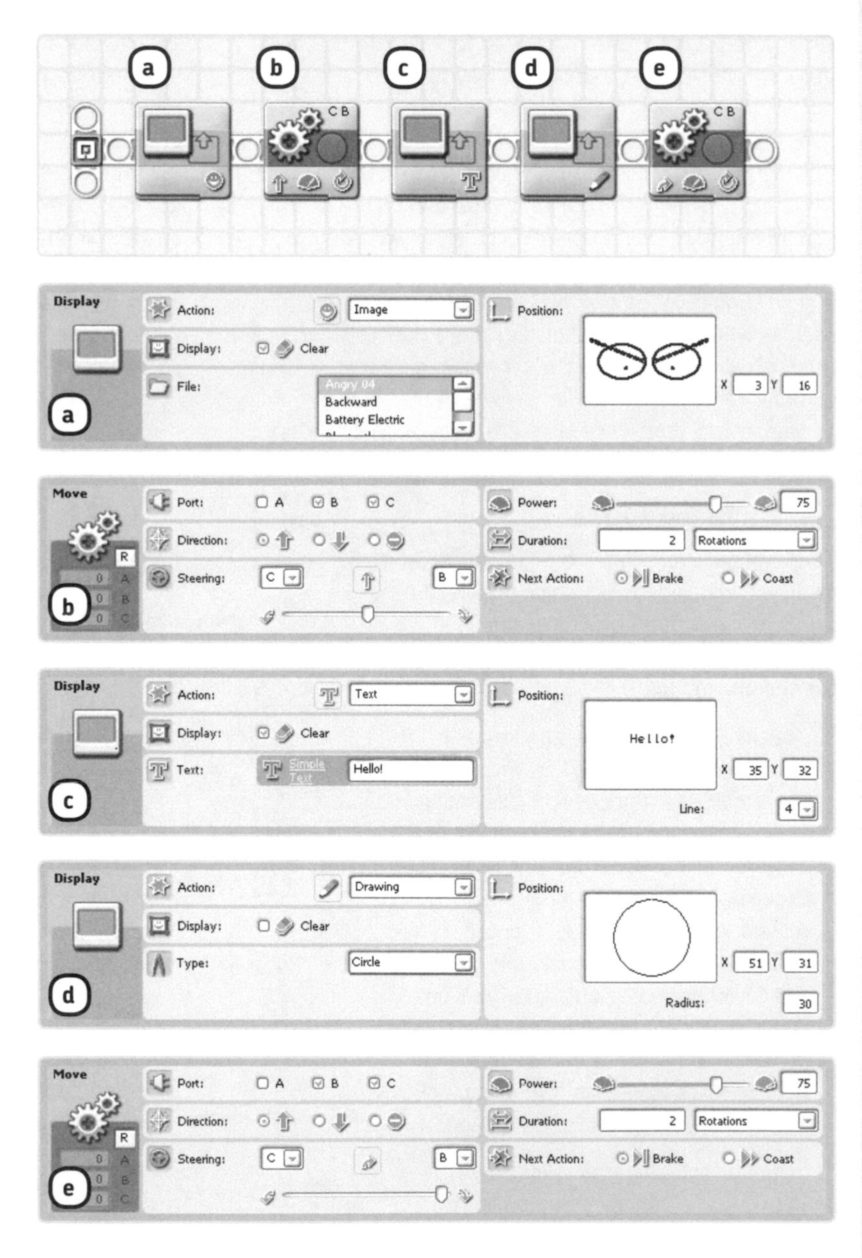

Figure 4-16: The configuration of the blocks in the Explorer-Display program

a circle on the screen. This block doesn't clear the screen in advance, so you'll see both the text and the circle on the screen. Finally, a Move block makes the robot turn to the right, and then the program ends.

further exploration

You've completed the basics of LEGO MINDSTORMS NXT programming. Congratulations! You should now know how to program robots to perform actions such as moving, making sounds, and displaying text and images on the NXT screen. Chapter 5 will teach you more about using programming blocks, including how to use blocks to pause a program and how to repeat a set of blocks.

But before you move on, try to solve some of the following discoveries to further fuel your programming skills, and then share your solutions with fellow book readers at the companion website (*http://www.discovery.laurensvalk.com/*).

DISCOVERY #7: SUBTITLES!

Difficulty: Easy
Create a program that uses three Sound blocks to say, "Hello, sir, thank you!" Use Display blocks to display what the robot says as subtitles on the NXT screen and to clear the screen each time the robot starts saying something new. Where do you place each of the Display blocks?

DISCOVERY #8: NAVIGATOR!

Difficulty: Medium
Create a program with Move blocks that make the robot drive in the pattern shown in Figure 4-17. While moving, the robot should display arrows on the NXT screen that show the direction of its movement. When finished, it should display a stop sign. In addition to displaying the directions, use Sound blocks to enable to robot to speak the direction it's moving in. How do you configure the Wait for Completion box in the Sound blocks?

HINT **You can find all the direction signs shown in Figure 4-17 in the list of images in the Display block's File box. Use Sound blocks to play sounds like "Forward" and "Left."**

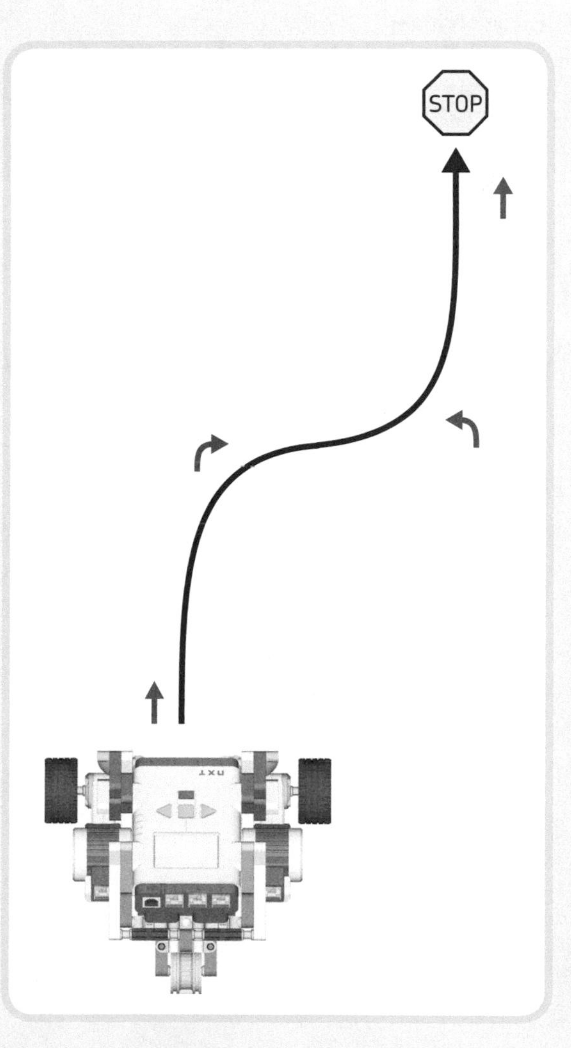

Figure 4-17: The driving pattern and the navigation images of Discovery #8

DISCOVERY #9: CIRCLE TIME!

Difficulty: Easy
Can you make the Explorer drive in a circular pattern with a diameter of about 1 meter (3 feet)? You'll need only one Move block to accomplish this. How do you configure the Duration and Steering settings? And how does changing the motor speed affect the circle?

DISCOVERY #10: W8 FOR THE EXPLORER!

Difficulty: Medium
Program the Explorer to drive in a figure eight, as shown in Figure 4-18. The robot should show a happy face on the screen as it moves.

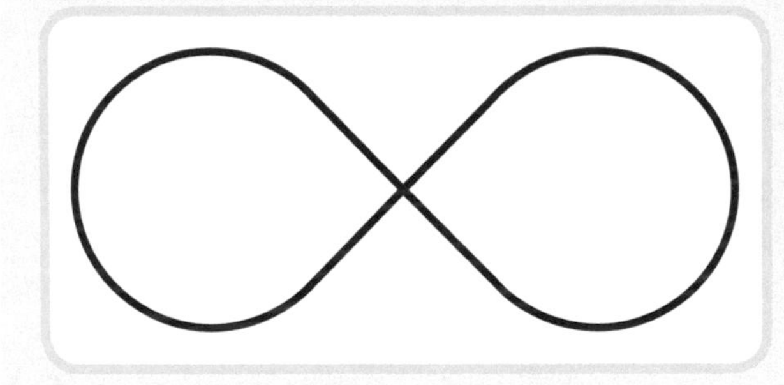

Figure 4-18: The drive track for Discovery #10

DISCOVERY #11: ROBODANCER!

Difficulty: Hard
Make the Explorer play musical beats (using Sound blocks) continuously while it dances in zigzagging movements (using Move blocks). After each movement, the robot should start playing a different sound.

HINT Experiment with the Repeat setting in the Sound blocks.

BUILDING DISCOVERY #1: EXPLORING ART!

In this building discovery you're challenged to expand the Explorer robot design. Using LEGO pieces, create an attachment for your robot to hold a pen. As the robot drives over a big piece of paper, it will draw lines and figures with the pen as it moves. As a start, you can make it draw the pattern in Discovery #11.

Advanced building: Attaching a fixed pen is certainly fun for drawing simple figures, but you won't be able to use a fixed pen to draw words since the pen is constantly on the paper. Use the third motor in your NXT kit to lift up the pen, and connect this motor to output port A with a cable. You can control this motor with a Move block, with the Port setting specified to control only motor A (not B and C). Can you make the robot write your name?

waiting, repeating, and other programming techniques

The previous chapter taught you how to program your robot to perform a variety of actions, such as moving. In this chapter, you'll learn several programming techniques that will allow you to do more with the blocks you used earlier, including how to pause a program with Wait blocks, how to repeat a set of actions with Loop blocks, how to run multiple blocks simultaneously, and even how to make your own so-called My Blocks.

the wait block

So far you've been using three different programming blocks to make the robot move, play sounds, or display something on its screen. Now you'll meet a block that does nothing more than pause the program for a given amount of time. This block is the *Wait block*, as shown in Figure 5-1.

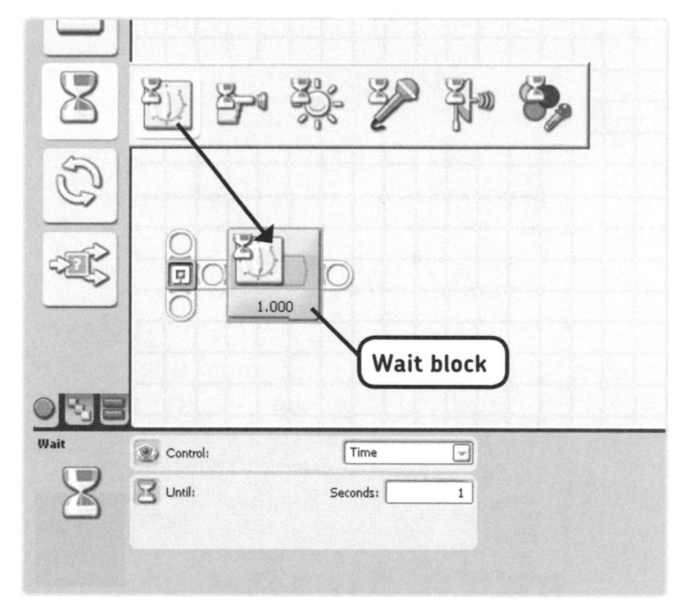

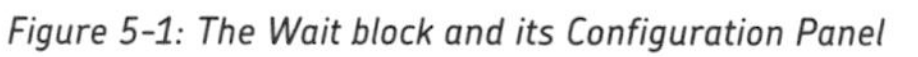

Figure 5-1: The Wait block and its Configuration Panel

understanding the wait block settings

You use the Wait block just like any other programming block. You place it on the Work Area and then configure its settings. The block can operate in two modes, as specified in the Control box of the Configuration Panel: Sensor or Time. You'll use only the Time option in this chapter.

When set to use the Time mode, the Wait block simply pauses the program for a certain amount time, such as five seconds. Once the time has elapsed, the program continues with the next programming block. To enter an amount of time for the block to wait, enter the time in the Seconds box as either an integer (such as 14) or a decimal (such as 0.5).

seeing the wait block in action

Why would you want to use a block that doesn't perform any actions? Here's an example of the Wait block in action. Create the **Explorer-Wait** program according to the instructions shown in Figure 5-2, and run it. This program will display two text lines on the NXT screen.

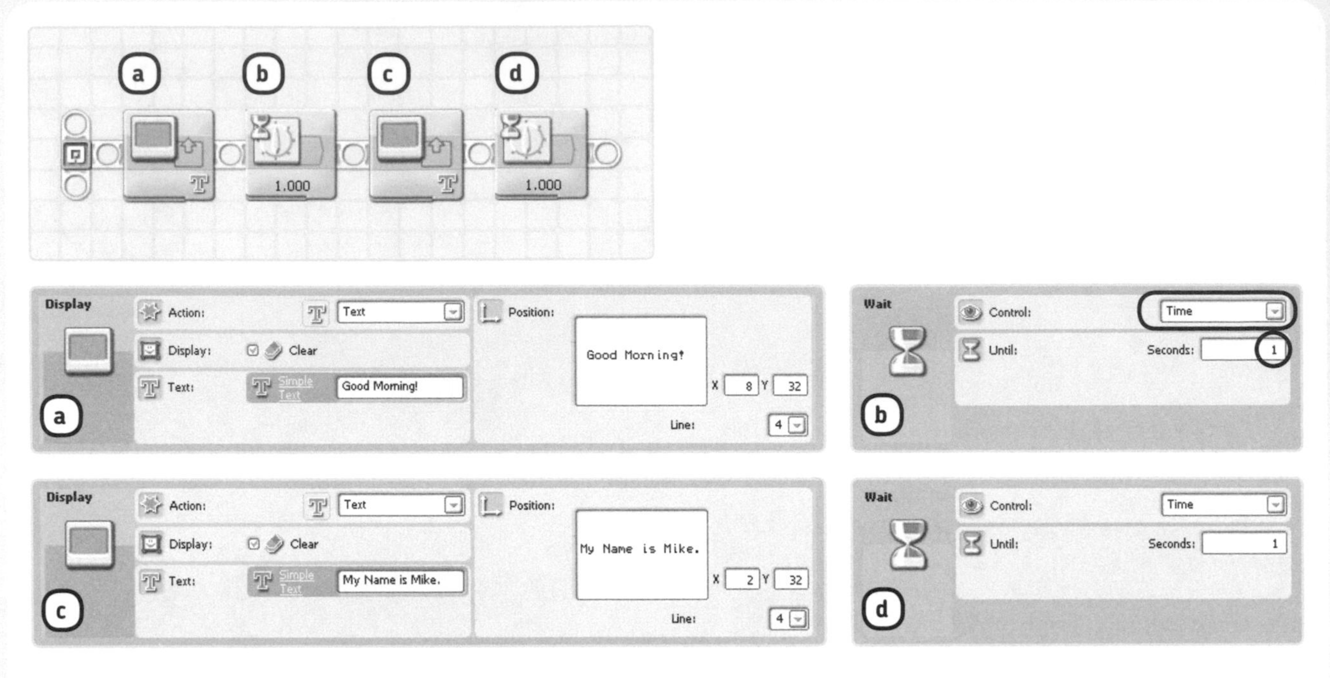

Figure 5-2: The configuration of the blocks in the Explorer-Wait program

understanding the explorer-wait program

When you run the Explorer-Wait program, the Wait blocks give you time to read what is displayed on the NXT screen. Had you not placed the Wait blocks in the program, the program would end immediately after the text was displayed on the NXT screen, making it impossible to read this text.

more on the move block: moving unlimited

The Move block comes in handy when you want to make the Explorer drive around, and you've used it in many of your programs so far. However, there is an important setting on its Configuration Panel that you have not used yet.

the unlimited option

The Move block's Duration box allows you to set how long the robot's motors will run in seconds, degrees, or rotations. Once the robot has moved for the specified duration, the

DISCOVERY #12: COUNT DOWN!

Difficulty: Hard

Create a program that makes the NXT display an exploding bomb after three seconds. The program should count down from 3 to 0, while displaying the remaining time on the screen. Create your program so that the display looks like Figure 5-3 after three seconds. For more fun, use Sound blocks to make the robot say how much time is left and have it shout when the bomb explodes.

Figure 5-3: The appearance of the NXT screen in Discovery #12

motors stop, and the next block in the program runs. When the Unlimited option is selected in the Duration box, the motors are simply switched on, and they run indefinitely. Once a Move block (defined as Duration Unlimited) has switched on the motors (which itself takes almost no time), the program moves on to the next block in the program while the motors are still turning.

To stop the robot, use another Move block set to stop the motors (by selecting the stop sign in the Direction box). The robot will also stop moving once the program has finished running all of its blocks. You can use the Unlimited option to create programs that allow your robot to make sounds while it moves, as you'll see in the following example program.

the duration unlimited setting in action

The next program, named **Explorer-Unlimited**, will help you understand exactly how the Unlimited setting works. This program will allow your robot to make a sound while it drives. Create the program as shown in Figure 5-4.

NOTE When configuring blocks in a program, you may find that you can't access some of the boxes on the Configuration Panel. This occurs when a setting is grayed out, meaning it's not one that you can use. For example, you cannot set the Steering in block *d* in this program because this block makes the robot stop, and you cannot steer while you're braking. When a setting in one of these images is grayed out, just ignore it. It's grayed out for a reason.

understanding the explorer-unlimited program

When you run the Explorer-Unlimited program, a Move block (block *a*) switches on the motors. Because block *a*'s Duration is set to Unlimited, the program instantly moves on to the next block, the Sound block (*b*). After two seconds, the sound stops, and a Wait block (*c*) makes the program wait three more seconds before moving on to the next block. Finally, another Move block (*d*) stops the motors, and a Sound block (*e*) triggers the NXT to play a tone.

problems with moving unlimited

When creating a program with just one Move block with the Duration set to Unlimited, you might think that the robot would go forward indefinitely, but that's not the case. This block only switches on the motors, and the program ends

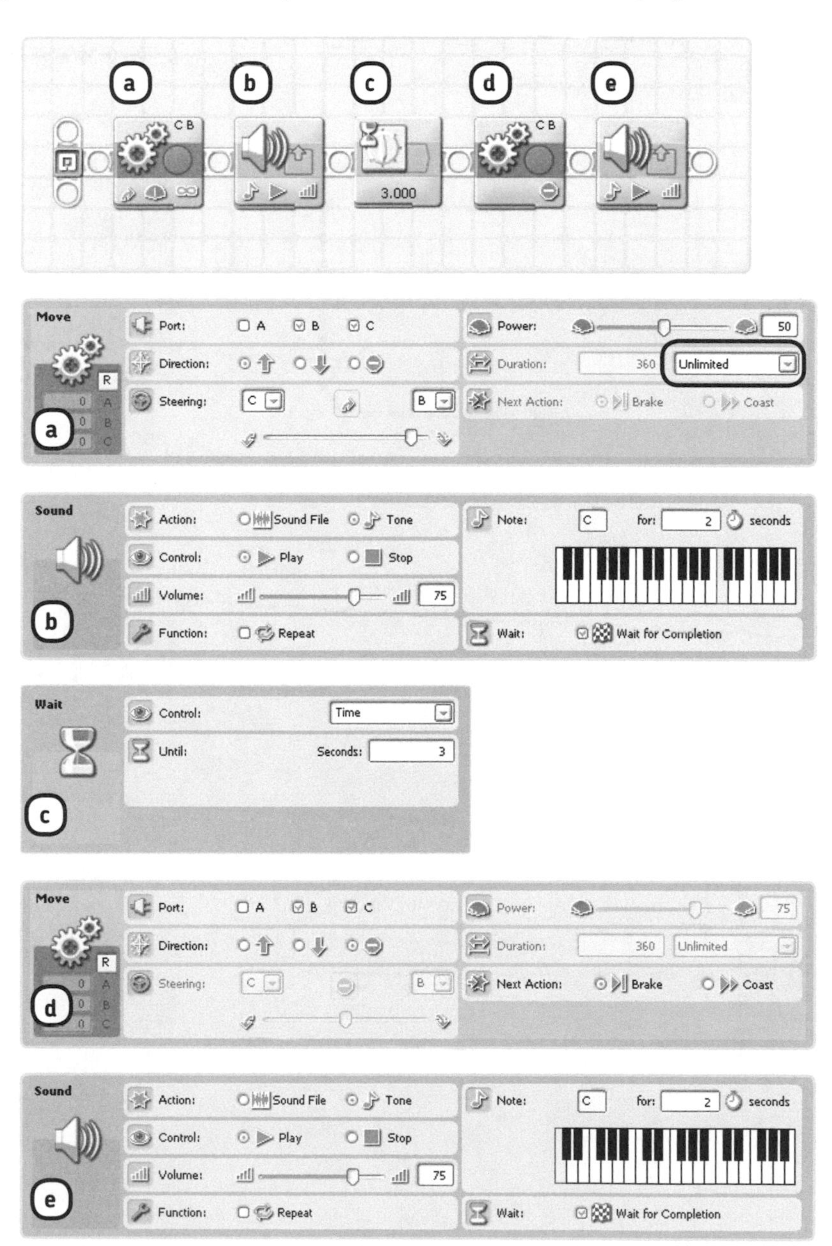

Figure 5-4: The configuration of the blocks in the Explorer-Unlimited program

because it has finished running all of its blocks; and when the program ends, the motors stop. (You'll learn how to create programs that run indefinitely in the next section.)

the loop block

Imagine you are walking on a square-shaped line, like the one shown in Figure 5-5. As you walk, you follow a certain pattern over and over again: Go straight, then go right, go straight, go right, and so on.

To create this sort of behavior with your robot, you could make your robot drive around with Move blocks so that it follows the same pattern. You could make it go straight and then right and use a Move block for each movement. To make your robot trace one complete square and return to the starting position, you would have to use each of these two blocks four times, for a total of eight blocks.

Rather than use eight Move blocks to create this program, it's much easier to use the *Loop block*, which lets you repeat sequences of blocks that are placed within it. Loop blocks are especially useful when you want to repeat certain actions many times. For example, to have your robot drive in a square, you would place the Move blocks to go forward and go right inside the Loop block, which would then run each block's action four times. You'll create just such a program in "Seeing the Loop Block in Action" on page 47.

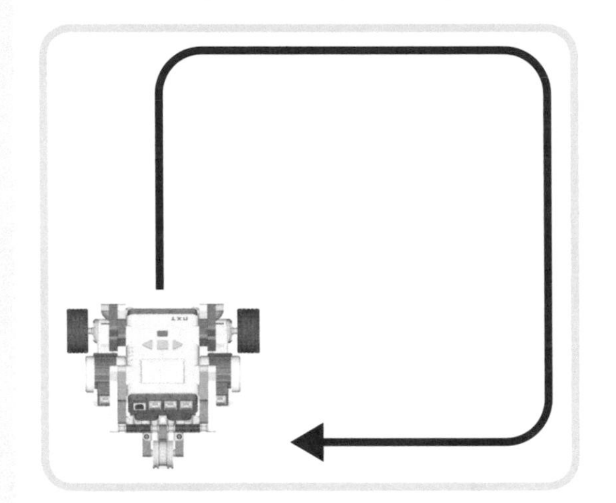

Figure 5-5: The Explorer moving in a square

using the loop block

Figure 5-6 shows the Loop block with its Configuration Panel, as well as how to place blocks inside a loop.

The flow of this program from beginning (left side) to end (right side) is slightly different when using Loop blocks, as shown in Figure 5-7.

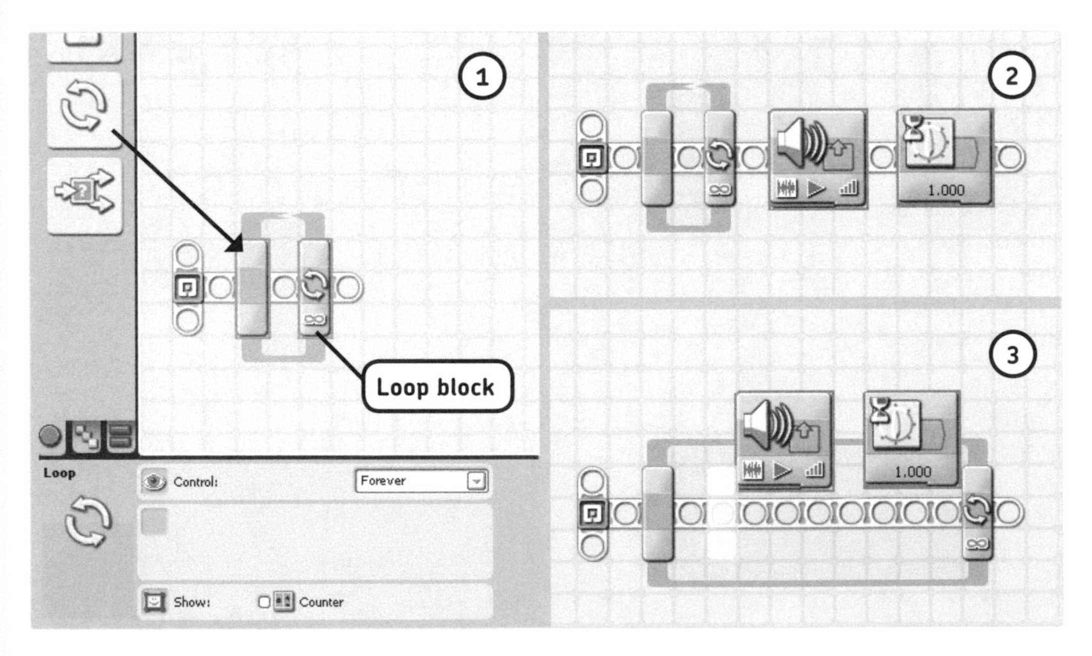

Figure 5-6: The Loop block and its Configuration Panel (1). To place blocks inside a Loop block, first place all required blocks on the Work Area (2). Next, select the blocks you want to move, and drag them into the loop (3). The Loop block should automatically enlarge to create space for the blocks as you try to drag them into it. Also, when you drag a Loop block around, its contents should remain inside it.

understanding the loop block settings

The settings on the Loop block's Configuration Panel control how many times the blocks inside the loop should be repeated. In the Control box, you can configure the block to repeat a set number of times (Count), to keep repeating for a certain amount of time (Time), or to repeat indefinitely (Forever).

* If you select Count, you can enter the number of repetitions in the Count box.
* If you select Time, you can enter the amount of time that the block should loop in the Seconds box.
* When Forever is selected, the Loop block keeps repeating the blocks placed inside indefinitely, unless you end the program by pressing the Exit button on the NXT brick. The Control box also offers settings for Sensor and Logic, which you'll learn about later in this book. You'll learn how to configure the Show setting on the Configuration Panel later as well.

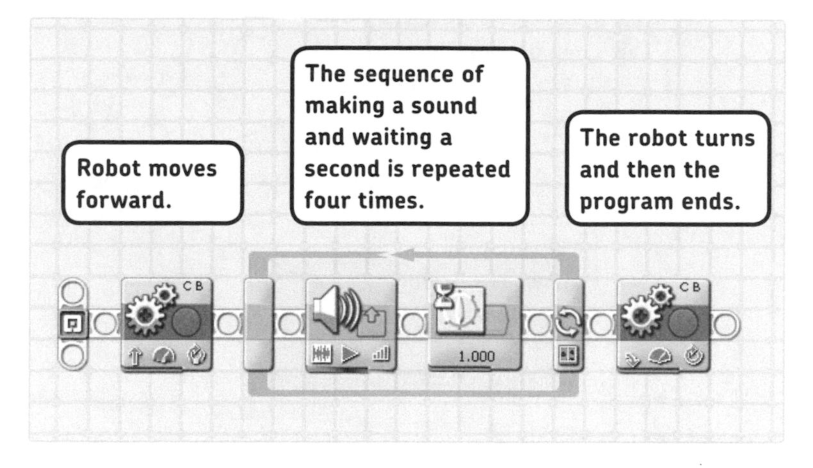

Figure 5-7: A program with a Loop block that loops four times. Once the Loop block has run the two blocks inside it four times (as specified by its Configuration Panel), the program continues with the next block, a Move block in this case.

seeing the loop block in action

You'll now create the program that I discussed earlier (Figure 5-5). This program will have the Explorer move in a square-shaped pattern. Create the **Explorer-Loop** program as shown in Figure 5-8.

NOTE If your robot doesn't make 90-degree turns when steering, try adjusting the number of degrees in Move block *c*, similar to what you did in Discovery #2 in Chapter 4.

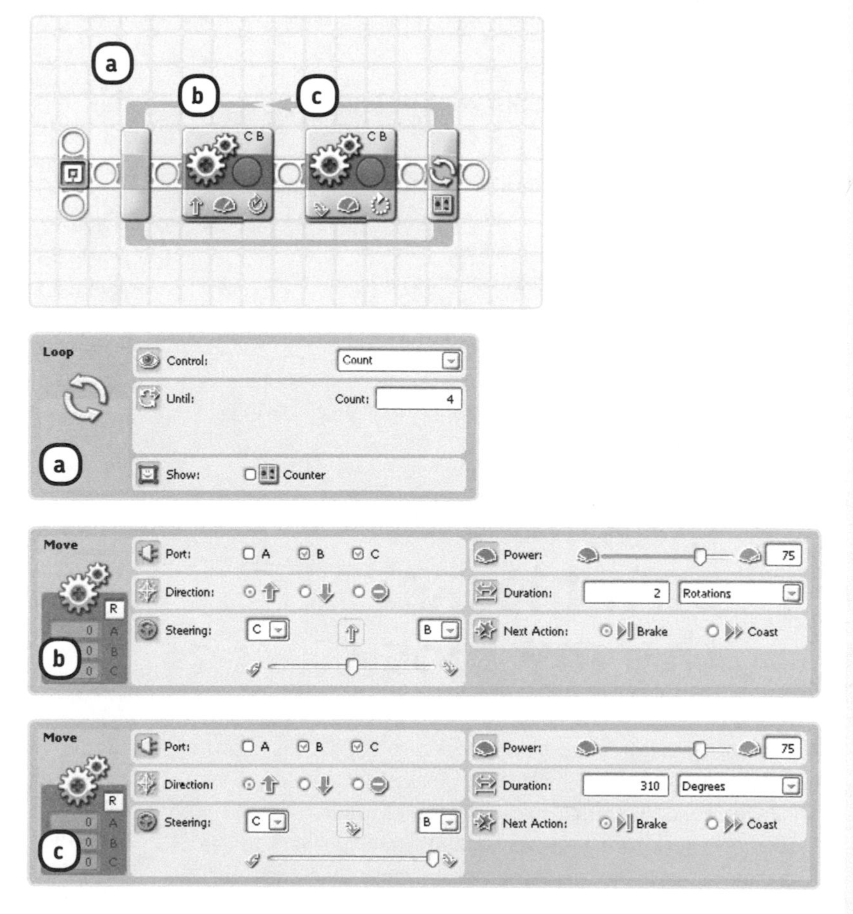

Figure 5-8: The configuration of the blocks in the Explorer-Loop program

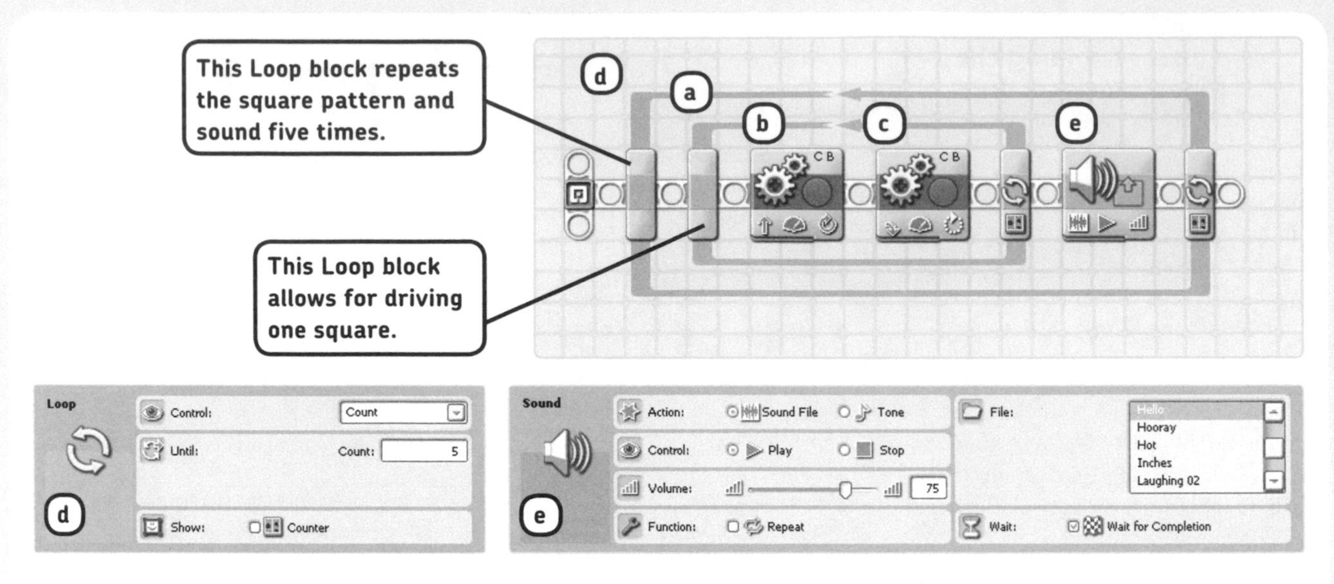

Figure 5-9: The Explorer-Square program shows you how you can have one Loop block inside another one. The inner Loop block makes the Explorer drive through a shaped pattern, and the outer loop repeats this behavior, as well as making a sound, five times.

using loop blocks within loop blocks

The structure of the Loop block with the two Move blocks in the Explorer-Loop program (Figure 5-8) makes the Explorer drive in a square. You can use a Loop block to repeat the square driving behavior so that the robot drives through, say, five complete square patterns. You'll configure the block to loop five times, and you'll drag the blocks from the previous program into it, as shown in Figure 5-9. You'll also add a Sound block to make Explorer say something after driving each square.

making your own blocks: the my block

In addition to using ready-made blocks, you can make your own blocks to meet specific needs. Blocks that you create are called *My Blocks* and consist of a set of programming blocks. My Blocks are especially useful when you want to use a specific set of blocks in your program more than once. For example, you could create a My Block to make the Explorer drive in a square-shaped pattern whenever you use that block.

DISCOVERY #13: GUARD THE ROOM!

Difficulty: Easy

Create a program to allow the Explorer to constantly move back and forth in front of your bedroom door, as if guarding it (see Figure 5-10). Use one Loop block configured to loop indefinitely (set to Forever), a Move block to move forward, and another Move block to turn around.

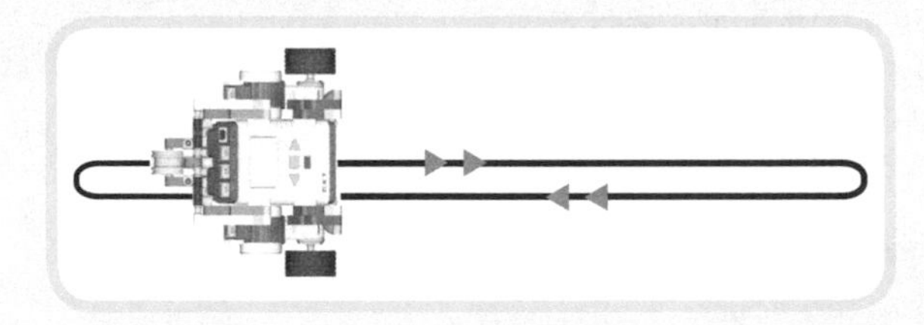

Figure 5-10: The path for the Explorer in Discovery #13

DISCOVERY #14: TRIANGLE!

Difficulty: Medium

You've created a program to drive in square-shaped patterns. How could you modify the Explorer-Loop program to drive in triangle-shaped patterns or in a six-sided figure? (Be sure to repeat each figure five times.)

Also, using My Blocks can help you keep your programs look organized, because you'll see fewer blocks on the screen. You'll see how to create and use My Blocks as you read on.

creating my blocks

To demonstrate the My Block functionality, you'll create a program that makes the Explorer drive in a square, turn around, make a sound, and then drive another square. Because Explorer drives the square-shaped figure twice, you'll create a My Block to perform this action, as shown in Figures 5-11 through 5-13. Once you've created your My Block, you can place it in a program whenever you want Explorer to drive in a square. You'll use the Explorer-Loop program as a foundation for this experiment.

NOTE If you didn't save the Explorer-Loop program earlier, create it again by following the directions in Figure 5-8, or download it from the companion website.

1. Select the blocks that you want to turn into a My Block, and click the **Create My Block** button, as shown in Figure 5-11.

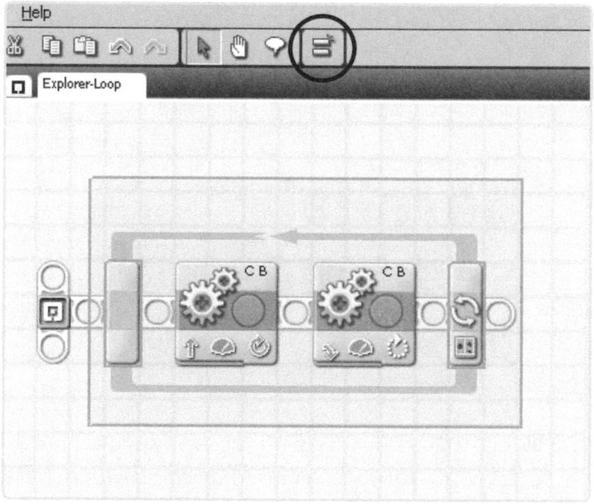

Figure 5-11: Selecting the blocks to be turned into a My Block

2. Enter a name for your My Block in the Block Name box, such as **Square**, as shown in Figure 5-12. Use the Block Description area to describe your block so that you'll remember how it works if you want to reuse it later, and then click **Next**.

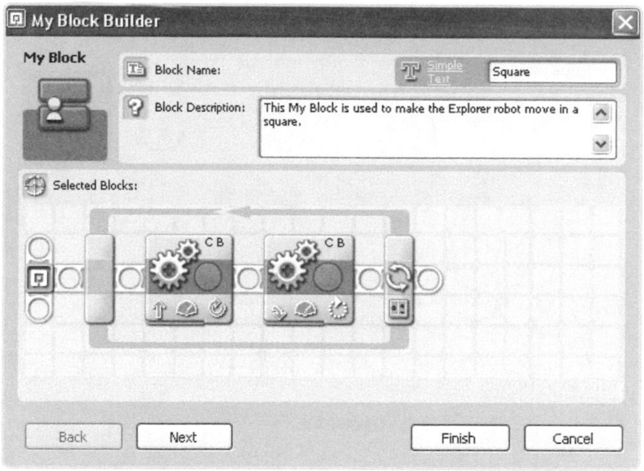

Figure 5-12: Enter a name and description for your My Block.

3. Drag icons to your block from the Icon Builder to give it a unique look (as shown in Figure 5-13), and then click **Finish**.

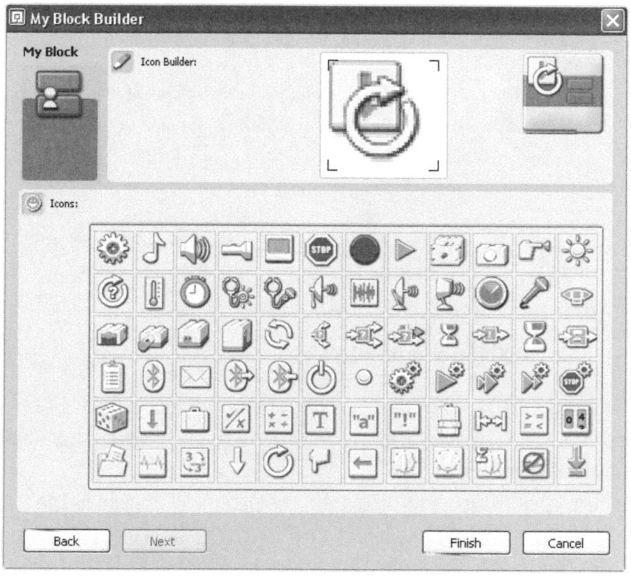

Figure 5-13: Adding icons to your My Block

using my blocks in programs

Once you've finished creating your My Block, it should appear on the Work Area and in the Custom Palette, which you can open as shown in Figure 5-14. (The figure also shows the Explorer-MyBlock program.)

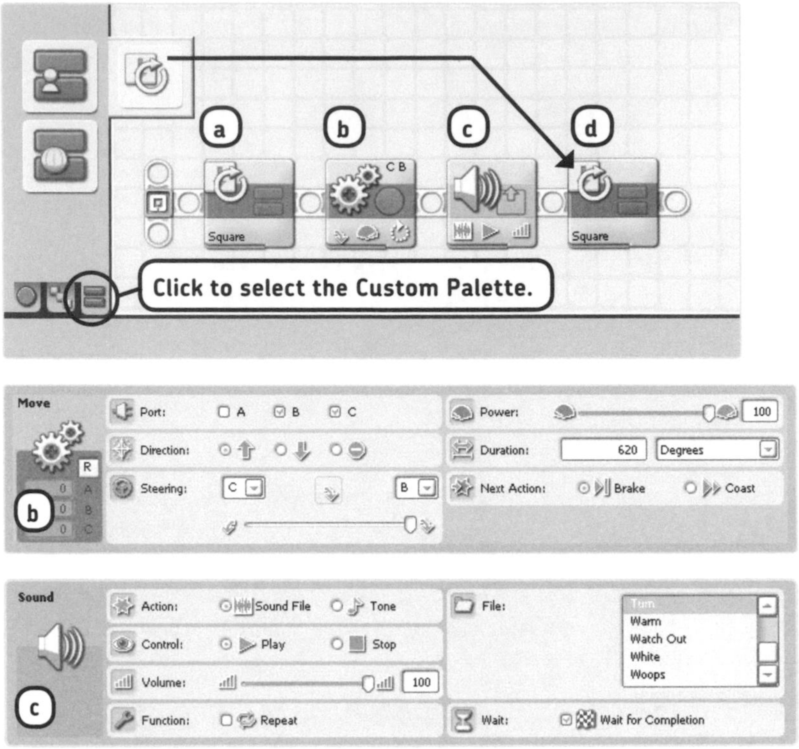

Figure 5-14: The configuration of the blocks in the Explorer-MyBlock program. You can find My Blocks on the Custom Palette. (The My Block's Configuration Panels are not shown because there are no settings to change on these panels.)

DISCOVERY #15: MY TRIANGLE!

Difficulty: Easy
Create a program like the Explorer-MyBlock program that drives in triangles instead of squares.

HINT Use the program you made in Discovery #14 to make the triangle My Block.

DISCOVERY #16: MY TUNE!

Difficulty: Easy
Remember the tune you made using Sound blocks in Discovery #6 in Chapter 4? Convert that sequence of blocks into a My Block so that you can easily use your favorite tune any time in your programs.

When you run the program, the robot should first move in a square (controlled by the square My Block) and then turn, make a sound, and drive another square.

editing my blocks

You can edit My Blocks after you create them. To do so, double-click a My Block on the Work Area to reveal its contents, and then edit it as you'd like. When you're finished, click **Save**, and return to the program that uses the My Block.

To change the icons that you added to your block, select the My Block on the Work Area, and select **Edit ▸ Edit My Block Icon** on the Toolbar.

parallel sequences of blocks

All the blocks that you've used so far are executed linearly in the order in which they were placed on the Sequence Beam. However, the NXT can execute multiple blocks at the same time, in parallel, using a *Parallel Sequence Beam*, as shown in Figure 5-15.

using parallel sequences in a program

To learn how Parallel Sequences work, you'll create the **Explorer-Parallel** program that makes the robot drive in circles as it makes sounds. Create the program shown in Figure 5-16.

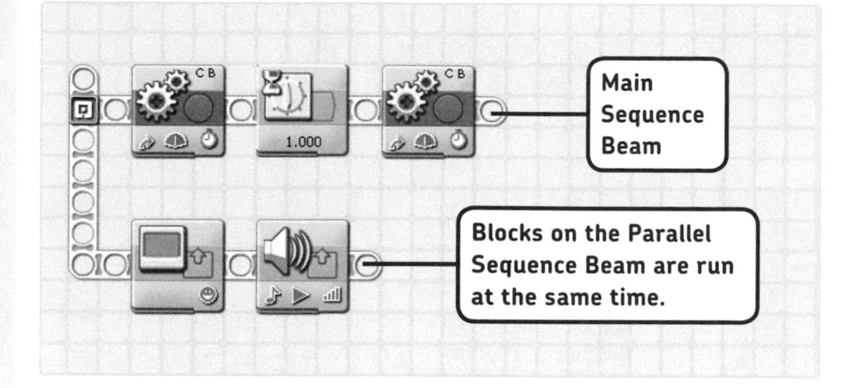

Figure 5-15: Blocks on the Main Sequence Beam and a Parallel Sequence Beam are run at the same time. Note that the Wait Block here causes only the sequence of blocks it appears in to pause; the blocks on the Parallel Sequence are not affected by the Wait Block.

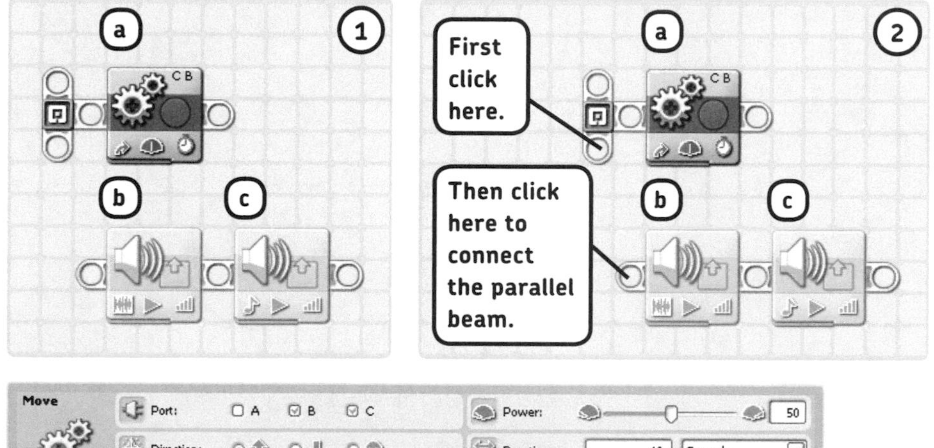

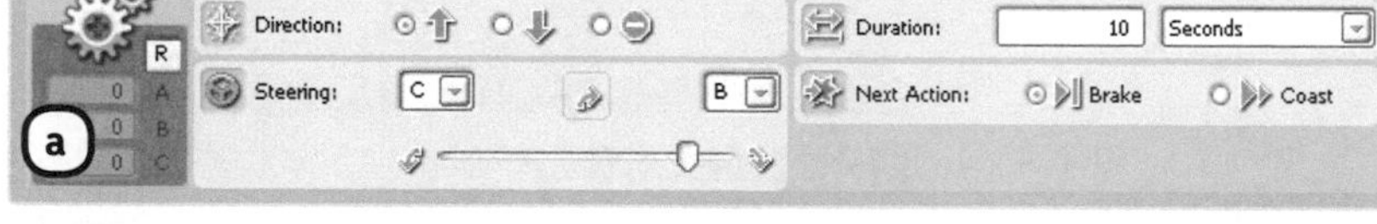

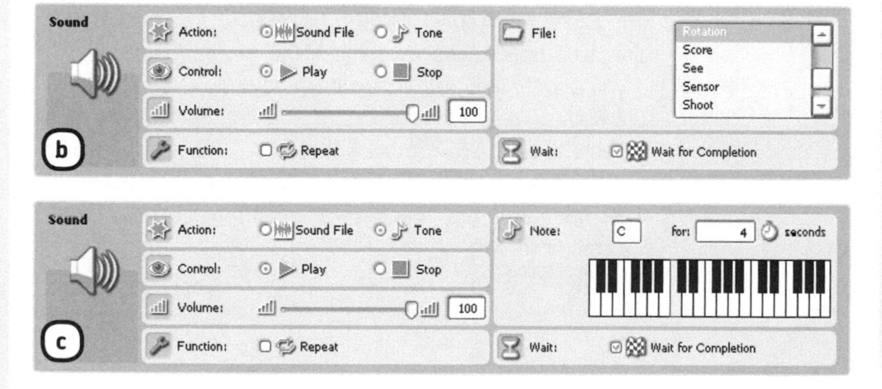

Figure 5-16: The configuration of the Explorer-Parallel program. Place the blocks required for this program on the Work Area (1). Next, connect the two Sequence Beams as shown in the figure (2).

DISCOVERY #17: LET'S MULTITASK!

Difficulty: Easy

Expand the Explorer-Parallel program (Figure 5-16) to make the robot drive in triangles indefinitely while at the same time playing a tune composed of different notes.

further exploration

Now that you've successfully gone through the first part of this book, you have a solid knowledge of several essential programming techniques. In this chapter, you learned how to use Wait and Loop blocks and how to work with My Blocks as well as parallel sequences of blocks.

In the next part of this book, you'll create robots that can interact with their environments via sensors. But before you do, practice a bit with what you've learned in this chapter by solving the following discoveries.

DISCOVERY #18: COMPLEX FIGURES!

Difficulty: Medium
Create a program that makes the Explorer drive in the pattern shown in Figure 5-17 while making different sounds.

HINT If you look carefully, you'll see that you can divide the track into four equal parts, so you have to configure a set of Move blocks for only one of these parts. Next, you'll place these blocks in a Loop block configured to loop four times.

Figure 5-17: The drive track for Discovery #18

BUILDING DISCOVERY #2: MR. EXPLORER!

In Building Discovery #1 (in Chapter 4), you found instructions to control a third NXT motor. This time you are challenged to use the extra motor to create a waving hand for your Explorer. Use other LEGO parts to further decorate your robot, and turn it into your own Mr. Explorer. Make him repeatedly wave his hand and play sounds like "Good morning!" at the same time.

PART II

building and programming robots with sensors

understanding sensors

The LEGO MINDSTORMS NXT 2.0 robotics kit includes three types of sensors: Ultrasonic, Touch, and Color. You can use these sensors to build a robot that makes sounds when it sees you or to build a vehicle that drives around while avoiding walls or that follows the black line on the Test Pad. This second part of the book will teach you what you need to know in order to create working robots with sensors.

To learn how to work with sensors, you'll upgrade the Explorer robot by adding several sensor attachments to it to create the Discovery robot shown in Figure 6-1. You'll learn to create programs for robots with sensors as you upgrade your robot with an Ultrasonic Sensor attachment. Once you have a good working knowledge of how to program with sensors, you'll continue creating more sensor attachments for this robot in Chapter 7.

Figure 6-1: The Discovery robot: an enhanced version of the Explorer equipped with an Ultrasonic Sensor to make the robot "see"

what are sensors?

LEGO MINDSTORMS robots can't actually see or feel the way humans do, but by adding *sensors* to them, they can collect and report information about the environment around them. Your programs can interpret sensor information in ways that will make your robot seem to respond to its environment as if it is experiencing it. For instance, you could create a program that makes the robot say "Blue" when one of its sensors sees a piece of blue paper.

understanding the sensors in the NXT 2.0 kit

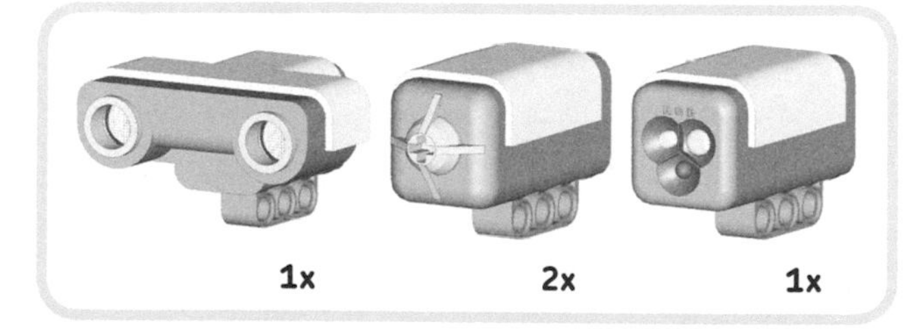

Figure 6-2: The NXT kit comes with an Ultrasonic Sensor (left), two Touch Sensors (middle), and a Color Sensor (right).

Your NXT kit contains three sensors, as shown in Figure 6-2. The *Ultrasonic Sensor* reads the distance to objects, the *Touch Sensors* detect button presses, and the

Color Sensor detects the color of a surface (among other things, as you'll learn in Chapter 7).

You connect the sensors to the NXT via *input ports*, numbered 1 through 4, as shown in Figure 6-3.

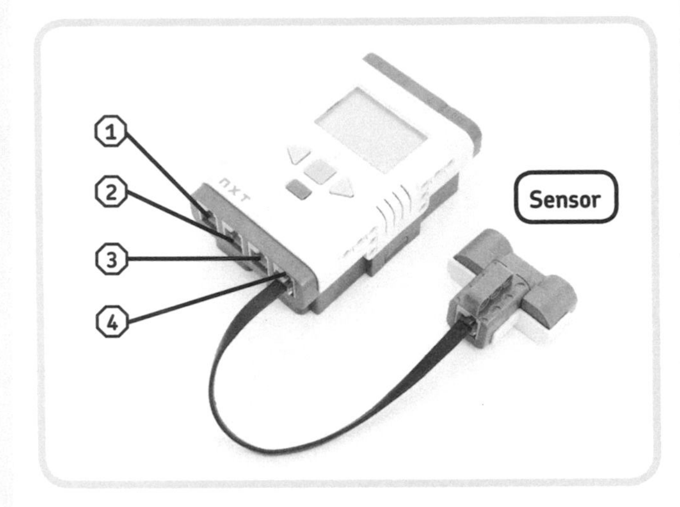

Figure 6-3: Sensors are connected to input ports.

In this chapter, you'll explore the Ultrasonic Sensor, and you'll take a more detailed look at the other sensors in Chapter 7. The programming techniques that you'll learn for the Ultrasonic Sensor can be used for all the sensors in the kit.

understanding the ultrasonic sensor

The Ultrasonic Sensor serves as your robot's eyes. To "see," the sensor measures the distance between it and other objects (as shown in Figure 6-4). The NXT retrieves the information from the sensor and uses its measurements in programs. For example, using input from this sensor, you could make your robot say "Hello" when the ultrasonic sensor reports an object in front of it that's nearer than 50 cm.

As you'll learn in this and the following chapters, you can use the Ultrasonic Sensor in a number of interesting ways. For example, you can use it to make a vehicle avoid walls (Chapter 6), create an intruder alarm and detect targets to shoot at (Chapter 8), find objects to grab (Chapter 13), and even detect a ceiling so that a vertical climber knows to go down again (Chapter 15).

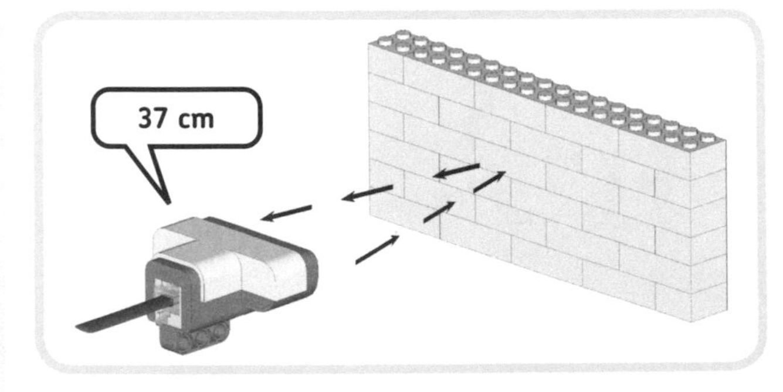

Figure 6-4: The Ultrasonic Sensor is used to detect objects by measuring the distance to them. It can see things up to 200 cm (80 inches) away, but the farther the object is, the harder it is for the sensor to see it. When the sensor doesn't see anything, it reports a value of 255 cm.

creating the ultrasonic sensor attachment

To begin, add an Ultrasonic Sensor to the Explorer robot as shown in the directions on the following page.

NOTE If you have trouble following these steps, try disconnecting the motor cables and then reattaching them after you connect the Ultrasonic Sensor. Be sure to connect the Ultrasonic Sensor to input port 4 on the NXT using a medium-sized cable.

polling sensors

Without having to do any programming, you can view the sensor readings on the NXT's *View menu*. Gathering information from a sensor is sometimes referred to as *polling*. To poll a sensor, follow these steps:

1. Turn on your NXT, navigate to the View menu (Figure 6-5), and select the sensor you want to poll.
2. Choose **Ultrasonic cm** (or **Ultrasonic Inch**).
3. Select the input port the sensor is connected to (port 4), and you should see the sensor's measurement, which is 37 cm in this case.

DISCOVERY #19: MIND YOUR HEAD!

Difficulty: Easy
How could you find out how far your robot is from the ceiling of a room? Use the View mode and the Ultrasonic Sensor to find out, making sure to point the sensor's "eyes" toward the ceiling while measuring. If you see only question marks on the screen, you might need to hold the sensor up a little closer to the ceiling.

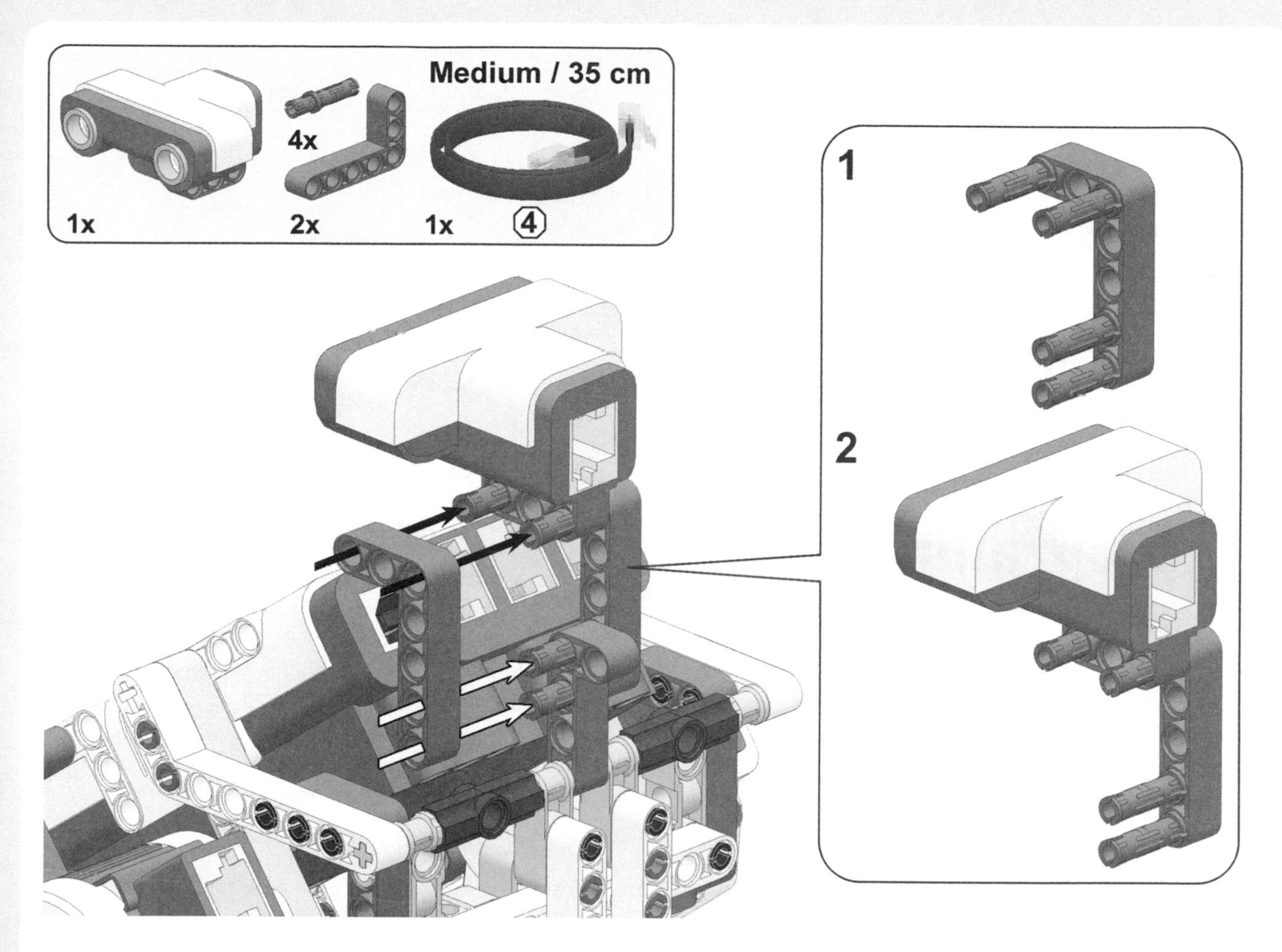
Medium / 35 cm
4x
1x
2x
1x
4
1
2

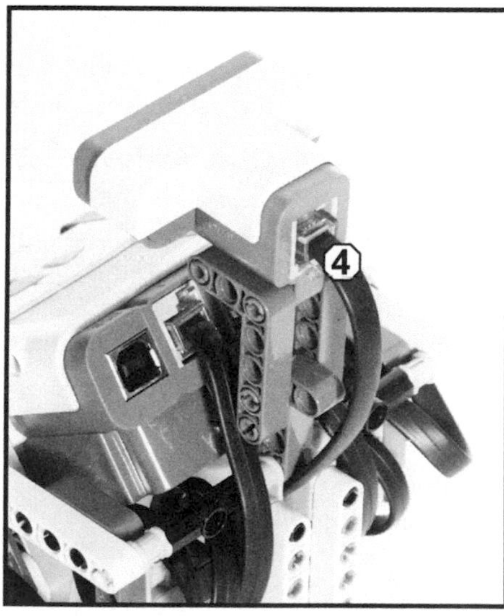
4

4

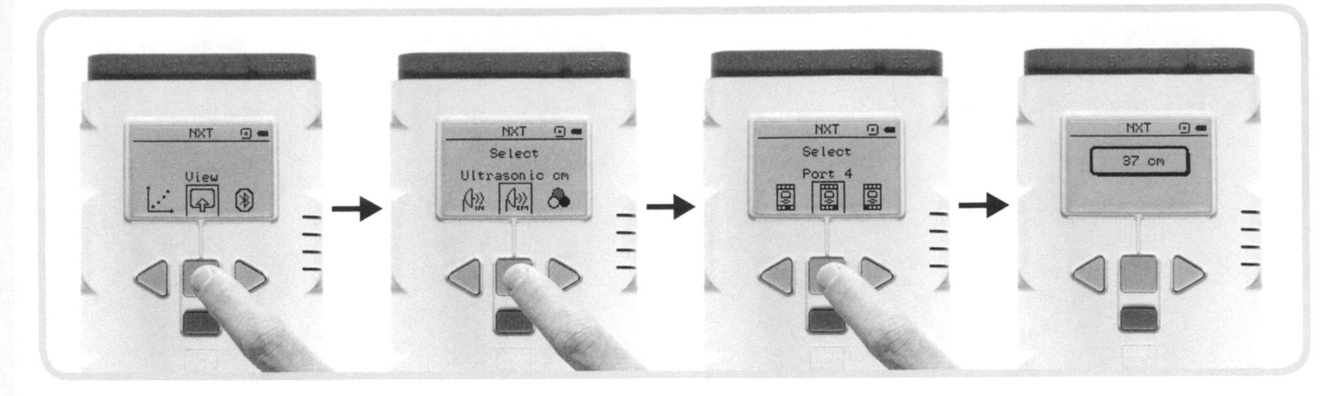

Figure 6-5: Polling a sensor with the View menu

programming with sensors

You've just seen how to poll a sensor yourself. Programs can also poll a sensor in order to use the sensor's data. As an example, you'll create a program that has the robot play a sound when the Ultrasonic Sensor sees something that's closer than 50 cm (20 inches), as shown in Figure 6-6.

You can use several programming blocks to poll sensors including the Wait and Loop blocks as well as the Switch block.

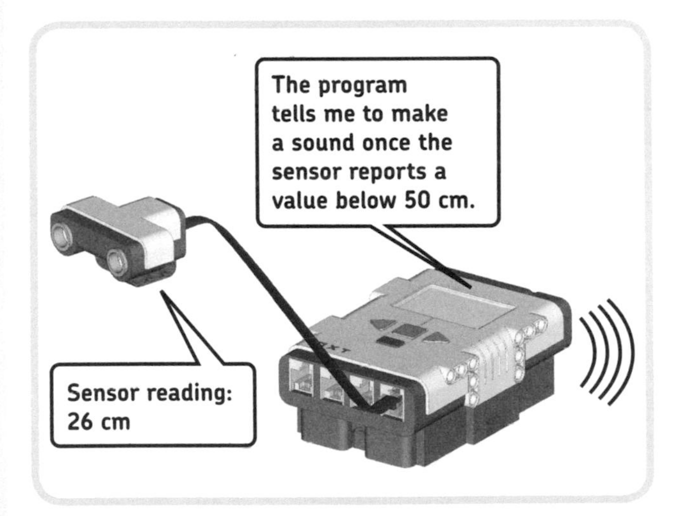

Figure 6-6: The program that runs on the NXT polls the sensor constantly. The program pauses until a sensor reading less than 50 cm is reported, at which point it plays a sound.

sensors and the wait block

You can use a Wait block to pause a program for several seconds, but also to pause until a sensor reading goes above or below a certain value. For example, Figure 6-7 shows a Wait block halting a program until the Ultrasonic Sensor detects a sensor value less than 50 cm. This value of 50 cm is called the *trigger value*. Once this trigger value is reached, the sensor is *triggered*, the Wait block stops waiting, and the next block in the program (i.e., a Sound block) runs.

using the configuration panel

You've just learned how a Wait block is used to poll a sensor. Now you'll have a look at the configurations of this block.

First select the Wait block configured to poll the Ultrasonic Sensor from the Programming Palette, as shown in Figure 6-7. (This is really just the Wait block you've been using with the Control setting set to Sensor instead of Time.)

* You use the Port setting to select the input port to which the sensor is connected.
* You set the trigger value in the Distance setting in the Until box by entering its value or by dragging the slider to the left (closer) or right (farther). You use the Distance setting to select whether the block should wait until a value above (>) or below (<) the trigger value is reached.
* You use the Show setting to choose Centimeters or Inches.

seeing the sensors and the wait block in action

Now you'll create the **Discovery-Wait** program that plays a sound when the Ultrasonic Sensor sees something closer than 50 cm (20 inches), as shown in Figure 6-8.

Before running this program, make sure that there is nothing in front of your robot. When you run the program,

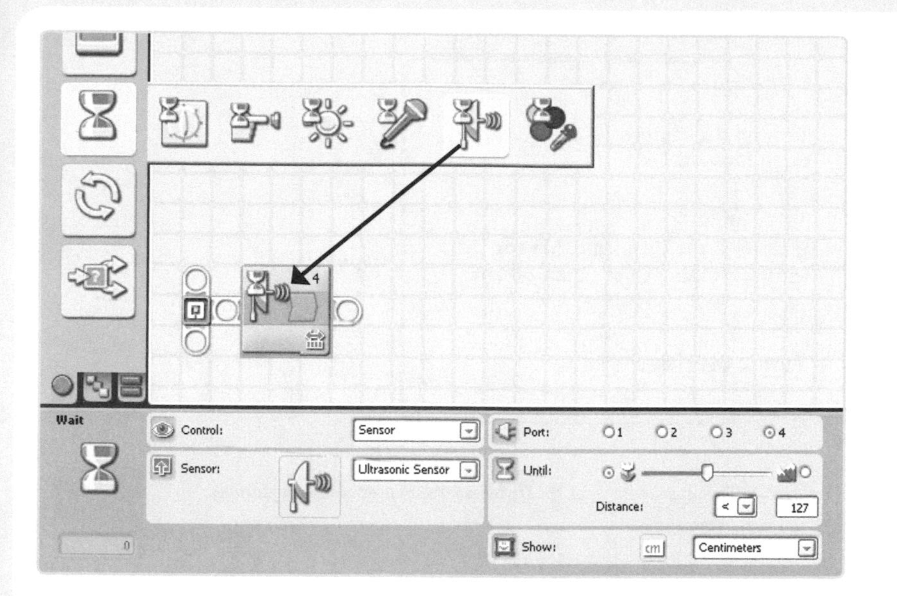

Figure 6-7: The Wait block configured to poll an Ultrasonic Sensor

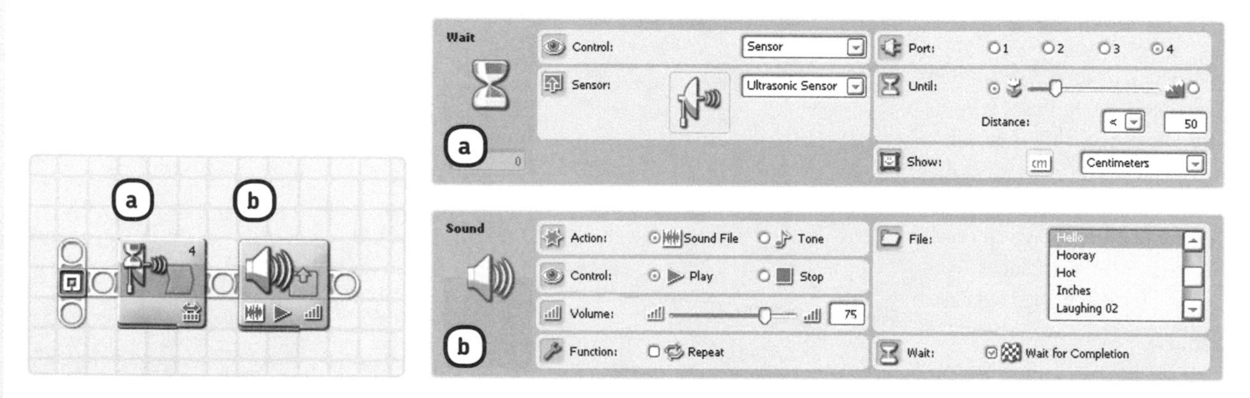

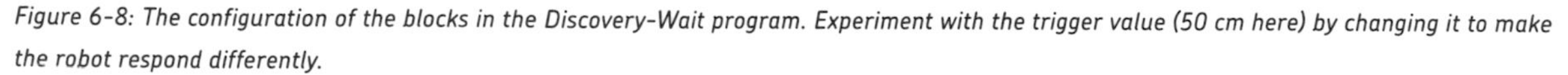

Figure 6-8: The configuration of the blocks in the Discovery-Wait program. Experiment with the trigger value (50 cm here) by changing it to make the robot respond differently.

nothing should happen at first, but as you move your hand in front of the Ultrasonic Sensor, the sensor should detect your hand once it's closer than 50 cm. Once your hand is detected, the Wait block should stop waiting, and your robot should play a sound.

Now, before you proceed with other sensor programming techniques, try to solve Discovery #20 to get a better understanding of how to work with sensors.

DISCOVERY #20: HELLO AND GOODBYE!

Difficulty: Medium

Can you create a program that has the robot say "Hello" when it detects your hand in front of the Ultrasonic Sensor and then "Goodbye" when you move your hand away?

HINT Place and configure the sequence of the two blocks in the Discovery-Wait program (Figure 6-8) twice. Configure the first Wait block to wait until it sees something closer than 50 cm, and configure the second Wait block to wait until the sensor sees something farther than 50 cm. Once you've confirmed that your program works, drag all the blocks into one Loop block so that your robot will repeatedly tell you when it detects your hand.

avoiding walls with the ultrasonic sensor

The next program, **Discovery-Avoid**, will make the Discovery robot drive around a room and turn around when it sees something in order to prevent it from bumping into an obstacle such as a wall. You can see an overview of the program in Figure 6-9.

Next you'll re-create the program with programming blocks. You can accomplish each action described in Figure 6-9 with one block. You'll use a Move block to turn on the motors with the Duration option set to Unlimited, and then you'll wait for the sensor to be triggered with a Wait block. (Note that while the program waits, the robot is still moving forward.)

Once the robot sees something, you use a Move block to turn it around (and stop its unlimited movement). For turning around, the block's Duration option is set to a specific number of rotations. After the robot turns around, the program returns to the beginning, which is why you must place the three blocks used inside a Loop block that's configured to loop forever.

Create the program now, as shown in Figure 6-10.

sensors and the loop block

As you learned in Chapter 5, you can configure a Loop block in many ways. You can configure Loop blocks to loop a certain number of times, loop for a specified amount of time, or loop forever. These conditions indicate when a Loop block should stop looping.

You can use a sensor to make a Loop block stop repeating. For example, you can repeat a sequence of Move blocks inside a Loop block until the Ultrasonic Sensor sees something that's closer than 25 cm (10 inches).

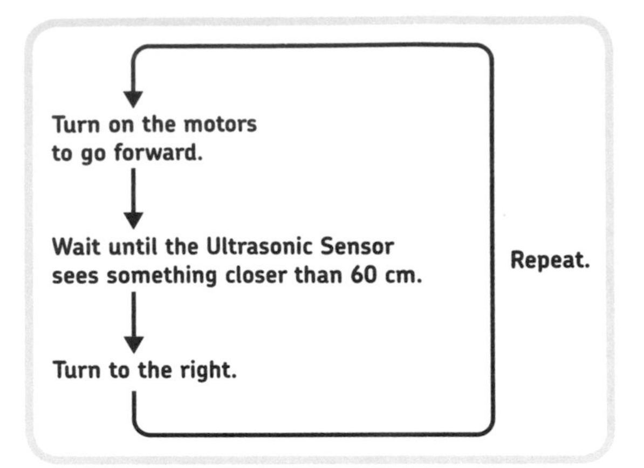

Figure 6-9: The program flow for the Discovery-Avoid program. After turning right, the program returns to the start, and the robot moves forward again.

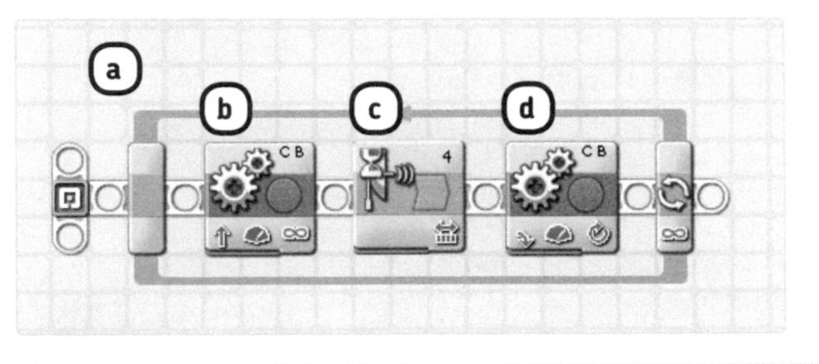

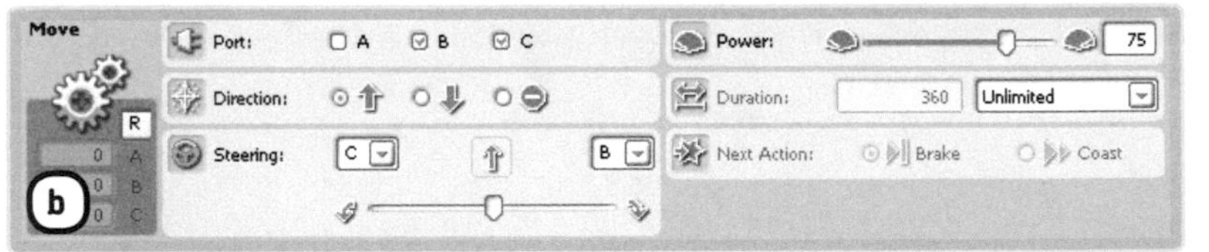

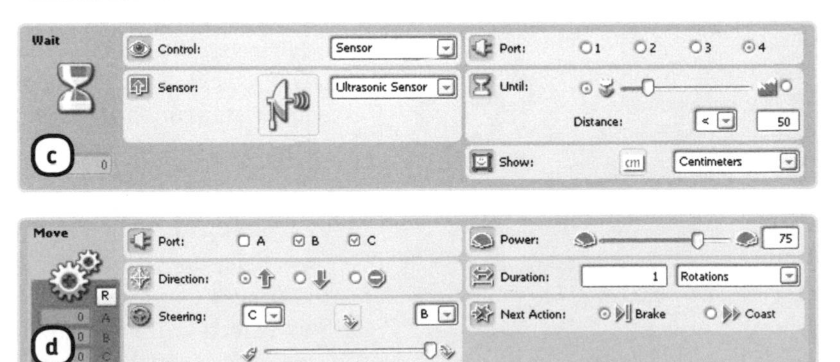

Figure 6-10: The configuration of the blocks in the Discovery-Avoid program

DISCOVERY #21: AVOID WALLS AND A BAD MOOD!

Difficulty: Easy

Expand the Discovery-Avoid program by making it display a happy face on the NXT's display as it moves forward and by making it display a sad face when it is turning around to avoid the wall.

HINT **To do so, try putting two Display blocks somewhere in the Loop block.**

DISCOVERY #22: FOLLOW ME!

Difficulty: Medium

Make the Discovery robot follow you in a straight line. When you place your hand in front of the robot, it should stop, but when you move your hand away, it should move forward until it sees your hand again.

TIP **Have one block wait for the hand to come close (and then use a Move block to stop the robot's movement) and another block wait for the hand to move away, and then continue the robot's movement. Put all the blocks in a Loop block.**

DISCOVERY #23: HAPPY TUNES!

Difficulty: Medium

Use a Loop block to have the robot play a tune until the sensor spots someone watching, at which point the robot should scream and turn its head the other way.

HINT **You can use the My Block that you made in Discovery #16 in Chapter 5 for your tune. If you have yet to create your own tune, simply select a sound file from the list in a Sound block.**

To use a sensor to control when a loop should stop, select **Sensor** in the Control box, and then select the sensor (**Ultrasonic**) you would like to use and the port (**4**) to which the sensor is connected. Set the trigger value in the Until box to specify exactly when the loop should stop. You can see how to make these configurations in the following sample program.

seeing the sensors and the loop block in action

Now make the **Discovery-Loop** program shown in Figure 6-11. When you run this program, the robot should move back and forth repeatedly until the sensor spots something closer than 25 cm (10 inches).

This program doesn't always respond to objects visible within 25 cm of the robot because it polls the sensor only once per repetition, just after all the blocks inside the loop complete. For this reason, when you quickly move your hand in front of the robot while the first block is running, the robot won't notice your hand.

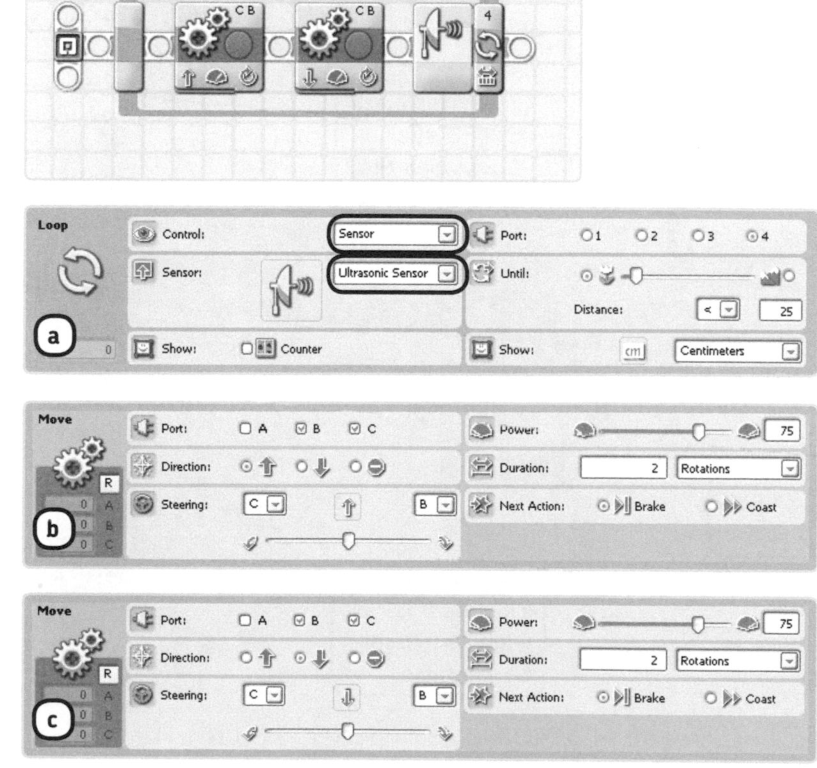

Figure 6-11: The configuration of the blocks in the Discovery-Loop program

sensors and the switch block

You can use Switch blocks to have a robot make decisions based on sensor data. Until now, your robots have been preprogrammed, meaning that each time you ran a program, the robot would behave in the same way. Switch blocks allow your robots to *choose* what to do based on a sensor reading. For example, you can make your robot move backward if the Ultrasonic Sensor sees something nearer than 50 cm, or you can have it say "Distance" if there is no object closer than 50 cm, as shown in Figure 6-12.

Similar to the question asked in Figure 6-12, the robot checks whether a given *condition* (a statement, such as "The sensor reading is less than 50 cm") is true or whether it is false, as shown in Figure 6-13.

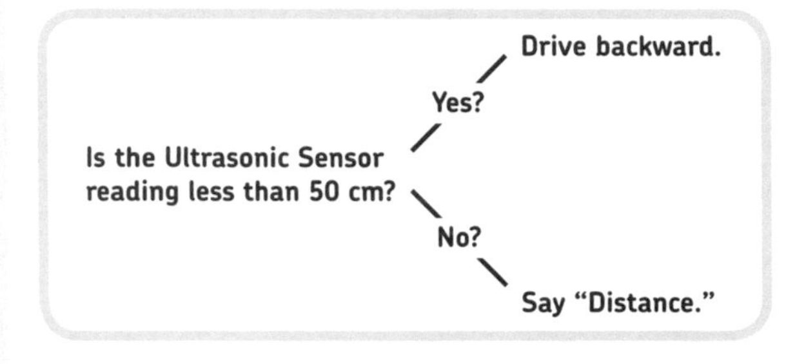

Figure 6-12: A robot can make decisions based on the Ultrasonic Sensor reading.

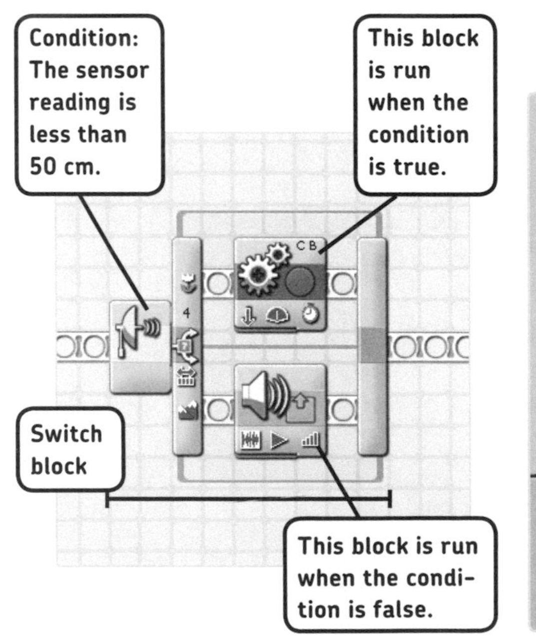

Figure 6-13: The Switch block checks whether the condition is true or false and runs the appropriate blocks. You specify the condition in the settings of the Switch block.

The Switch block in this program contains two blocks in separate parts of the switch; the switch decides which of the two blocks to run. If the condition is true, the block in the upper part of the switch is run, and the robot moves backward; if the condition is false, the lower block is run, and you should hear a sound.

configuring a switch block

Figure 6-14 shows the Switch block's Configuration Panel. By changing the Switch block's settings, you define the condition (like the one in Figure 6-13), and the program will check whether it's true. If it is true (the sensor reading is less than 50 cm), the upper blocks in the switch should run. If the condition is false (the reading is 50 or greater), the blocks in the lower part of the switch should run.

You select the Sensor option in the Control box to indicate that you want to make a decision about a sensor reading. In the Sensor box, you select the sensor to poll, which is the Ultrasonic Sensor in this case. On the right half of the block's Configuration Panel, you specify the condition. For example, you could set the block to check whether the sensor sees something that's farther away than 25 cm (10 inches).

seeing the switch block in action

The **Discovery-Switch** program that you'll now create has the robot drive forward for four seconds. Then, if the robot sees something closer than 50 cm, it goes backward for a short while. If the robot does not see an object closer than

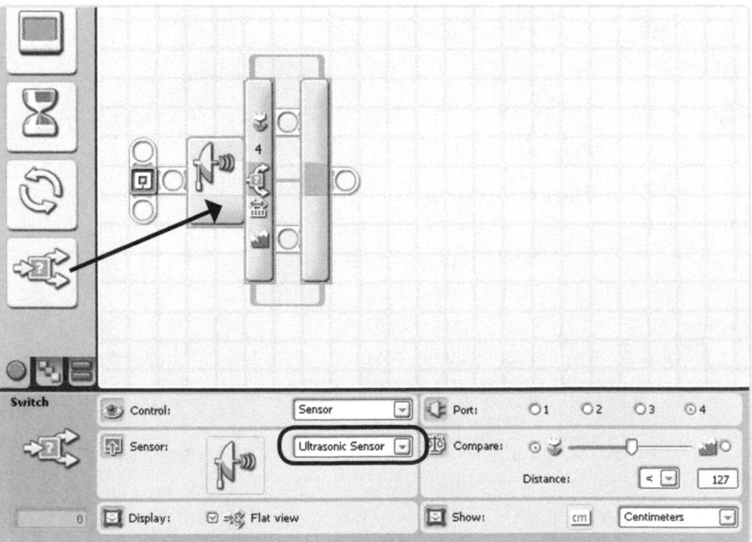

Figure 6-14: The Configuration Panel of a Switch block configured to poll the Ultrasonic Sensor

50 cm, Discovery will say "Error." Finally, regardless of the Switch block's decision, the robot will play a tone.

Create the program now as shown in Figure 6-15.

Run this program multiple times to determine when you must keep your hand in front of the sensor to make the robot go backward. Doing so should demonstrate that when you use the Switch block in your programs, the robot polls the sensor only once and compares the sensor value to the trigger value to see whether the particular condition is true. In this program, the sensor value is measured after the robot finishes going forward (when the Move block finishes).

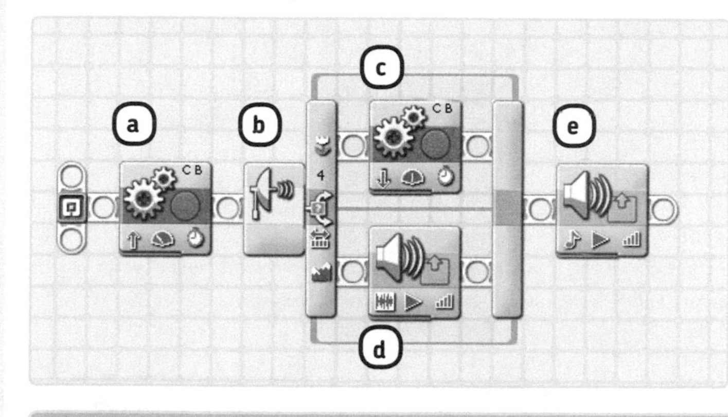

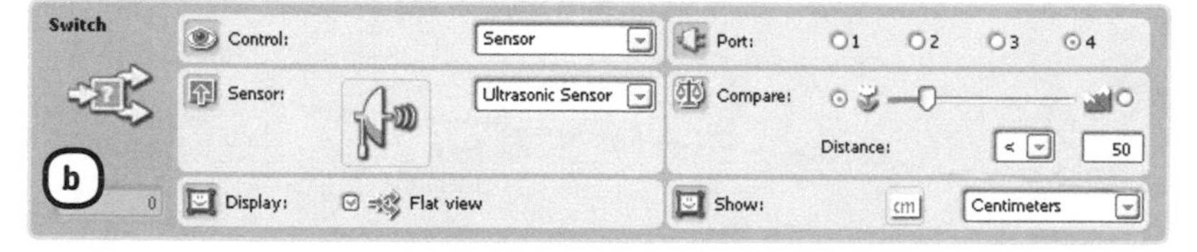

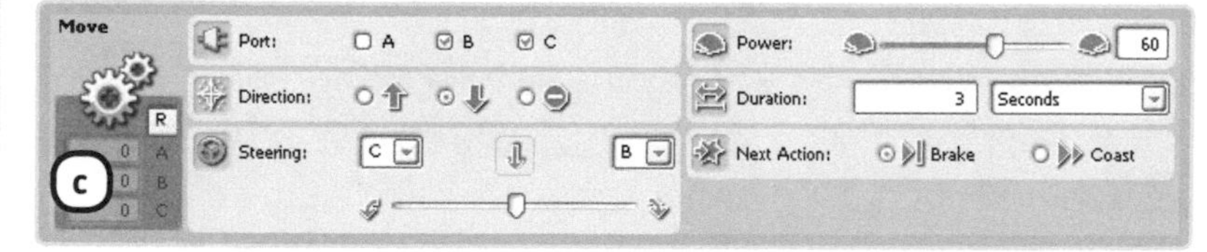

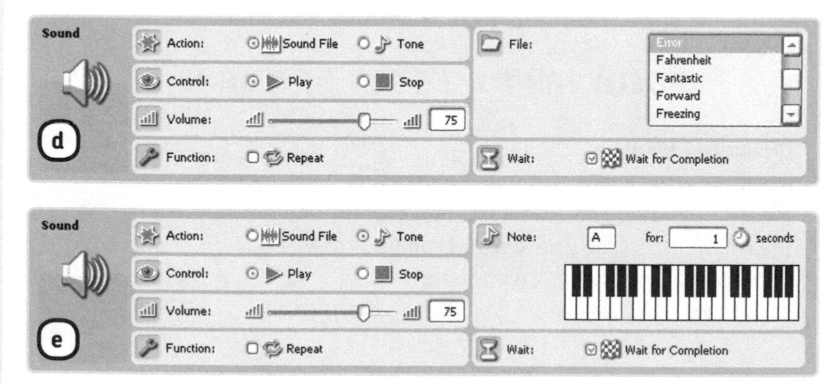

Figure 6-15: The Discovery-Switch program has the robot decide what do based on a sensor reading.

DISCOVERY #24: SEE THE DISTANCE!

Difficulty: Easy

Let's practice with the Switch block! Try to create a program to implement the decision tree shown in Figure 6-16. How do you configure the Switch block, and why do you have to put a Wait block at the end of the program?

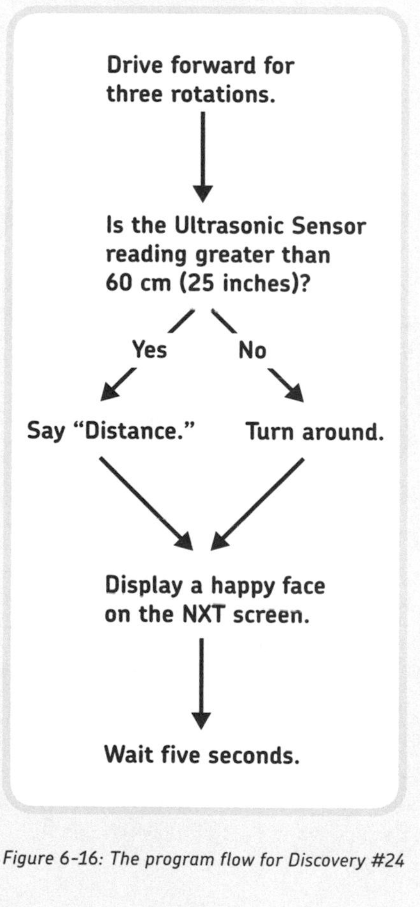

Figure 6-16: The program flow for Discovery #24

When either block *c* or *d* finishes running, the program continues, and the next Sound block is activated to play a tone.

adding blocks to a switch block

There's no limit to the number of blocks you can place inside a Switch block. If one part of a switch has multiple blocks, they're simply run one by one, as shown in Figure 6-17. You can also leave one of the two parts of a Switch block empty, as shown in the figure.

Run this modified program to see what happens. If the condition is true (the robot sees you), the robot should move backward and say "Error," and the program should continue by playing the tone. If the condition is false (the robot does not see you), the program should find no blocks in the lower part of the switch and instantly move on to the Sound block after the switch (block *e*), which plays the tone.

using the flat view option

The Switch block's Flat View option in the Configuration Panel displays the complete Switch block on the Work Area. When you make large programs containing Switch blocks, it's easy to lose track of how your program actually works. In such cases, uncheck the **Flat View** checkbox to decrease the size of the Switch block, as shown in Figure 6-18. Both parts of the switch are still in the program, but they're on separate tabs, which you can open by clicking them.

You won't need to use Flat View often with smaller programs, but as you continue creating programs for your robot in the next chapter, you'll learn that it can be useful to disable Flat View, especially when part of a switch contains no programming blocks.

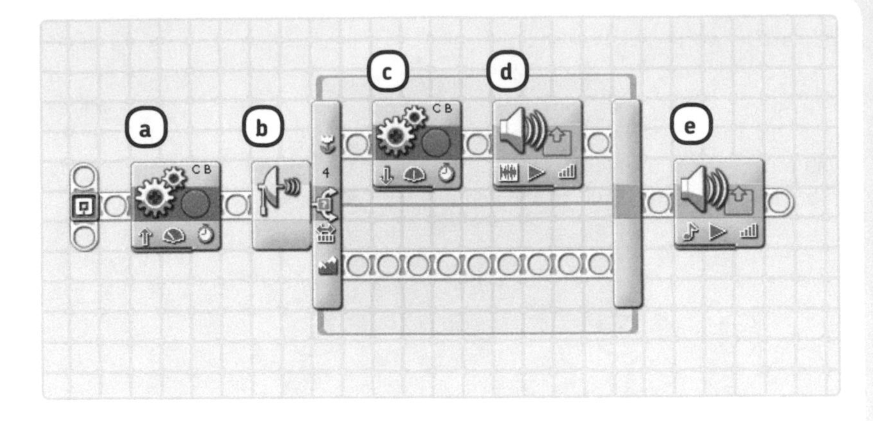

Figure 6-17: A modified version of the Discovery-Switch program. You move Sound block d *formerly in the lower part of the switch (which runs when the condition specified in the Switch block is false) to the upper part of the switch (which runs when the condition is true).*

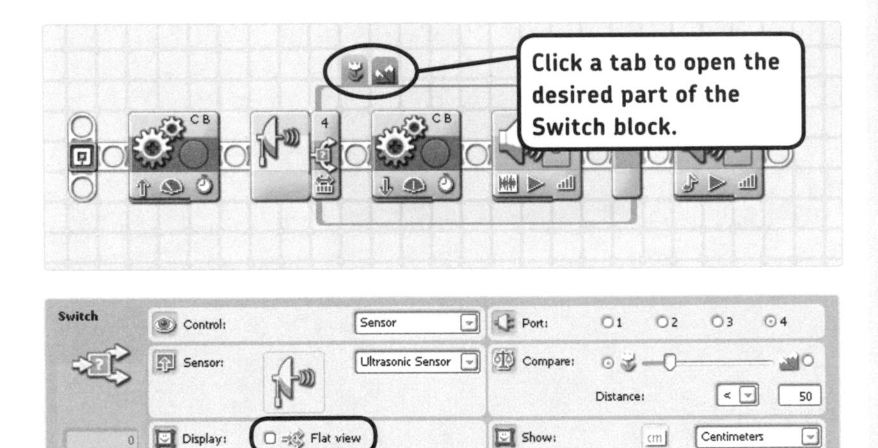

*Figure 6-18: Decrease the size of a Switch block by deselecting the **Flat View** option in its Configuration Panel. This options changes your view of the block only; it won't affect the way the program works.*

DISCOVERY #25: STOP OR TURN?

Difficulty: Medium
In this discovery, you'll continue working on the program that you made in Discovery #24. Modify the program to make it do nothing if the condition is true (the sensor reading is greater than 60 cm), and if the condition is false, have the robot turn around and then drive back to where it was when you started the program.

repeating switches

Switch blocks poll the Ultrasonic Sensor once and compare the measured value to the trigger value. Then, depending on how a condition is configured, blocks in the true or false part of the switch are run.

To have a robot make decisions more than once, you can drag a Switch block into a Loop block. For example, you could program a robot to say "Yes" when it sees something closer than 100 cm (40 inches) and say "No" otherwise. If you place a Switch block with this configuration in a Loop block, the robot will continue to say "Yes" and "No" based on the sensor reading.

Create the **Discovery-Repeat** program shown in Figure 6-19 now.

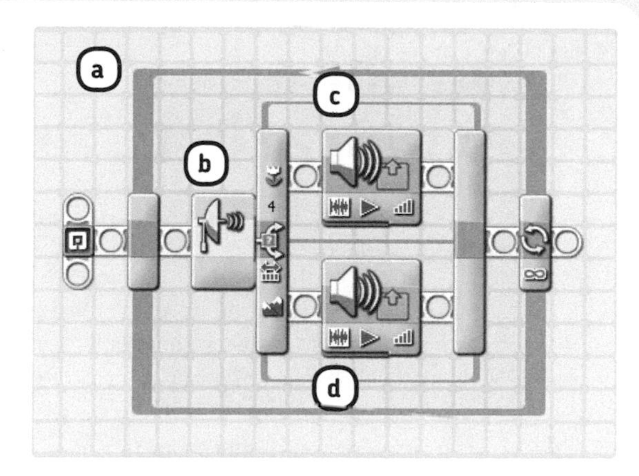

Figure 6-19: The configuration of the blocks in the Discovery-Repeat program

further exploration

Robots use sensors to gather input from their environments, and in response to this input you can create an NXT-G program to make a robot do different things. To make a program for a robot that uses sensors, you use Wait, Loop, and Switch blocks, each configured to control a sensor. Your choices will be based on what your program should do.

You've been using only the Ultrasonic Sensor so far, but you can use all the programming techniques you've learned in this chapter when working with each of the other sensors. In Chapter 7, you'll play with the Touch Sensors and the Color Sensor and use them in the Discovery robot.

DISCOVERY #26: INTRUDER ALARM!

Difficulty: Easy
Create a program that makes a beeping alarm sound when someone enters your room. Use a Wait block (configured as an Ultrasonic Sensor) to sense an opening door, and then follow it with Sound blocks in a Loop block to create the alarm. What trigger value do you need in the Wait block?

DISCOVERY #27: ULTRASOUND!

Difficulty: Hard
Use Switch blocks to determine the measured distance of the Ultrasonic Sensor. If the distance is 0 cm to 10 cm, a low tone should be played with a Sound block. If the distance is between 11 cm and 20 cm, the tone should be higher, and so on. How do you configure the Switch blocks, and where do you put each of them in your program?

BUILDING DISCOVERY #3: RAILROAD CROSSING!

As in Chapters 4 and 5, attach a third motor to the Discovery robot, and control it with a Move block that's configured to control the motor on output port A. Use this motor to move a self-made barrier that stops cars from crossing the railway if a train passes. Use the Ultrasonic Sensor to spot when a model train approaches and when the barriers should be lowered, as well as when they should be raised again.

using the touch, color, and rotation sensors

Chapter 6 showed you how to use the Ultrasonic Sensor, together with the Wait, Loop, and Switch blocks, to make your robot interact with its environment. In this chapter, you'll learn how to use the Touch, Color, and Rotation Sensors, as well as the NXT buttons. You'll use these sensors to allow the Discovery robot to avoid walls with bumpers, follow a line, play different sounds based on sensor readings, and even tell you which color its sensor detects. Once you've finished this chapter, you'll be ready to build two cool robots that use sensors in their own way in Chapters 8 and 9.

You'll begin by upgrading Discovery with a dual bumper attachment that use two Touch Sensors, as shown in Figure 7-1.

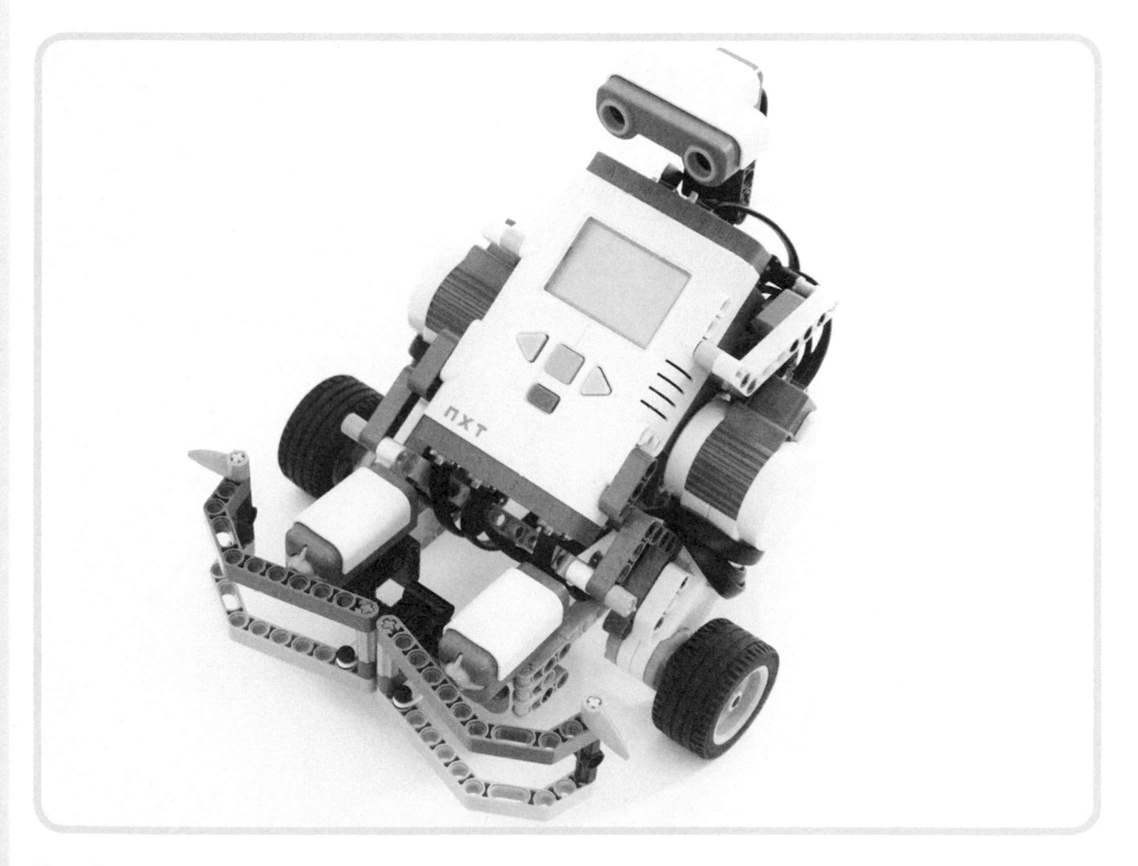

Figure 7-1: Discovery with a bumper attachment that uses two Touch Sensors

the touch sensor

The *Touch Sensor* can make your robot feel and respond to its environment by detecting whether its orange button is *pressed* or not (*released*). By combining the readings from the Touch Sensor with, for example, a Wait block, you can make the robot respond to a press of the Touch Sensor, as shown in Figure 7-2. The sensor can also detect whether it's bumped, meaning the sensor was quickly pressed and then released.

You can use Touch Sensors in many different ways. For example, in this chapter, you'll use them to create bumpers for your robot so that it can back up and turn around when it strikes an obstacle. In Chapter 8 you'll use Touch Sensors to create remote control buttons, and in Chapter 9 they'll act as antennas for an animal robot. In Chapter 13 you'll see how to use a Touch Sensor to perform a specific mechanical function, such as detecting when a grabber has lifted its load to the maximum height.

creating the bumper attachment with touch sensors

You'll now create an extension for your Discovery robot that uses two Touch Sensors as bumpers (Figure 7-1) to allow your robot to feel objects that it runs into. Once it determines which bumper was pushed, the robot will back up and turn away from this object. Create the bumpers now, as shown in the instructions on the next pages.

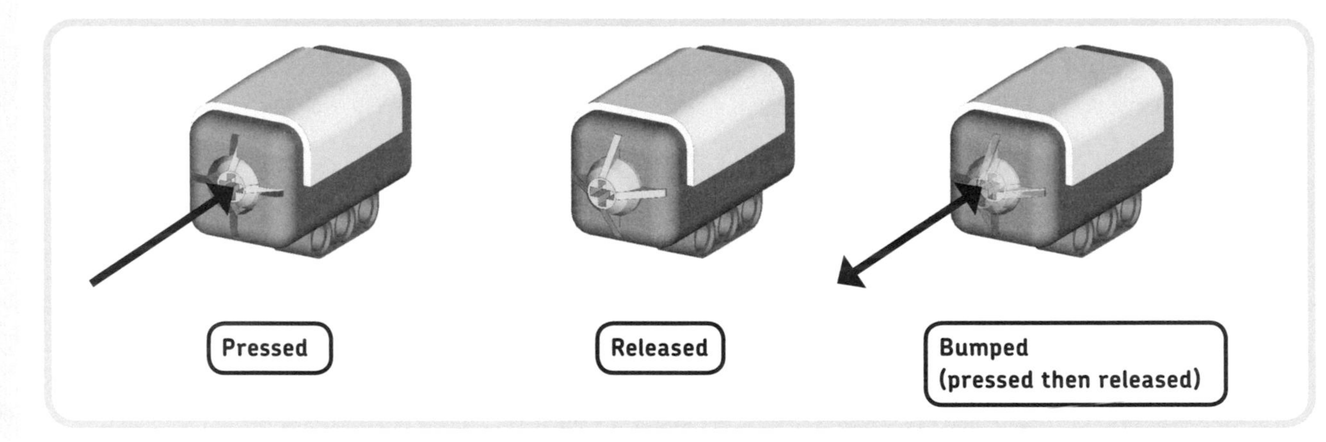

Figure 7-2: The Touch Sensor detects three actions: pressed, released, and bumped.

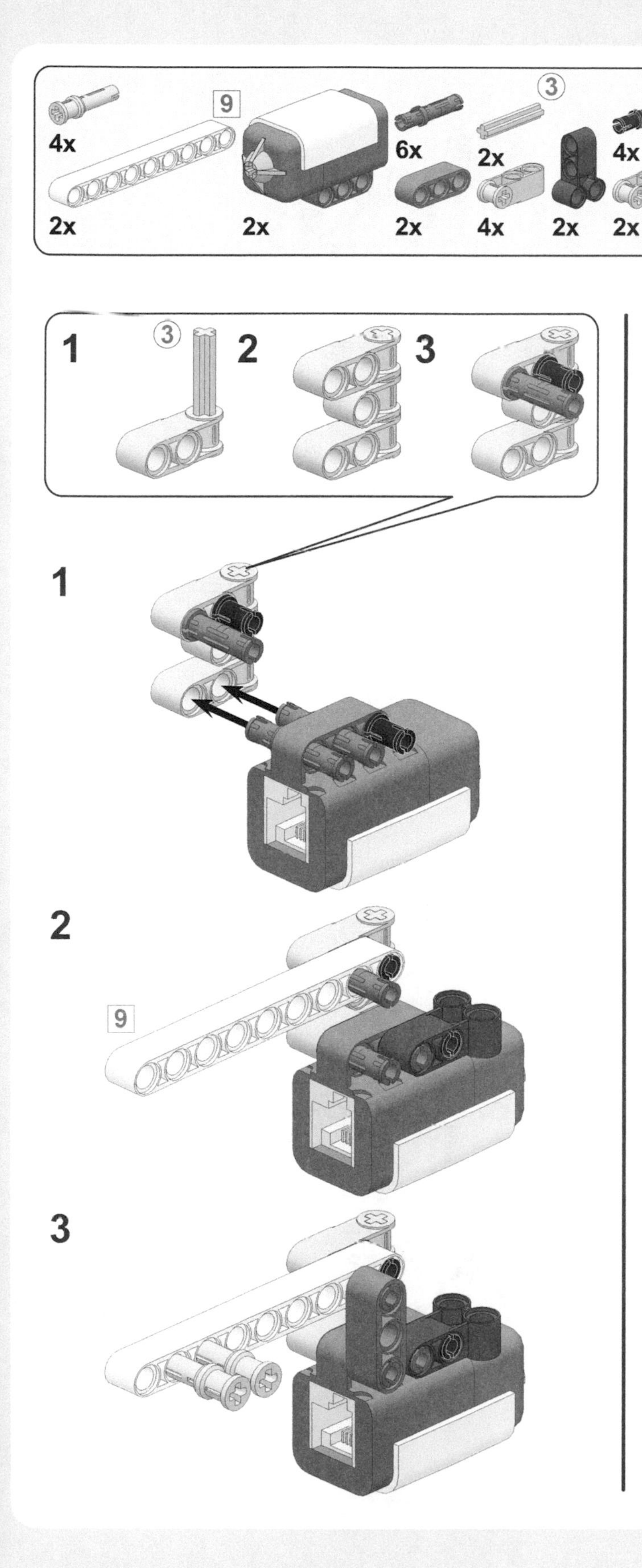

4x
9
6x
3
2x
4x
2x
2x
2x
4x
2x
2x
1
3
2
3
1
2
9
3

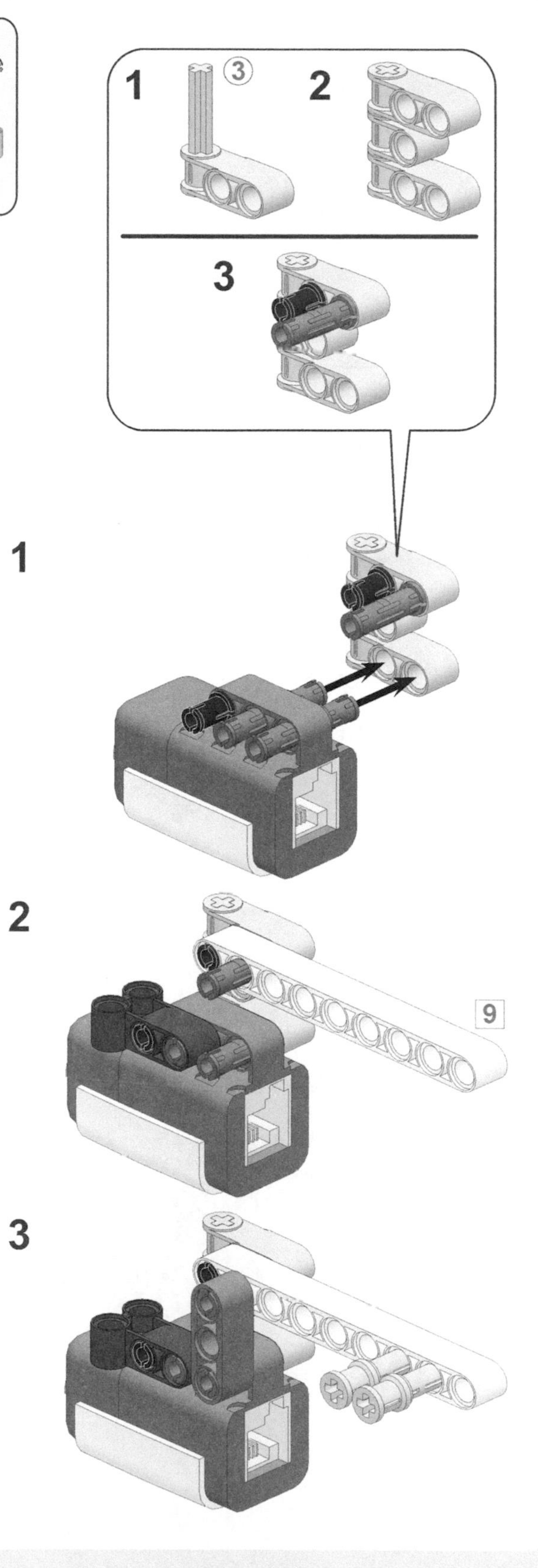

1
3
2
3
1
2
9
3

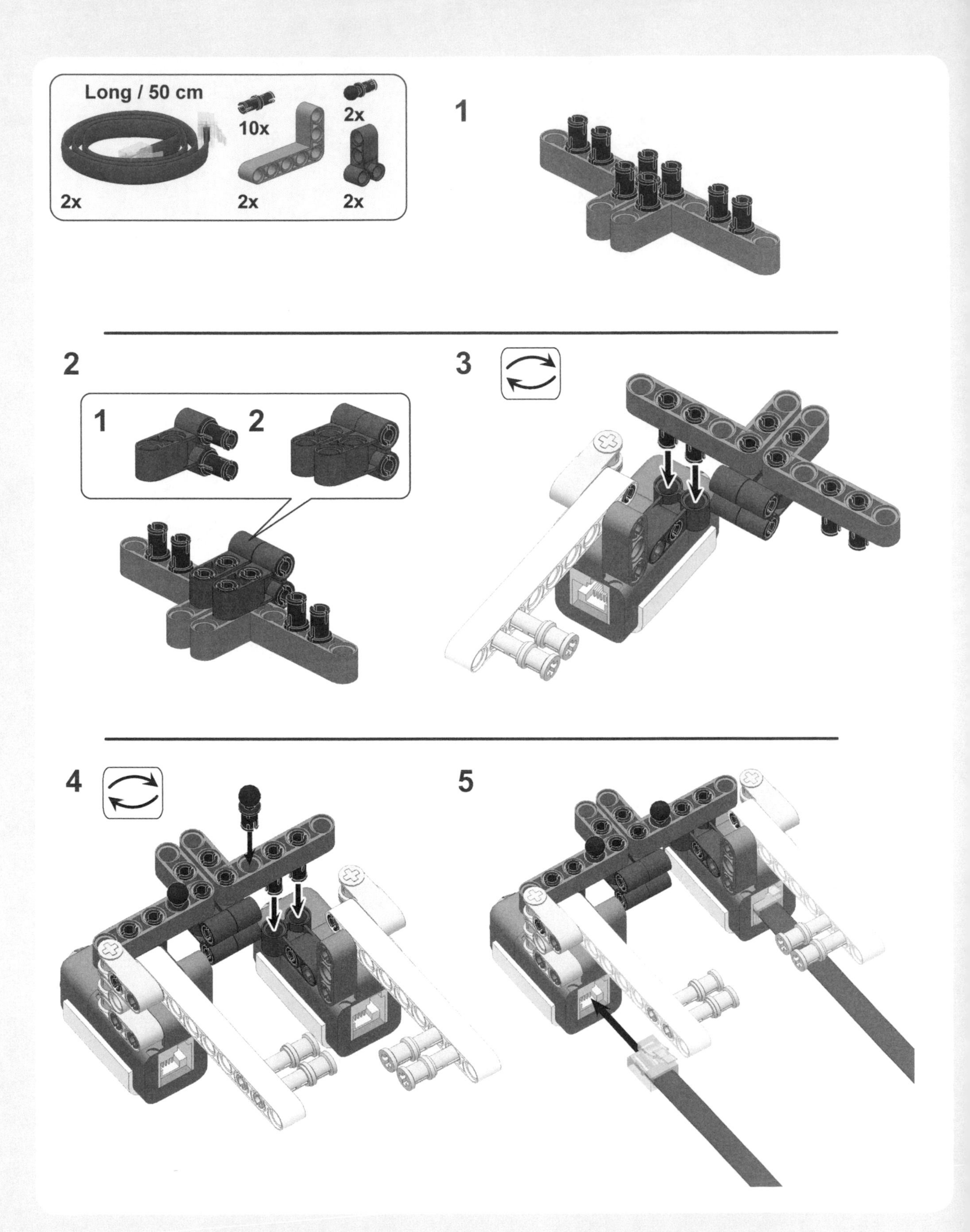
Long / 50 cm
2x
10x
2x
2x
2x
1
2
1
2
3
4
5

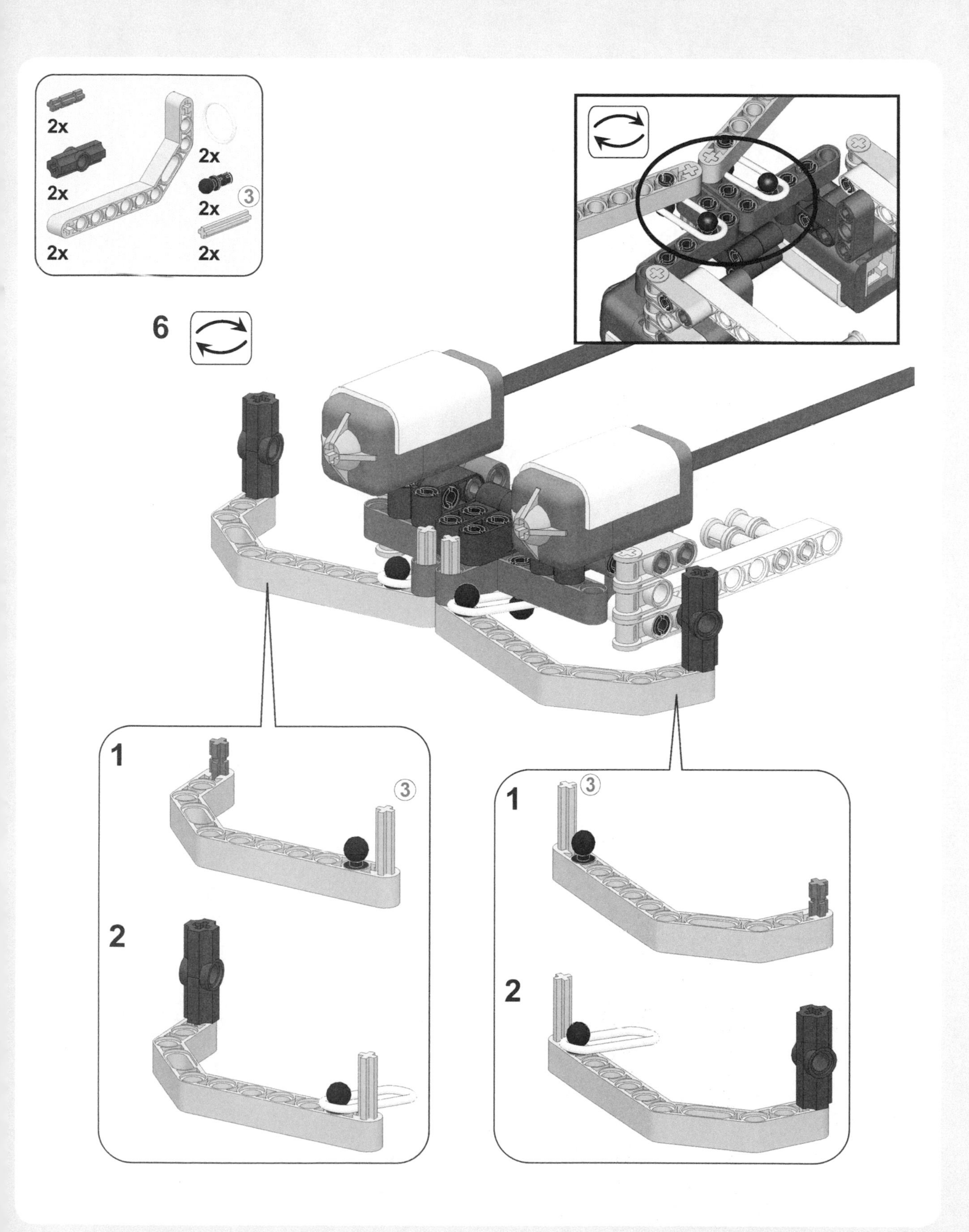
2x
2x
2x
2x
2x
3
2x
2x
6
1
3
2
1
3
2

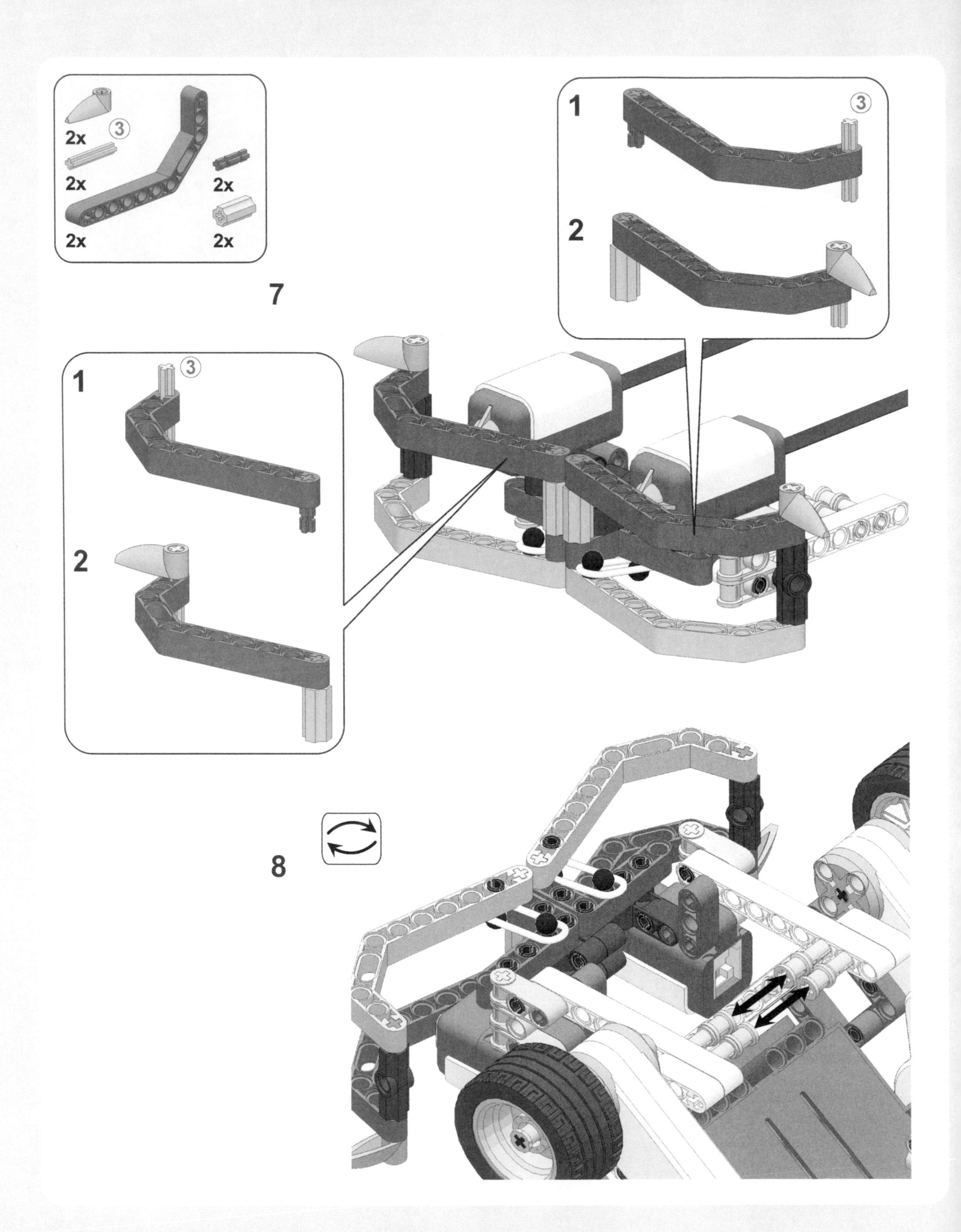
2x
3
2x
2x
2x
2x
2x
7
1
3
2
1
3
2
8

connecting the cables

Now connect the cables that you've already added to the Touch Sensors to the NXT, as shown in Figure 7-3.

From now on I'll refer to the Touch Sensor connected to port 1 as the *Right Touch Sensor* and the Touch Sensor connected to port 2 as the *Left Touch Sensor*.

Figure 7-3: Connect the Touch Sensors to the appropriate input ports on the NXT with long cables. To prevent the cables from dragging on the ground or interfering with the wheels, wrap each cable around a motor, as shown here.

programming with the touch sensor

Let's look at the Configuration Panel settings that are specific to the Touch Sensor, as shown in Figure 7-4.

Use the Port setting to select to which port the sensor is connected.

Use the Action setting to control which action will trigger the sensor (Pressed, Released, or Bumped). When the action occurs, this Wait block stops pausing the program.

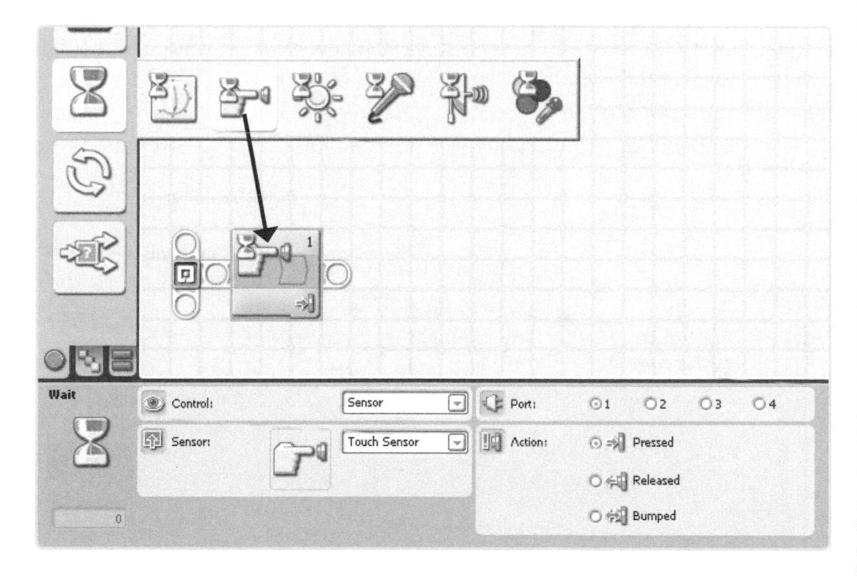

Figure 7-4: The Configuration Panel of a Wait block, configured to poll a Touch Sensor. On the right half of the panel are the specific settings for the Touch Sensor. The Configuration Panels of Loop and Switch blocks that control a Touch Sensor contain the same settings.

creating a test program for the touch sensor

The **Discovery-Touch** program you'll create in this section will make the robot play a sound when you press the Right Touch Sensor. Create the program as shown in Figure 7-5, run it, and then create modified versions with each of the other two Action settings (Released and Bumped) so you can see how they work.

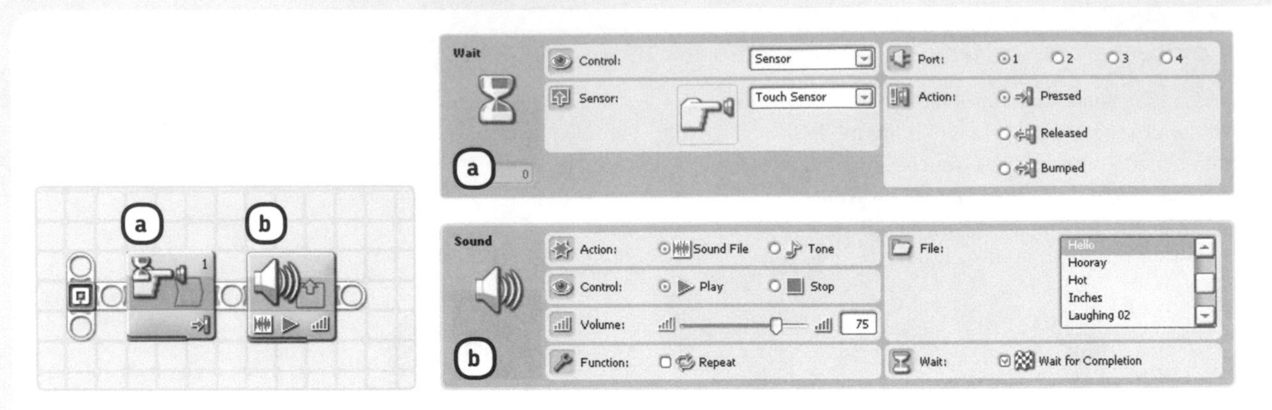

Figure 7-5: Configuration of the blocks in the Discovery-Touch program

NOTE The Wait blocks on the Programming Palette are basically all the same. The difference between them is that each block has preset Control and Sensor settings to make programming easier.

avoiding walls with touch sensors

The Discovery robot can easily avoid walls with its Ultrasonic Sensor, but it can do the same thing using the Touch Sensors. Although it can't use the Touch Sensors to sense objects from a distance, of course, it will be able to feel smaller objects than it might otherwise be able to sense with the Ultrasonic Sensor. Another advantage of the two bumpers is that they allow the robot to determine in which direction it should turn after running into something.

When creating your program to use the Touch Sensors, the first thing you need to do is to switch on the motors. Next, you check to see whether a sensor has been pressed. If the Left Touch Sensor is pressed, Discovery should back up, turn right, and then continue its path; if the Right Touch Sensor is pressed, it should back up and turn left. To find out which sensor is pressed, you'll use Switch blocks.

A Switch block can poll only one sensor at the time, so you'll start by checking to see whether the Right Touch Sensor is pressed. If it is, you go left. If it's not, you'll see whether the Left Touch Sensor is pressed. If it is, you turn right. If neither sensor is pressed, you essentially do nothing different and keep moving forward. All of these actions and decisions repeat continuously, as shown in the schematic overview of the Discovery-Bumper program in Figure 7-6.

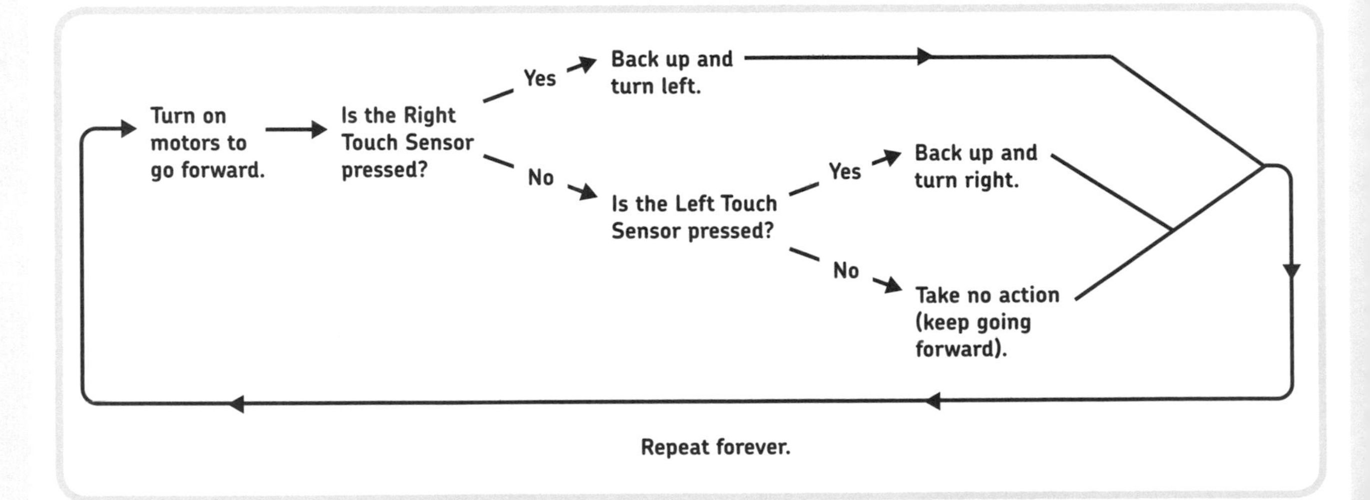

Figure 7-6: An overview of the Discovery-Bumper program. If the answer to the first question is yes, the robot turns left, but if the answer is no, another question is asked. The programming blocks that you use to make the second decision are placed inside the lower part of the Switch block that you use to make the first decision.

creating the discovery-bumper program

You'll now create a wall-avoidance program. First, create a new program called **Discovery-Bumper**, and place and configure the blocks as shown in Figure 7-7 and Figure 7-8.

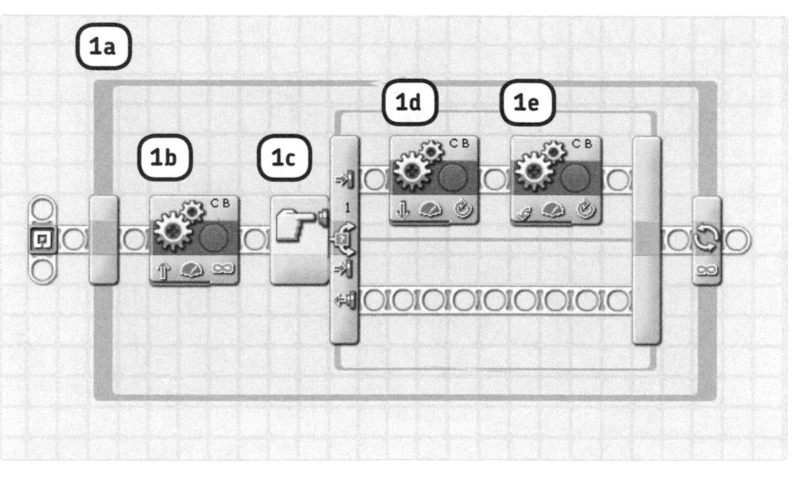

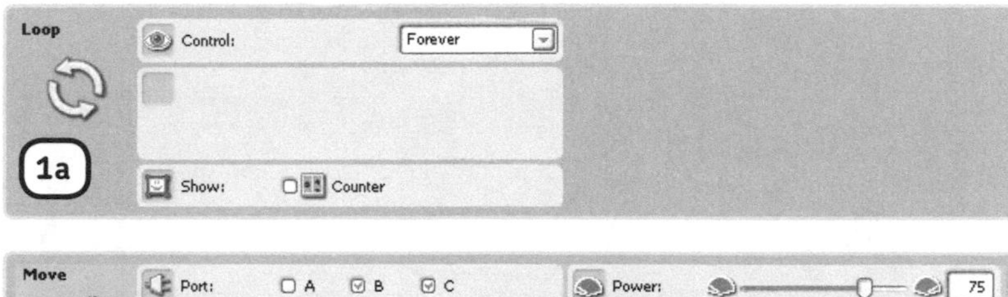

Move
Port: A B C
Direction:
Steering: C B
Power: 75
Duration: 360 Unlimited
Next Action: Brake Coast
1b

Switch
Control: Sensor
Sensor: Touch Sensor
Display: Flat view
Port: 1 2 3 4
Action: Pressed Released Bumped
1c

Move
Port: A B C
Direction:
Steering: C B
Power: 75
Duration: 0.5 Rotations
Next Action: Brake Coast
1d

Move
Port: A B C
Direction:
Steering: C B
Power: 75
Duration: 1 Rotations
Next Action: Brake Coast
1e

Figure 7-7: ***Step 1:*** *The blocks shown here turn the motors on and handle the first decision with a Switch block. The complete sequence of blocks is placed inside a Loop block configured to loop forever.*

DISCOVERY #28: ONLY TWO IS ENOUGH!

Difficulty: Easy
Create a program that shows a happy face on the display only if you press both bumpers simultaneously. If none or only one sensor is pressed, a sad face should be displayed. You'll need two Switch blocks to program this behavior.

DISCOVERY #29: SMART DECISIONS!

Difficulty: Hard
Can you make a wall-avoidance program that uses both the Touch Sensors and the Ultrasonic Sensor? Expand the Discovery-Bumper program with an extra Switch block to determine whether the Ultrasonic Sensor sees anything closer than 20 cm. Where in the program do you place this Switch block?

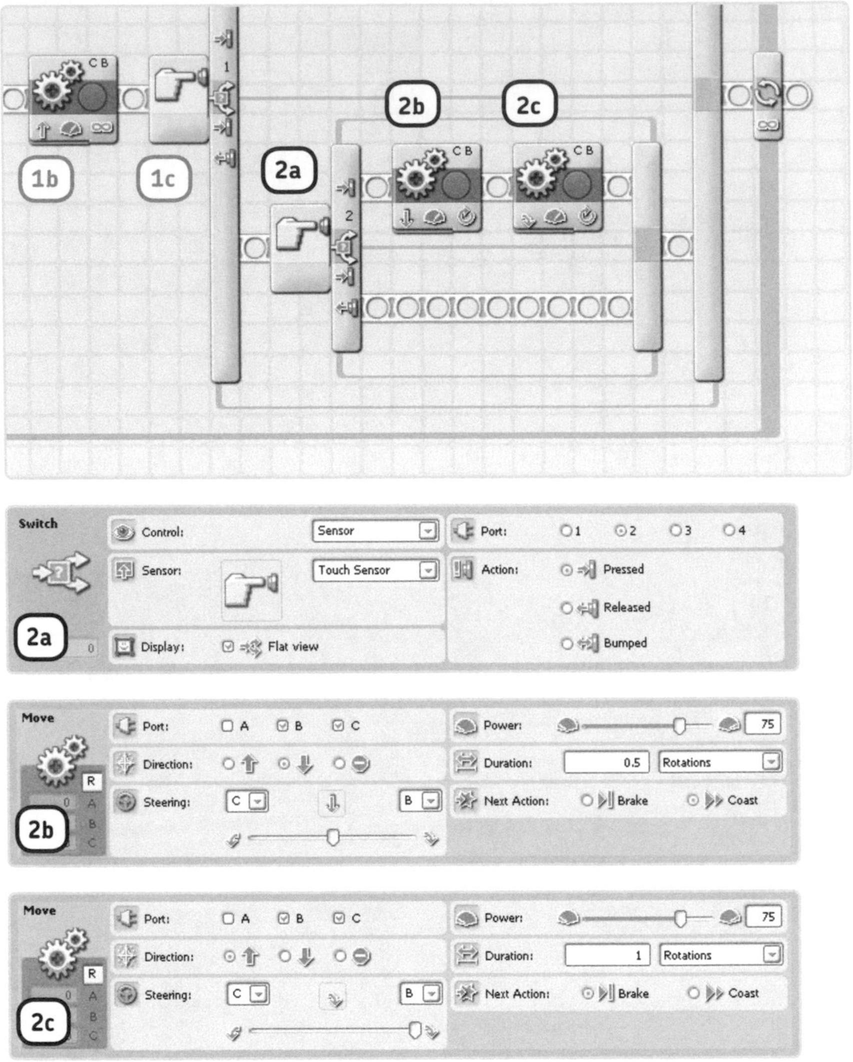

Figure 7-8: **Step 2:** The blocks you add in this step are run when the Right Touch Sensor (on port 1) is not pressed. The blocks enable the robot to make the second decision. If neither sensor is pressed, the robot doesn't take any action, and it returns to the beginning of the program to see whether a sensor is being pressed.

the color sensor

The *Color Sensor* detects the color of a surface, the brightness of a light source, and the intensity of light reflected by a surface. Figure 7-9 shows the Color Sensor being used to detect the color of LEGO bricks. The color sensor can also act as a colored lamp, emitting a bright light colored red, green, or blue. You'll use the color sensor's ability to see color in this chapter and learn to use its other features in Chapters 8 and 9.

Because the Color Sensor has so many functions, it's useful in many applications. For instance, you can use it to make a vehicle follow lines (Chapter 7), as a Color Lamp to indicate that a certain action is taking place (Chapter 8), as a light detector (Chapters 8 and 9), to sort colored LEGO bricks (Chapter 14), and even as a stabilization sensor (Chapter 15).

Figure 7-9: The Color Sensor can sense the color of surfaces such as the color of LEGO bricks. It can identify black, blue, green, yellow, red, and white.

creating the color sensor attachment

In this section, you'll enhance Discovery by creating a sub-module with the Color Sensor. But before you do, detach the bumpers by disconnecting their cables and the gray pins that connect them. (Don't take them apart, though; you'll need them for some of the discoveries at the end of this chapter.)

Once you've completed the Color Sensor attachment, connect it to input port 3 on the NXT using the short cable, as shown in Figure 7-10.

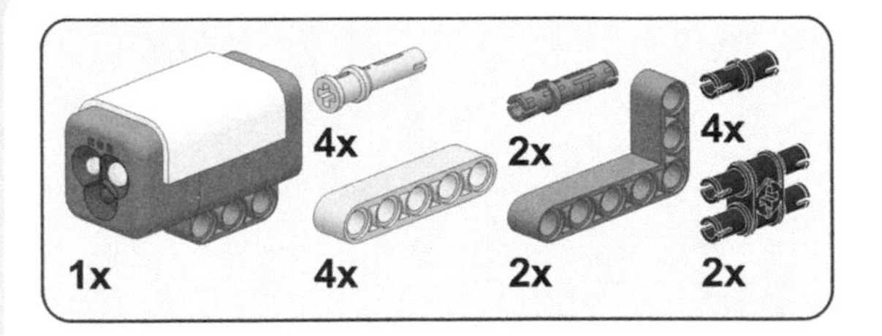
4x
2x
4x
2x
1x
4x
2x
2x

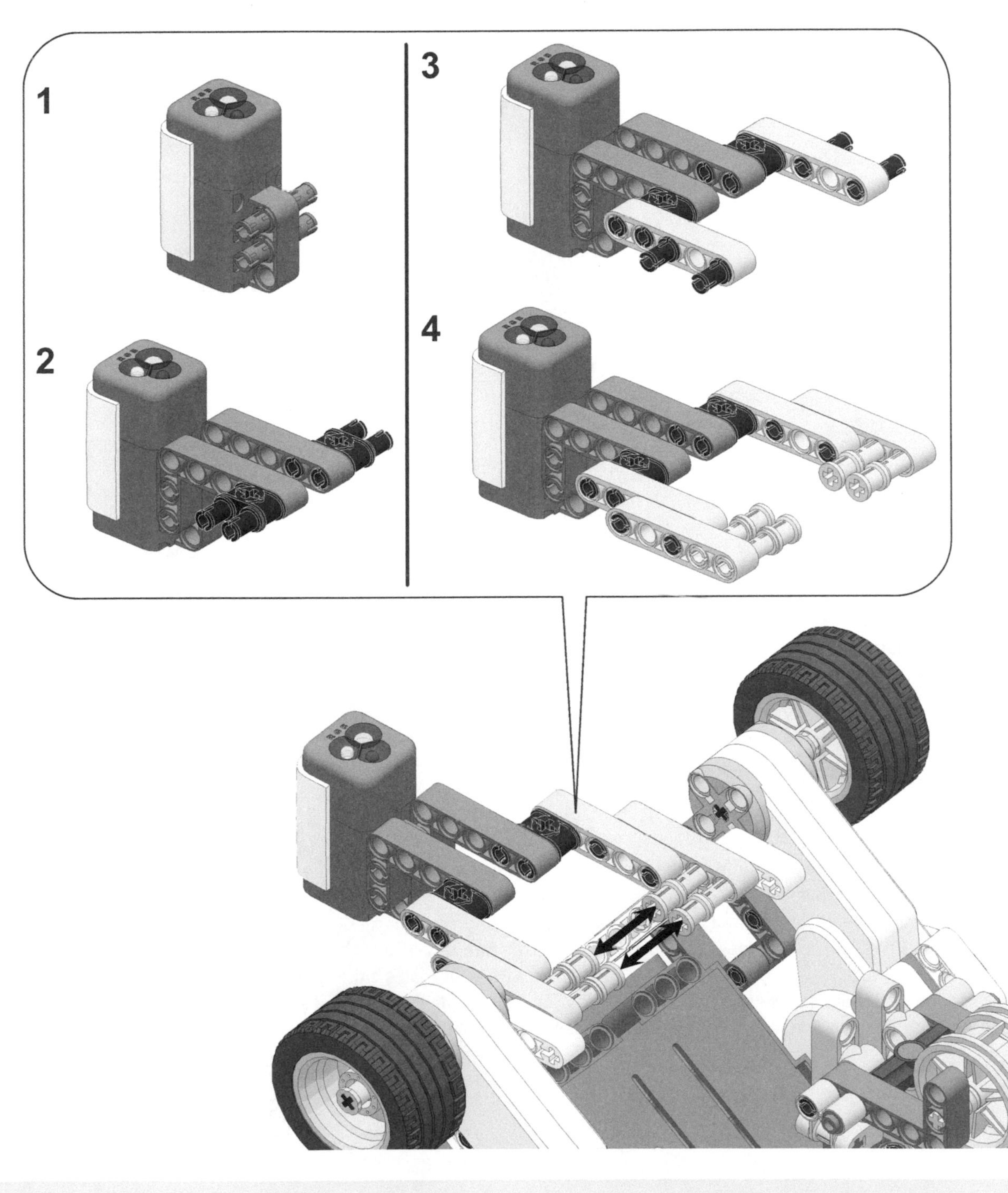
1
2
3
4

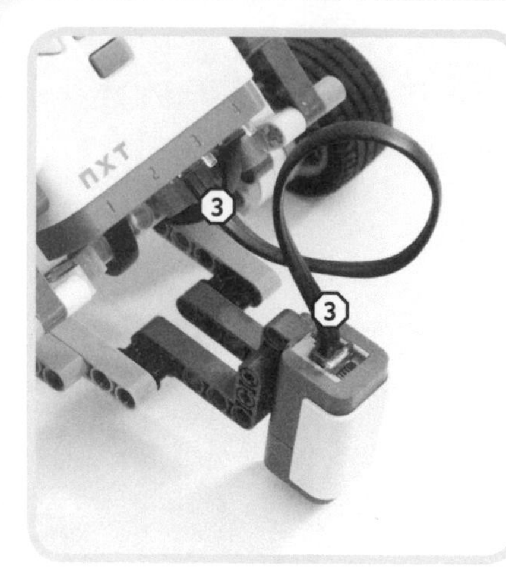

Figure 7-10: Connecting the Color Sensor to the Discovery robot using the short cable

Figure 7-11: The placement of the Discovery robot and the Color Sensor on the Test Pad

using the view mode to poll the color sensor

To poll the Color Sensor, select **View ▸ Color ▸ Port 3** on your NXT, place your robot on the Test Pad, and point the sensor at the colored line, as shown in Figure 7-11. Notice how the color name displayed on the NXT screen changes as you move the sensor over the line. The sensor can recognize six colors this way: black, blue, green, yellow, red, and white.

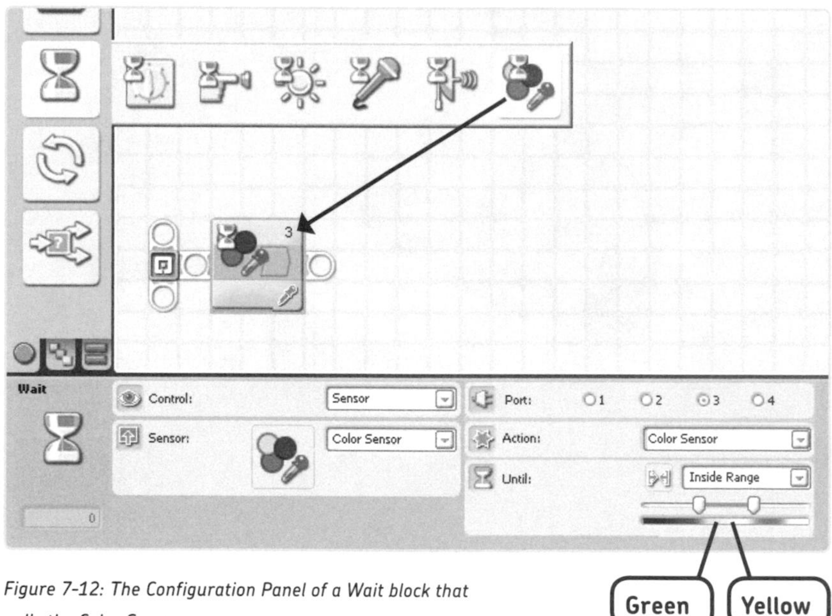

Figure 7-12: The Configuration Panel of a Wait block that polls the Color Sensor

programming with the color sensor

In this section, you'll create programs for the Discovery robot that use the Color Sensor with Wait and Switch blocks, just as you did with the Ultrasonic and Touch Sensors. Figure 7-12 shows a Wait block that polls the Color Sensor and its Configuration Panel.

In the Port area, select the input port the sensor is connected to (port **3**). In the Action box, choose **Color Sensor**. In the Until box, specify what the sensor should see in order for the Wait block to stop waiting, by selecting a range of colors, such as green and yellow as shown. With **Inside Range** selected, the robot will wait until the sensor reports a color that is either green or yellow. If you were to select ***Outside Range***, the block would wait until the sensor spots something that's not green or yellow.

staying inside a colored line

Your next program, **Discovery-Circle**, will demonstrate how you can use the Color Sensor as a line detector. Once your program is loaded, you'll place the Discovery robot in the circle on the Test Pad and have it drive around without leaving the circle. Figure 7-13 shows the program flow for this behavior; Figure 7-14 shows how to create the program.

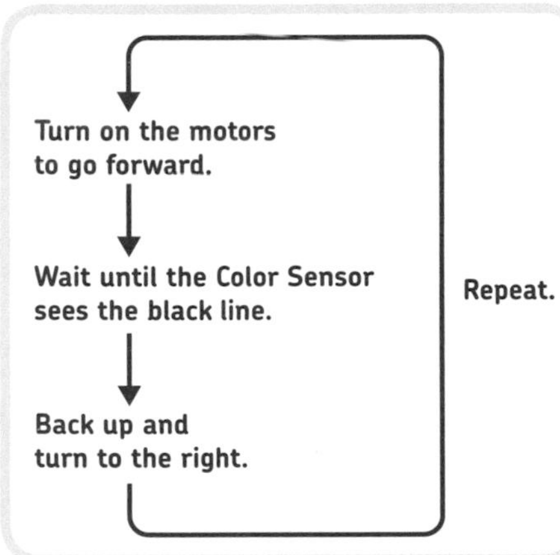

Figure 7-13: The program flow of the Discovery-Circle program

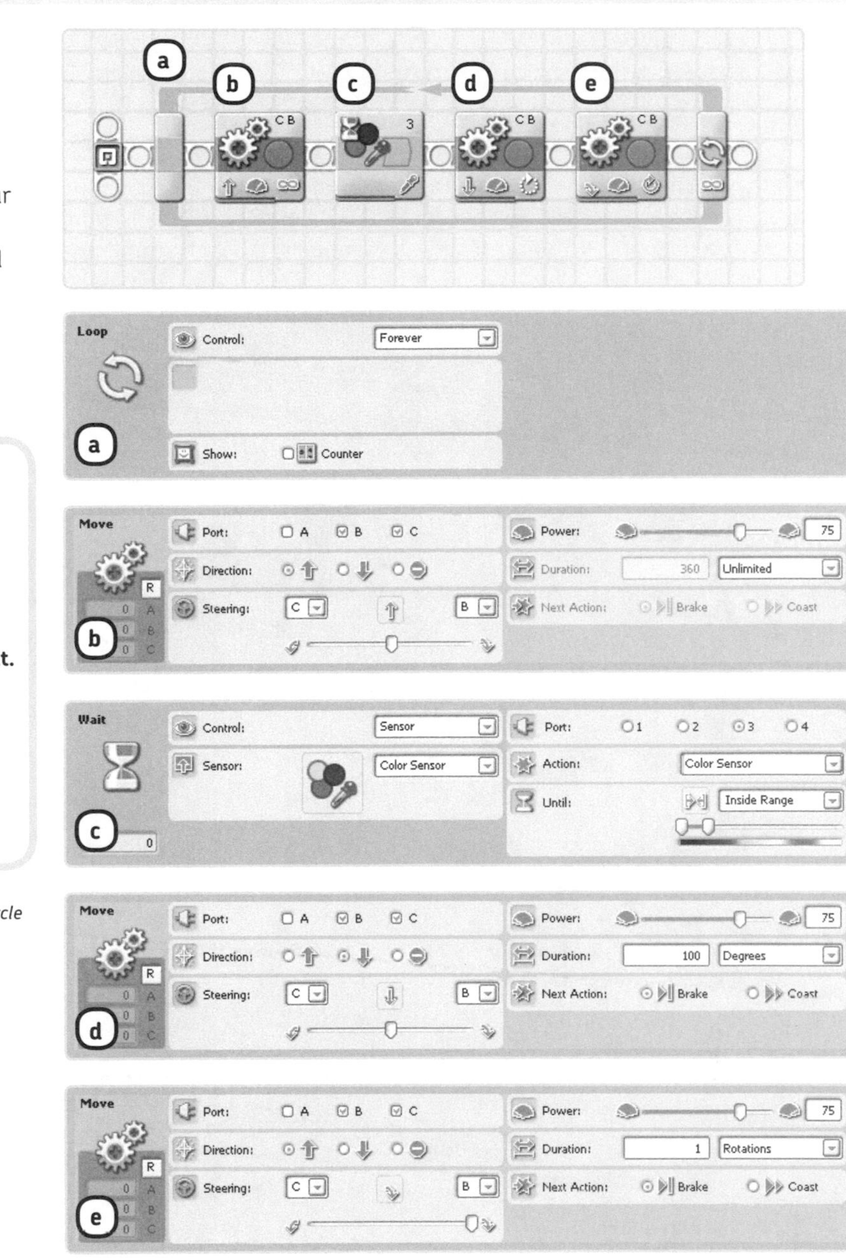

Figure 7-14: The configuration of the blocks in the Discovery-Circle program. The Wait block is configured to wait until the sensor sees something that is black as specified in the Until setting (using the sliding bars).

BUILDING DISCOVERY #4: CLEAN THAT TEST PAD!

With the Discovery-Circle program the robot constantly moves through the circle. If you put some LEGO pieces in the circle (Figure 7-15), your robot will eventually push them out, although before it can do that, you need to give Discovery a bulldozer blade. Can you build such a blade with your LEGO pieces?

TIP **Connect the bulldozer blade to the Color Sensor attachment, and use the Discovery-Circle program.**

Figure 7-15: The setup for Building Discovery #4: Using a bulldozer blade, Discovery can sweep the mess out of the circle. Can you build it?

DISCOVERY #30: TELL ME WHAT YOU SEE!

Difficulty: Hard

You learned earlier that you can use multiple Switch blocks in a program and that the color sensor can detect up to six different colors. Can you create a program that will make the robot say the name of the color it sees, using sound files? To begin, look at the program flow scheme shown in Figure 7-16, and then test your program using the colored bar on the Test Pad. When you're ready, turn the entire series of blocks (except for the Loop block) into a My Block called **Say Color**.

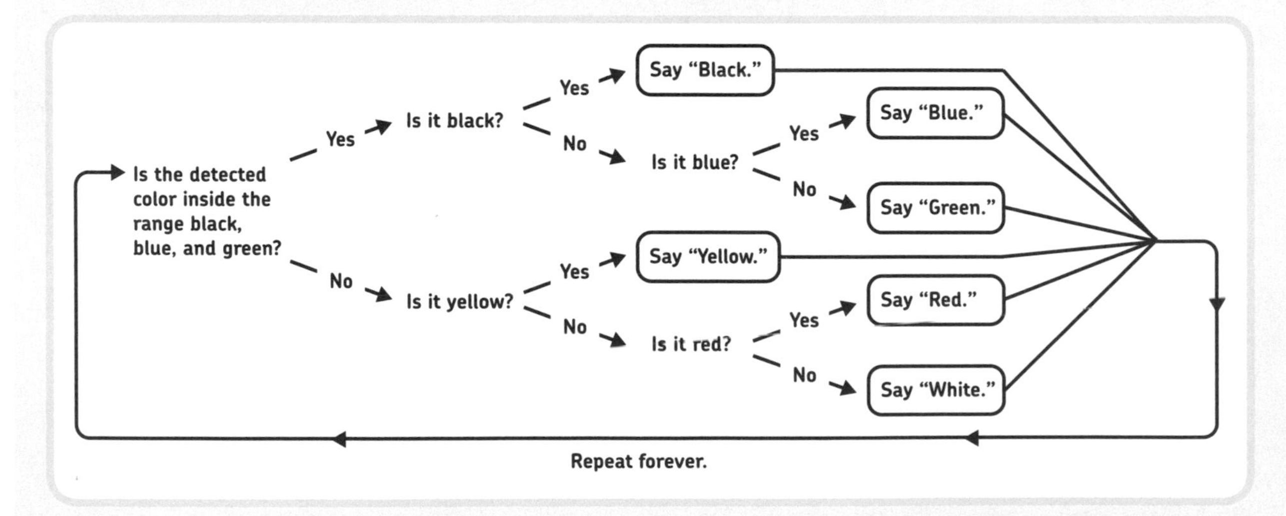

Figure 7-16: A model for a program to make the robot say which color it sees

following a line

In your next project, you'll use the Color Sensor to create a line-following robot, which means the robot will follow a colored track on a mat, such as the black line on the Test Pad. Let's look at the strategy behind this program.

When following a black line on a white mat, there are always only two possibilities: the sensor measures either white or black. Therefore, when creating a line-following program for a black-and-white environment, you'll use a Switch block, which looks for the color black. When the sensor sees black, the Switch block will trigger a Move block to perform one movement; if it sees another color (white), it will perform a different movement, as shown in Figure 7-17.

If Discovery sees white, it cannot determine on which side of the line the color lies, so you need to make sure it will always stay on only one side of the line; otherwise, it will stray off the line into the white area. You do this by always driving Discovery right when it sees black and left when it sees white. Figure 7-18 shows the line-following program.

NOTE **Be sure when configuring the Move blocks not to drag the Steering slider all the way to the right (in block *c*) or to the left (in block *d*), or the robot will turn in place rather than moving forward. Also, be sure that before you start the program, you place the robot on the mat so that the sensor is pointed at the line, with the outside of the circle to the robot's right (see Figure 7-17*a*).**

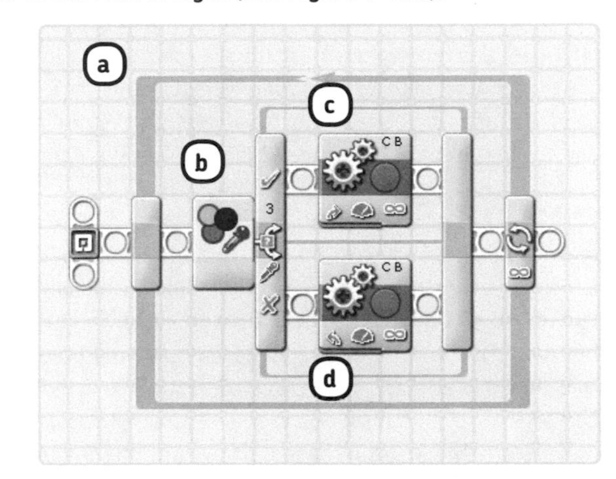

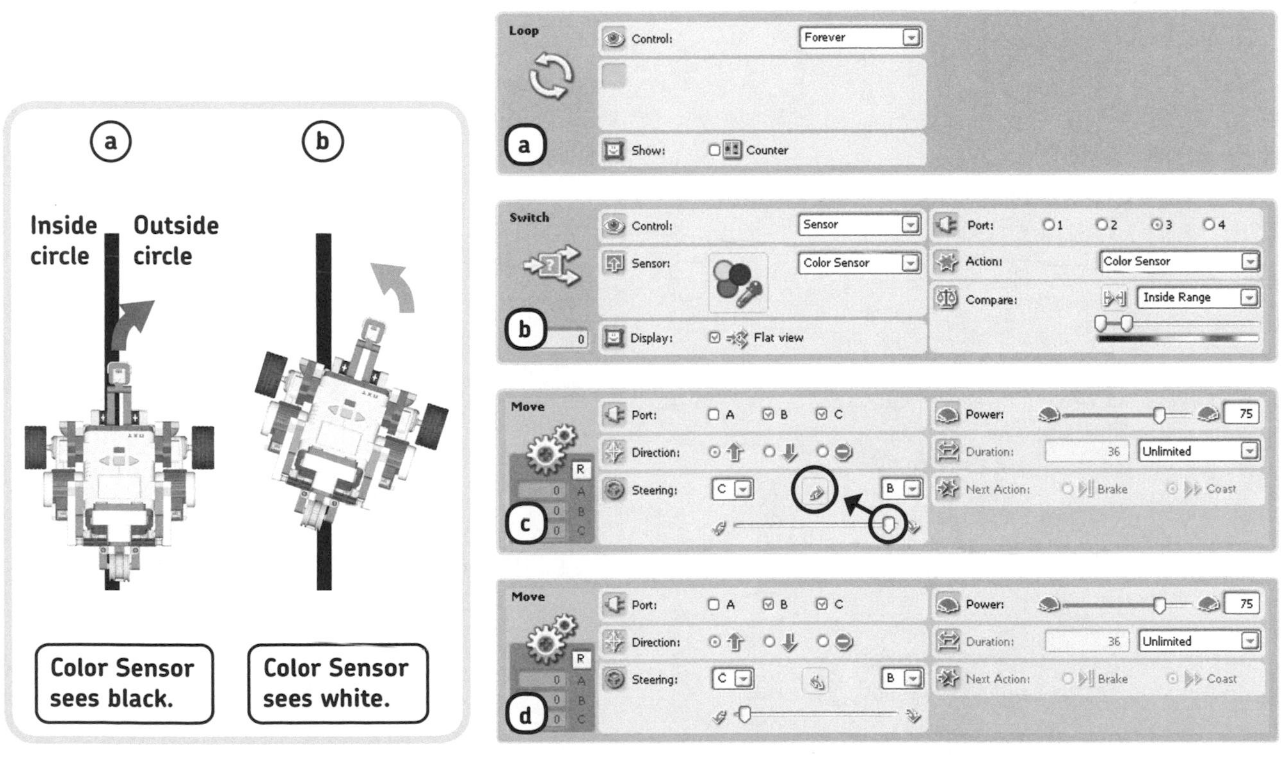

Figure 7-17: Discovery steers right if it sees the black line (a) and steers left if it sees the white area (b). As Discovery steers, it moves forward, so if you repeat this behavior, you end up with a line-following robot.

Figure 7-18: The configuration of the Discovery-Line program. Note that the Duration settings in the Move blocks are set to **Unlimited**. Once the robot starts to turn, it instantly goes back to the beginning of the program to see whether a different color has been detected or whether it should keep turning in the same direction. Unlimited Move blocks just switch on the motors and have the program continue.

using the NXT buttons as sensors

In addition to the Ultrasonic, Color, and Touch Sensors, the NXT contains its own sensors: the *NXT buttons*. You can use both the Enter and Right Arrow and Left Arrow buttons on the NXT just like you use Touch Sensors. For example, you can make the robot turn around if you press the Right Arrow button.

To use the NXT buttons, you'll configure a Wait, Switch, or Loop block set to control a sensor and select **NXT Buttons** in the Sensor box. In the Button box, choose the **Enter Button**, **Right Button**, or **Left Button**. Finally, in the Action box, you choose whether the button should be pressed, released, or bumped, just as with the Touch Sensor. You can see these configurations in the next sample program.

Let's practice using these buttons with the **Discovery-Button** program, as shown in Figure 7-19. Create this program now.

DISCOVERY #31: EXPERT LINE FOLLOWING!

Difficulty: Expert
When you ran the Discovery-Line program, you may have noticed that Discovery actually followed the *outside edge* of the line. You can make it follow the *inside edge* of the line by running the program when the robot is on the line but with the inside of the circle on the robot's right. As it does so, it will pass four colored squares on the Test Pad each time it goes around the circle. Make sure that it stops when it detects one and says which color it sees, and then make it continue following the line. When following the inside edge, your robot might have problems following the line. How can you modify the program to resolve this?

DISCOVERY #32: WHICH BUTTON DID YOU PRESS?!

Difficulty: Medium
Create a program that says "Left" when the Left Arrow button on the NXT is pushed and "Right" when the Right Arrow button is pressed. If you press the **Enter** button, the NXT screen should show a happy face until you release the button.

HINT **Use Switch blocks to determine which NXT button is being pressed, and use a Wait block to wait until the Enter button is released.**

DISCOVERY #33: SOUNDBOT!

Difficulty: Medium
Create a program that enables you to play music on your Discovery. Program it so that each bumper and NXT button triggers a different tone using Switch blocks to determine which buttons are pressed.

HINT **You can make the program even more interactive if you implement the Ultrasonic Sensor in your program. How can you use it as a sound trigger?**

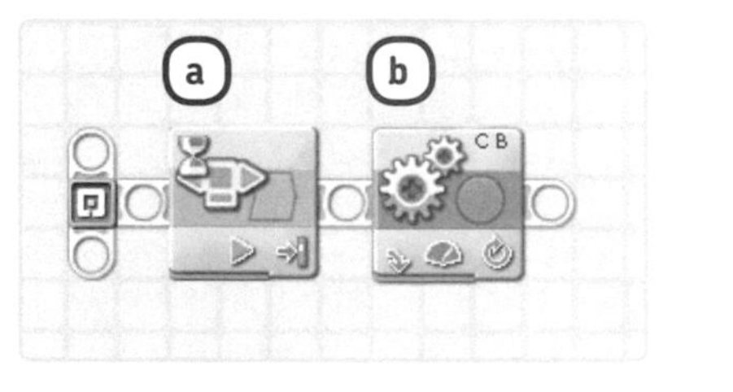

Figure 7-19: The configuration of the blocks in the Discovery-Button program. When you run this program, the robot should turn right when you press the Right Arrow button.

the rotation sensors

When you tell the robot to move forward for three rotations with the Move block, the vehicle knows that it should stop moving when the wheels have each made three revolutions. The robot knows this because the *Rotation Sensor* in each NXT motor tells the NXT how much the motors have turned.

You can use the information from this sensor in your programs, for example to create a program that repeatedly says "Hello" until you turn one wheel by hand. The sensor tells you how much (either in degrees or rotations) a motor has turned since you started the program, as well as in which direction the motor turned, as shown in Figure 7-20.

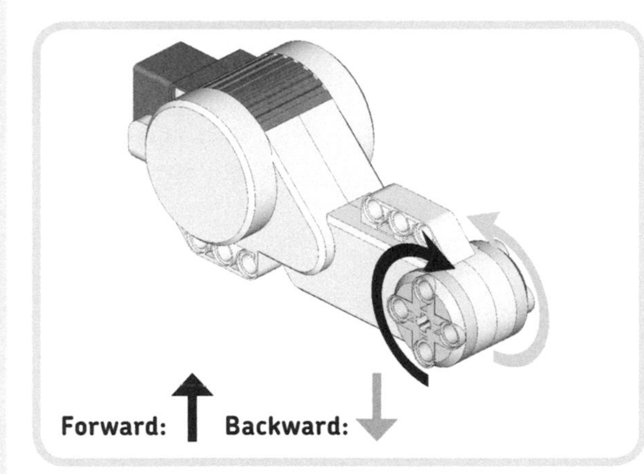

Figure 7-20: The Rotation Sensor inside the motor tells the NXT program how many degrees or rotations the orange part of the motor has turned since you activated a program and in which direction. If a program is running and you spin the wheel around twice in the black direction, the sensor will report two rotations (or 720 degrees) in the forward direction. If you then turn it half a rotation in the gray direction, the sensor reports one and a half rotations in the forward direction.

using the view mode to poll the rotation sensor

You can use the motor's Rotation Sensors in your programs like normal sensors, but since they are inside the NXT motors, they are *always* connected to output ports.

To use the View mode on the NXT to read the sensor value reported by a Rotation Sensor, turn on the NXT, select **View**, choose **Motor Degrees**, and then select port **B** or **C**.

Now, as you rotate the motor that you selected by hand, the value on the display should change. Positive degree values mean that you turned the wheel forward, and negative ones represent reverse rotations. Instead of Motor Degrees, you can select **Motor Rotations** to make the NXT screen show how many complete revolutions the wheel has made.

DISCOVERY #34: CIRCLING DEGREES!

Difficulty: Easy

Do you remember how you tried to make your robot do an accurate 90-degree turn in Chapter 4? The question was how many degrees you had to enter in the Move block's Duration setting to accomplish this turn. You've just learned how to measure how many degrees the wheels have turned. Now go to View mode, poll the Rotation Sensor on port B, and manually turn the robot in place for 90 degrees by slowly rotating each wheel. Use the value that appears on the NXT display, and enter it in a Move block's Duration setting. Was your measurement accurate?

making programs with rotation sensors

As you've seen in this chapter, you can use Wait, Loop, and Switch blocks to control a Rotation Sensor. For instance, a program with a Loop block can repeatedly say "Hello" until the motor on port B has turned 180 degrees in the forward direction. The boxes in the right half of the Configuration Panel are used to configure the sensor, as shown in Figure 7-21, and you use the Port box to select to which output port the motor you want to poll is connected. (You'll learn more about the Reset function in the Action box in "Resetting the Rotation Sensor" on page 84.)

In the Until box you configure the *trigger value* (the condition that makes a Loop block stop looping) by using the orange arrows to specify whether it should look for forward or backward rotation. Select a number of degrees or rotations, as well as whether the sensor reading should be greater (>) or less (<) than this number in order to trigger the block and make the Loop block stop Looping.

Now create the sample program **Discovery-Rotation** shown in Figure 7-21 to perform the behavior described.

NOTE **The list of sensors in the Wait, Loop, and Switch blocks may contain both Rotation Sensor and !Rotation Sensor. The !Rotation Sensor is there for compatibility with older versions of the NXT-G software, but you'll always use the Rotation Sensor option (the one without the exclamation point) in this book.**

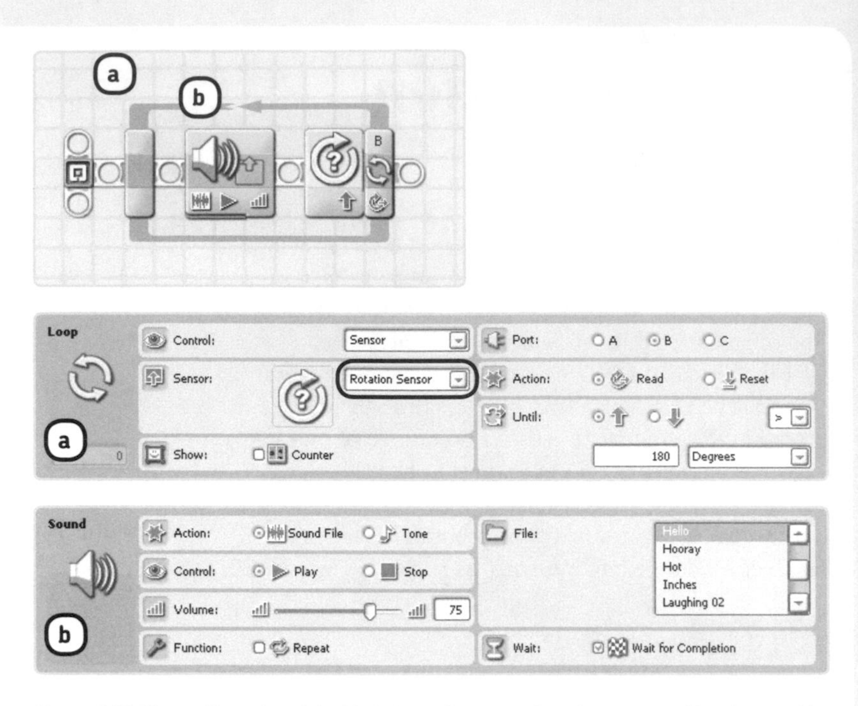

Figure 7-21: The configuration of the blocks in the Discovery-Rotation program. Experiment with this sensor by changing the settings in the Until box of the Loop block to see what each setting really does.

resetting the rotation sensor

As a program runs, the Rotation Sensor value changes as you turn a motor by hand or if a Move block makes the robot move. Sometimes, though, it is useful to reset this value to zero. For example, you might want to modify the Discovery-Rotation program to reset the sensor after it confirms that the motor has turned 180 degrees. With a Loop block and the Action setting specified to **Read**, the Rotation Sensor will repeat the

DISCOVERY #35: ROTATIONAL MUSIC!

Difficulty: Medium

The program shown in Figure 7-22 plays different tones depending on how the left and right wheels of Discovery are rotated. If a wheel is turned forward, the robot plays a different tone than when it is turned backward. Can you figure out how this program works and how it is configured?

HINT **The hidden tabs of the Switch blocks also contain Sound blocks.**

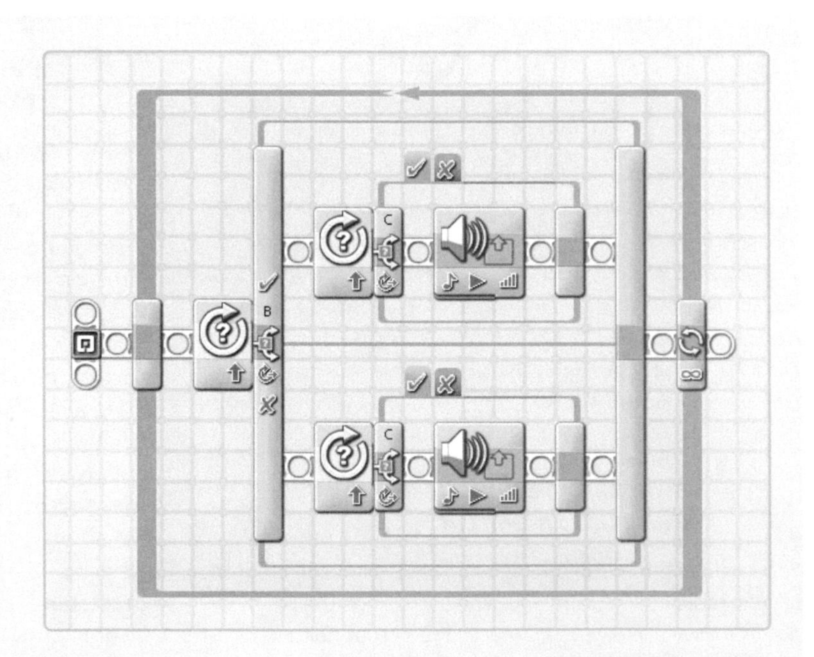

Figure 7-22: Can you re-create the program shown here? How do you think the blocks should be configured to make the robot play four different tones depending on how the wheels have turned?

blocks in it until the trigger value is reached. When **Reset** is selected, the sensor value is reset to zero after each loop. Therefore, a Loop block with this setting stops only if the trigger value is reached *during* one loop.

Now modify the Discovery-Rotation program. Change the Action setting to **Reset**, and run the modified program to see how this works.

NOTE If you select Reset in a Switch block, the value is reset to zero *after* the sensor value is compared to the trigger value. You cannot reset the sensor value with a Wait block.

further exploration

Now that you've learned how to work with the NXT sensors, you should be able to create robots that interact with their environments. Discovery is, of course, only one example. As you continue reading this book, you'll build several robots with sensors, each of which will use sensors differently.

By now you've learned to use the components that are essential to create a working robot: the NXT, the motors, the sensors, and the NXT-G software. The following chapters will explore each of these subjects in more detail so that you'll be able to create increasingly sophisticated (and fun!) robots.

The following discoveries will help you explore more possibilities with the sensors. Be sure to post your ideas and solutions to the book's companion website (*http://www.discovery.laurensvalk.com/*)!

DISCOVERY #36: COLOR THE BALL!

Difficulty: Hard
For this discovery you'll have to disassemble the Color Sensor attachment so that you can mount the sensor elsewhere on the robot. Connect the sensor so that it looks upward. The NXT 2.0 robotics kit comes with a set of small colored balls; if you hold one in front of the Color Sensor, the sensor should be able to identify the color of the ball. Can you program the Discovery robot to perform different actions such as moving and making sounds when it sees these colored balls?

DISCOVERY #37: ULTRASONIC LINE FOLLOWING!

Difficulty: Expert
You used the Discovery-Line program to follow the black line on the Test Pad, but you can have your program do more than one thing at the time. Place a book upright somewhere along the line on the Test Pad, and then expand the program to follow the line until the Ultrasonic Sensor spots the book. When Discovery spots the book, it should turn around and follow the line again, but in the other direction.

HINT Instead of a Loop block configured to loop forever, configure it to loop until it sees the book. How do you turn around to follow the line in the other direction?

BUILDING DISCOVERY #5: AUTOMATIC HOUSE!

Long before you got the LEGO MINDSTORMS NXT 2.0 robotics kit, you might have built houses from regular LEGO bricks. Now that you know how to work with motors, how to use sensors, and how to make working programs, how about trying to build a robotized house with the NXT?

IDEAS Use motors to automatically open the door when someone presses the doorbell (the Touch Sensor), and set an intruder alarm that sounds when the Ultrasonic Sensor sees someone. Use another motor to close and open the shutters when you hold a colored ball in front of the Color Sensor.

shot-roller: a robotic defense system

Now that you've learned how to control motors, use sensors, and program the NXT, you can begin making more complex machines. This chapter will show you how to build and program the Shot-Roller (Figure 8-1). This three-wheeled construction can shoot balls in any direction.

NOTE Enjoy the Shot-Roller, but remember to never shoot at people! As long as you respect the safety of others, this is not a dangerous robot.

The Shot-Roller can operate in either autonomous or remote-control mode. In *autonomous mode*, it does everything by itself. You can program it to look around and shoot at targets detected by the Ultrasonic Sensor or to respond to light signals using the Color Sensor. In *remote-control mode*, you use two Touch sensors to completely control the robot's actions. The sensors are used as buttons to activate the shooter's movements.

Unlike the Explorer and the Discovery robots that you built earlier in this book, the Shot-Roller uses three motors, as you can see in Figure 8-1. The *Turn motor* enables the robot to spin around in order to look for targets; the *Turret motor* moves the turret up and down;

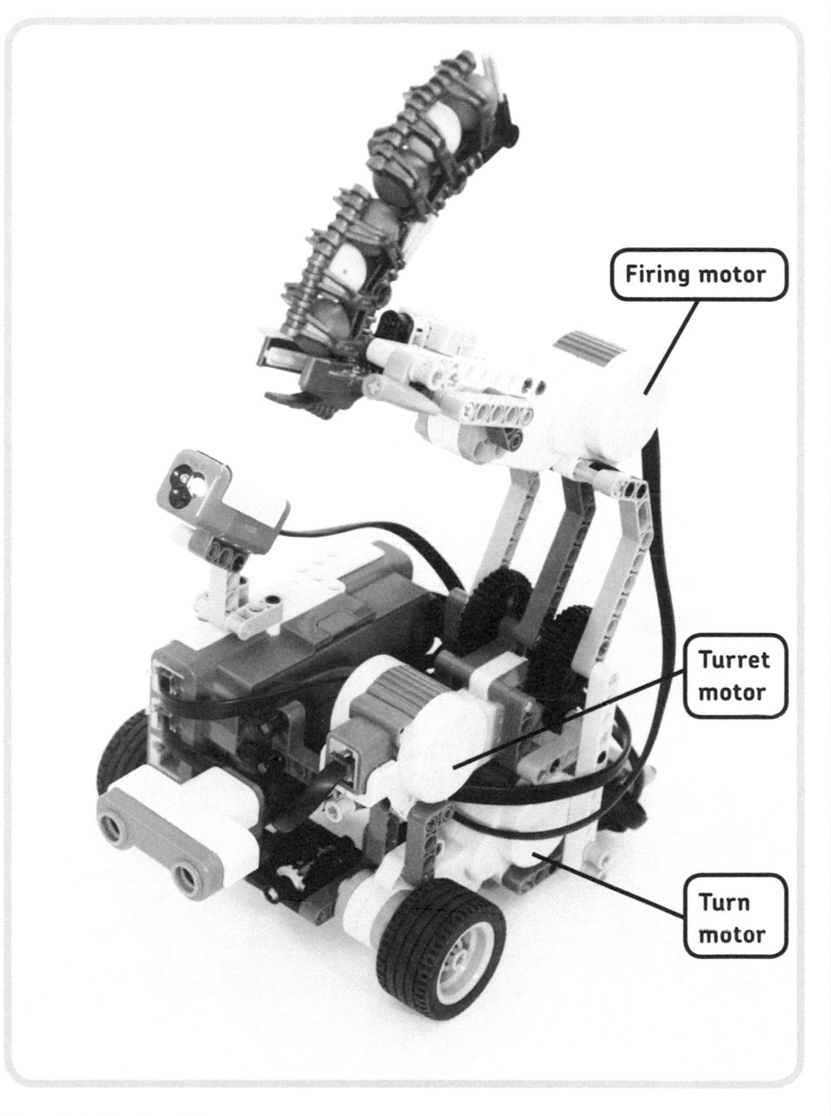

Figure 8-1: The Shot-Roller

and the *Firing motor* is the shooter, which can shoot balls at high speed. The combination of these three motors allows the robot to fire in any direction!

In addition to building the Shot-Roller, you'll learn some new programming techniques. You'll also get to use some new programming blocks such as the Color Lamp block and the Motor block.

building the shot-roller

Now that you've learned a bit about Shot-Roller's functionality, you're ready to build it. To do so, follow the directions on the next pages, but first select the pieces you need as shown in Figure 8-2.

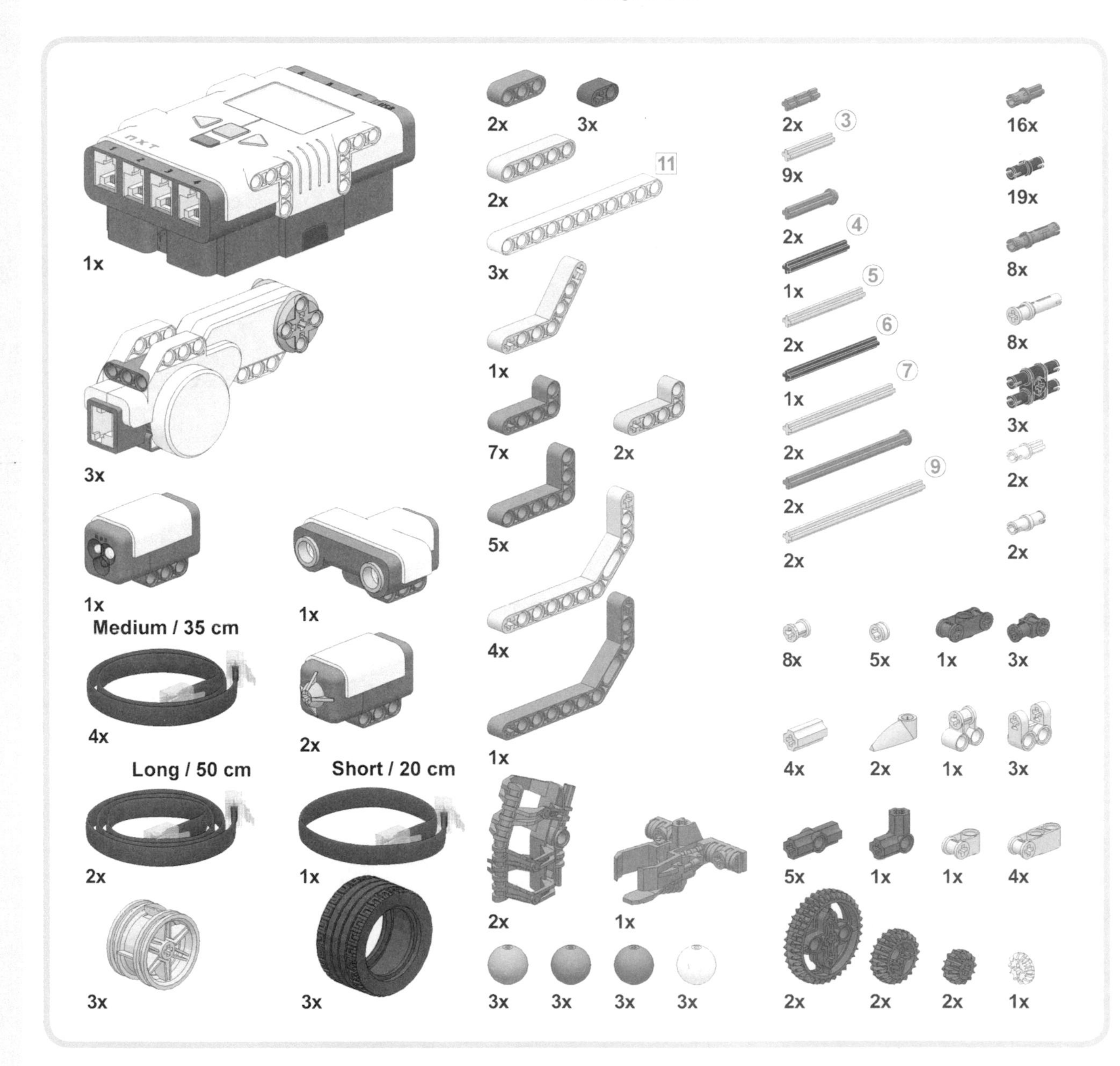

Figure 8-2: The required pieces to build the Shot-Roller

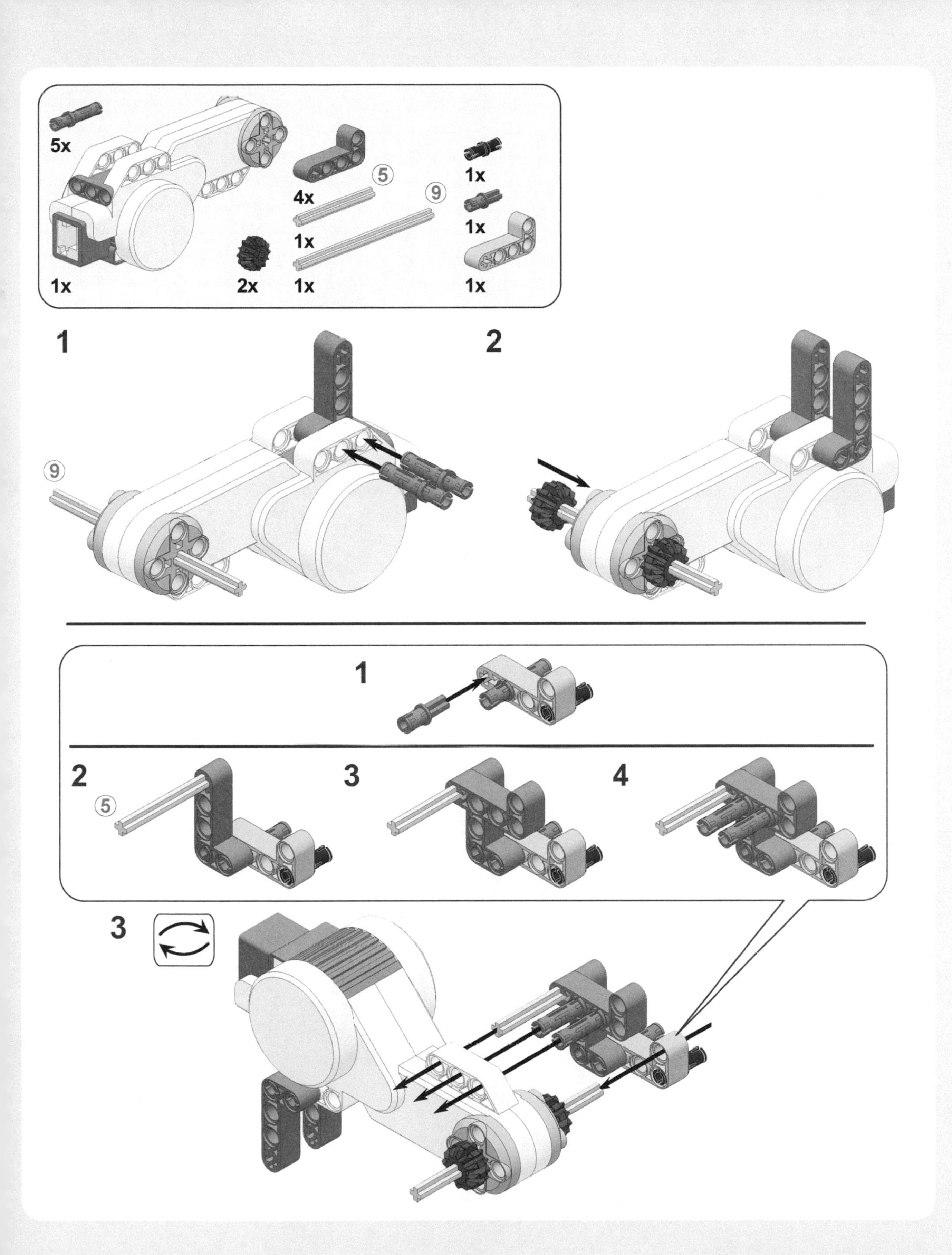
5x
4x
5
9
1x
1x
1x
1x
1x
2x
1x
1x
1
2
9
1
2
5
3
4
3

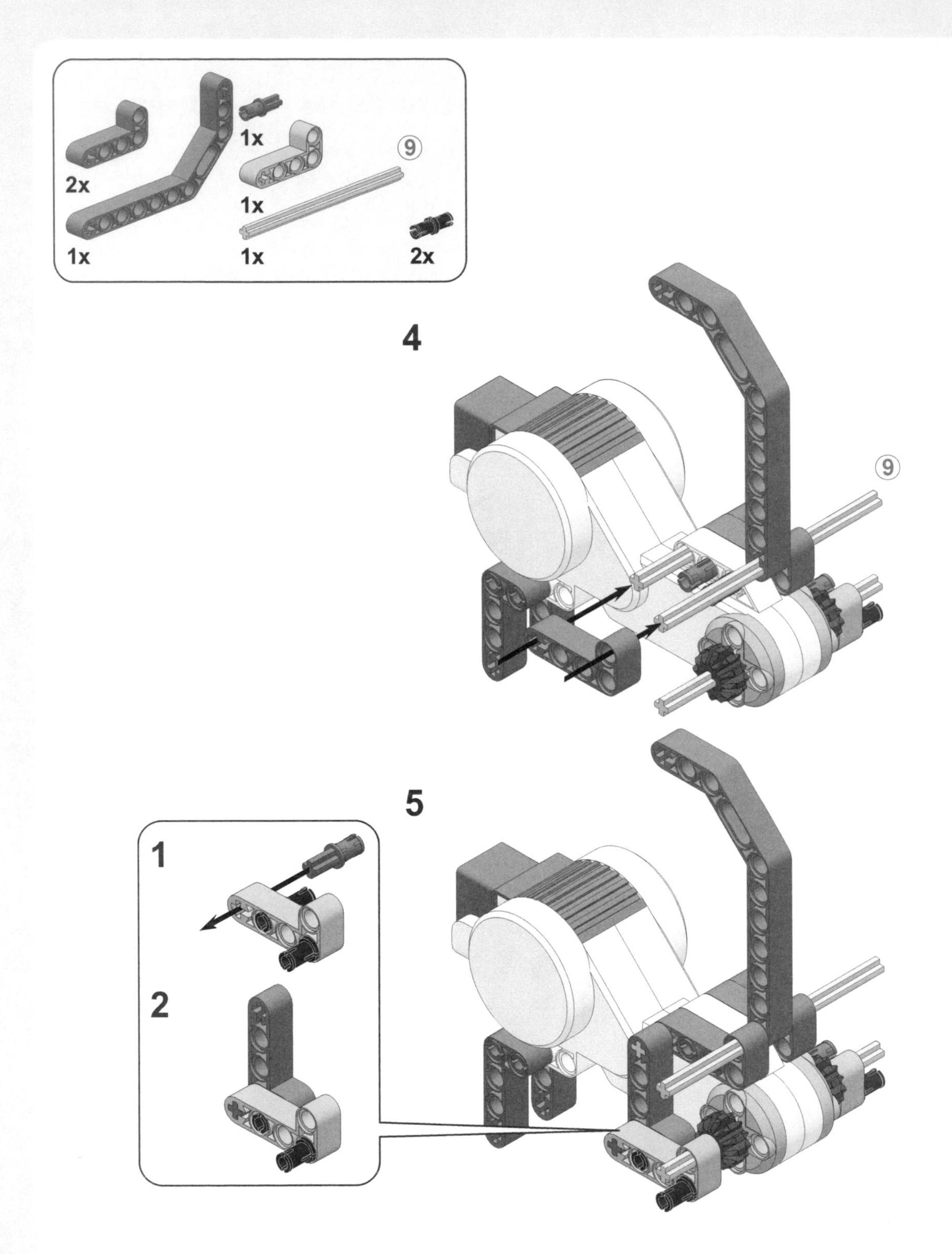

2x
1x
1x
1x
1x
9
2x
4
9
5
1
2

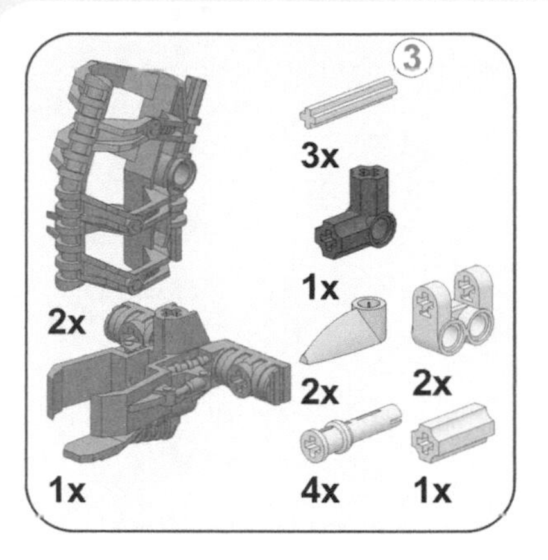
3
3x
1x
2x
2x
2x
1x
4x
1x

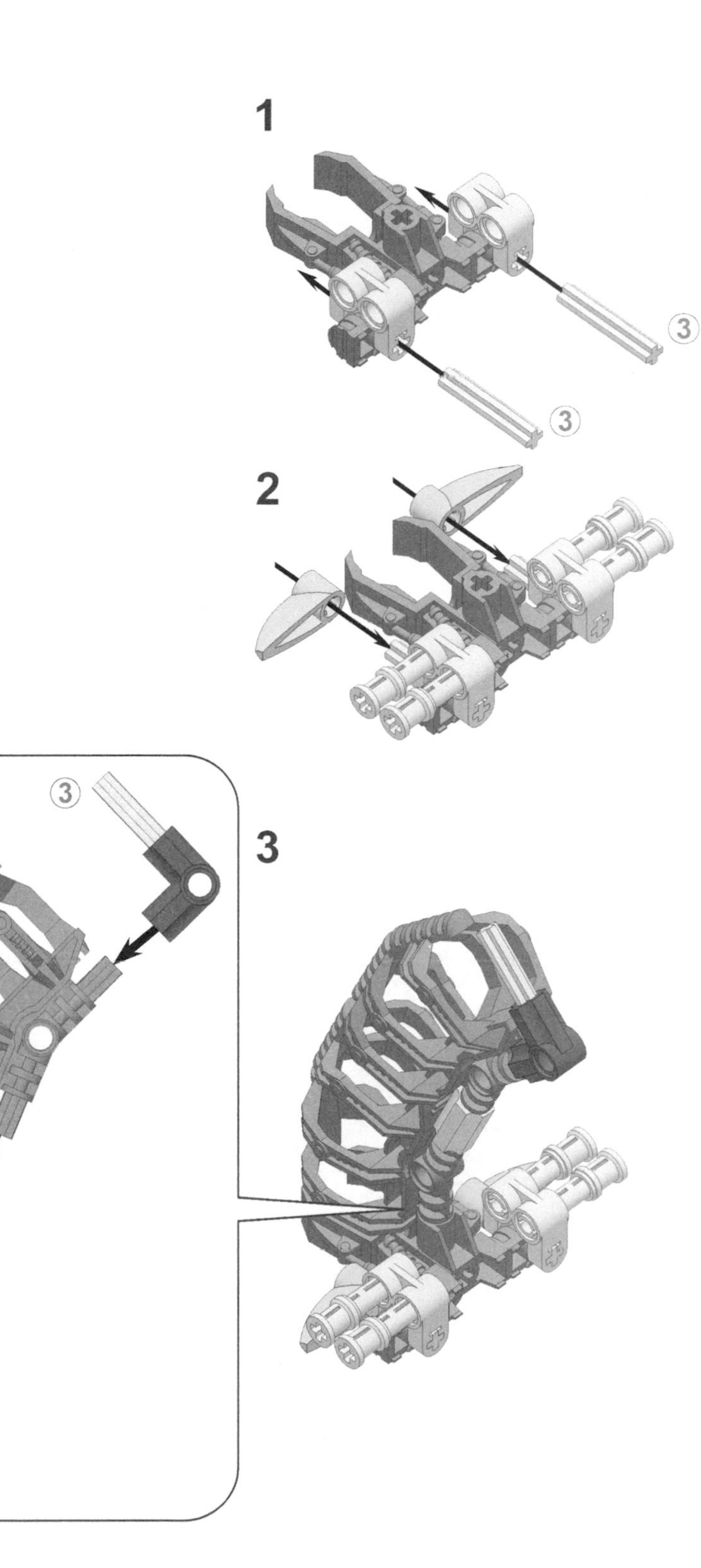
1
3
3
2
3
3

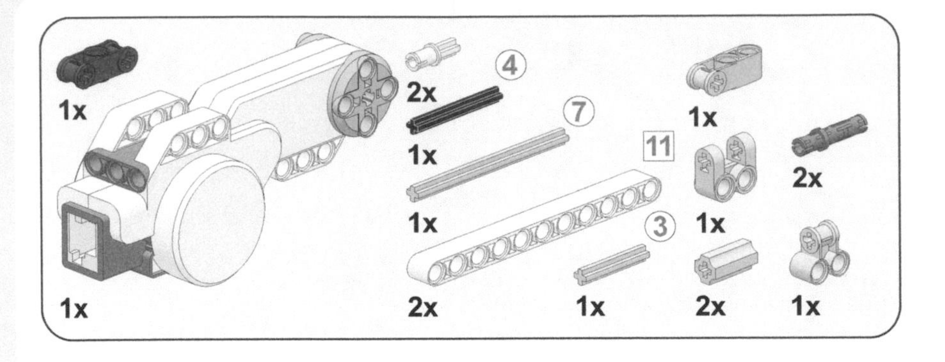
1x
2x
4
1x
7
1x
11
2x
1x
1x
3
1x
1x
2x
2x
1x
2x
1x

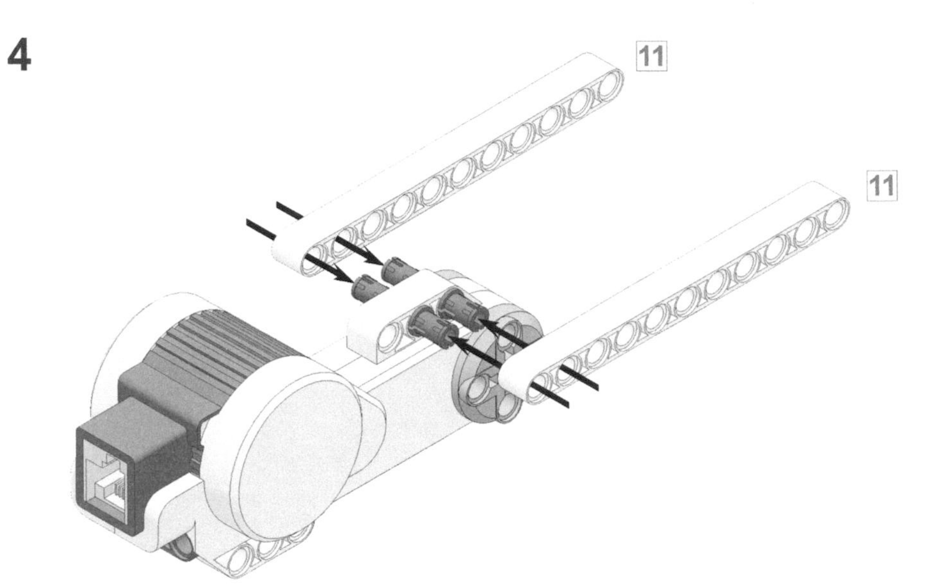
4
11
11

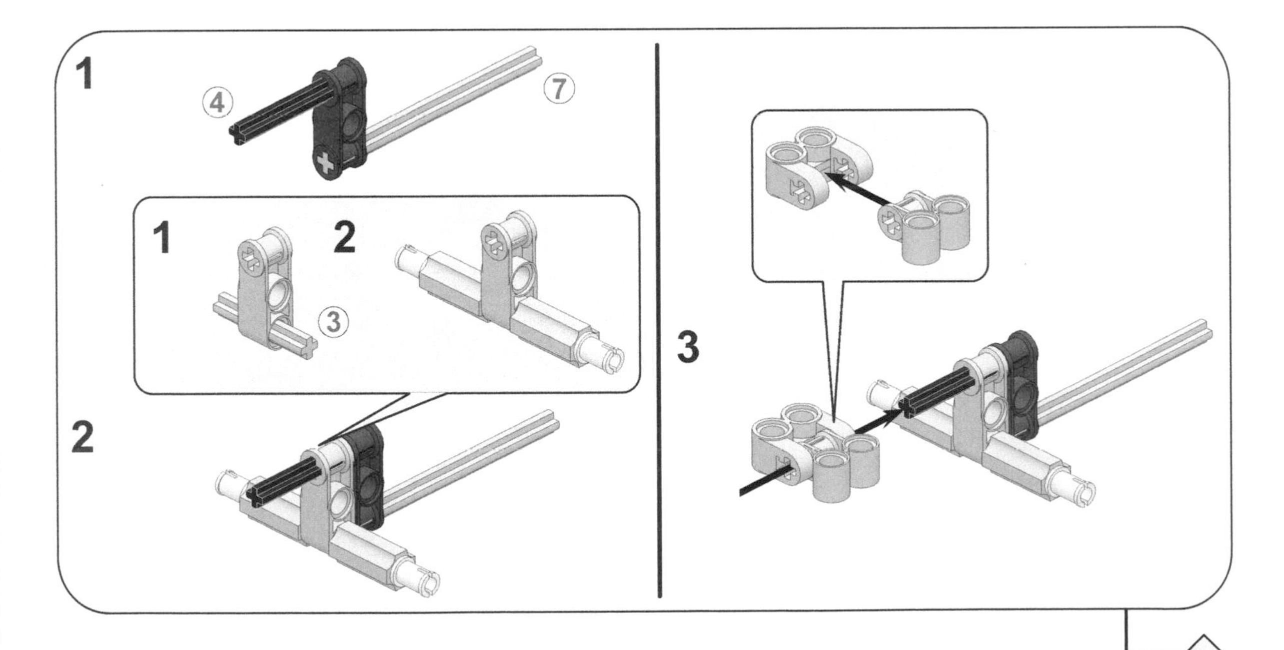
1
4
7
1
2
3
2
3

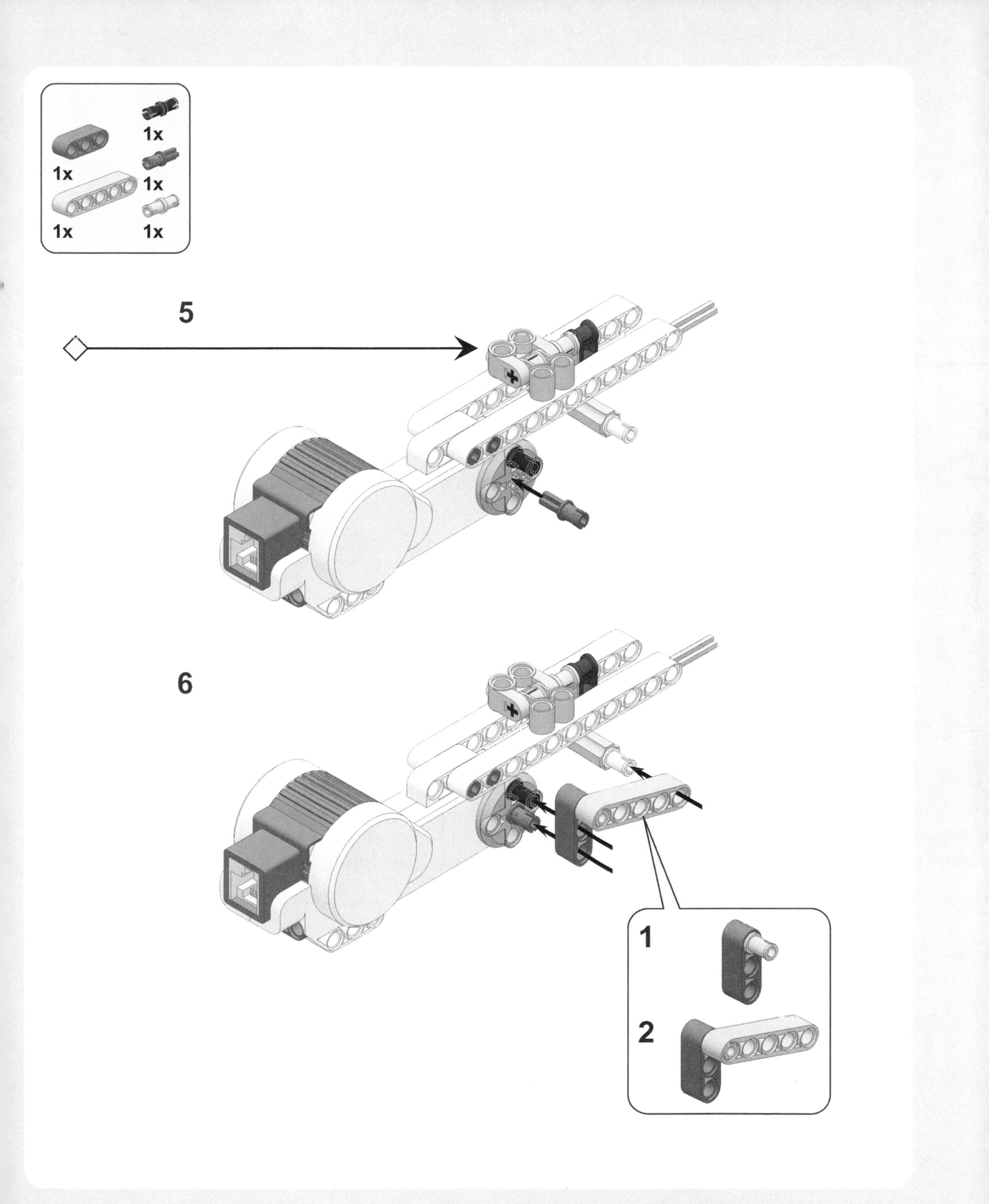
1x
1x
1x
1x
1x
1x
5
6
1
2

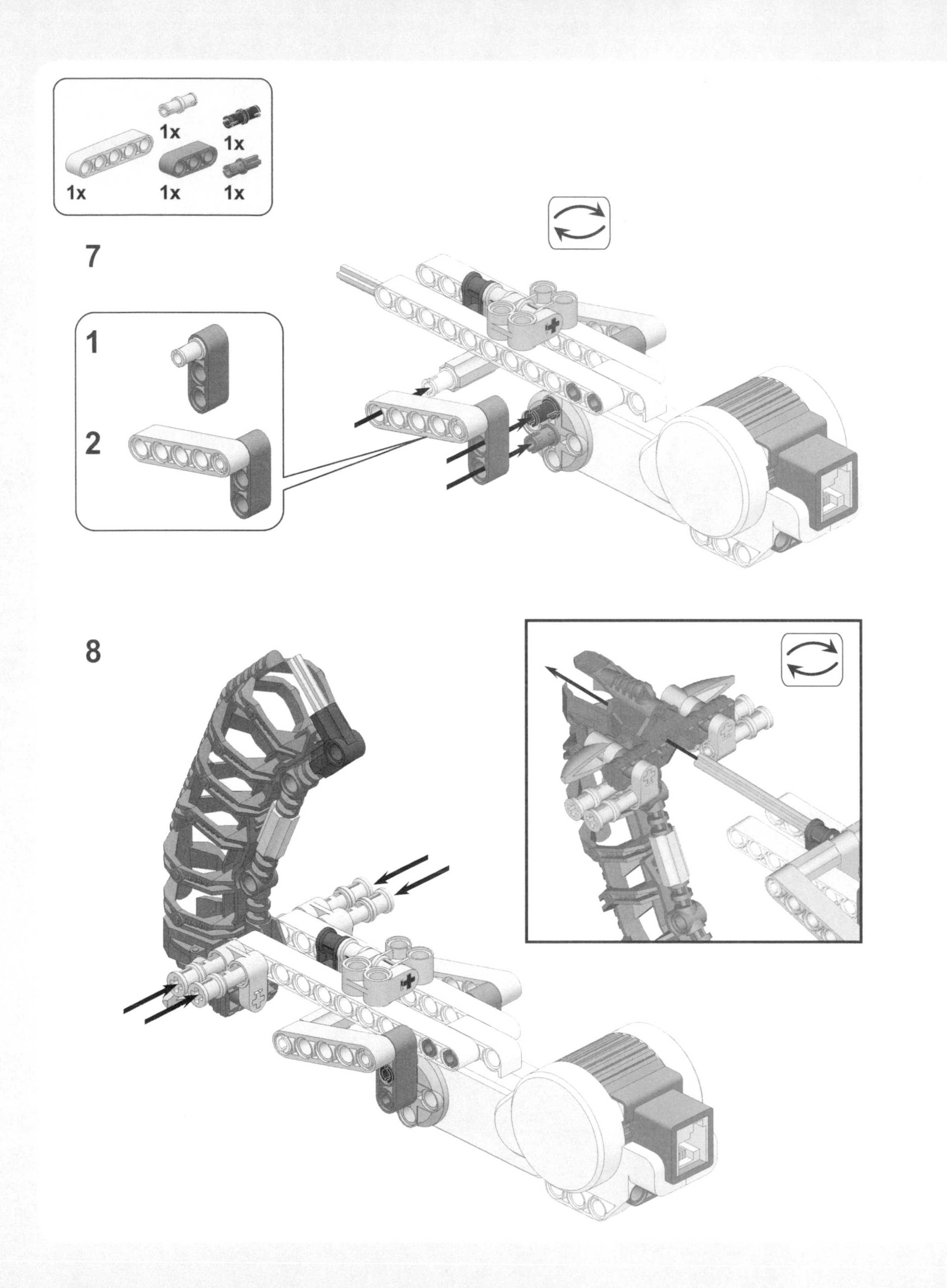
1x
1x
1x
1x
1x
7
1
2
8

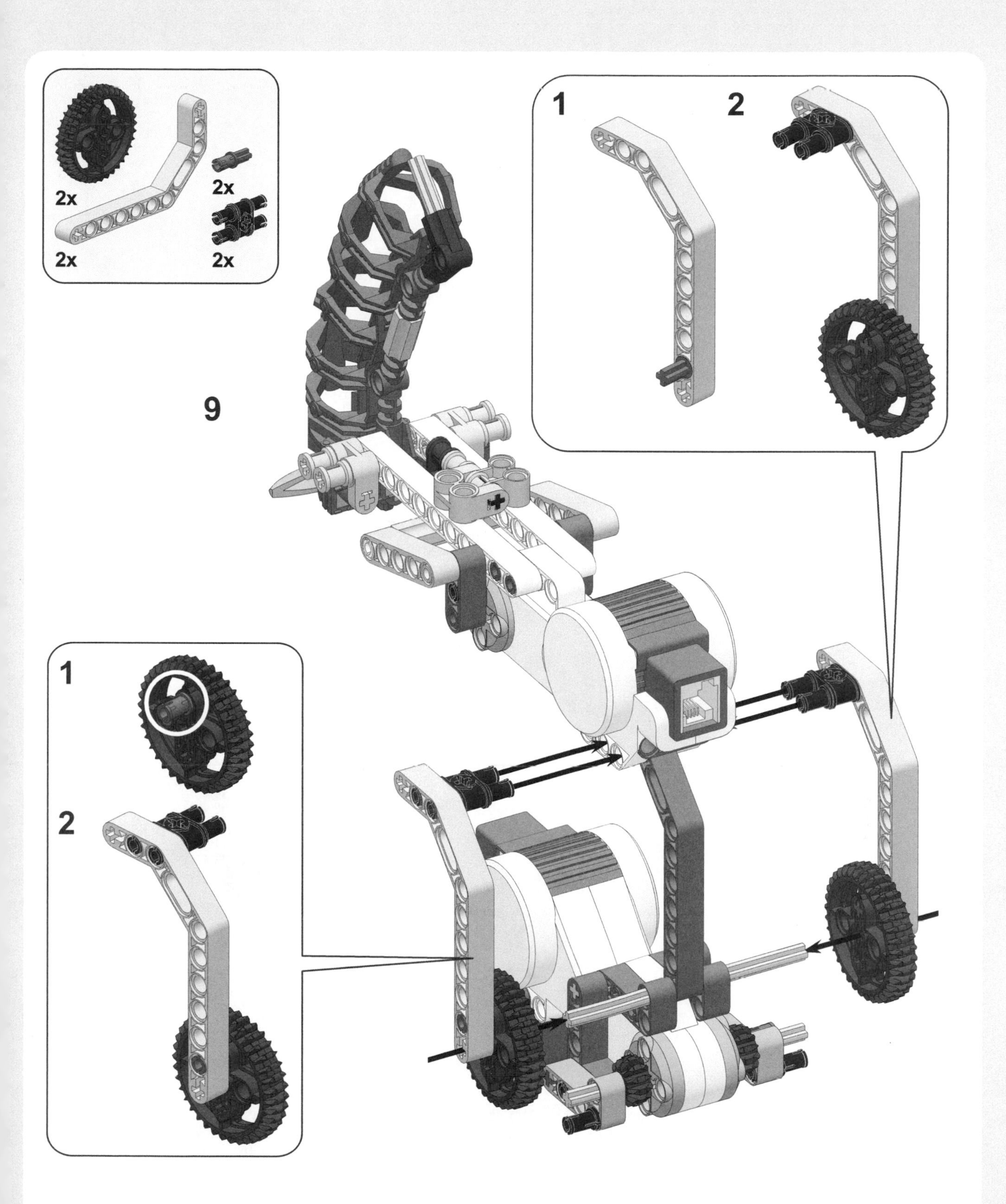
2x
2x
2x
2x
1
2
9
1
2

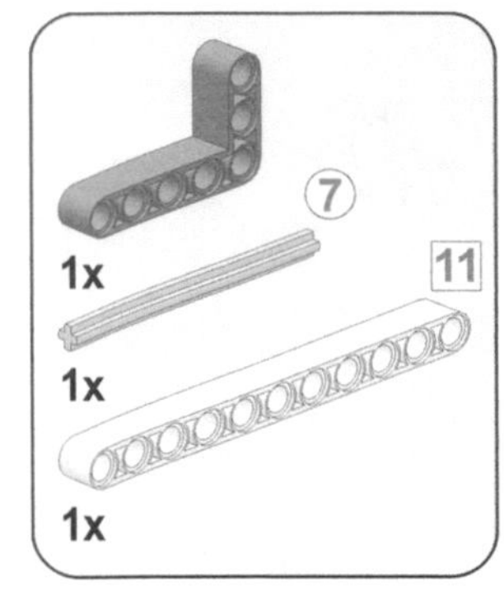

10

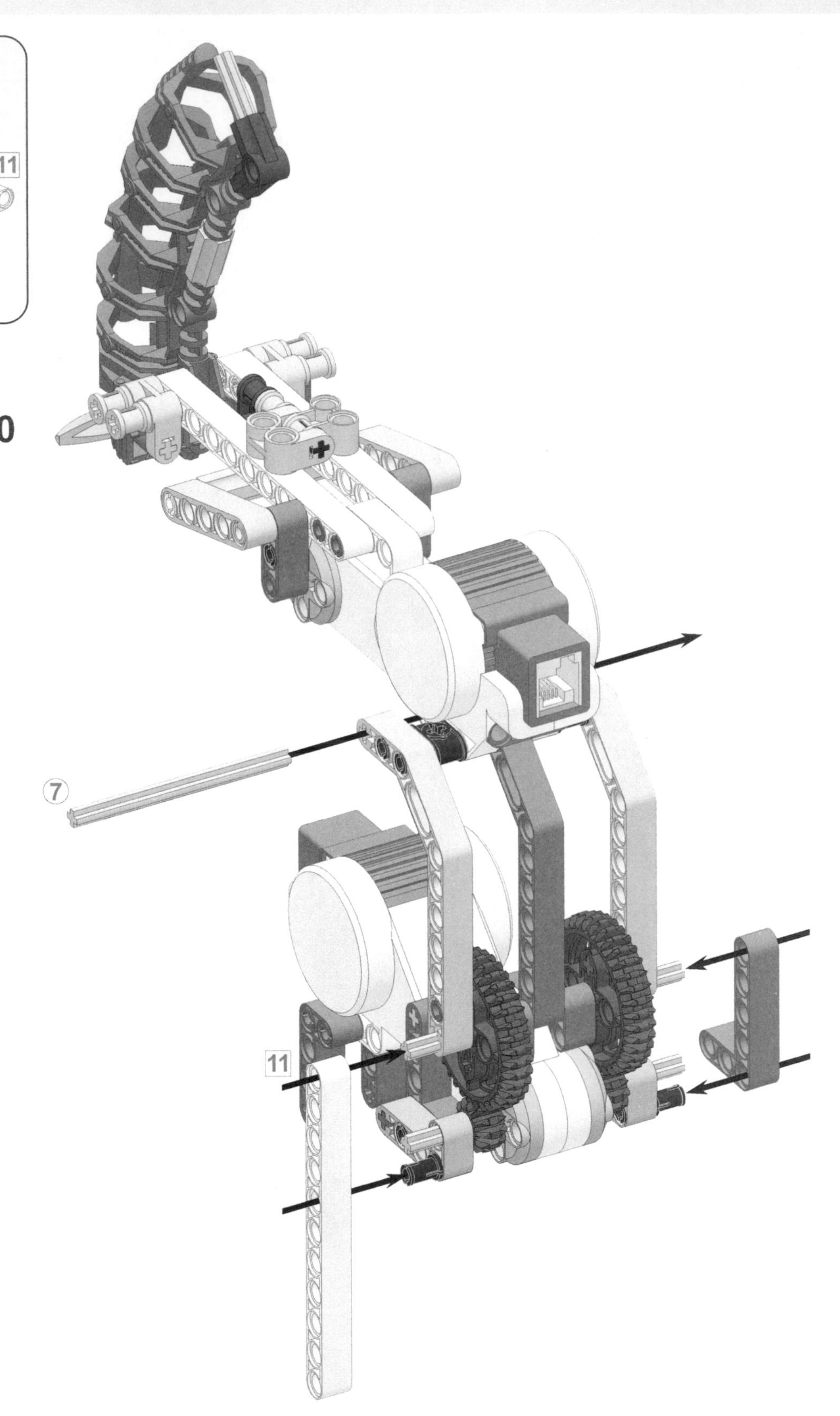

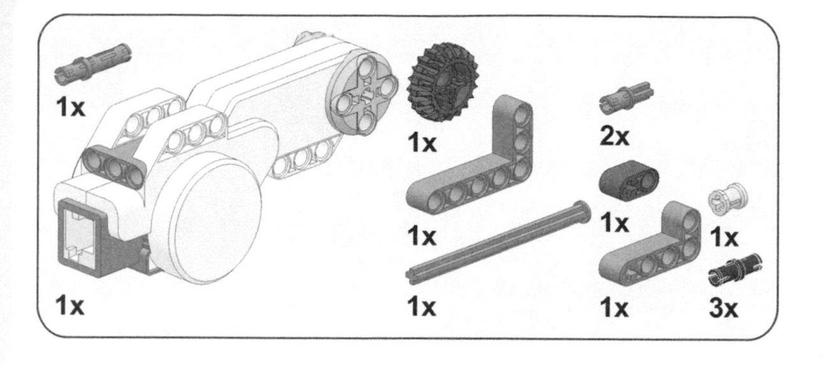
1x
1x
2x
1x
1x
1x
1x
1x
1x
3x

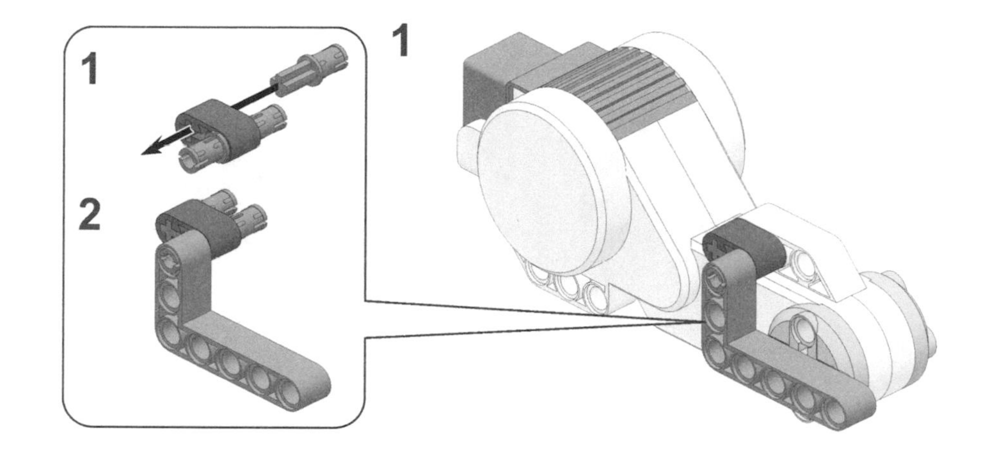
1
1
2

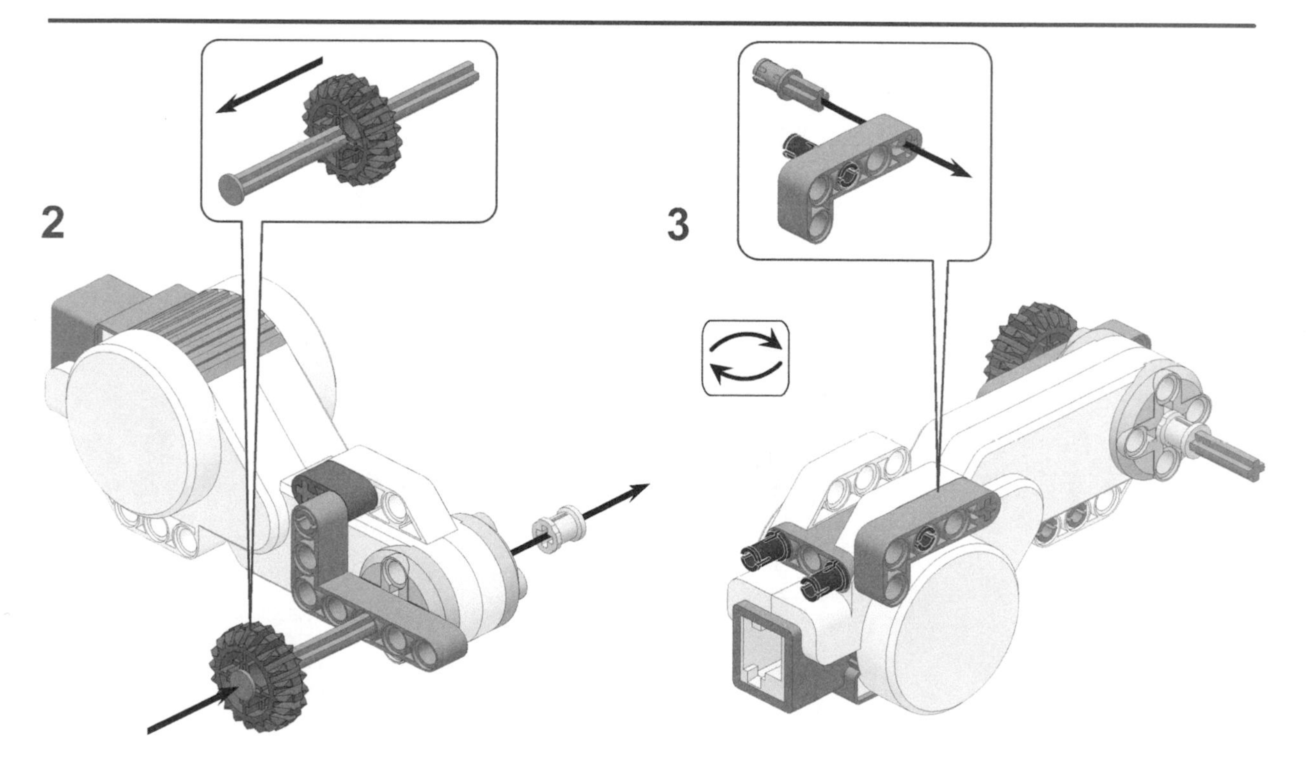
2
3

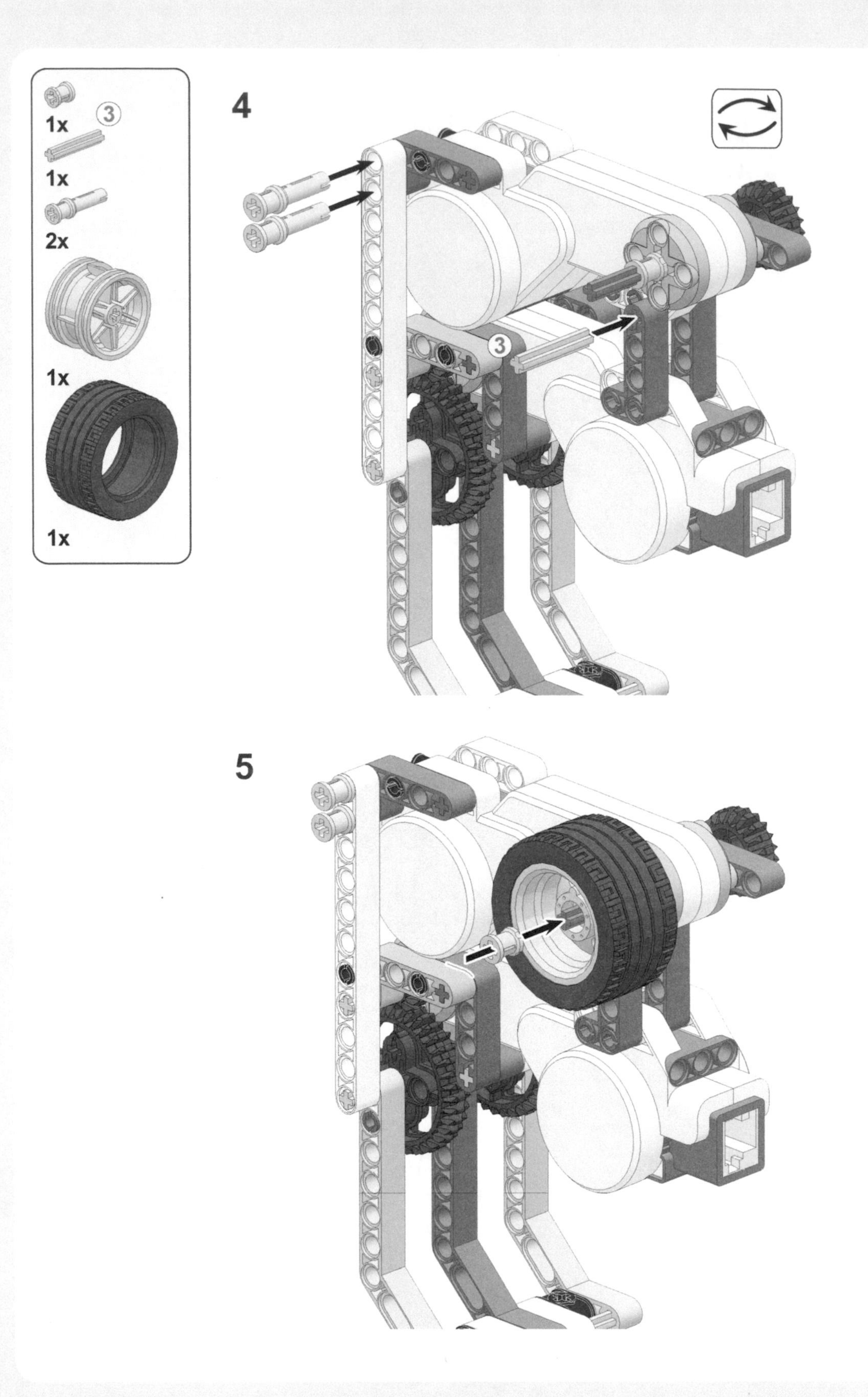
1x
3
1x
1x
2x
1x
1x
4
3
5

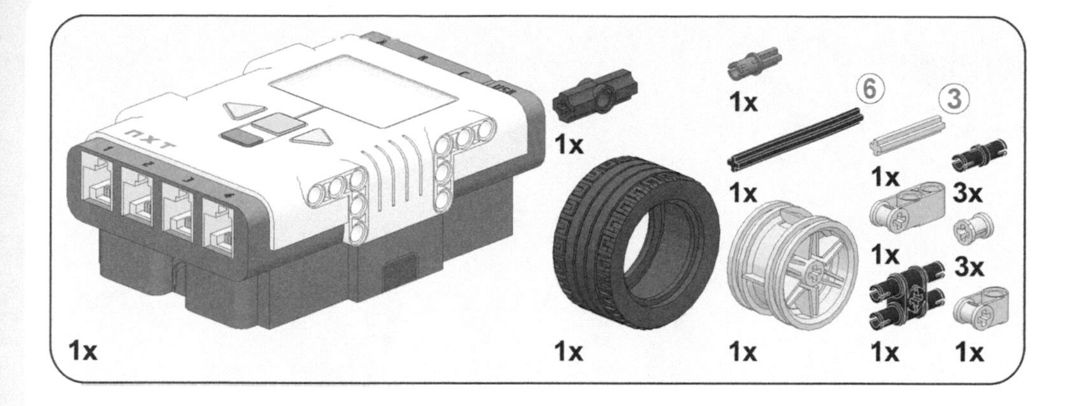
1x
1x
1x
6
3
1x
1x
1x
3x
1x
3x
1x
1x
1x
1x

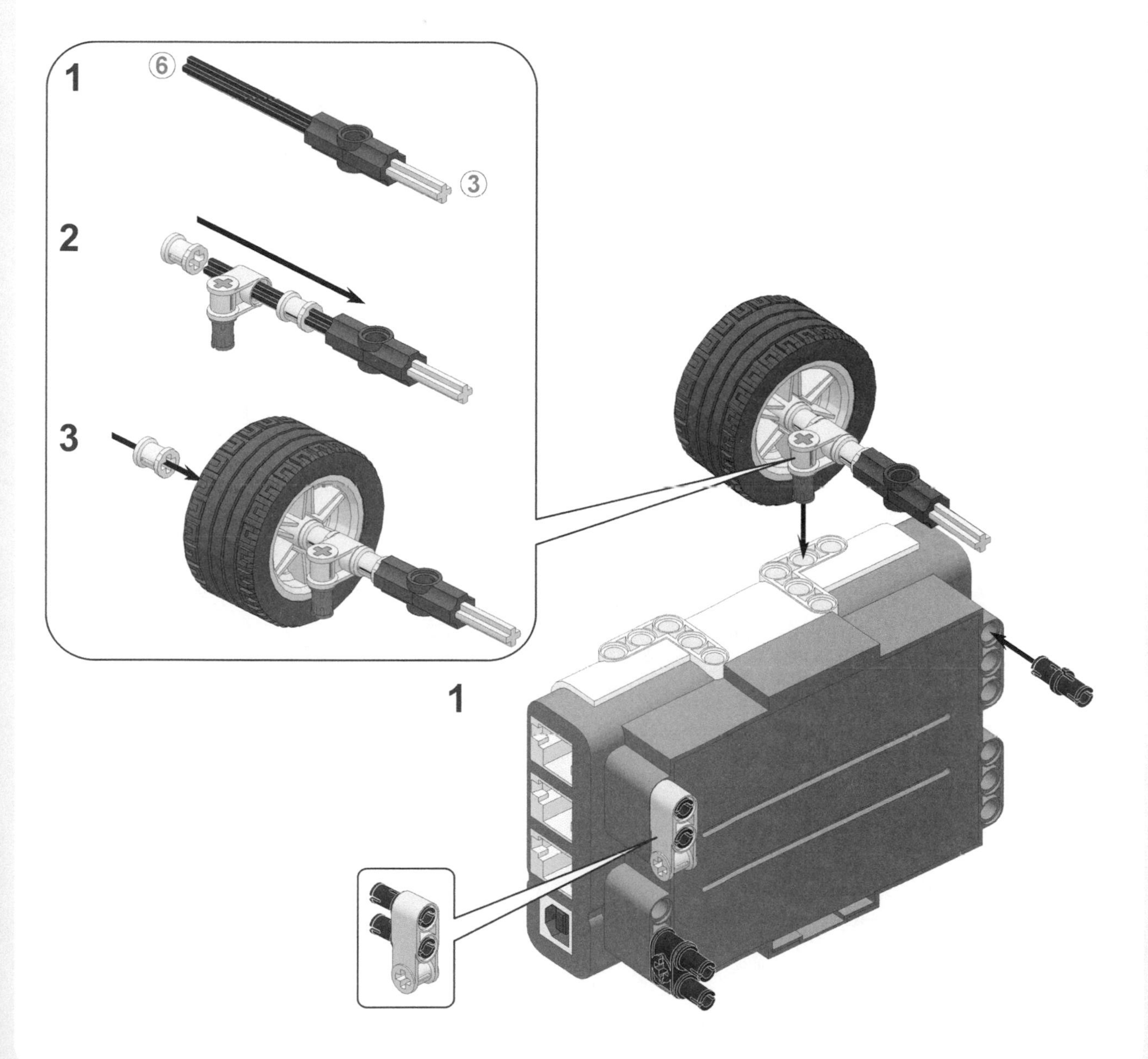
1
6
3
2
3
1

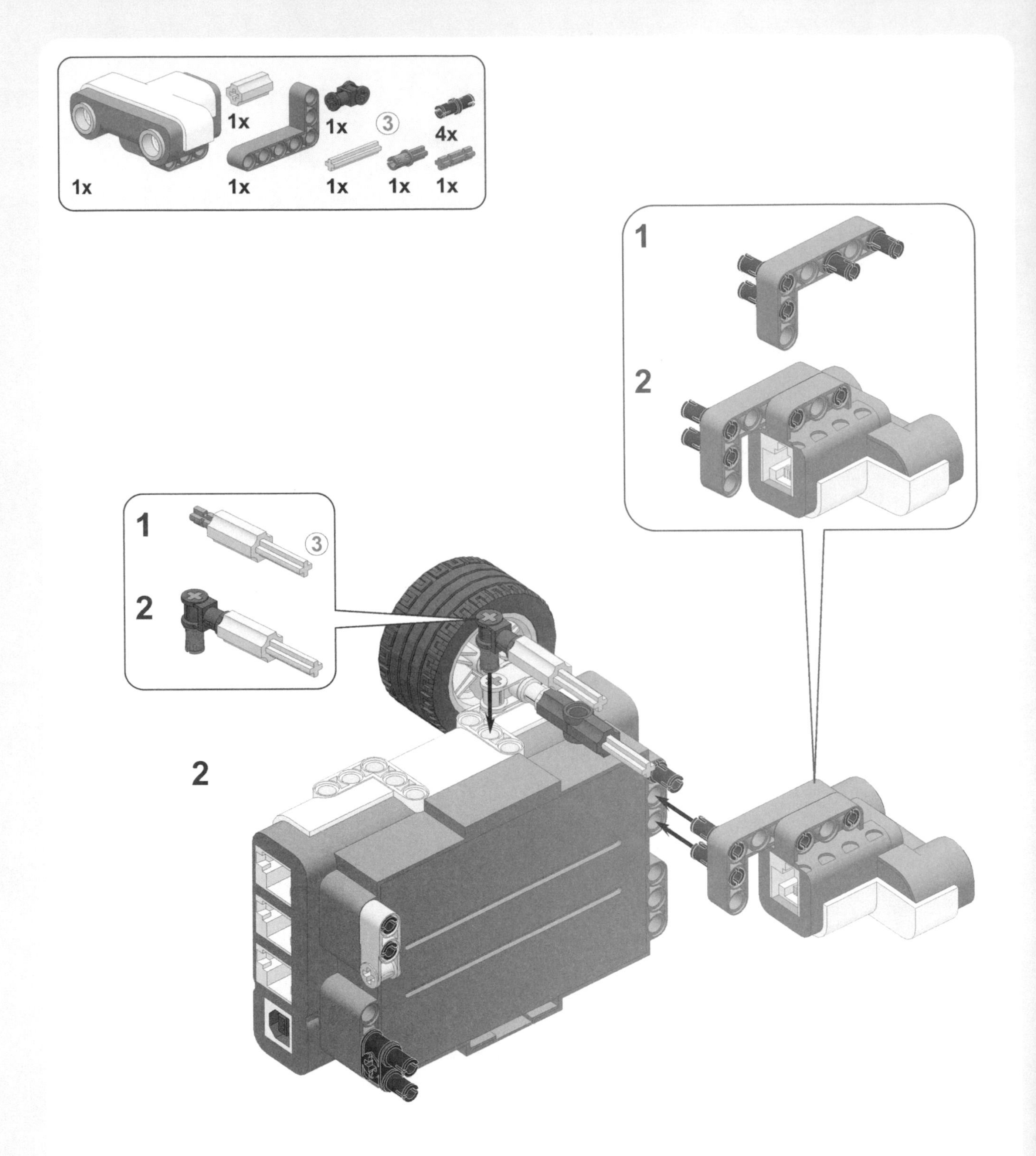
1x
1x
1x
3
4x
1x
1x
1x
1x
1x
1
2
1
3
2
2

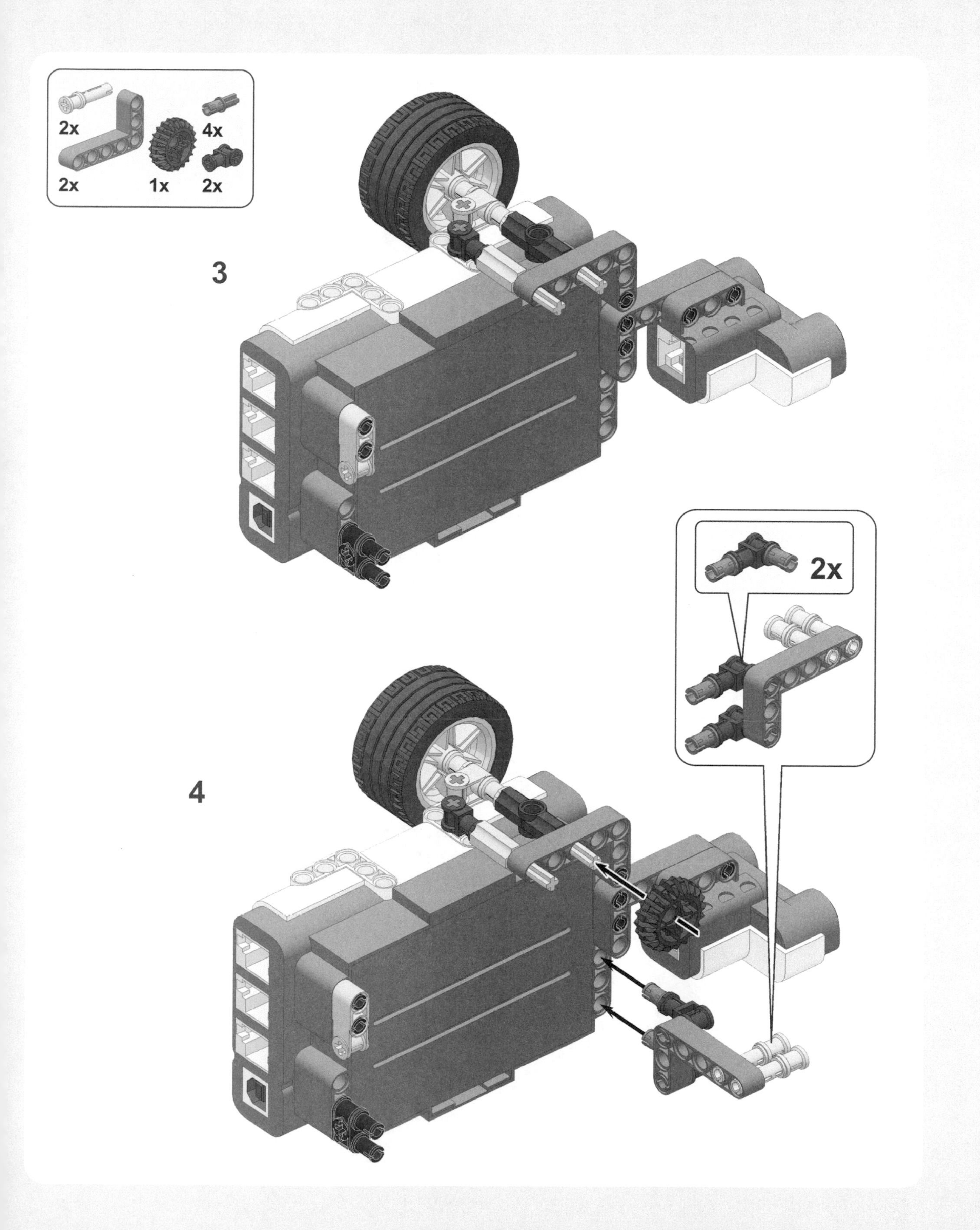
2x
2x
1x
4x
2x
3
2x
4

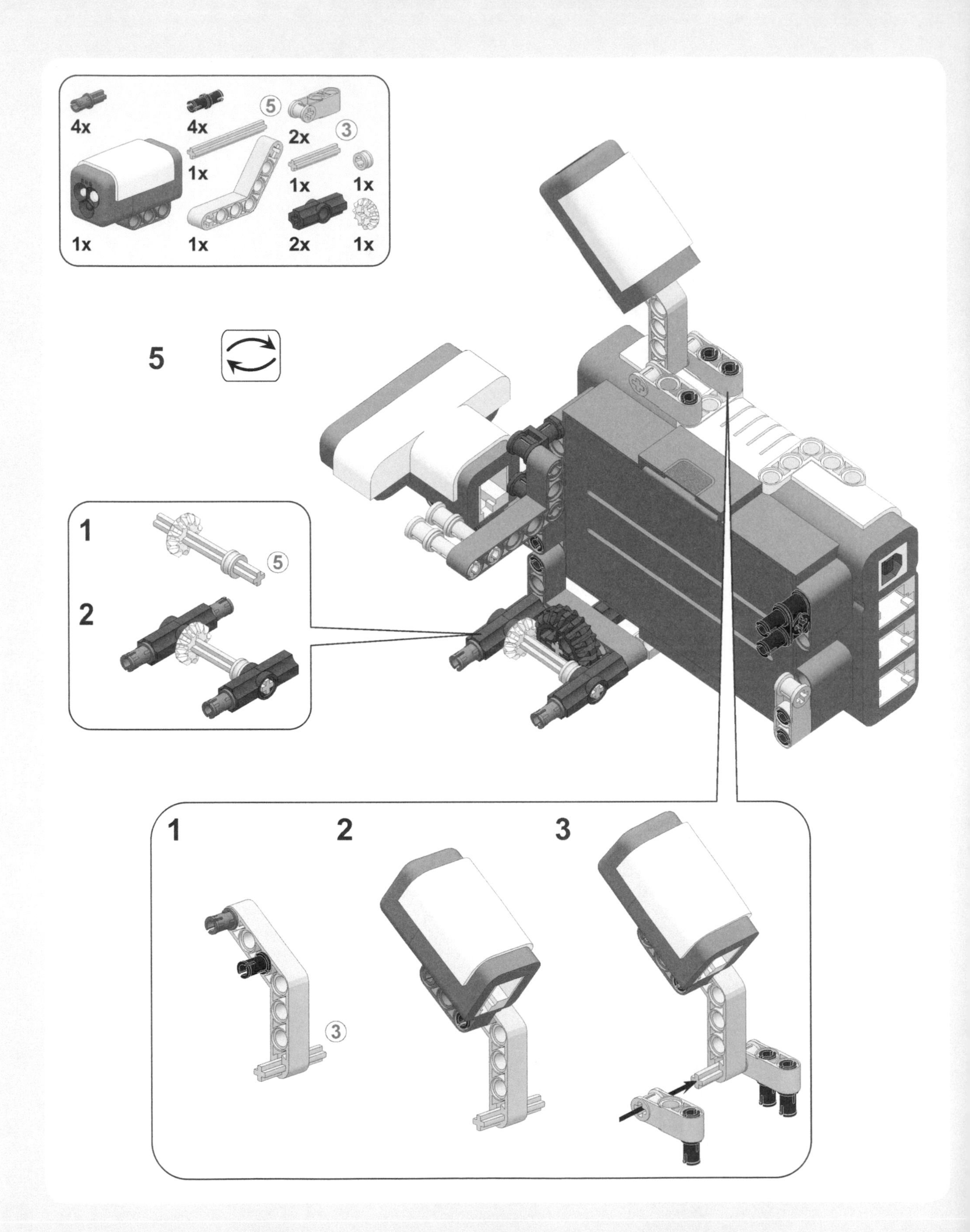
4x
4x
5
2x
3
1x
1x
1x
1x
1x
1x
2x
1x
5
1
5
2
1
2
3
3

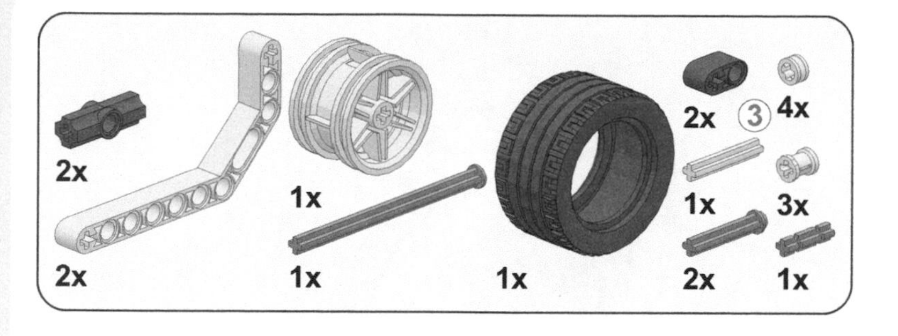
2x
2x
1x
1x
1x
2x
3
4x
1x
3x
2x
1x

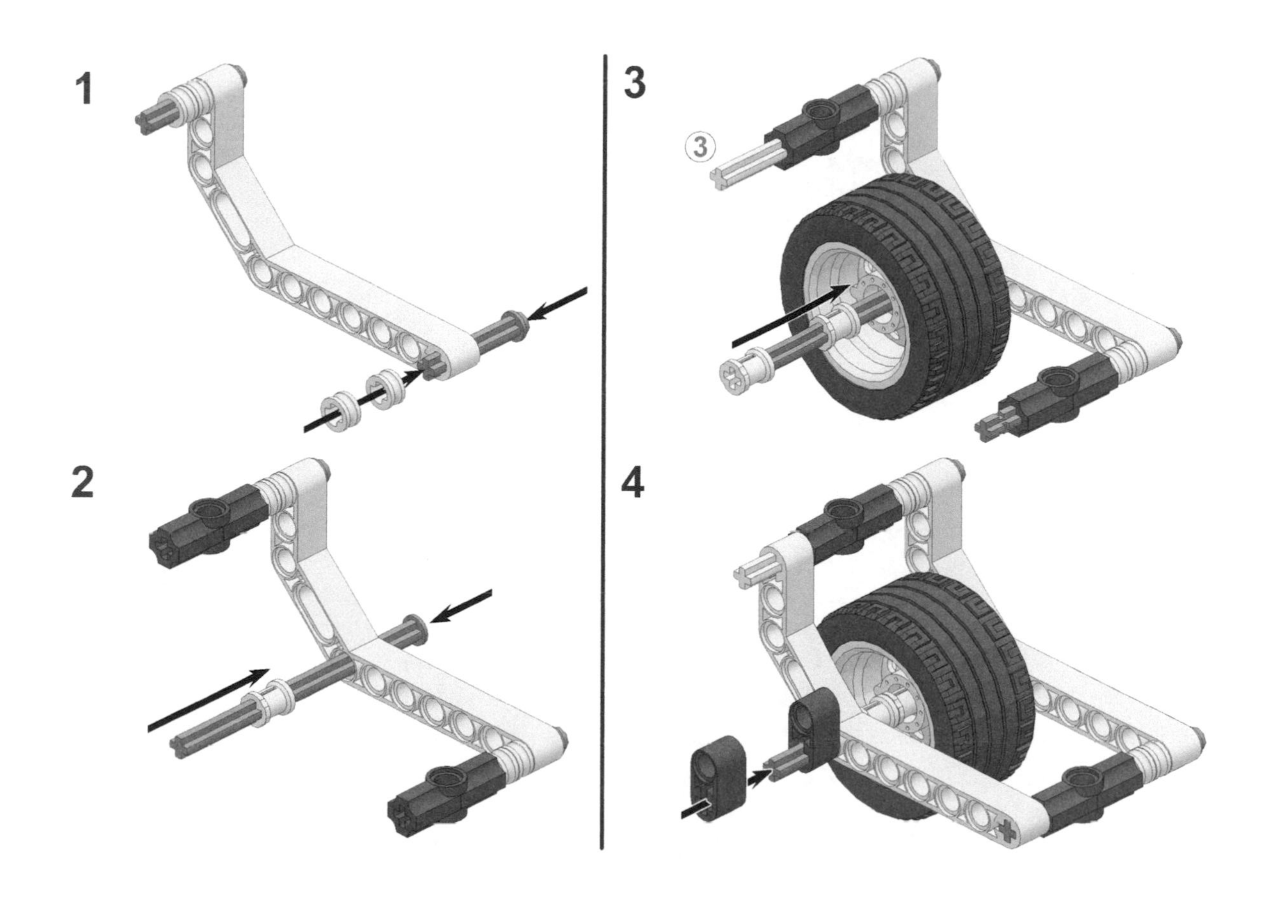
1
2
3
3
4

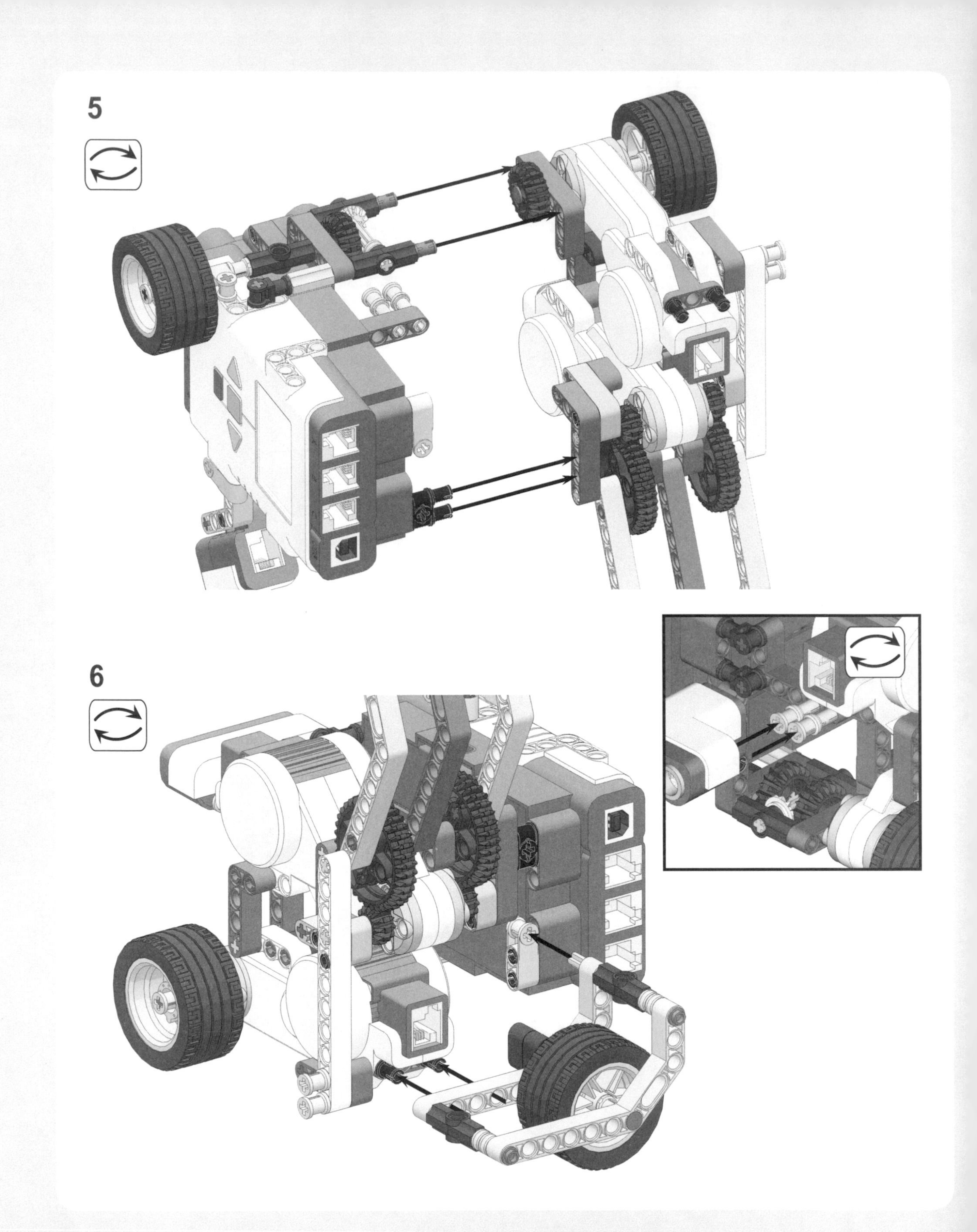
5
6

connecting the cables

Connect the sensors and motors to the NXT brick according to Table 8-1, making sure that they do not interfere with the wheels. One way to accomplish that is to wind the cables around several LEGO pieces on the robot.

Make sure that your robot can turn around smoothly and move the turret up and down easily, without having the cables blocking these movements. Check this by moving the turret and the wheels with your hands.

table 8-1: cable placement for the shot-roller

From motor/ sensor	To NXT brick port	Cable length
Turret motor	Output port A	Medium (35 cm/15 inches)
Turn motor	Output port B	Medium
Firing motor	Output port C	Medium
Color Sensor	Input port 3	Medium
Ultrasonic Sensor	Input port 4	Short (20 cm/8 inches)

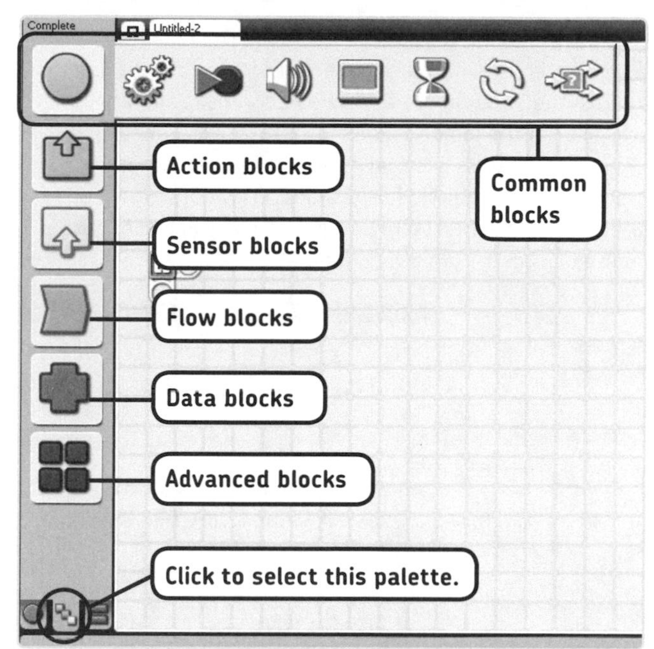

Figure 8-3: The Complete Palette. To open this palette, select the appropriate tab at the bottom of it. Hover your mouse pointer over the colored icons to see the programming blocks in each category.

programming the shot-roller

Before you program your robot, let's look at some new programming blocks that you'll need for your programs.

the complete palette

The NXT software has three different Programming Palettes. So far, you've used only the Common Palette (for most blocks so far) and the Custom Palette (for the My Blocks). Now you'll use the *Complete Palette*, as shown in Figure 8-3. This palette contains all blocks that can be used in an NXT program, except for the My Blocks that you create yourself.

Each colored icon on the Complete Palette represents a certain type of block in the categories Common, Action, Sensor, Flow, Data, and Advanced blocks.

* *Common blocks* are the blocks from the Common Palette. The Common blocks are just a collection of frequently used blocks. Since blocks such as the Display block and the Sound block are Action blocks, you'll also find them among this category; there is no difference between the Sound block in the Common category and the Sound block in the Action category.
* *Action blocks* are blocks to make the robot perform an action, such as turning motors, playing a sound, or displaying a line of text on the NXT screen.
* *Sensor blocks* (colored yellow) are blocks to read values from sensors for use in your programs. These blocks differ from the blocks that you've used so far to poll sensors, such as the orange Wait block. (I'll show you how to use the yellow Sensor blocks in Chapter 10.)
* *Flow blocks* (like the Wait block) are typically used to change a program's flow. For instance, some blocks may need to be repeated (with a Loop block), or a decision may need to be made (with a Switch block).

I'll discuss some of the Data and Advanced blocks later in this book.

the color lamp block

In Chapter 7 you learned that you can use the Color Sensor to determine the color of a surface, but you can also use it as a red, green, or blue lamp. To use it as a lamp, you use the *Color Lamp block* in the Action blocks.

To use the Color Lamp block, open the Configuration Panel, select the port it's connected to, then choose whether to turn the lamp on or off (in the Action box), and finally set the color of the lamp.

Let's create a program that turns the sensor into a disco lamp that quickly flashes different colors. Start a new program called **TestColorLamp**, and place and configure four blocks in it as shown in Figure 8-4.

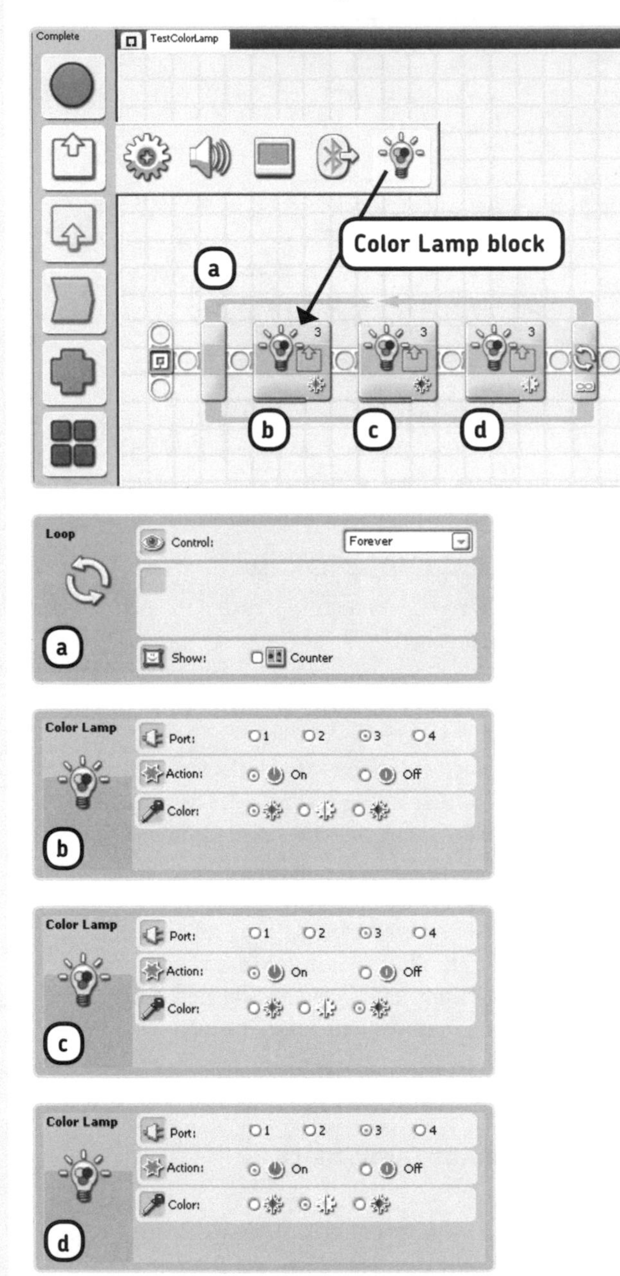

Figure 8-4: The configuration of the blocks in the TestColorLamp program

When you run TestColorLamp, you should see colored flashes coming from the sensor. You see this effect because you don't use a Wait block to tell the program to pause. Immediately after the first Color Lamp block instructs the sensor to display a red light, the second block requests a blue light and then a green light. This sequence is repeated over and over with the Loop block.

DISCOVERY #38: A COLORED VOICE!

Difficulty: Easy
Can you make the robot say which lamp color is turned on? Expand the TestColorLamp program with three Sound blocks. How do you configure the Wait for Completion option in the Sound blocks?

the motor block

Like the Move block, the *Motor block* in the Action blocks controls a motor. The important difference between the two is that the Motor block has extra features for controlling *individual* motors, while the Move block is perfect for vehicles with two wheels like the Discovery we used in earlier chapters. Because the Shot-Roller uses three motors, each for a different function in the robot, you'll use Motor blocks to control them individually.

Some of the configurations in a Motor block are the same as those in a Move block. For instance, you can use the Motor block's Configuration Panel to select to which output port the motor is connected, in which direction the motor should turn, the motor power, and the time of rotation (Duration); you can also tell the robot what to do when the motor finishes rotating (Next Action).

using the control motor power option

One unique setting in the Motor block is the Control Motor Power option. Normally, when you set a motor to move at a certain power level (such as 50), the motor turns more slowly when you try to stop it with your hands because the power to the motor stays constant. However, when you use the Control Motor Power option, the NXT will automatically apply more power to the motor when there's a load on it so that the motor continues turning at a constant speed. You can see a Motor block with its Configuration Panel in the following sample program.

To illustrate the Control Motor Power setting, you'll shoot two balls with the Shot-Roller's Firing Motor. To shoot a ball, the motor has to make one full rotation. You'll launch the first ball with a Motor block specified to control the motor power and with the second one without the power control, as shown in Figure 8-5.

When you run the program, the first ball should be released at high speed because the motor power is controlled, while the second ball is stuck in the ball magazine: there is not enough power to push it out.

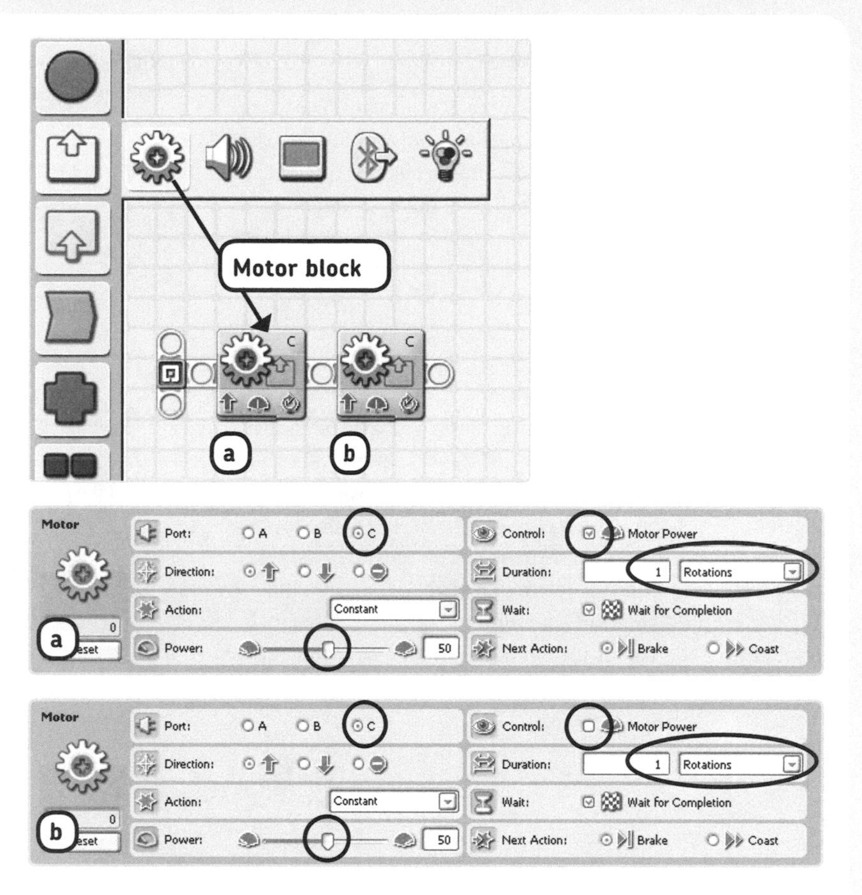

Figure 8-5: The configuration of the blocks in the MotorControlTest program

DISCOVERY #39: A MOTOR BLOCK TASK!

Difficulty: Easy

Before creating big programs for a robot, it is good to test the robot's mechanical functions to give you a better understanding of how they work and to make it easier to localize problems. For example, if running the Turn motor doesn't make the robot turn, some LEGO pieces were probably misaligned while constructing the robot. Create a program that makes each motor on the Shot-Roller robot move forward and then backward for a while. Try different motor speeds, and test the Control Motor Power function. Note that the Turret motor cannot spin infinitely, so be careful when configuring its block.

NOTE It's not always obvious in which direction a motor has to turn in order to make a certain movement. For example, to raise the turret in this robot, motor A has to turn backward. The direction depends on both the orientation of the motors and the gears in your robot. To determine the direction your motors need to turn in order for your robot to make a certain movement, simply try both directions using a small test program with one Motor block.

autonomous mode

Now that you know how to use the Motor and Color Lamp blocks, you can start making better programs for the Shot-Roller! You'll begin by creating some programs that make the robot do everything by itself, without your help. In this autonomous mode, the NXT controls the motors and the robot's actions based on the running program and input from the sensors.

In your next program, you'll use the Color Sensor as a Color Lamp, so in effect, the only real sensor you'll use is the Ultrasonic Sensor. Your program will instruct the Shot-Roller to turn around while looking for targets. If the robot sees a target, it will raise the turret. If the target is closer than 25 cm (10 inches), it will fire two balls; otherwise, it will fire just one. Once the balls have been fired, the robot lowers its turret, and the robot returns to the beginning of the program, looking for targets again. The Color Lamp indicates the state of the shooter: scanning for targets (blue light), aiming with the turret (green light), and firing (red light).

NOTE **Before starting the program, make sure that the shooter is parallel to the ground.**

creating the program

Create a new program called **Shot-Roller-Autonomous**, and then follow the instructions in Figures 8-6 through 8-10.

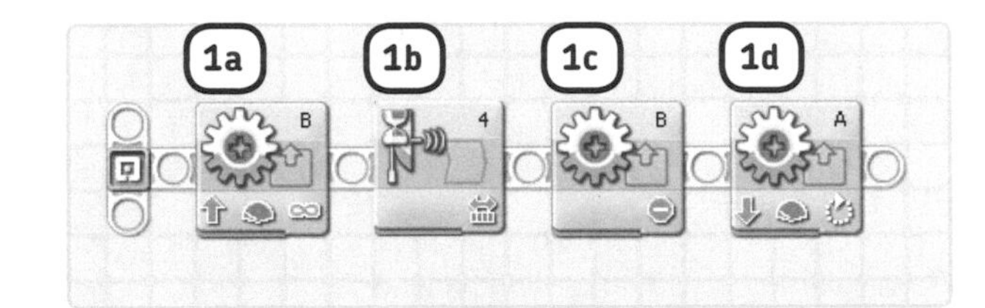

Motor
1a
Port: A B C
Direction:
Action: Constant
Power: 20
Control: Motor Power
Duration: 1080 Unlimited
Wait: Wait for Completion
Next Action: Brake Coast

Wait
1b
Control: Sensor
Sensor: Ultrasonic Sensor
Port: 1 2 3 4
Until:
Distance: < 45
Show: cm Centimeters

Motor
1c
Port: A B C
Direction:
Action: Constant
Power: 20
Control: Motor Power
Duration: 1080 Unlimited
Wait: Wait for Completion
Next Action: Brake Coast

Motor
1d
Port: A B C
Direction:
Action: Constant
Power: 20
Control: Motor Power
Duration: 80 Degrees
Wait: Wait for Completion
Next Action: Brake Coast

Figure 8-6: ***Step 1:*** *You make the Turn motor move by activating motor B with the Duration option set to* ***Unlimited****. As the Shot-Roller turns right, a Wait block makes the Ultrasonic Sensor look for targets in sight. Once the robot sees a target closer than 45 cm (18 inches), motor B is turned off, and the turret is raised by spinning motor A backward.*

NOTE In addition to the newly placed programming blocks, Figure 8-7 also shows some of the blocks placed previously. You don't need to configure these blocks again; they just help you see where to put the new block. You may also see such blocks from previous steps in other figures in this book.

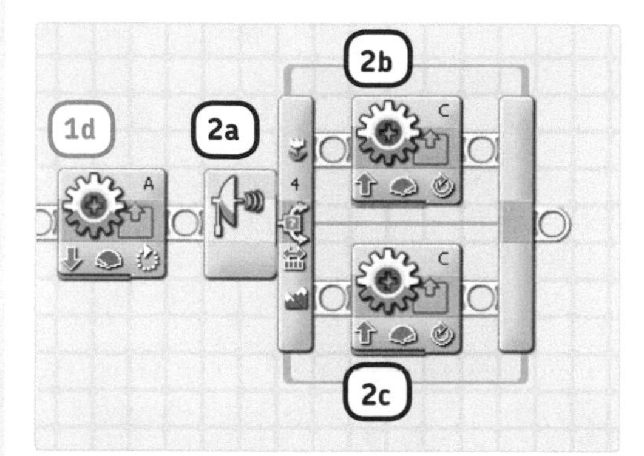

Switch (2a)

Control:	Sensor	Port:	○1 ○2 ○3 ⊙4
Sensor:	Ultrasonic Sensor	Compare:	Distance: < 25
Display:	☑ Flat view	Show:	cm Centimeters

Motor (2b)

Port:	○A ○B ⊙C	Control:	☑ Motor Power
Direction:	⊙↑ ○↓ ○	Duration:	2 Rotations
Action:	Constant	Wait:	☑ Wait for Completion
Power:	100	Next Action:	⊙ Brake ○ Coast

Motor (2c)

Port:	○A ○B ⊙C	Control:	☑ Motor Power
Direction:	⊙↑ ○↓ ○	Duration:	1 Rotations
Action:	Constant	Wait:	☑ Wait for Completion
Power:	100	Next Action:	⊙ Brake ○ Coast

Figure 8-7: ***Step 2:*** *The Wait block in step 1 waited until the sensor saw something closer than 45 cm, but it couldn't determine exactly how close a target was. To find out, you use a Switch block. If the detected object is closer than 25 cm (10 inches), you shoot two balls, and just one otherwise.*

NOTE When you pick a Wait block from the Programming Palette (among the Flow blocks), it is by default configured to poll the Touch Sensor. To make it wait for a certain number of seconds, set the Control box to *Time* (Figure 8-8).

Figure 8-8: ***Step 3:*** *After the robot shoots the balls, it waits for half a second, lowers the turret, turns slightly, and then starts the whole program over again—all to keep it from shooting at the same target repeatedly.*

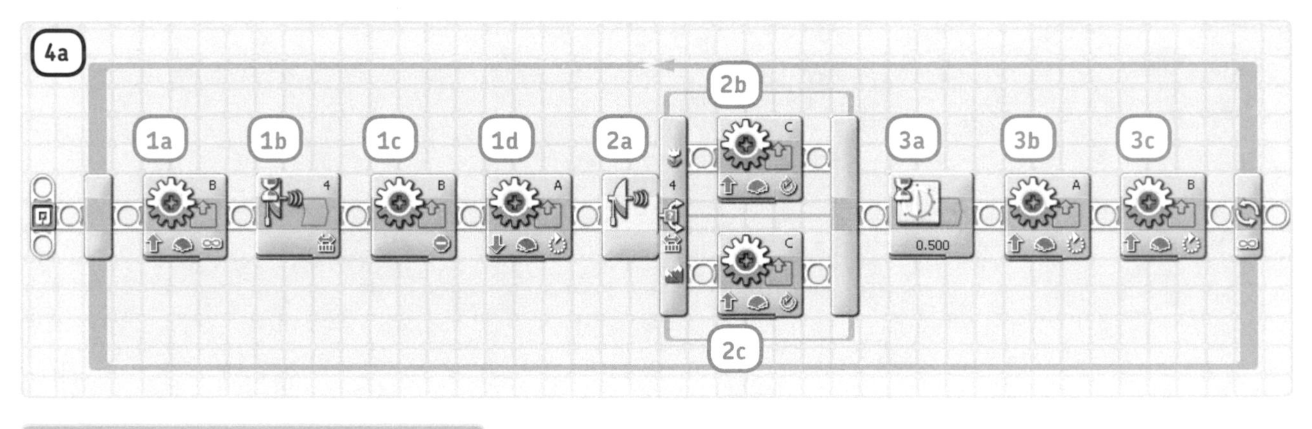

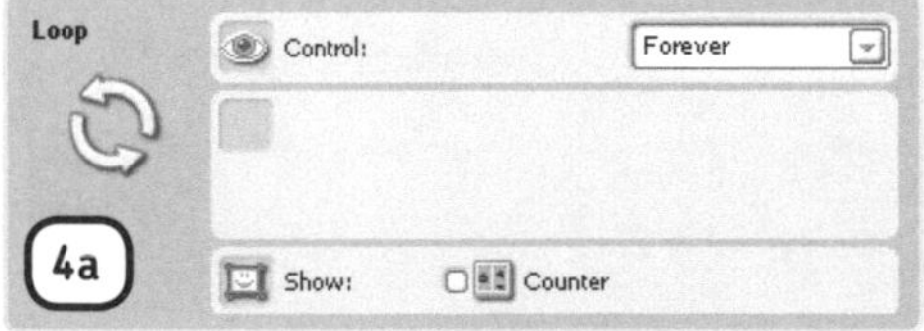

Figure 8-9: ***Step 4:*** *You use a Loop block to repeat the sequence of blocks by placing a Loop block at the start of your program (before block 1a) and then selecting the rest of the blocks and dragging them into the Loop block. The result is the program shown in this figure.*

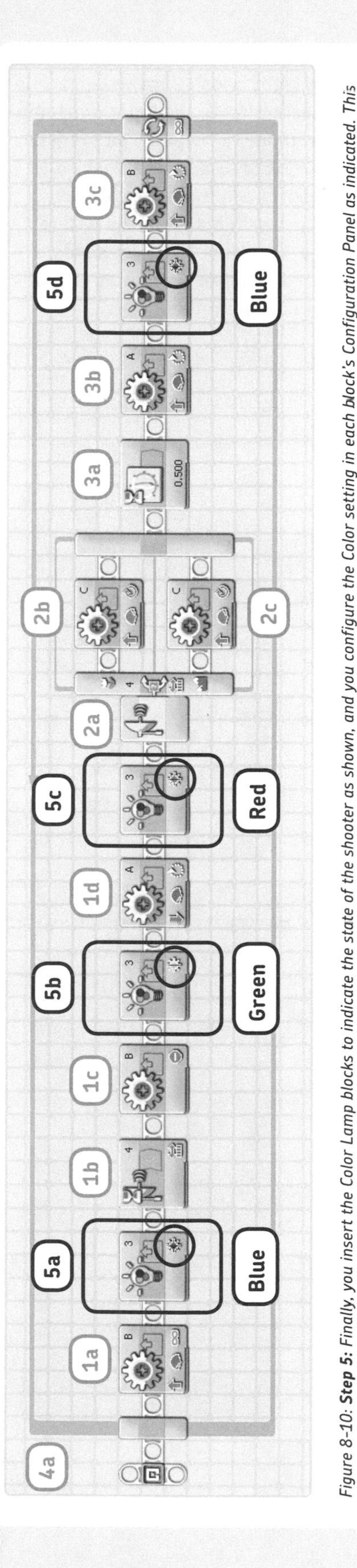

Figure 8-10: ***Step 5:*** *Finally, you insert the Color Lamp blocks to indicate the state of the shooter as shown, and you configure the Color setting in each block's Configuration Panel as indicated. This figure shows the final program.*

Congratulations! You've finished creating the program. Now you can download it to the Shot-Roller and run it!

DISCOVERY #40: DANGEROUS INTRUDER ALARM!

Difficulty: Medium
Create a program to turn the Shot-Roller into an intruder alarm. Use the Ultrasonic Sensor to sense when the door opens. Once the door opens, the alarm should not go off just yet, because the intruder would still be too far away to be hit by one of the Shot-Rollers balls. After a Wait block pauses the program for a few seconds, the Shot-Roller should fire balls quickly, while making loud sounds using Sound blocks on a Parallel Sequence Beam.

light sensor mode

In addition to using the Color Sensor as a color detector and a colored lamp, you can use it to measure the light intensity in a particular area. For example, it can measure the difference between bright areas covered in sunlight and a dark closet. The sensor values range from 0 to 100. A sensor value of 0 says that the sensor sees no light, while a value of 100 indicates that it sees very bright light. You can see the settings of a Wait block configured to poll the Color Sensor in Light Sensor mode in Figure 8-11.

Use the Until box to specify which value the sensor should measure in order for the Wait block to stop waiting. The block that you see here waits until the sensor reports a light value brighter (greater) than 50.

When measuring light intensity, you can choose to turn on the Color Lamp by selecting the **Light** checkbox in the Function box and specifying the color you want to see.

Wait
Control: Sensor
Sensor: Color Sensor
0
Port: 1 2 3 4
Action: Light Sensor
Until: > 50
Function: Light

Figure 8-11: The configuration of a Wait block polling the Light Sensor. To configure it, pick a Wait block from the Programming Palette, select ***Color Sensor*** *from the list of sensors, and then select* ***Light Sensor*** *in the Action box.*

defending a territory with the shot-roller

The next program needs a dark room, such as a windowless bathroom with the lights turned off. Once you've programmed the Shot-Roller, put it in this room, and you'll have 30 seconds to leave the room and to carefully close the door. Alone in its darkened room, the Color Sensor's measure of light intensity should be at or close to 0. Now if you open the door, and once the measured value exceeds 5 (the trigger value), the Shot-Roller robot notices that someone has opened the door and will defend its territory using its shooter. (Don't forget to close the toilet lid!)

Start a new program called **Shot-Roller-Light**, and then place and configure the blocks as shown in Figure 8-12.

NOTE **When I refer to the Light Sensor, or the Light Sensor value, I'm actually referring to the Color Sensor functioning in Light Sensor mode. When using a programming block to control the Light Sensor, first select *Color Sensor* from the list of sensors, and then specify that you want to use the *Light Sensor* function in the Action parameter of the Configuration Panel. In other words, don't select Light Sensor in the sensor list, because that's an older sensor that's not included in the 2.0 kit.**

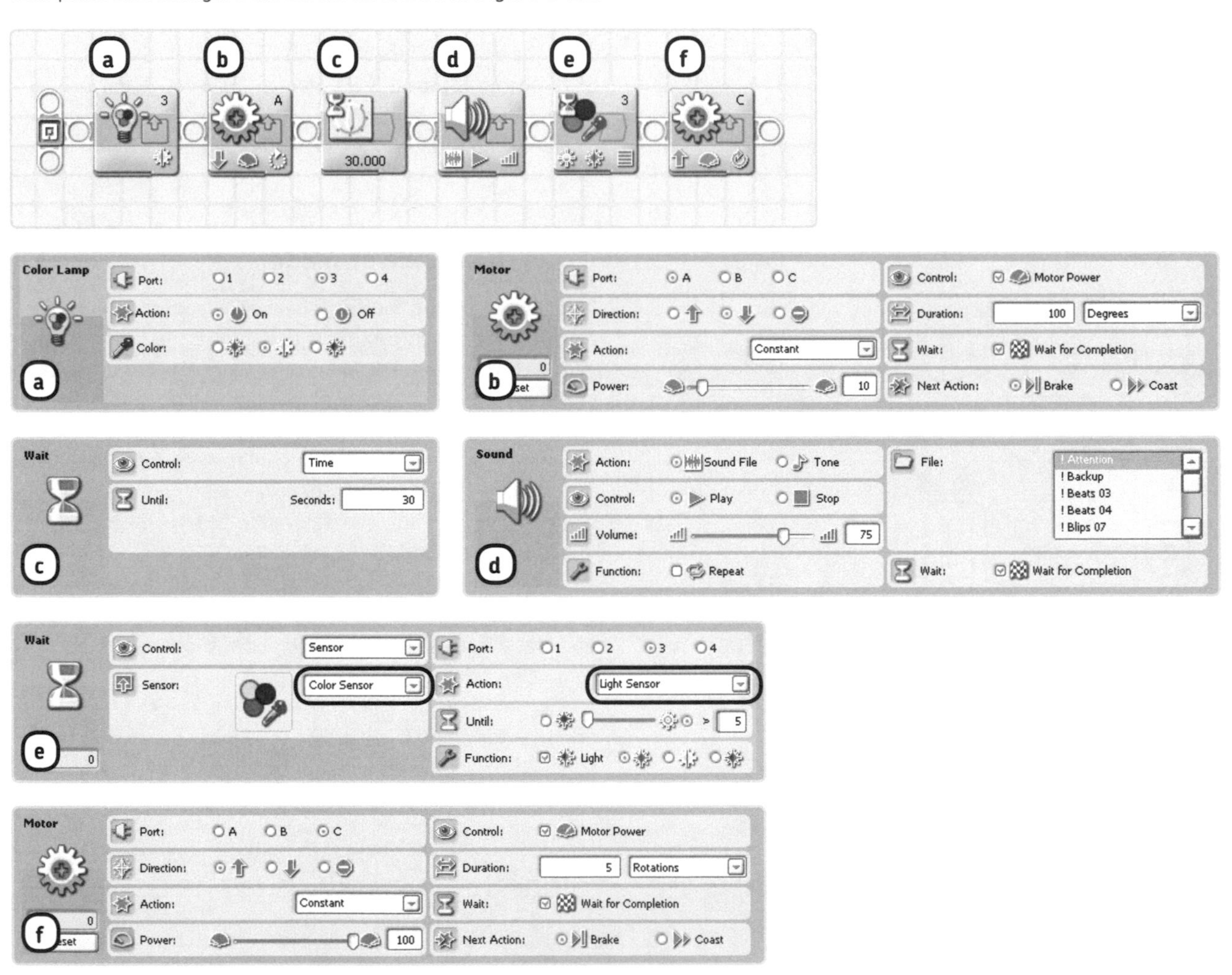

Figure 8-12: The configuration of the blocks in the Shot-Roller-Light program

troubleshooting your program

If your robot doesn't run the program as expected, the trigger value of 5 in Wait block *e* may be too low; the room may not be completely dark. Try a higher trigger value, such as 10. (You'll learn about a more solid solution to this problem in Chapter 9.)

remote-control mode

In autonomous mode, the robot handles all the actions itself, but it is also fun to control a machine like the Shot-Roller remotely. In Chapter 3, you saw that NXT-G has a built-in function to control robots remotely, but it's suitable only for vehicles like the Explorer. Therefore, you'll create a program that *responds* to the remote that you'll use: two Touch Sensors connected to input ports 1 and 2 on the NXT with long cables, as shown in Figure 8-13.

You'll use the Touch Sensor on port 1 to control the Turn motor and the sensor on port 2 to control the Turret motor. When you press both buttons at the same time, you'll

DISCOVERY #41: COMBINING SENSOR POWER!

Difficulty: Hard
Now that you've mastered the techniques to control the Shot-Roller, it is your turn to be creative. Make a program that uses both the Light Sensor and the Ultrasonic Sensor to detect intruders, and use multiple Switch blocks to determine which sensor has been triggered. For instance, you could make a Shot-Roller that shoots only when there is an intruder at night: The Light Sensor detects darkness, and the Ultrasonic Sensor detects the intrusion. Use Sound and Display blocks to describe the sensor values. For example, you could make your robot say "Dark" as the Light Sensor value drops below 20. Post your program to the companion website to show others what you've made!

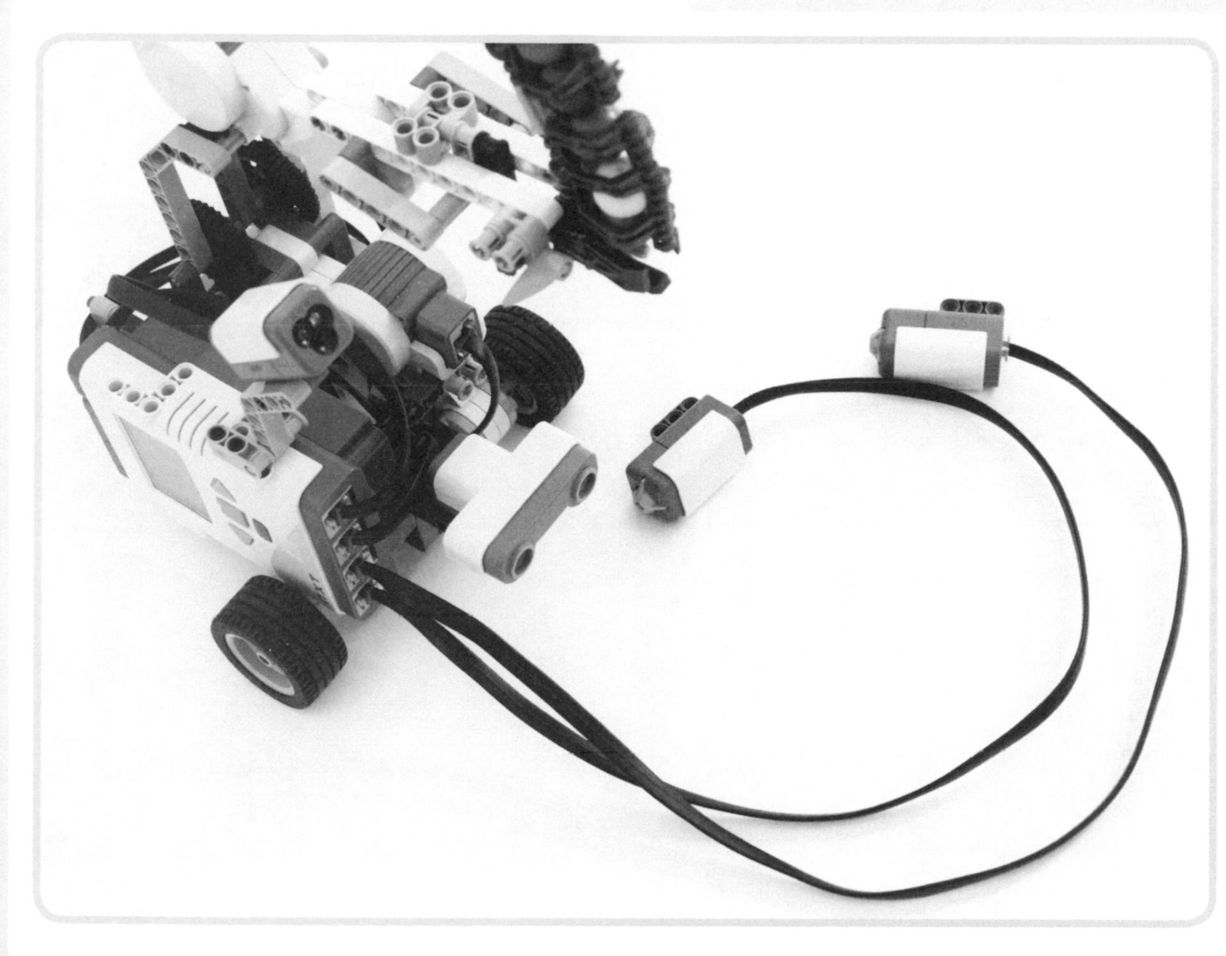

Figure 8-13: Two Touch Sensors are used to control the Shot-Roller remotely.

make the Shot-Roller fire a ball. Each of these three control actions are placed on a separate Sequence Beam. But how do you control the Turn motor and the Turret motor with just one button each?

You can solve this problem as follows: The first time you press a Touch Sensor, the controlled motor moves *forward* until you release the button. Then, when you press the button again, the motor moves *backward* until you release the button. The next time you press the button, the motor spins forward again, and so on.

Because you want to be able to use the Turret and Turn motors simultaneously, you place the blocks to handle these motors on two separate Sequence Beams so that they can run at the same time.

Create the **Shot-Roller-Remote** program as shown in Figures 8-14 to 8-20.

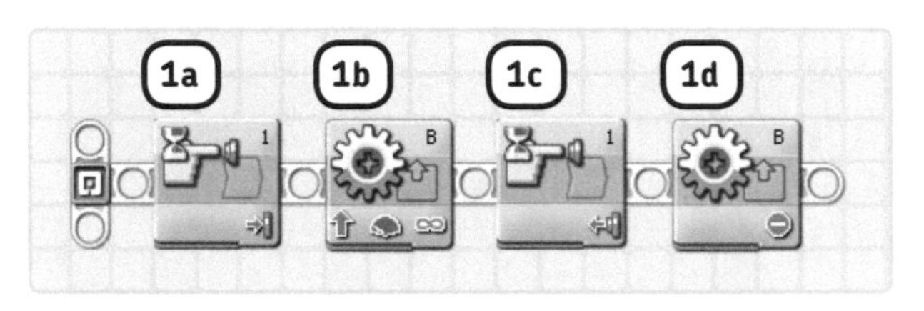

Wait
Control: Sensor | Port: 1 2 3 4
Sensor: Touch Sensor | Action: Pressed / Released / Bumped
1a

Motor
Port: A B C | Control: Motor Power
Direction: | Duration: 72 Unlimited
Action: Constant | Wait: Wait for Completion
Power: 20 | Next Action: Brake / Coast
1b

Wait
Control: Sensor | Port: 1 2 3 4
Sensor: Touch Sensor | Action: Pressed / Released / Bumped
1c

Motor
Port: A B C | Control: Motor Power
Direction: | Duration: 72 Unlimited
Action: Constant | Wait: Wait for Completion
Power: 10 | Next Action: Brake / Coast
1d

Figure 8-14: ***Step 1:*** *When you press the Touch Sensor on port 1, the Turn motor moves forward, and the Shot-Roller spins to the right. When you stop pressing the sensor, a Motor block makes the motor stop.*

Figure 8-15: **Step 2:** These blocks work like the ones placed in step 1, except that they make the Turn motor spin backward, causing the Shot-Roller to turn left.

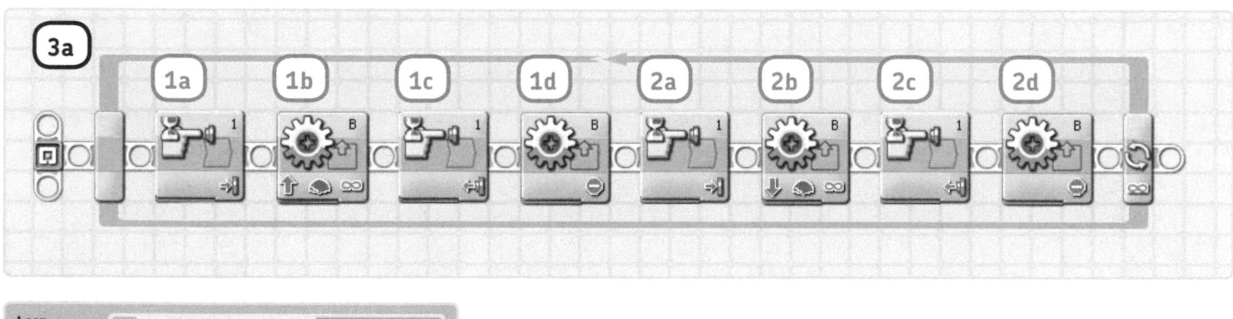

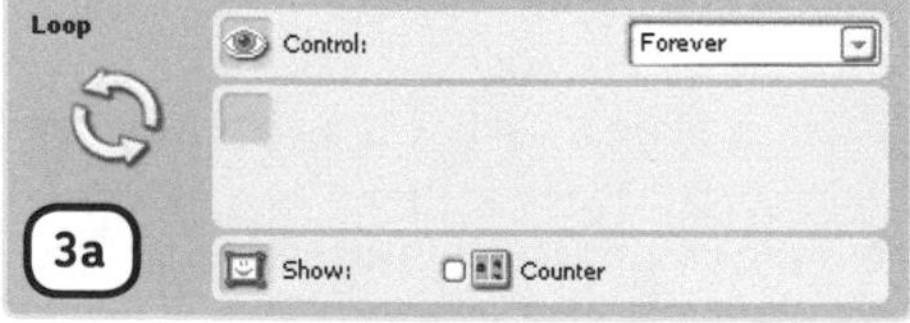

Figure 8-16: **Step 3:** Place a Loop block on the Work Area, and then select all the other blocks and drag them into the loop. Because of the Loop block, the program will repeatedly wait for button presses and therefore repeatedly control the Turn motor based on this input.

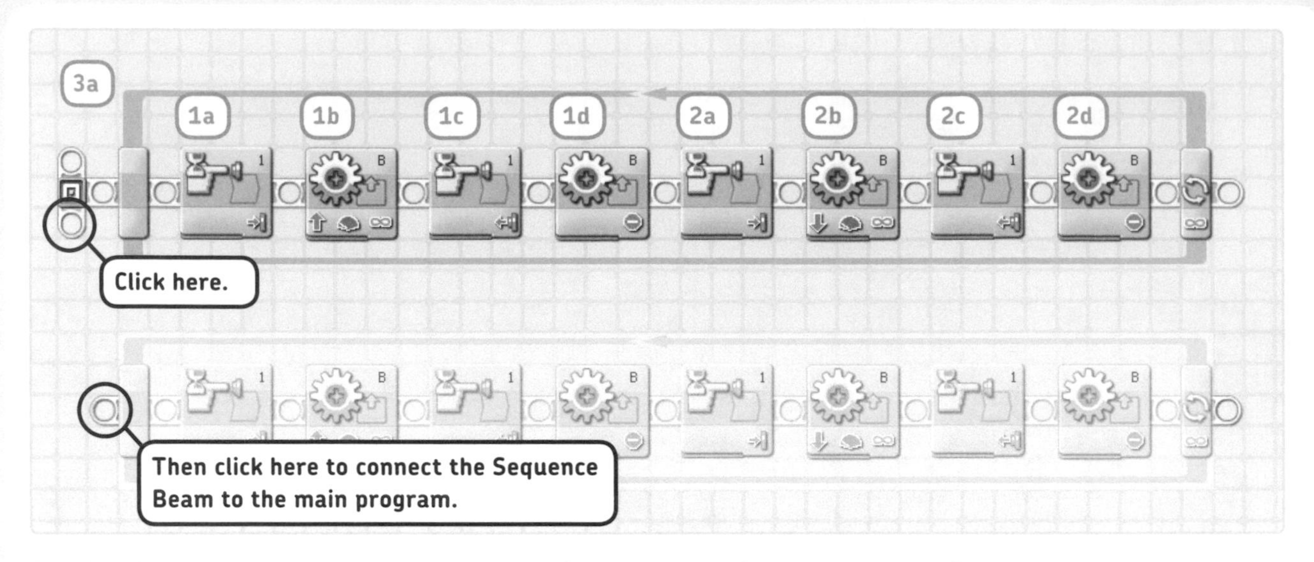

Figure 8-17: ***Step 4:*** *Select the Loop block that you placed in step 3, and then press the* ***Copy*** *button and then the* ***Paste*** *button on the toolbar to duplicate the loop and the blocks in it. Drag the new loop to the position shown in the figure, and then connect the new blocks to the main program as shown here.*

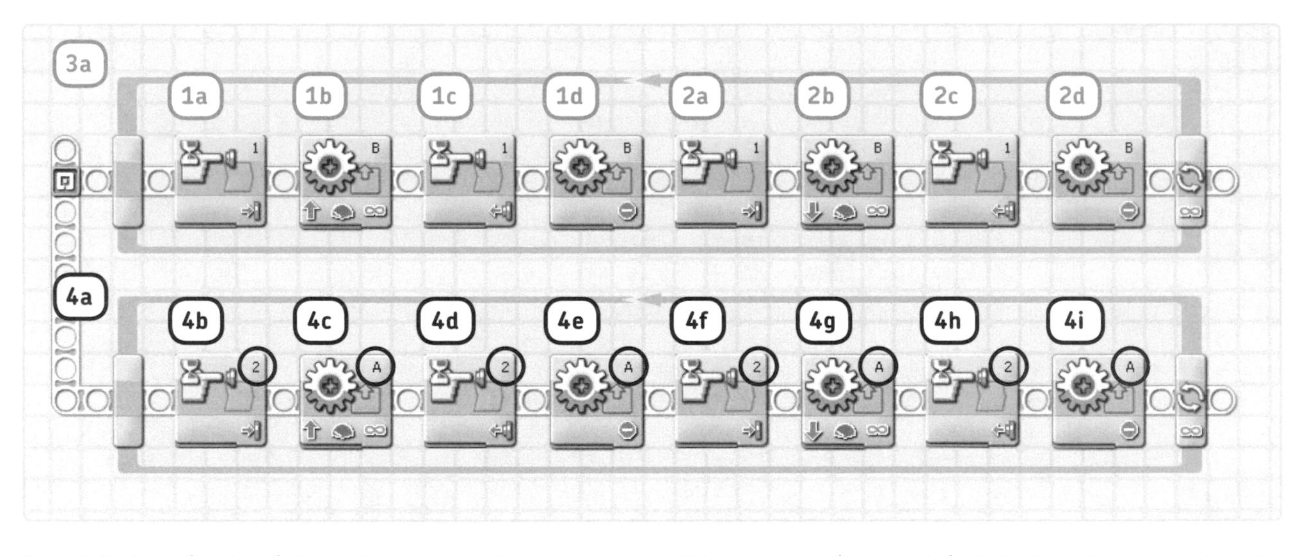

Figure 8-18: ***Step 4 (continued):*** *The blocks you just copied will be used to control the Turret motor (output port A) with the Touch Sensor on input port 2. Therefore, you change the Configuration Panels of the Wait blocks to wait for button presses of the Touch Sensor on port 2, and you change the Motor blocks to control the Turret motor (port A). Your Work Area should look like the one shown here.*

Finally, you need to program the robot to shoot when both Touch Sensors are pressed simultaneously. To do so, you'll use two Switch blocks, each configured to check whether a sensor is pressed. When both sensors are pressed, the Turret and Turn motors are switched off, and the Firing motor shoots until you release the Touch Sensors.

The Switch blocks are displayed in compact mode (Flat View is not checked in the Configuration Panels) to make the programs more readable. There are no blocks hidden in the false condition tabs of the switches because in this part of the program no action is taken when one of the Touch Sensors is not pressed.

The two switches are placed inside a Loop block so that the robot continuously checks to see whether both sensors are pressed. Because the robot needs to perform this check constantly, you place the blocks that control the Firing Motor on a Parallel Sequence Beam.

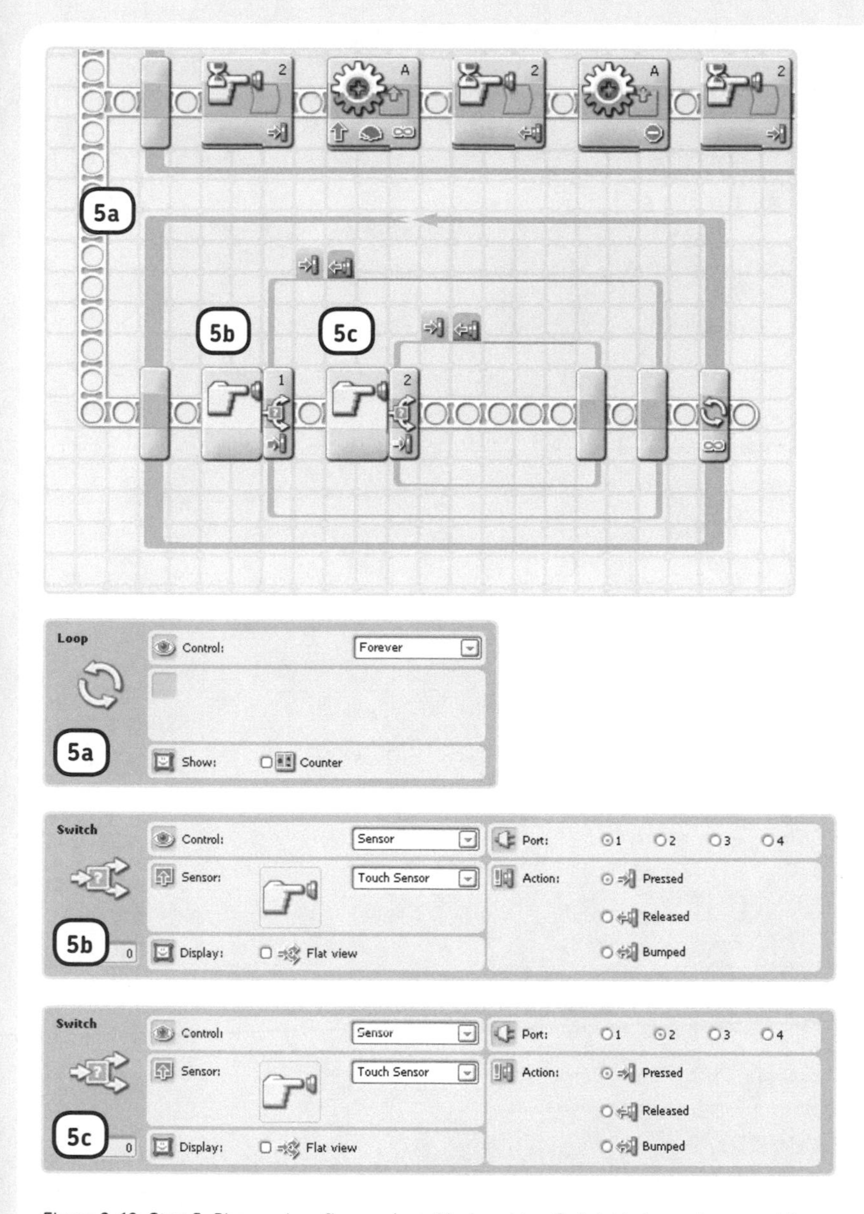

Figure 8-19: ***Step 5:*** *Place and configure a Loop block and two Switch blocks as shown, and then connect the blocks to the main program by connecting the Sequence Beams, as you did in step 4. Because this is the second Parallel Beam, hold the* SHIFT *key while making the connection.*

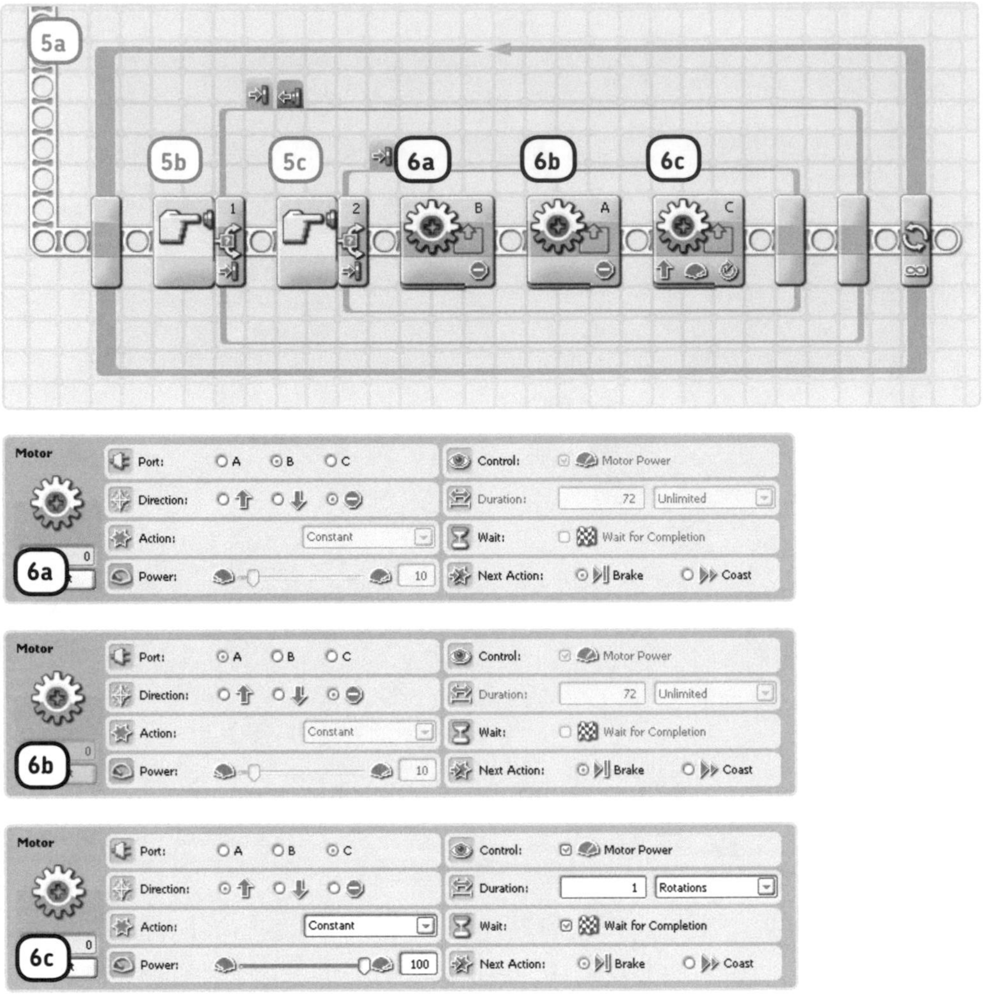

Figure 8-20: **Step 6:** Place and configure three Motor blocks as shown. These blocks are run when the condition of both Switch blocks is true; in other words, when both Touch Sensors are pressed. The Turn and Turret motors are turned off, and the Firing motor starts to fire balls.

Congratulations, you finished creating the remote-control program! Download the program to the Shot-Roller, and have fun!

further exploration

In this chapter, you had a chance to build, program, and play with a prebuilt robot. This is, of course, a lot of fun, but it is even more fun to create your own robot designs. For example, you could take the shooter from this robot and mount it on your self-built car or tank, or you could turn it into a dangerous creature, such as a ball-shooting insect. Don't worry if you don't succeed on the first try; you will gain more and more building experience as you continue to try new designs.

DISCOVERY #42: RESEARCH WITH THE NXT!

Difficulty: Medium
How far can the Shot-Roller shoot? The distance that a ball travels will depend on the angle that the Turret motor makes with the ground. Which angle makes the ball go the farthest? Can you make it go even farther by modifying the shooter? Investigate the effect that the turret angle and the motor speed have on the distance of the fired ball. Post your findings on the book's companion website (*http://www.discovery.laurensvalk.com/*)!

BUILDING DISCOVERY #6: LOOK BEFORE YOU SHOOT!

Can you modify the Shot-Roller design to mount the Ultrasonic Sensor on the turret, just below the ball magazine? This way, not only can you shoot in any direction, but you can also look in any direction so that the Shot-Roller has a better view of its targets.

BUILDING DISCOVERY #7: CATAPULT!

The Shot-Roller's shooting mechanism makes it easy to launch the balls in the NXT robotics kit, but it won't let you fire other LEGO parts. Can you make a construction like a catapult that fires LEGO pieces when it sees you?

HINT **Think of how you would make a catapult out of something other than LEGO. You might use rubber bands to launch something or a plastic spoon that you bend with your hands. Build such a mechanism with the MINDSTORMS system, and use an NXT motor to, for example, release the rubber band or the spoon in order to launch something when the robot sees you. When using rubber bands, use ones other than those found in the NXT 2.0 kit, or you'll likely break them.**

strider: the six-legged walking creature

You can make interesting models that move on wheels, but it is also possible to make robots that walk. Such creations are slightly more challenging to make, but in this chapter you'll find instructions to build Strider, a six-legged walking creature, as shown in Figure 9-1. Once you finish building it, you'll program it to walk around and respond to human interaction.

Figure 9-1: Strider

The Strider robot uses three identical motor assemblies to walk, each of which controls one pair of legs. The leg modules are interconnected with a triangle-shaped frame, which also carries the NXT with several sensors attached. Two Touch Sensors on the Strider act like antennas, detecting touches from objects or people in Strider's environment (they are not used as bumpers to prevent the robot from walking into something). The Ultrasonic Sensor allows the walker to measure the distance of nearby objects, while the Color Sensor in Light Sensor mode can detect whether it is dark or light outside.

building strider

We'll now move on to building the robot by following the instructions on the subsequent pages. Before you start building, select the pieces you'll need to complete the robot as shown in Figure 9-2.

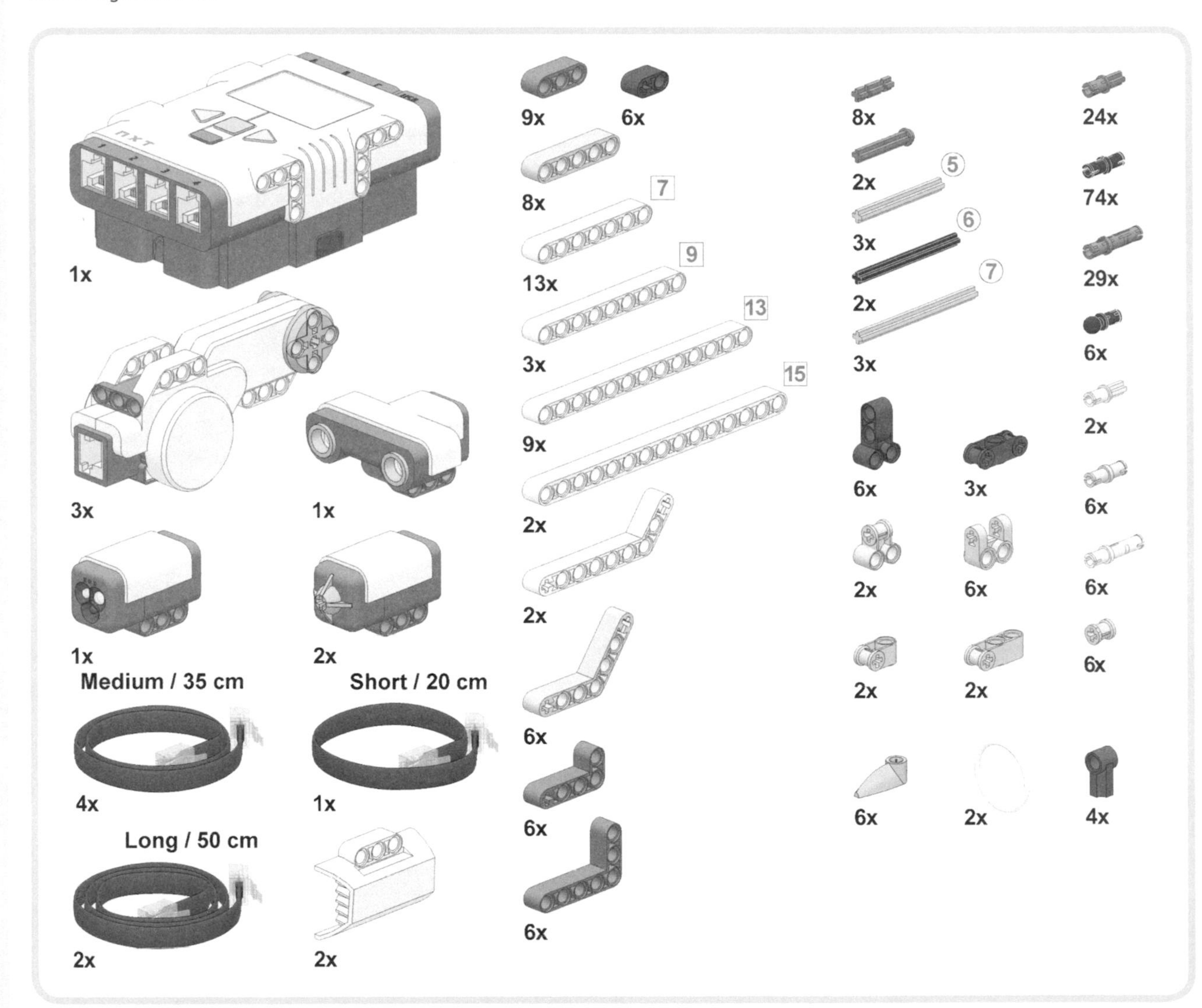

Figure 9-2: The required pieces to build the Strider

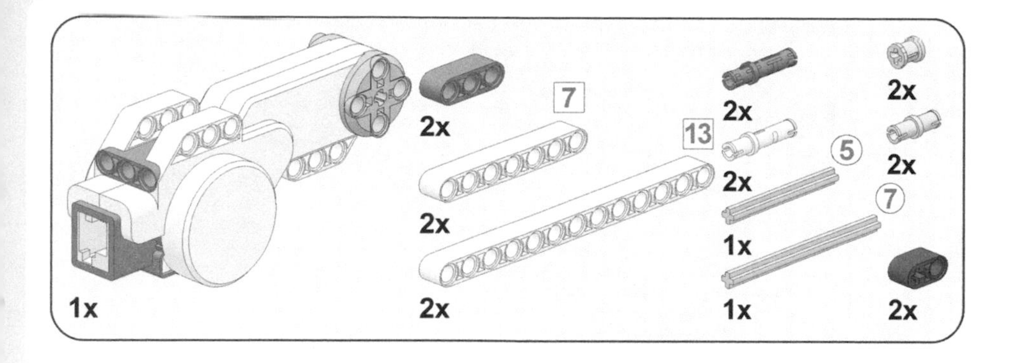
1x
2x
7
2x
13
2x
2x
2x
2x
5
1x
7
1x
2x
2x
2x

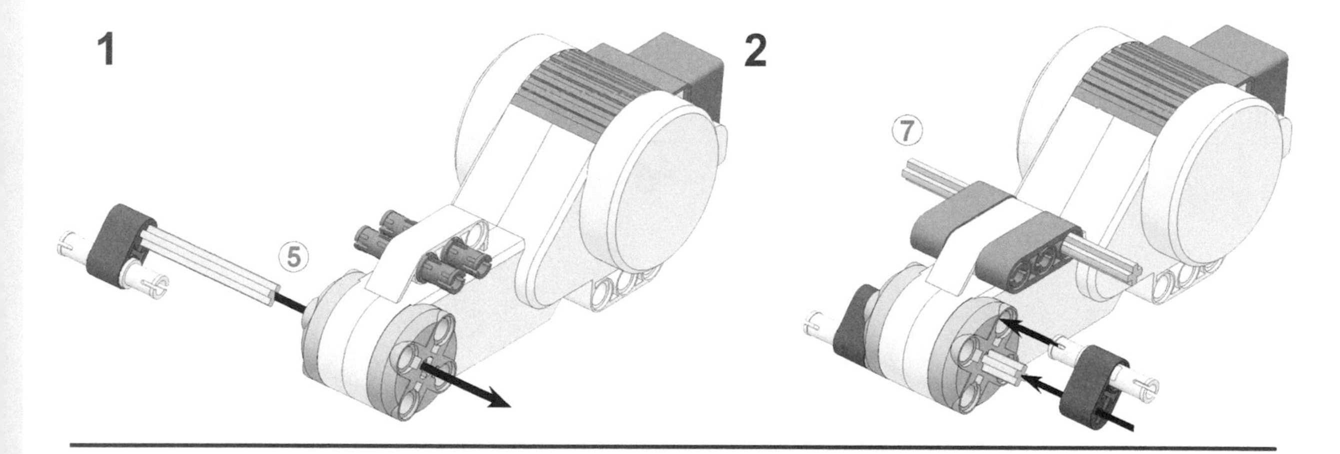
1
5
2
7

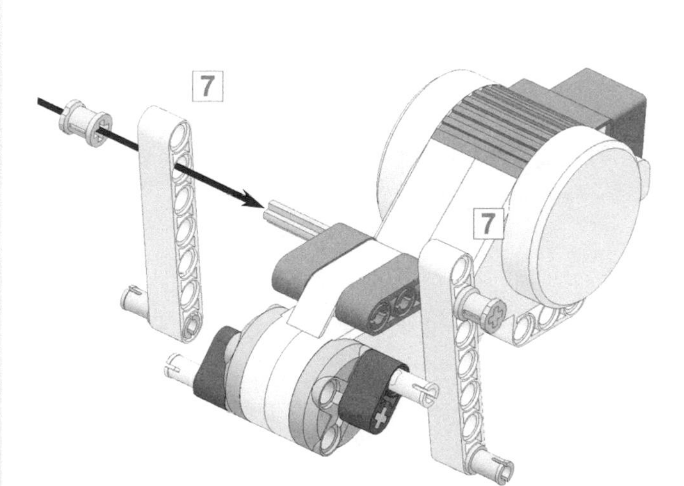
3
7
7

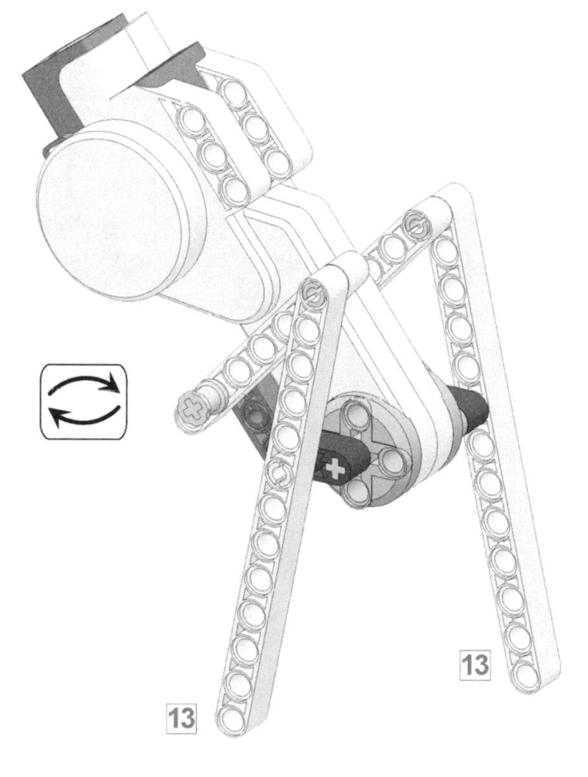
4
13
13

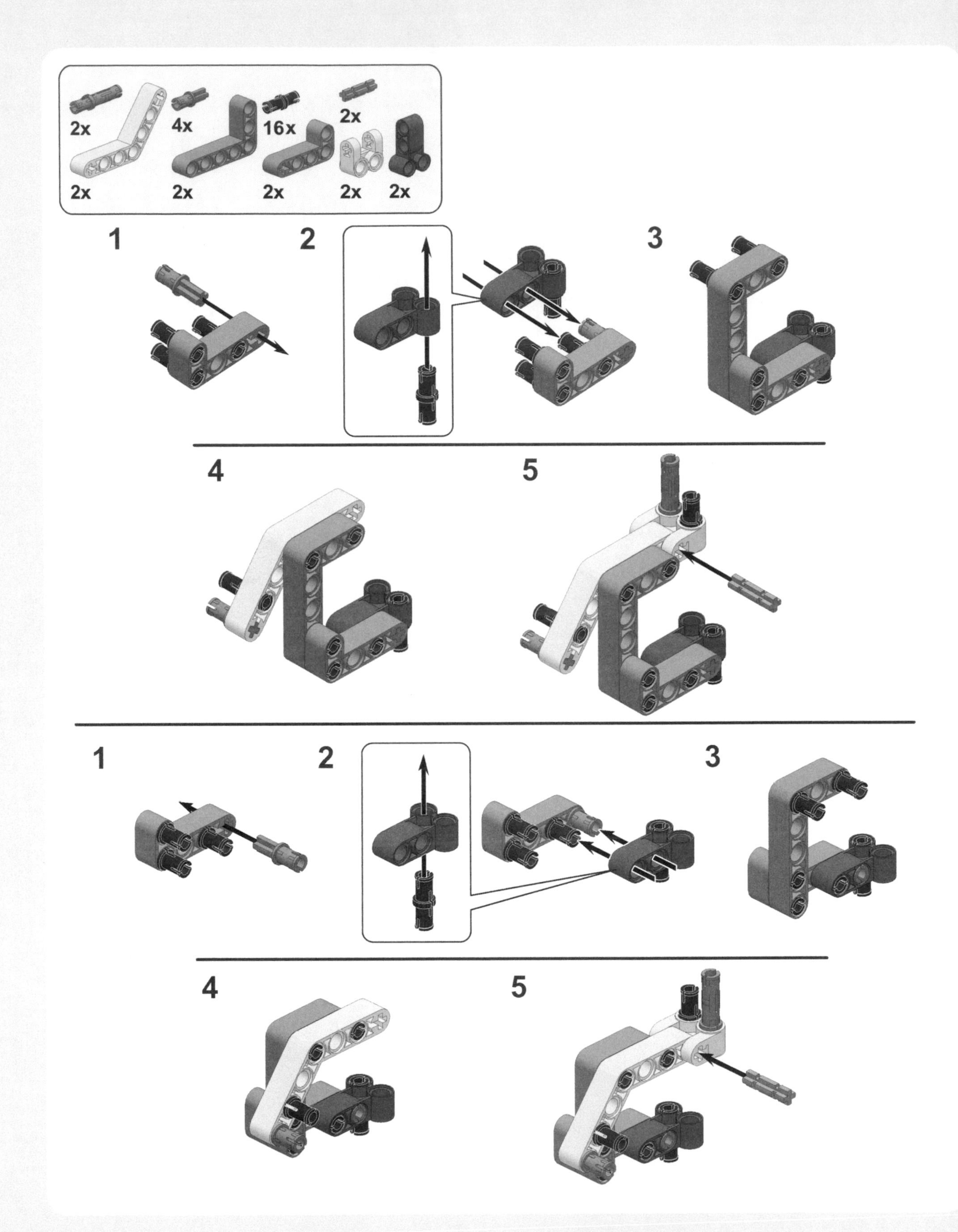
2x
4x
16x
2x
2x
2x
2x
2x
2x
1
2
3
4
5
1
2
3
4
5

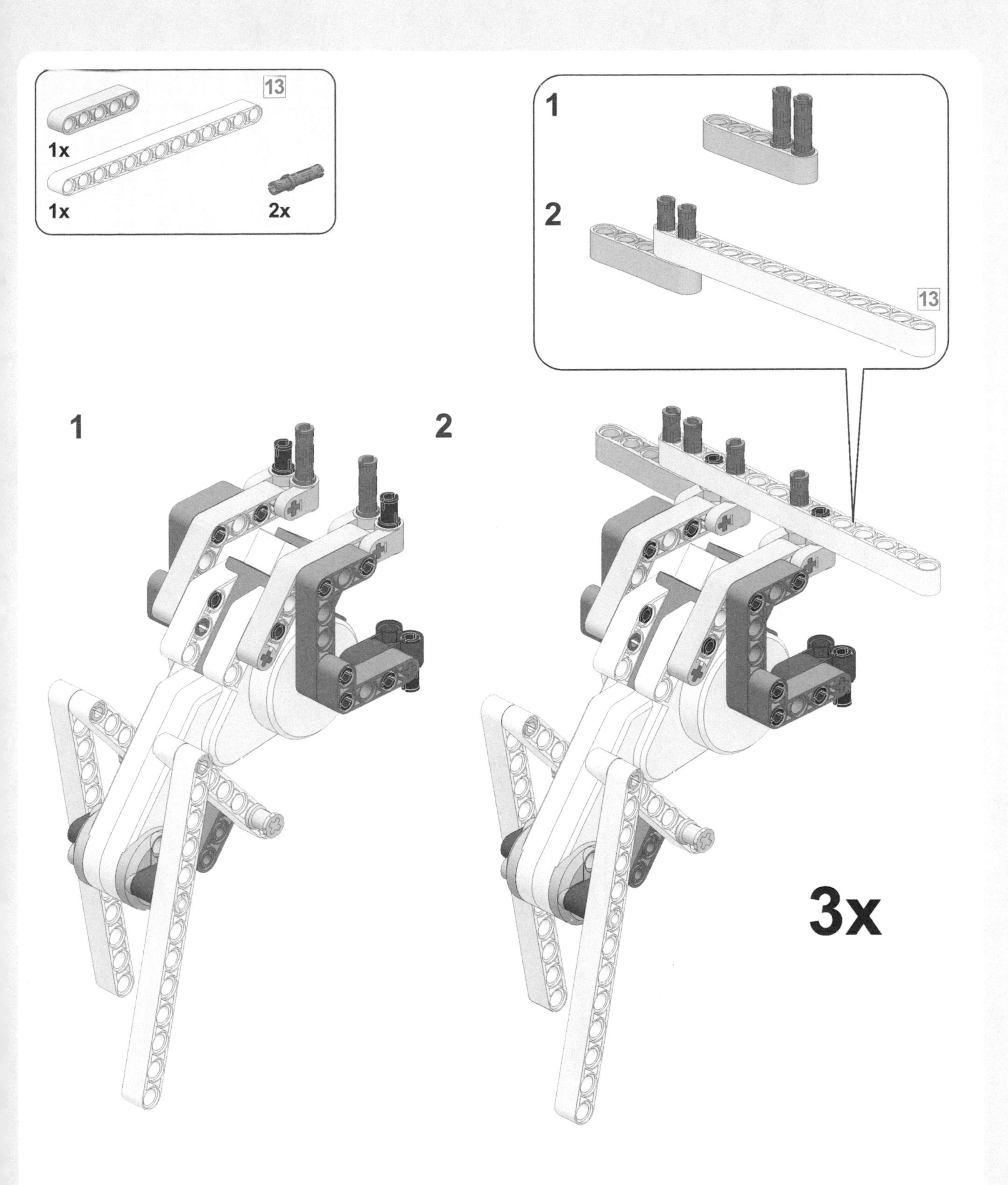
13
1x
1x
2x
1
2
13
1
2
3x

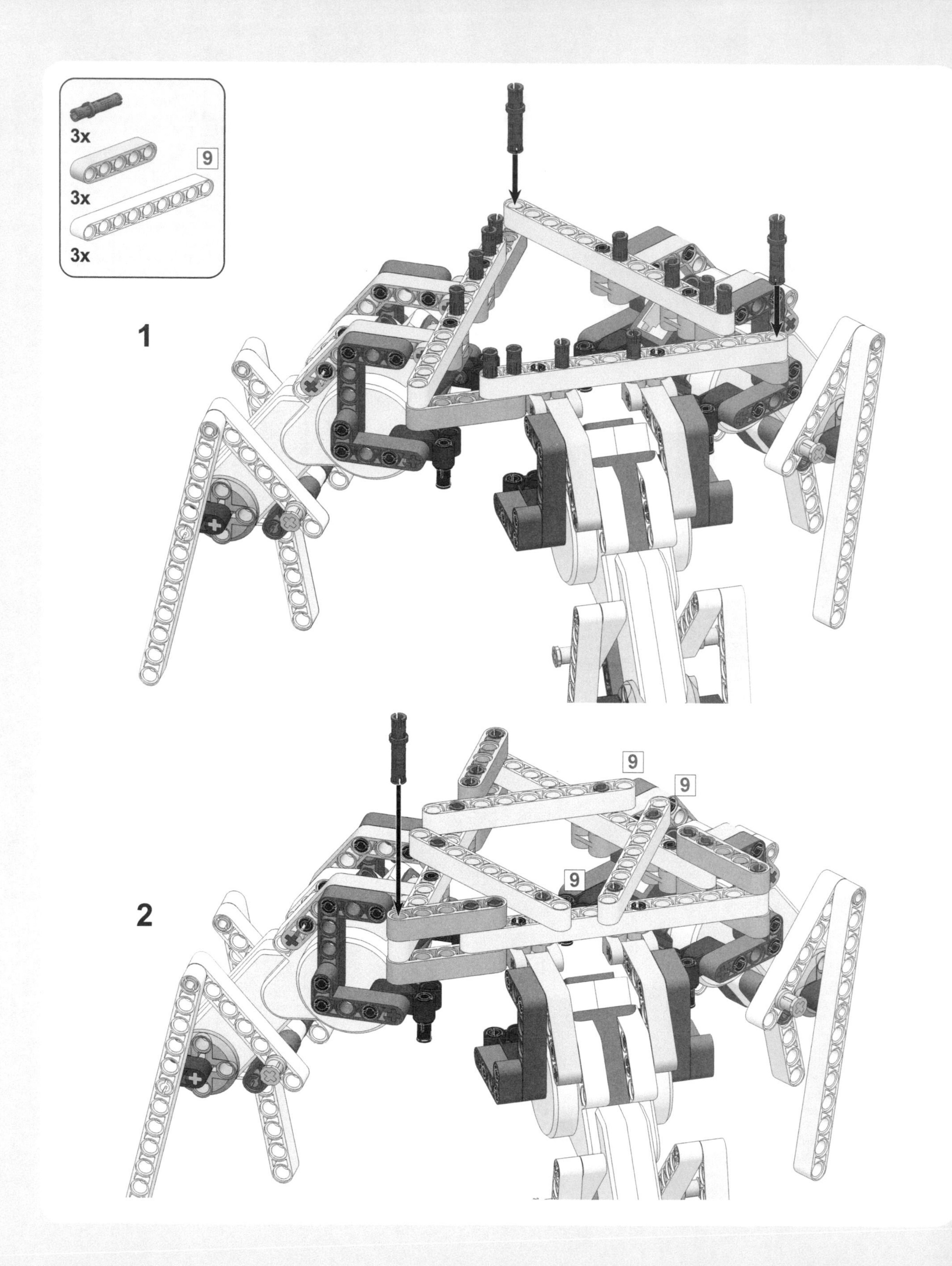
3x
3x
9
3x
1
2
9
9
9

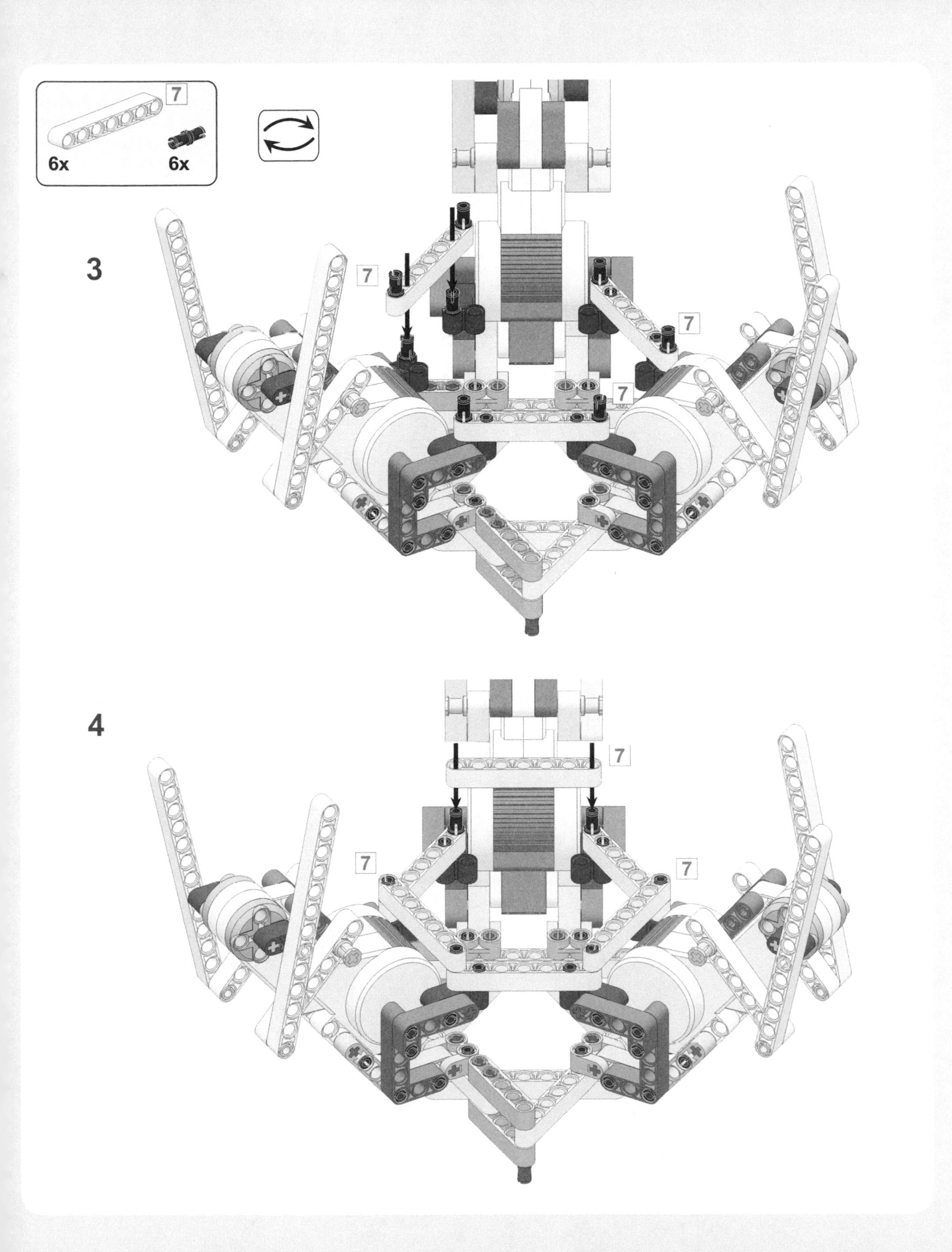

7
6x
6x
3
7
7
7
4
7
7
7

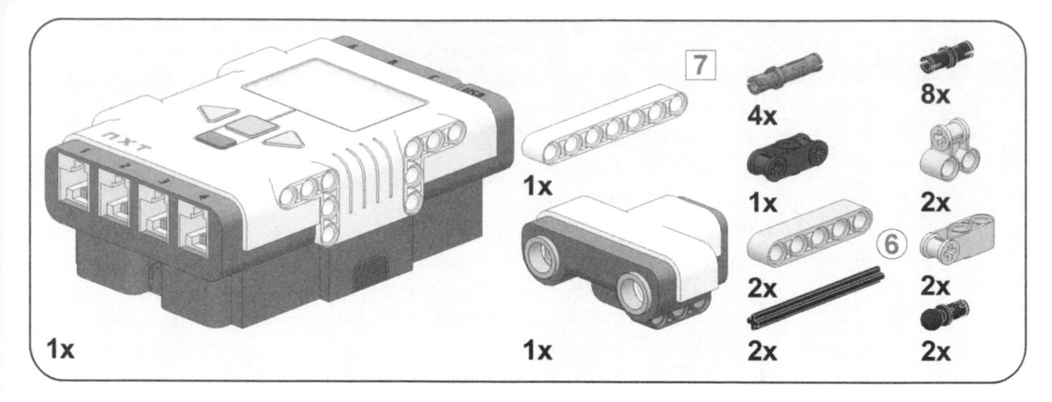
1x
7
1x
1x
4x
1x
2x
6
2x
8x
2x
2x
2x
2x

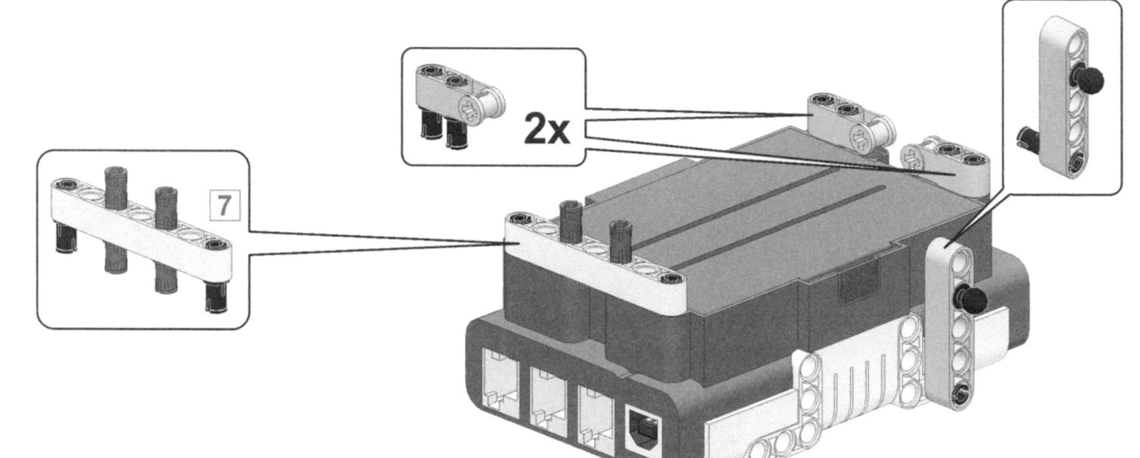
2x
7

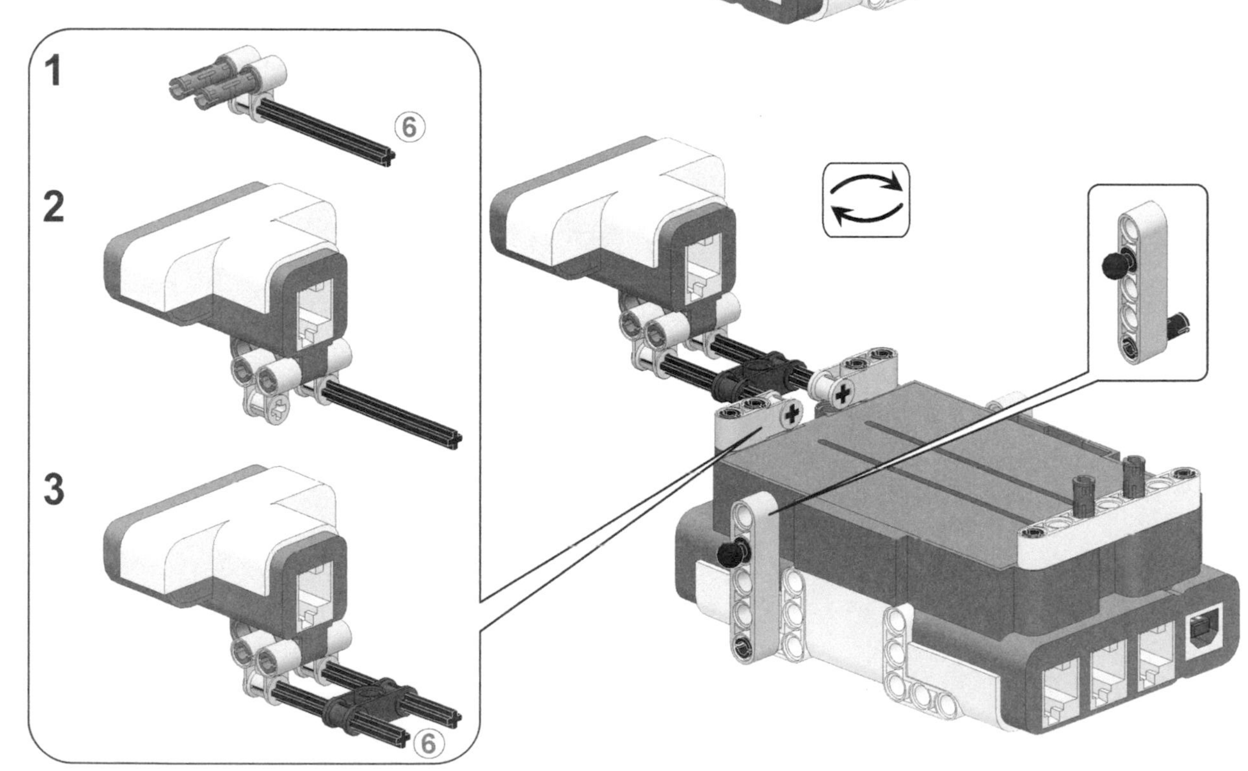
1
6
2
3
6

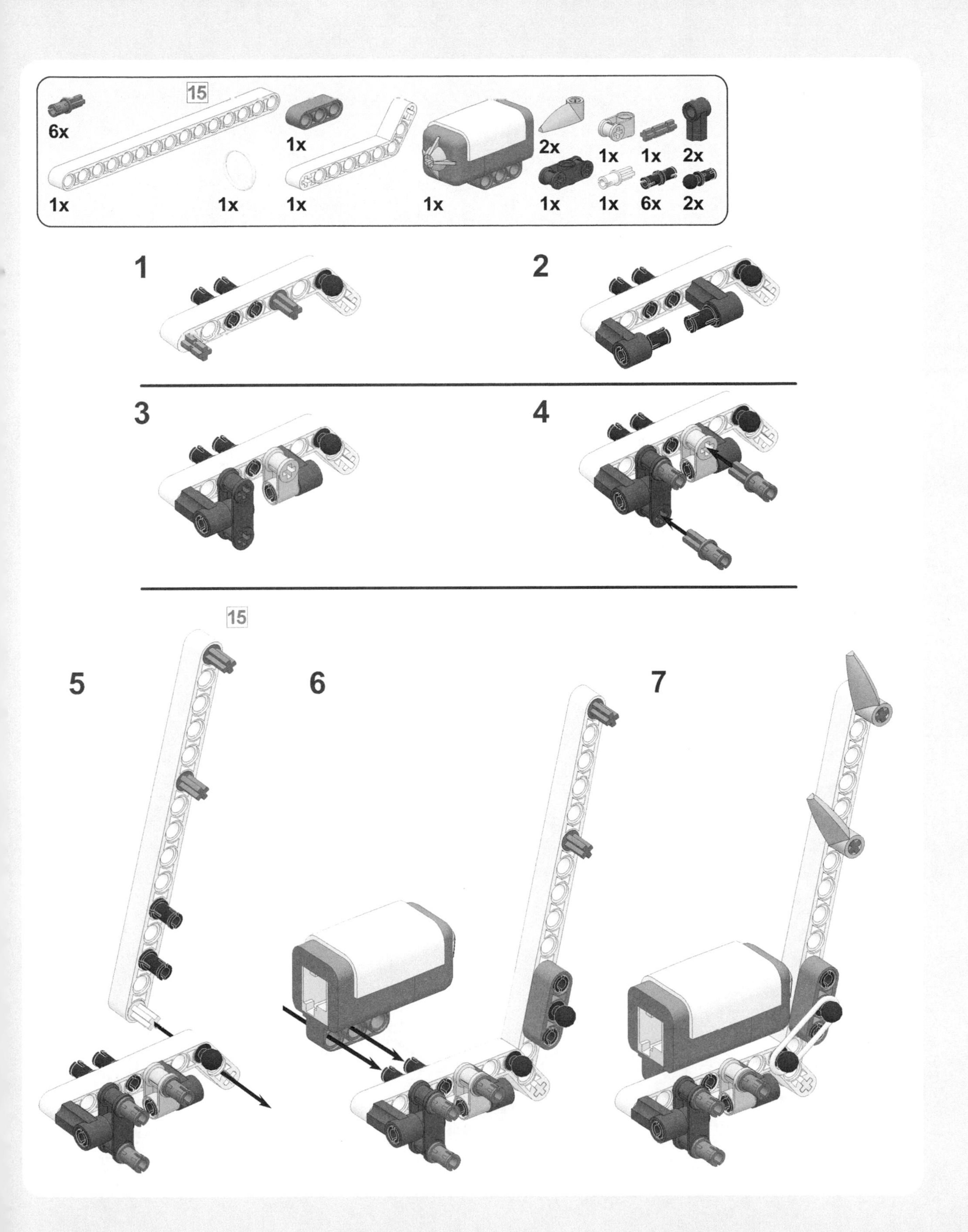
15
6x
1x
1x
1x
1x
1x
2x
1x
1x
2x
1x
1x
6x
2x
1x
1
2
3
4
15
5
6
7

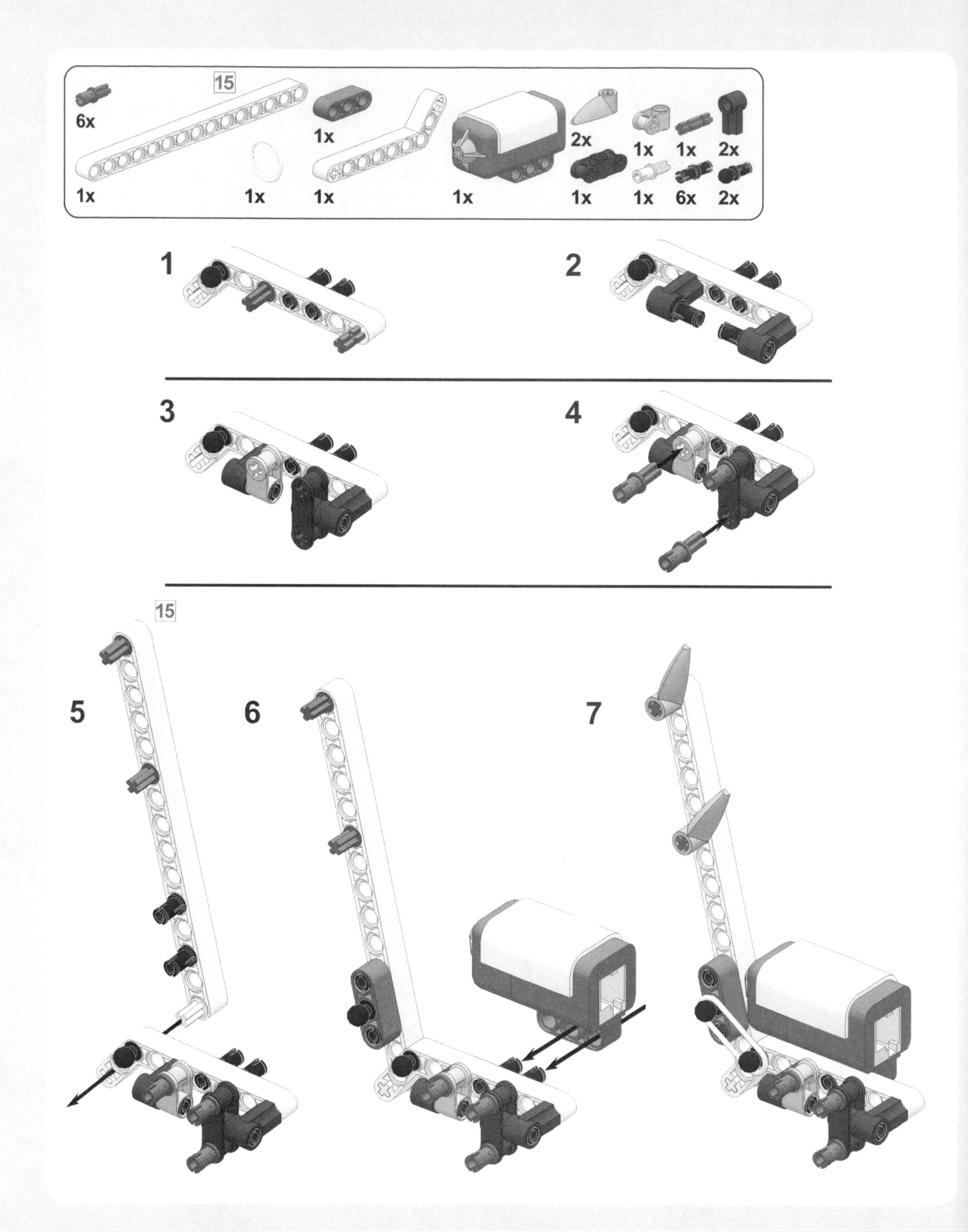
6x
15
1x
1x
1x
1x
1x
2x
1x
1x
1x
1x
2x
1x
6x
2x
1
2
3
4
15
5
6
7

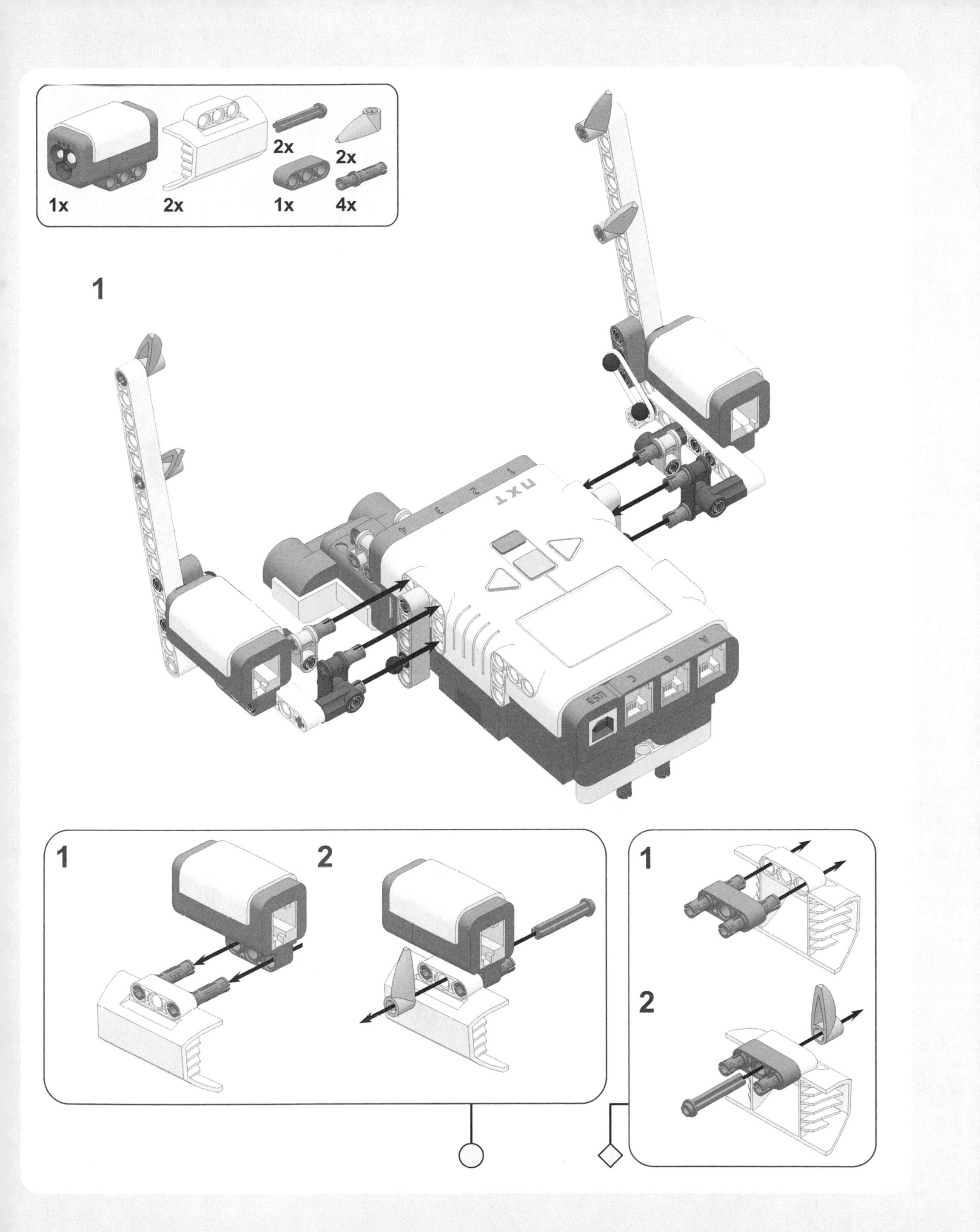
1x
2x
2x
1x
2x
4x
1
NXT
USB
C
B
A
1
2
1
2

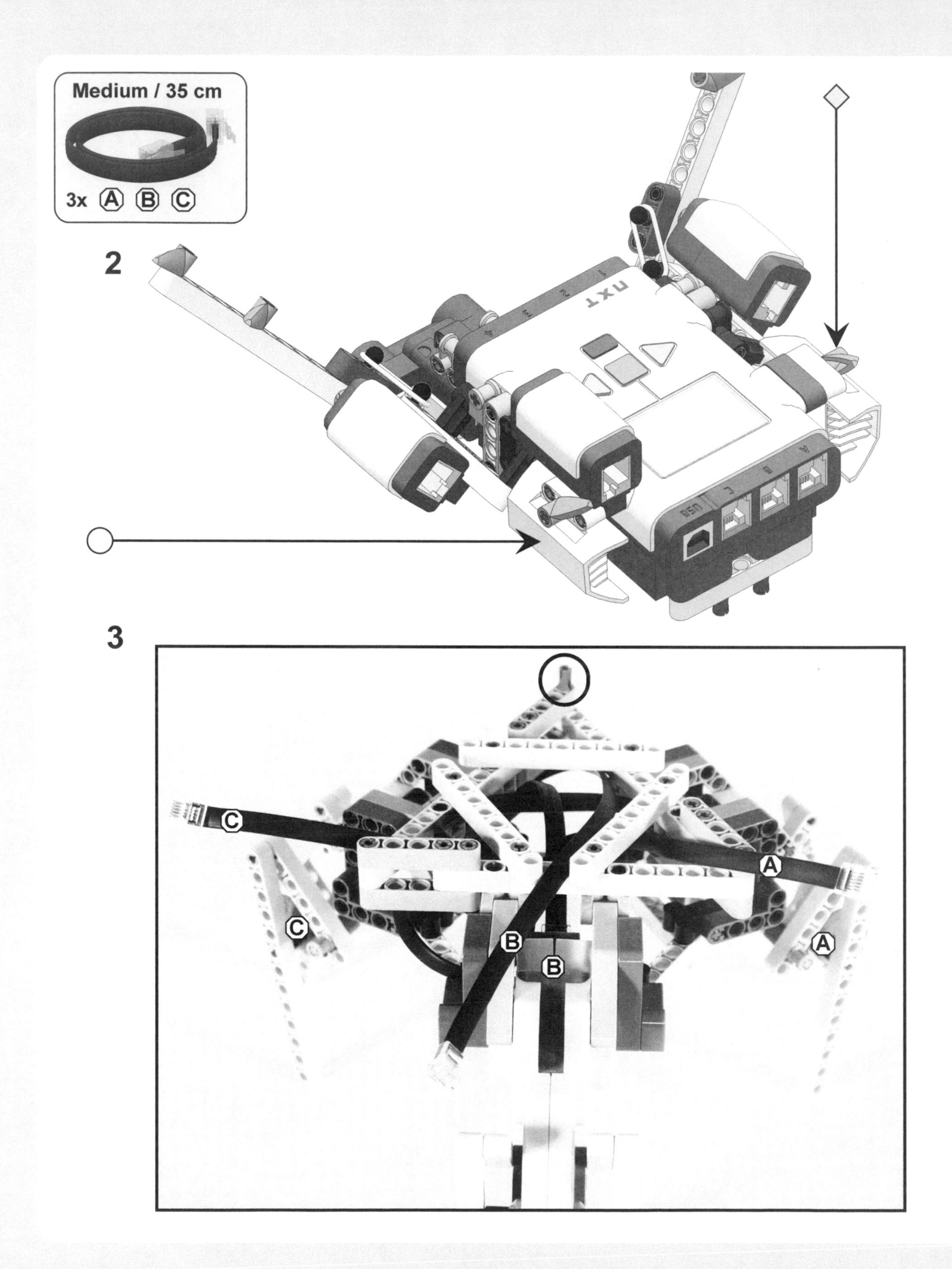
Medium / 35 cm
3x A B C
2
3
C
A
C
B
B
A

4

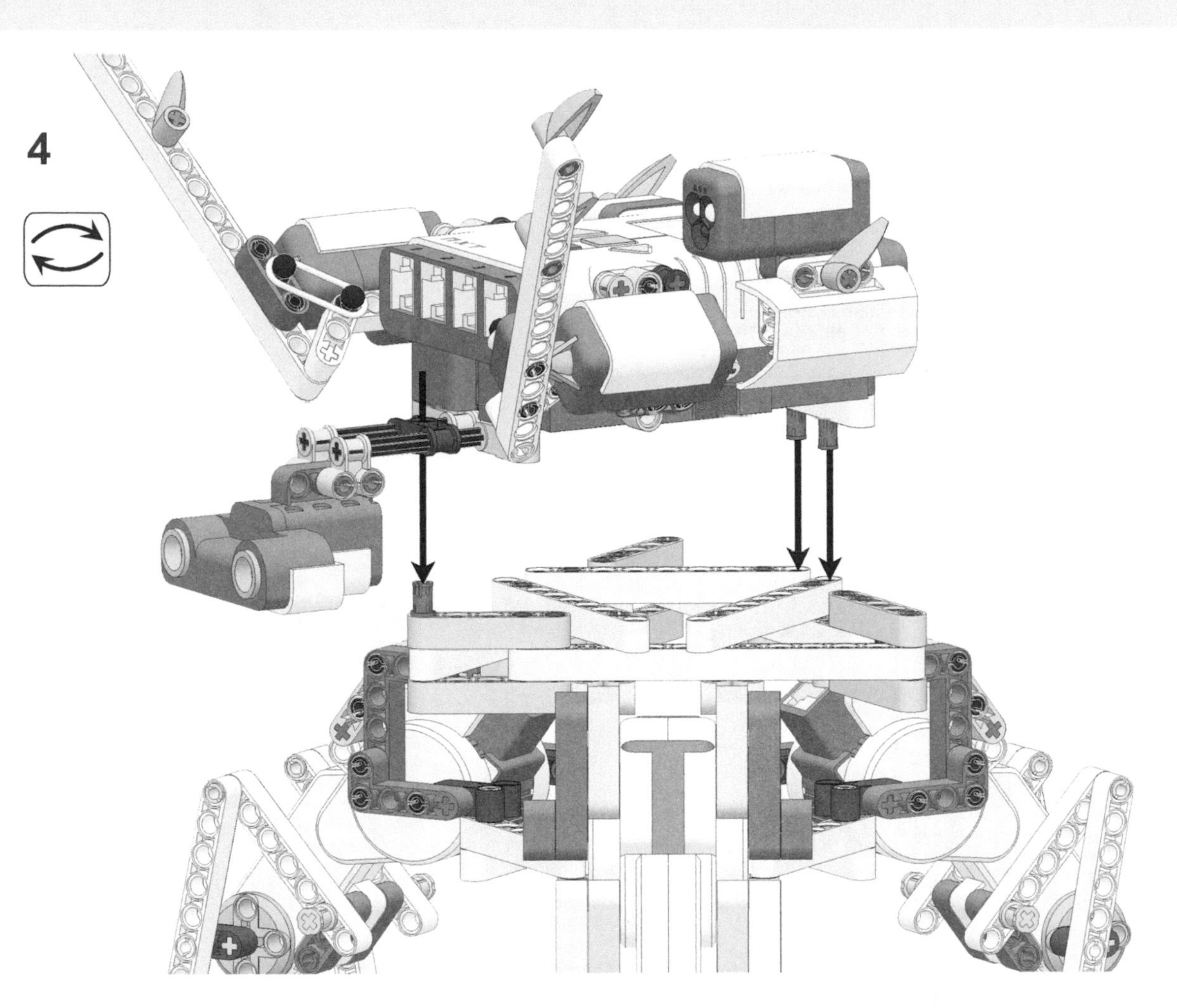

5

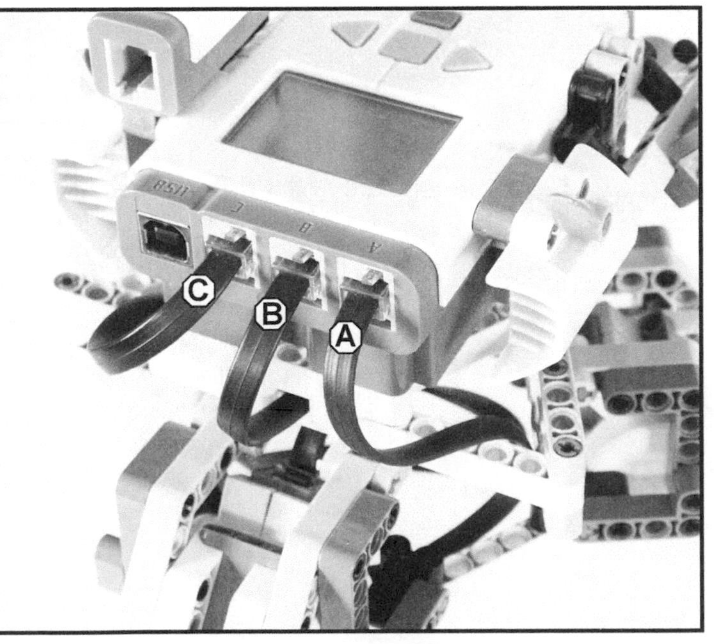

connecting the sensor cables

You connected the motor cables on pages 132 and 133. Table 9-1 and Figure 9-3 will show you how to connect the sensors. When connecting the cables, guide them through the space under the NXT so that they don't stick out.

table 9-1: the cable placement for strider

From motor/ sensor	To NXT brick port	Cable length
Right Touch Sensor (1)	Input port 1	Long (50 cm/20 inches)
Left Touch Sensor (2)	Input port 2	Long
Color Sensor	Input port 3	Medium (35 cm/15 inches)
Ultrasonic Sensor	Input port 4	Short (20 cm/8 inches)

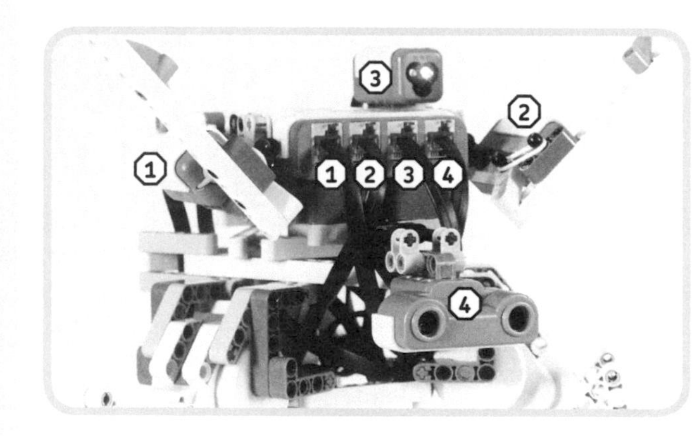

Figure 9-3: Connect the cables as shown.

understanding strider's walking technique

Before you can program Strider, you need to understand how it walks. Figure 9-4 shows how three minifigures can move a heavy object forward.

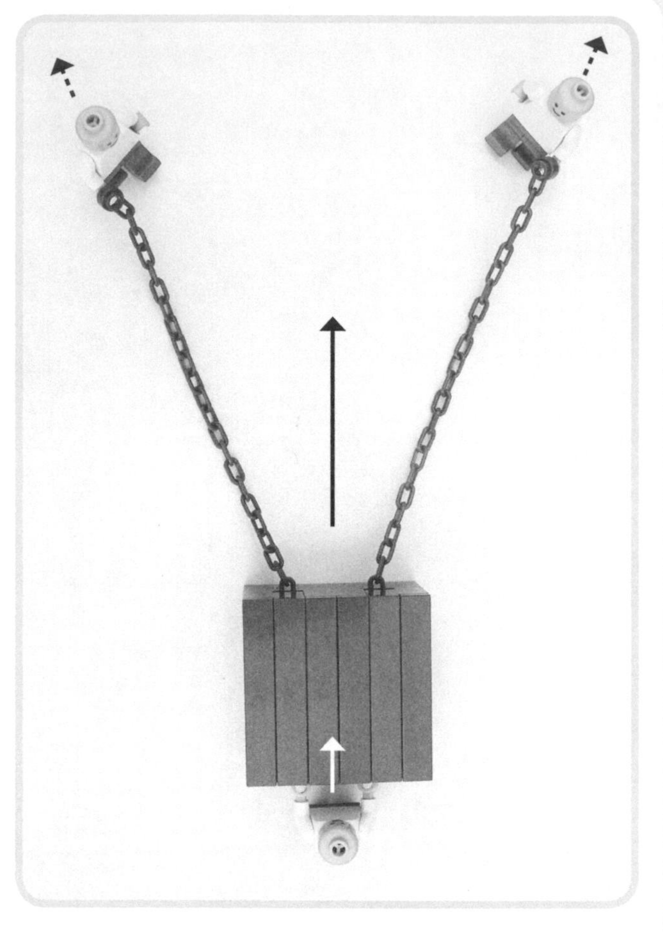

Figure 9-4: Two minifigures pull the heavy object (dashed arrows), and one pushes (white arrow). As a result, the object moves forward (solid black arrow).

The minifigures in Figure 9-4 represent each of Strider's motor compartments, and the heavy object represents the NXT, as shown in Figure 9-5.

Each of the Strider's three leg pairs functions a bit like your own legs. As a motor rotates, its attached leg pair walks by repeatedly putting one leg in front of the other. Whether a leg pair pushes or pulls the robot in a certain direction depends on whether the motor is configured to turn forward or backward (Figure 9-6).

When a leg's motor spins backward, it has the same effect on the Strider as the minifigure pulling an object. When the leg's motor rotates forward, it acts like the minifigure pushing an object. The combination of motor directions makes up the direction that Strider eventually walks in.

Figure 9-5: Motors A and C pull the robot forward (dashed arrows), and motor B pushes (gray arrow). As a result, the Strider walks forward (solid black arrow). (The sensors have been removed here to make the motors more visible.)

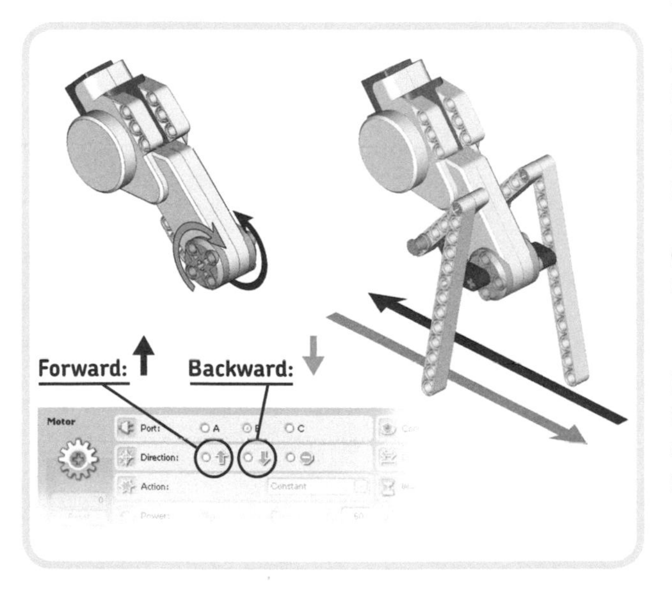

Figure 9-6: When NXT motors are specified to turn forward, the corresponding leg pair moves in the black direction. The gray denotes backward movement of the motor.

NOTE The Strider can only walk on a really smooth surface, such as tiles, a desk, or a smooth wooden floor. If you put it on carpet or a rough floor, the legs will probably break!

programming strider

Now that you understand the basics of making Strider walk, you'll create three My Blocks to make it walk. Following that, you'll use these blocks to create larger programs.

creating the walk-forward my block

You'll begin by creating the Walk-Forward My Block. Because motors A and C will pull the robot, those motors will rotate backward. Motor B will be configured to turn forward since its leg compartment pushes the robot. The Duration settings are all set to **Unlimited** so that the My Block will just switch on the motors. You'll use other blocks (like Wait blocks) to control how long the Strider walks in a specific direction.

To begin, create a new program, pick three motor blocks from the Programming Palette, and configure them as shown in Figure 9-7. Motor B spins slightly faster than the other two because the corresponding leg pair pushes the robot along in the direction it's walking (Figure 9-5); this makes the robot more stable as it walks.

NOTE For a refresher on how to create My Blocks, see Chapter 5 for step-by-step instructions.

Select these three Motor blocks, turn them into a My Block called **Walk-Forward**, and then add a Wait Block to the program (Figure 9-8). Run the program, and the Strider should walk forward for 10 seconds and then stop when the program ends.

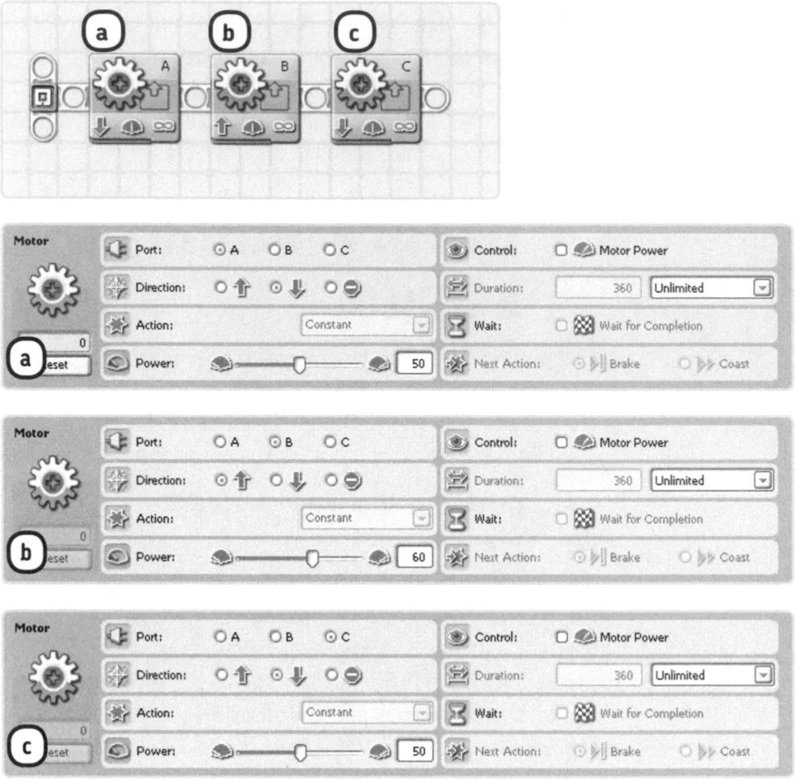

Figure 9-7: The configuration of the blocks in the Walk-Forward My Block

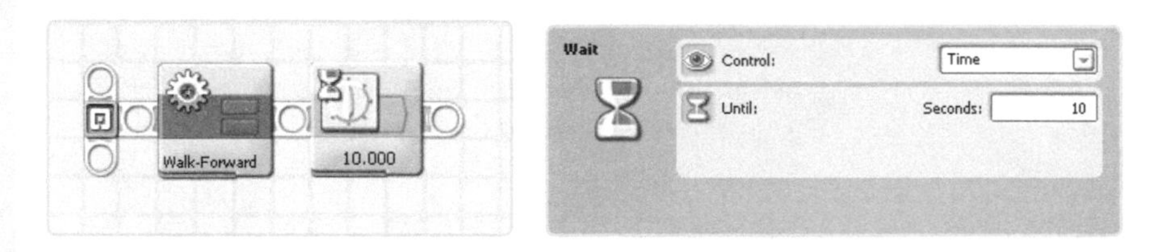

Figure 9-8: A quick test program to make the Strider walk forward

creating the walk-left and walk-right my blocks

Now you'll create two more My Blocks to make Strider walk left and right. The blocks are essentially the same as the one you just made, except that the Direction and Power settings now make the robot move in a different direction, as shown in Figure 9-9. Create these two My Blocks just like the one you made in the previous section, and see Table 9-2 for their names and motor Direction and Power settings.

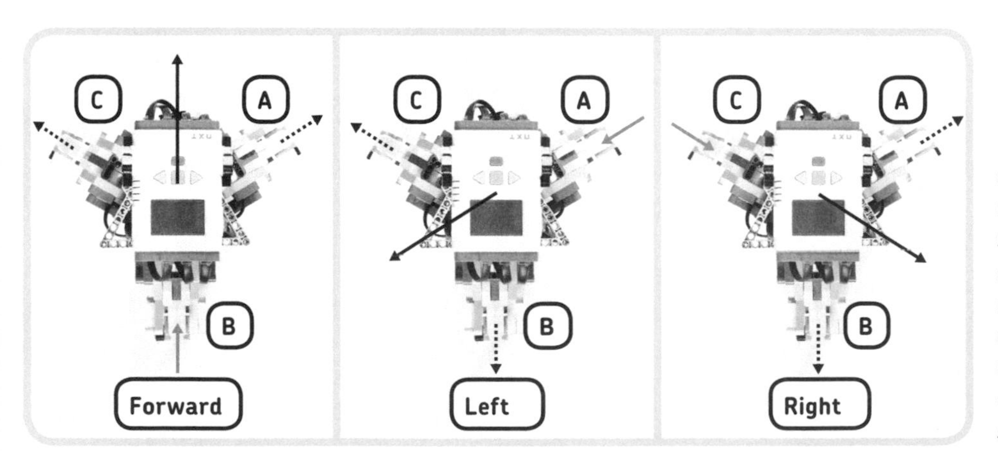

Figure 9-9: By changing the direction in which its motors turn, the Strider can walk forward, left, or right. In each case, two motors pull the robot in one direction (dashed arrows), and the third motor pushes the robot (gray arrow). The resulting direction is indicated with a solid black arrow.

table 9-2: the motor direction and power settings for the motor blocks

My Block name	Motor A	Motor B	Motor C
Walk-Forward	Backward, 50	*Forward, 60*	Backward, 50
Walk-Left	*Forward, 60*	Backward, 50	Backward, 50
Walk-Right	Backward, 50	Backward, 50	*Forward, 60*

DISCOVERY #43: TRIANGLE TIME, AGAIN!

Difficulty: Easy

You know how to drive in a triangle pattern using the Explorer robot, but how do you do it with the Strider robot? Use the three My Blocks you've just made, and set three Wait Blocks to wait for five seconds (put one Wait Block after each My Block). For more fun, put all of these blocks in a Loop Block and add some Color Lamp effects or sounds.

NOTE If you're experiencing that the Strider has trouble walking or if its legs come off, try making it walk on a smoother surface.

using the my blocks in an interactive program

Now that you know basically how to control the Strider, you can create bigger programs. As you noticed in Discovery #43, the Strider cannot turn around, which means it will always look in the same direction with the Ultrasonic Sensor. However, it is still capable of going everywhere when you combine the Walk-Forward, Walk-Left, and Walk-Right My Blocks properly.

The next program you'll create will have the Strider walk forward until you press one of its antennas, at which point it will walk left or right, depending on which sensor you press (programming steps 1 and 2). You'll use sounds and the NXT screen to give feedback about the sensor readings (steps 3, 4, and 5). You can see an overview of the program in Figure 9-10.

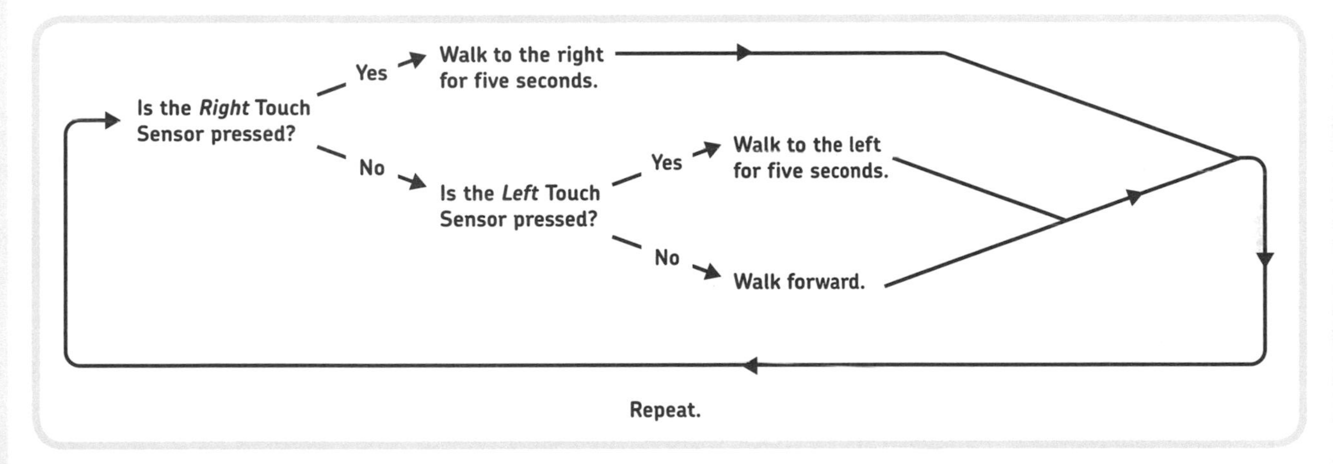

Figure 9-10: The program flow for the Strider-Touch program. While walking sideways for five seconds, the robot also makes sounds and displays text on the NXT screen to say which direction it is going (left or right).

creating the program

Start a new program, and save it as **Strider-Touch**. Then follow the instructions in Figure 9-11 and Figure 9-12 to create the program.

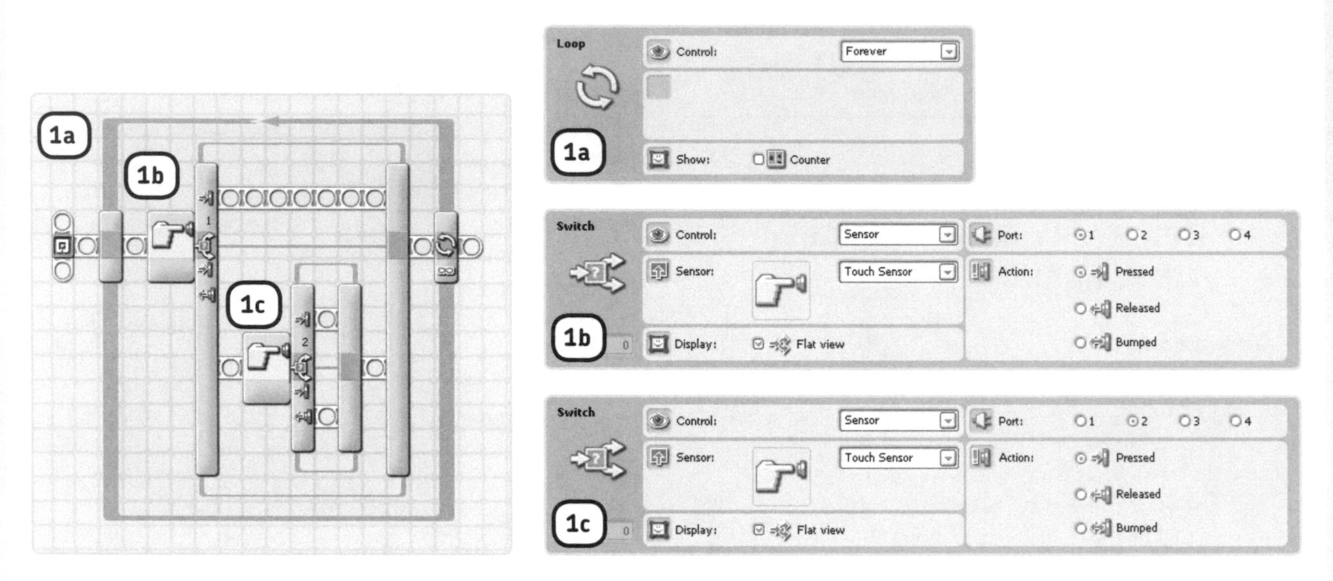

Figure 9-11: ***Step 1:*** *These blocks form the program's main structure. The program should keep running until you manually abort it, so you use a Loop Block. Inside the loop, you place two Switch Blocks to tell whether a sensor is pressed.*

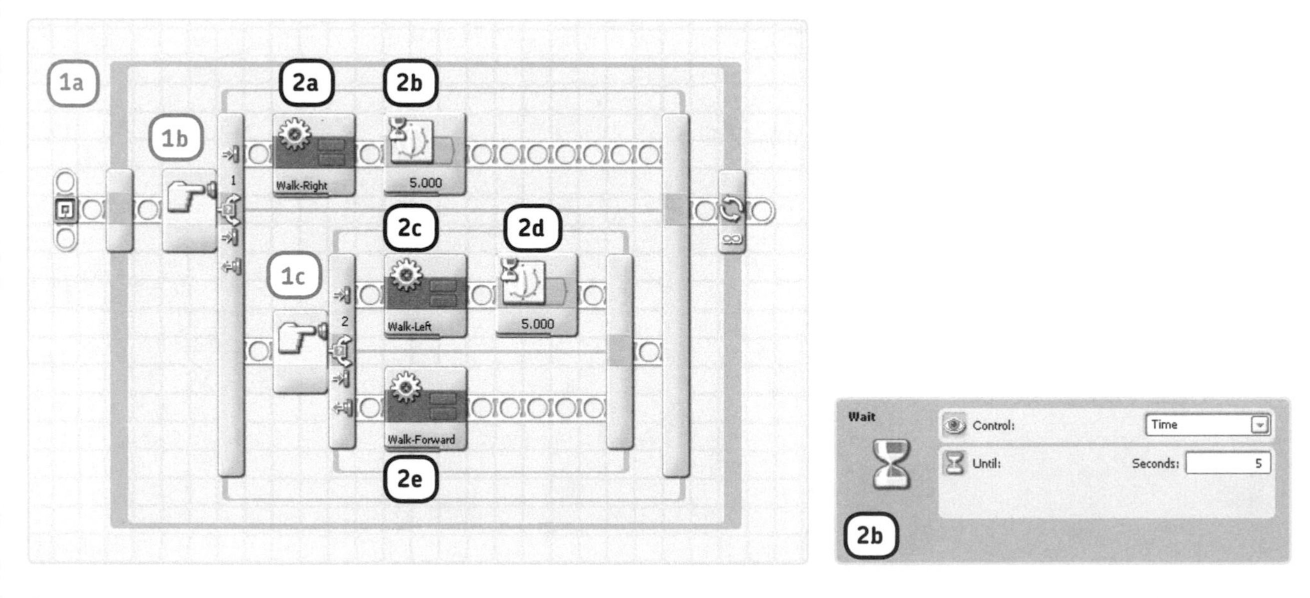

Figure 9-12: ***Step 2:*** *You configure the robot's movements by placing your My Blocks in the Switch Blocks. For example, when you press the Touch Sensor connected to port 1, the Strider robot should start to walk to the right and (because of the Wait Block) keep going in this direction for five seconds. Next, it should go back to the beginning of the program to see whether a sensor is pressed. If no sensor is pressed at that point, the Strider should continue to walk forward.*

The blocks you configured in steps 1 and 2 form the basis of this program. At this point, download the program to the Strider to test it. To make the program more fun and interactive, you'll add Display, Sound, and Color Lamp Blocks in steps 3, 4, and 5 (Figures 9-13, 9-14, and 9-15).

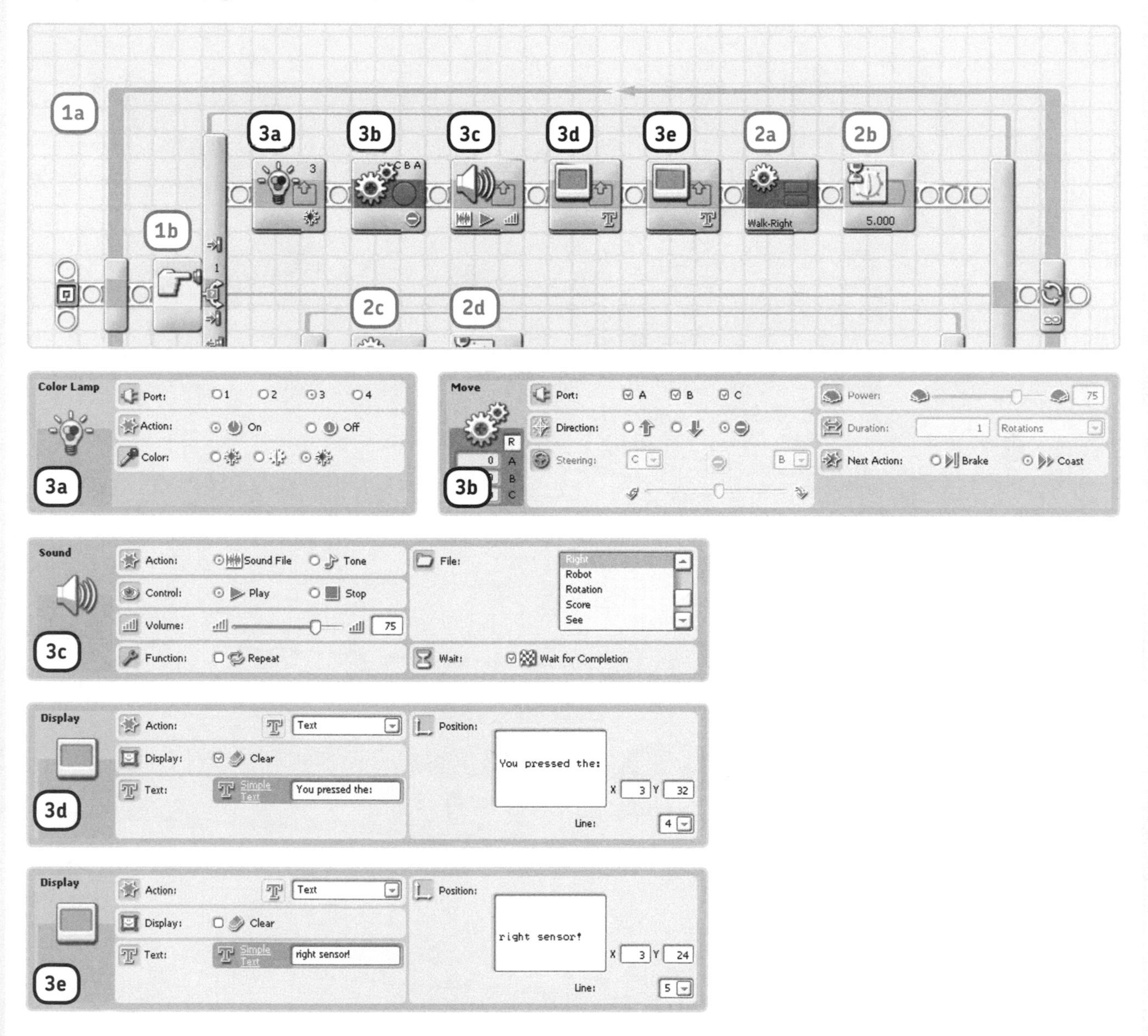

Figure 9-13: ***Step 3:*** *Here you add the blocks that should be run when the right Touch Sensor is pressed. When the sensor is pressed, a blue light should be switched on, the robot should stop walking, and it should say "Right" and display "You pressed the right sensor!" on the NXT screen. Next, the Strider robot should walk to the right because of the Walk-Right My Block that you placed earlier.*

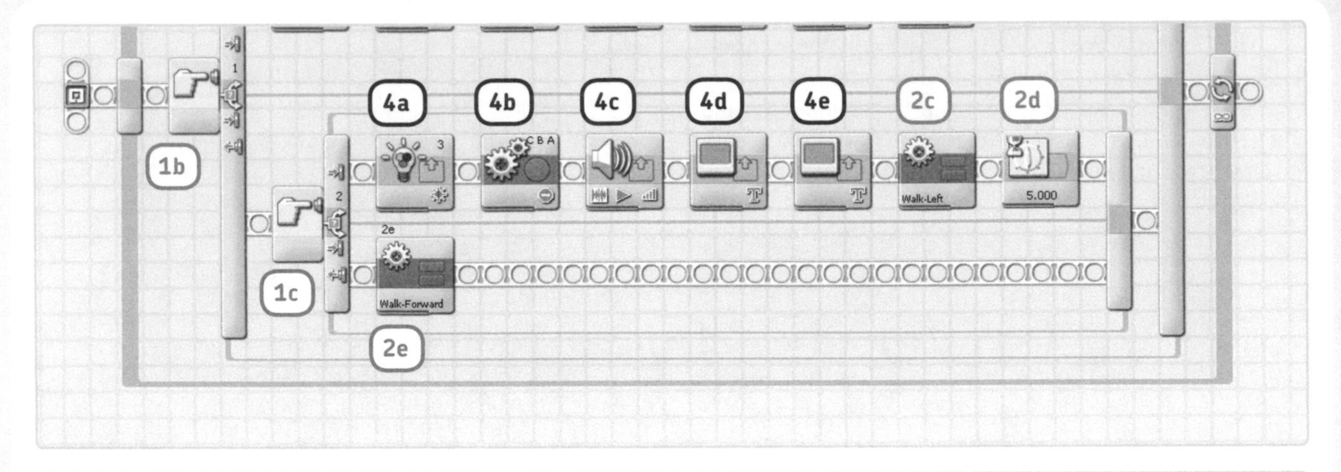

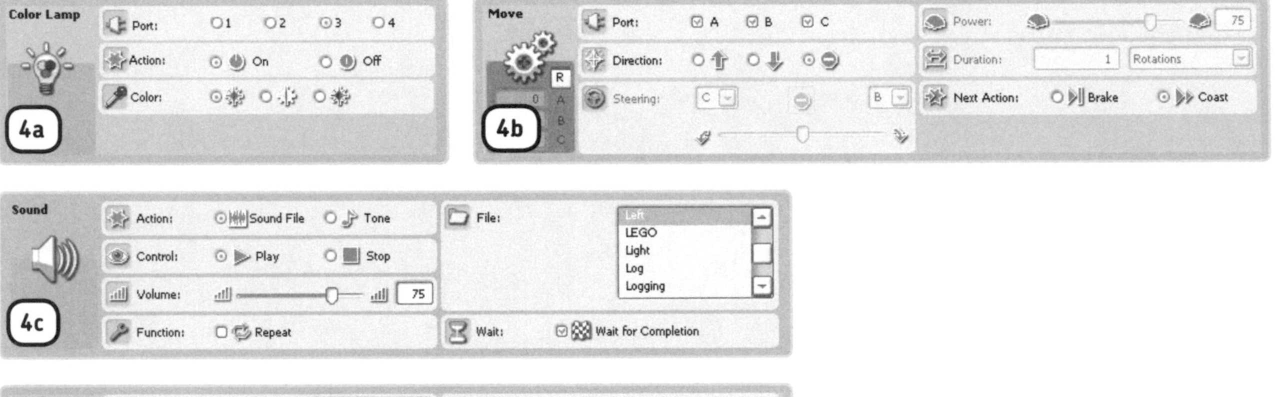

Sound
Action: Sound File Tone
Control: Play Stop
Volume: 75
Function: Repeat
File: Left LEGO Light Log Logging
Wait: Wait for Completion
4c

Display
Action: Text
Display: Clear
Text: Simple Text You pressed the:
Position: You pressed the: X 3 Y 32
Line: 4
4d

Display
Action: Text
Display: Clear
Text: Simple Text left sensor!
Position: left sensor! X 3 Y 24
Line: 5
4e

Figure 9-14: ***Step 4:*** *The blocks you place here are similar to those in step 3, except that they are run when the left Touch Sensor is pressed. Consequently, the Sound and Display Blocks are configured to tell that the robot is walking to the left.*

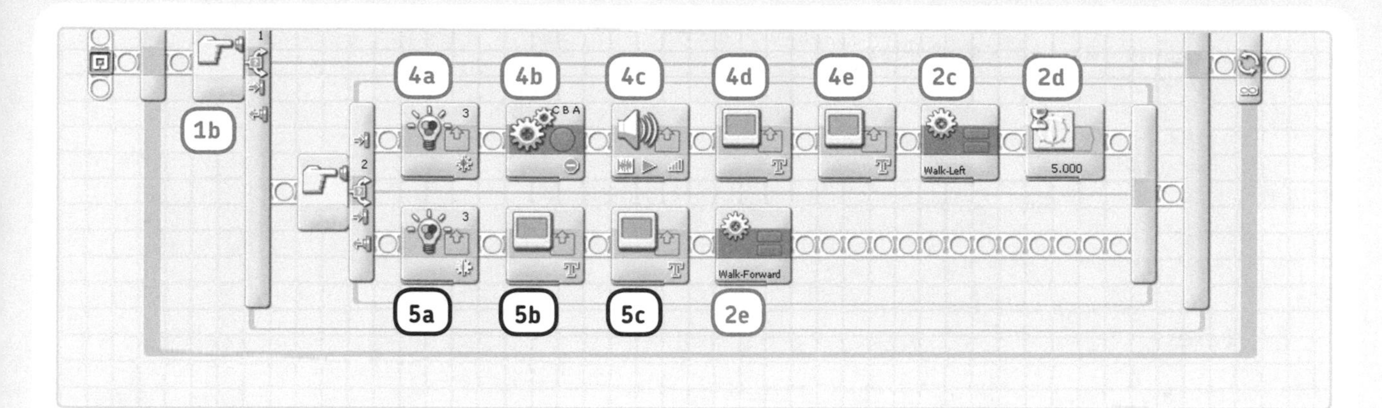

Figure 9-15: ***Step 5:*** *The blocks here run when no antennas are touched, so the NXT screen will display "You pressed no sensors!" After configuring the Strider to walk forward with the Walk-Forward My Block, the program goes back to see whether any sensors are pressed, so you don't need a Wait Block here.*

Congratulations, you can now download and run the program to make Strider walk!

DISCOVERY #44: WALKING IN SIX DIRECTIONS!

Difficulty: Easy
Take a look at the Walk-Forward My Block. Can you create a new My Block that uses the same Motor Blocks but that flips all motor directions? Which way does Strider walk now? Do the same thing for the other two My Blocks so that Strider can walk in six different directions. Once you've created the extra My Blocks, build a program that uses each of them.

creating the scared strider program

In the next program, you'll have Strider walk forward until someone turns on a light, at which point it will shout and immediately sit still until the light is switched off. But before you create this program, you need to learn two new programming tricks: Feedback Boxes and thresholds.

polling sensors with feedback boxes

The Configuration Panels of certain blocks contain a section with a fixed value called *Feedback Boxes*. Feedback Boxes (as shown in Figure 9-16) display sensor values when the robot is connected to the computer, either through USB or through Bluetooth.

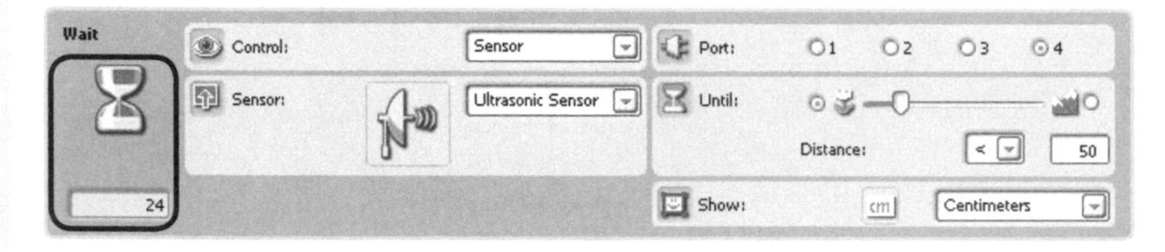

Figure 9-16: A Feedback Box reports sensor values. The Wait Block Configuration Panel shown here is configured to poll the value of the Ultrasonic Sensor.

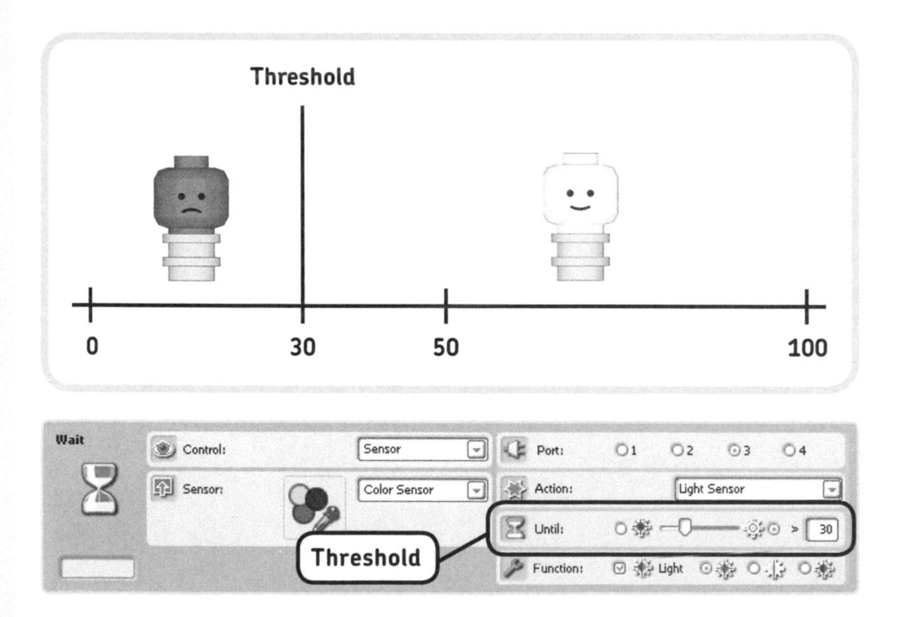

Figure 9-17: You use a threshold (trigger value) to set the edge between light and dark. You set this value in the Configuration Panel of a block set to work with a sensor, such as the Wait Block shown here. In this case, any value greater than 30 will be considered light, and values 30 and less will be dark. As a result, this block makes the robot wait until the Light Sensor reports a value greater than 30 before the program continues.

setting thresholds

It's easy to tell whether a Touch Sensor is pressed or not because it's either pressed or released, but how can you use a Light Sensor to tell when it is dark or light? Unlike a mechanical sensor that is either switched on or switched off, light can have many measured values. On the NXT, the values from the Color Sensor in Light Sensor mode range from 0 (darkest) to 100 (brightest). To tell your robot what you consider to be light or dark, you'll define light and dark in your program using a trigger value, also known as the *threshold* (Figure 9-17). You'll consider a measured light sensor value greater than this threshold as *light* and a measurement lower than the threshold as *dark*.

Threshold values are unique to every situation and will likely vary depending on the light conditions in the room. To set the threshold value, you begin by deciding how you want to define the light conditions in your program (a room with the lights on and the same room with the lights off), and next you measure the light value in each case.

Because the NXT's View menu won't let you poll the Light Sensor, you'll use a Wait Block's Feedback Box configured as a Color Sensor in Light Sensor mode (Figure 9-18) to poll it. For dark conditions, I measure a value of 4, and for bright conditions I measure a value of 30 with the Feedback Box method. Your measurements may vary.

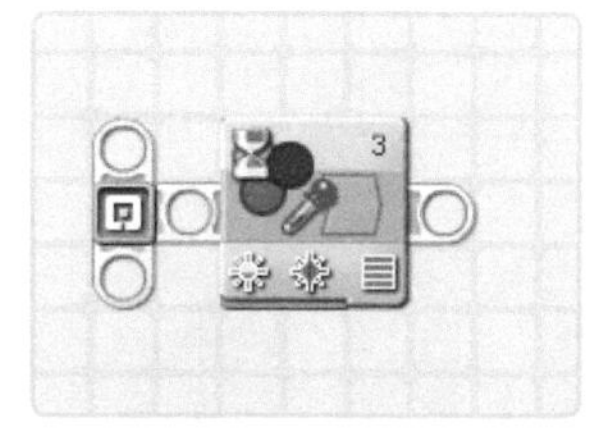

Figure 9-18: Using the Feedback Box to poll the Light Sensor value

Once you have the two measured values (one for the dark room and one for the light one), you calculate the average of the two, as shown in Figure 9-19. Your measured values may differ; be sure to calculate your own thresholds using your own measurements.

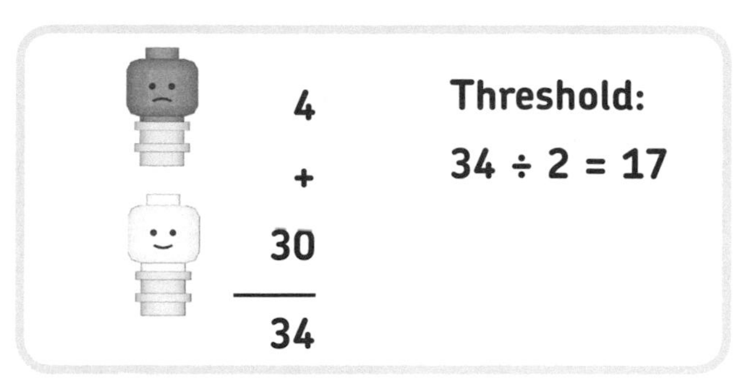

Figure 9-19: The threshold is the average of the Light Sensor value found in the dark room (a low number) and the one found in the room with the lights on (a bigger number). To calculate the average, add both values, and divide the total by 2.

creating the program

You'll now create the program that has the Strider robot move forward, shout when a light comes on, and stop moving until the light turns off. Create a new program, store it on your computer as **Strider-Scared**, and follow the instructions shown in Figure 9-20.

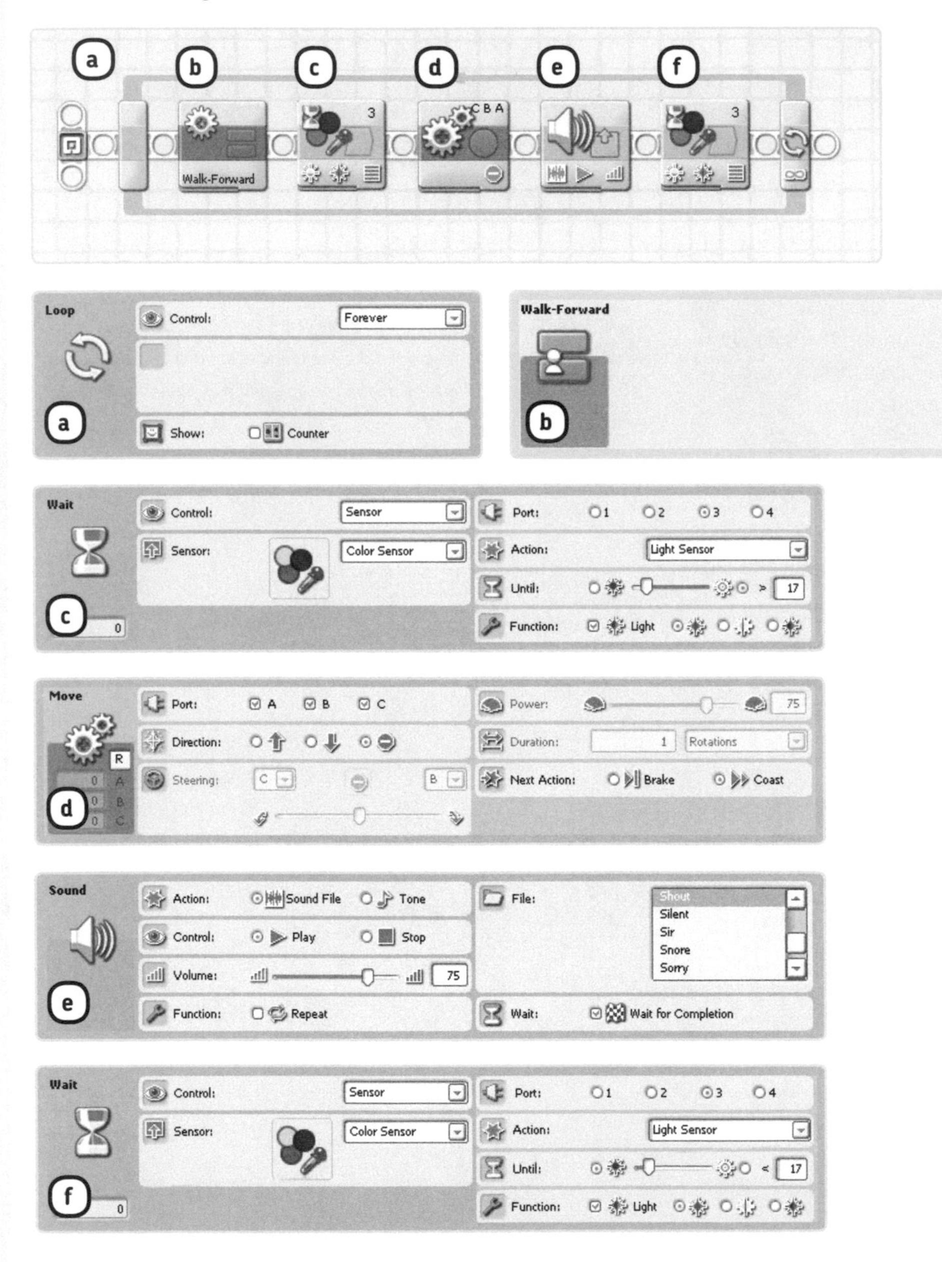

Figure 9-20: After making the Strider move forward, the program waits until the measured light intensity exceeds the threshold (set to ***17*** *here). Once the threshold is exceeded, all motors stop, and the robot makes a loud sound. The last block waits until the sensor sees that it is dark again, at which point the loop restarts and Strider starts walking again.*

DISCOVERY #45: WALKING AT THE SPEED OF LIGHT!

Difficulty: Hard
With your My Blocks, Strider walks at the speed specified in the Motor Blocks, but you can make it walk faster or slower by configuring these blocks differently. Create a program that makes the robot walk faster, based on its Light Sensor readings. If the Light Sensor value is less than 33, make Strider go slow; if this value is between 34 and 66, make it walk faster (around 50 percent motor power); and if the value is greater than 66, make it run as fast as possible without breaking its legs! Use Switch Blocks to determine the Light Sensor value. You could point a flashlight at the sensor from a distance to make the robot go faster.

further exploration

It's not easy to design walking robots, so if you want to create your own walker, try building another one according to a set of directions first to get a better sense of how walkers are designed. On the back of the LEGO MINDSTORMS NXT 2.0 box, you'll find a small picture of Manty, another six-legged creature of mine. This robot's walking technique is very different from Strider's because Manty has two sets of three legs and because it can turn in place. You can download building and programming instructions for this NXT 2.0 robot from *http://www.discovery.laurensvalk.com/*.

Now, before you move on to your next robot, take the time to improve your building and programming skills in Building Discoveries #8 and #9 and Discovery #46.

BUILDING DISCOVERY #8: TIRED OF WALKING?!

You may have noticed that Strider's base structure is a solid triangle, an uncommon construction for a LEGO MINDSTORMS robot. What happens if you put this triangle on wheels? Remove the legs from the motor compartments, and attach wheels (without the rubber tires) to them, as shown in Figure 9-21. Try to run the programs you made for the Strider robot. Do they still work? What happens if you attach the rubber tires?

DISCOVERY #46: REMOTE CONTROL!

Difficulty: Medium
Do you remember the remote control you made for the Shot-Roller? In this discovery, you'll do something similar with the Strider. Remove the antennas from the Strider robot, and use the Touch Sensors with long cables as remote buttons. How do you program Strider to walk in multiple directions?

BUILDING DISCOVERY #9: A PAIR OF EYES IN THE BACK!

When you built the Discovery robot with the two bumpers in Chapter 7, it almost never got stuck because it could always turn away from the walls it ran into. This is not the case for Strider because it cannot turn around. To make sure the robot never gets stuck, Strider needs to be able to sense the walls from any direction. It can already see things ahead of it with the Ultrasonic Sensor. Now, remove the antennas from the Strider, and create special bumpers on the two other sides of Strider. Can you use Touch Sensors to build bumpers so that this robot will not crash into objects?

Figure 9-21: Strider on three wheels

PART III

creating advanced programs

10 using data hubs and data wires

In this third part of the book, you'll learn how to use data hubs and data wires to create more advanced programs for your robots. For example, you'll use data hubs and data wires to create programs that display sensor readings on the screen (Chapter 10), do mathematics with the NXT (Chapter 11), or remember things such as high scores of games (Chapter 12).

In earlier chapters you configured each programming block by entering the desired settings in the Configuration Panel. One of the fundamental concepts in this chapter is that blocks can configure each other. For example, one block can instruct a Motor block to run a motor at a certain power level. Blocks transfer information such as the power level using *data hubs* and *data wires*. To make this a bit more tangible, think of this as like a person who uses a (wired) phone to ask someone to set the radio to half of its maximum volume, as shown in Figure 10-1. The two people here represent programming blocks. The actual phone is the data hub, and the phone wire is the data wire.

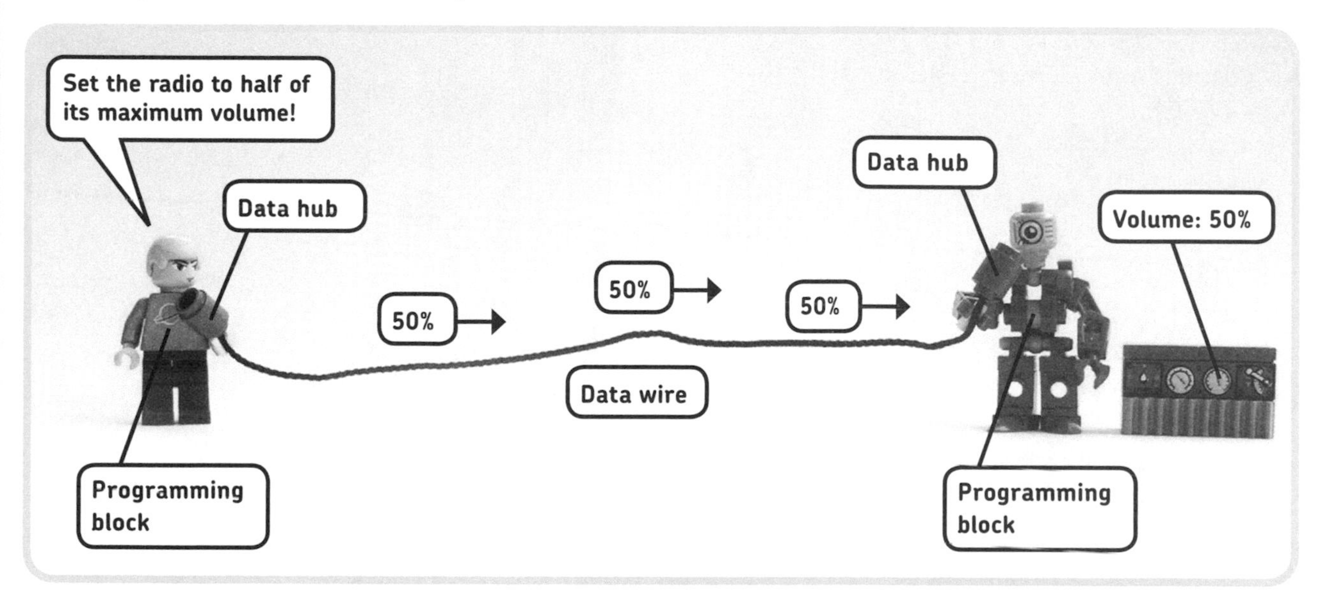

Figure 10-1: You use data hubs (the phones) and a data wire (the phone wire) to carry a value (the radio volume) from one programming block (the person on the left) to the other (the person on the right). The second block uses the value to turn on the radio at the requested volume level.

The "block" on the left basically passes the desired radio volume to the "block" on the right, which can then set the radio to the appropriate volume. For the volume value to move from the first to the second programming block, it goes from the block through a data hub (the phone on the left), through a data wire (the phone wire), and through another data hub (the phone on the right); when it reaches the second block, it specifies its (volume) setting.

To summarize this concept, data wires carry values from one block to another to configure one of this block's settings. A data hub, which is part of a programming block, allows a block to pass values into the wire (the data hub on the left of

Figure 10-1) and retrieve values from it (the data hub on the right of Figure 10-1).

This chapter teaches you how to make programs that use data hubs and data wires. You may find this a bit difficult at first, but as you go through the sample programs and the discoveries, you'll master all of these programming techniques!

building SmartBot

To make it easier to learn all these new features, you'll build a small platform with two motors, some sensors, and the NXT, called SmartBot, as shown in Figure 10-2. This robot will help you better understand how the advanced programs that you'll make really work.

Now build SmartBot by following the instructions on the next few pages, but first select the required pieces, as shown in Figure 10-3.

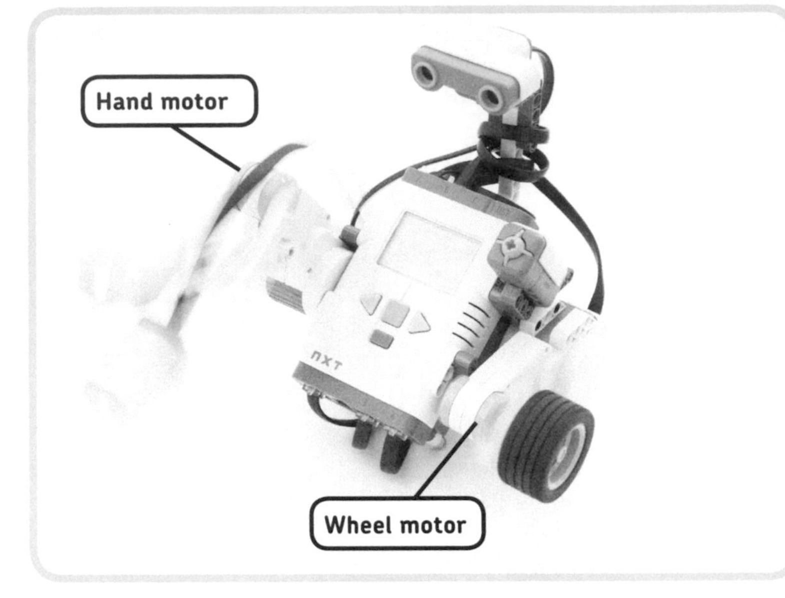

Figure 10-2: You'll use the SmartBot robot to learn many new programming techniques. I'll refer to the motor with the Color Sensor as the Hand motor *and the one with the wheel as the* Wheel motor.

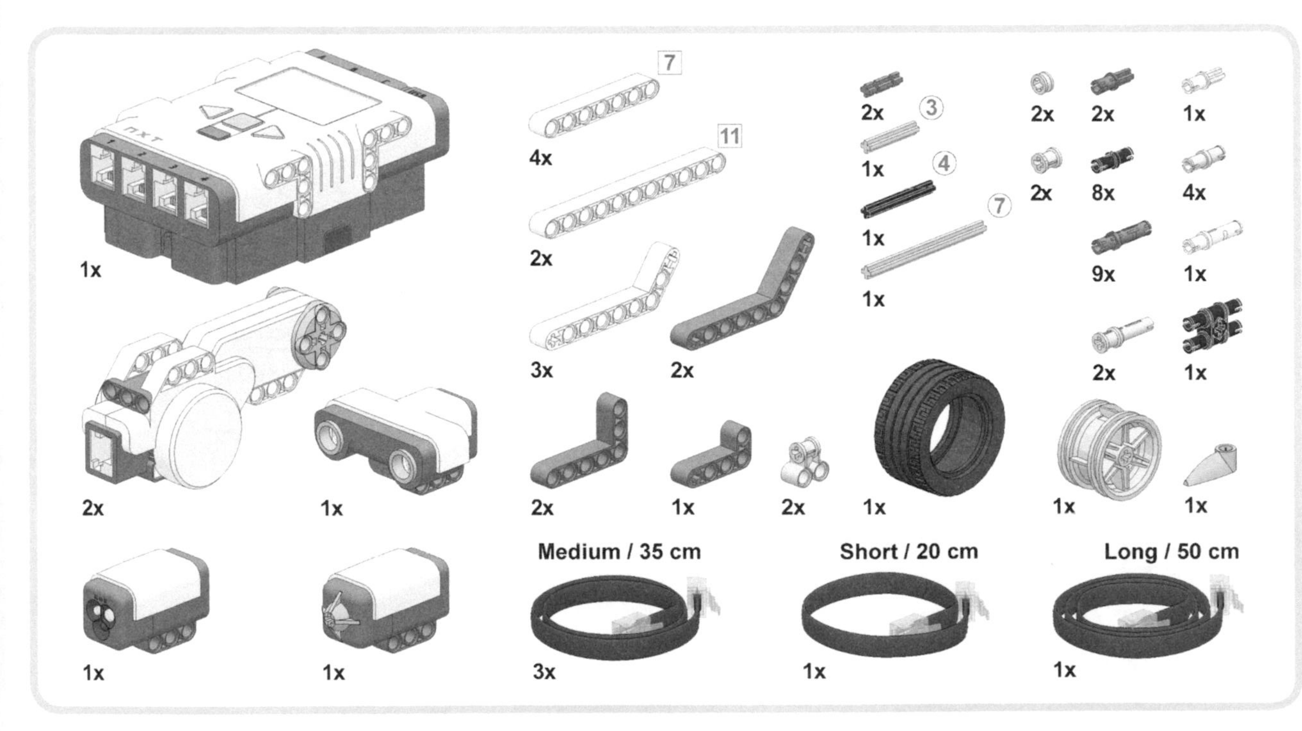

Figure 10-3: The required pieces to build SmartBot

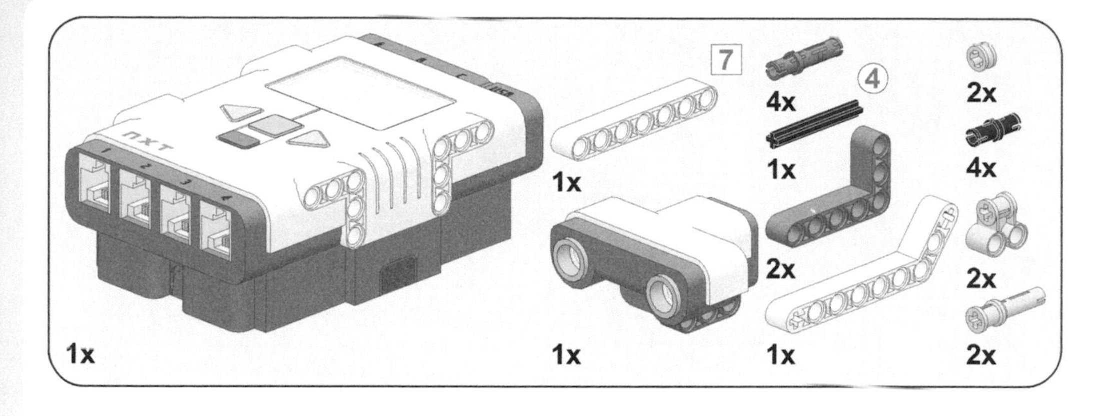
7
4x
4
2x
1x
4x
1x
1x
2x
2x
1x
1x
1x
2x
1x
1x
2x

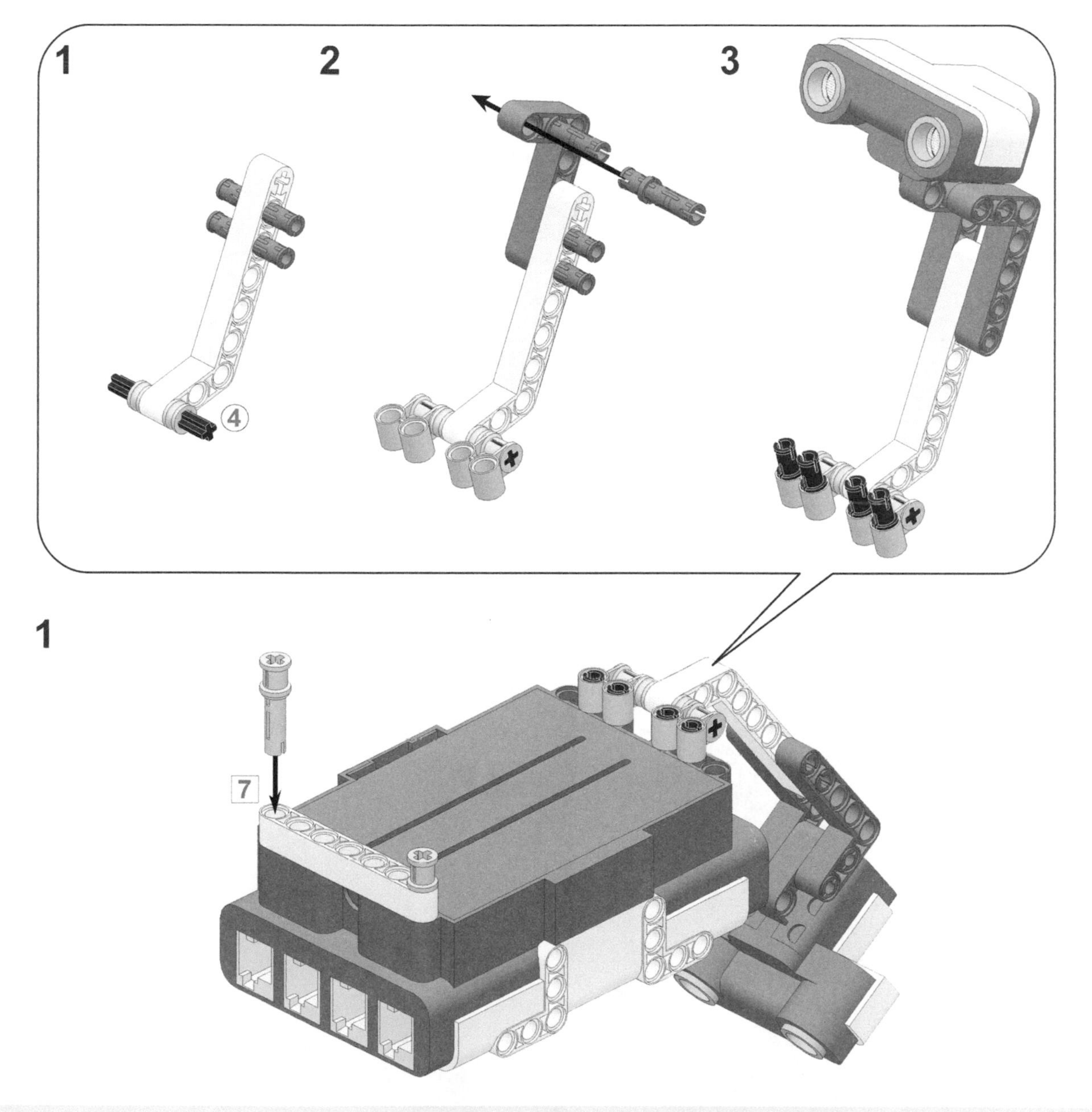
1
2
3
4
1
7

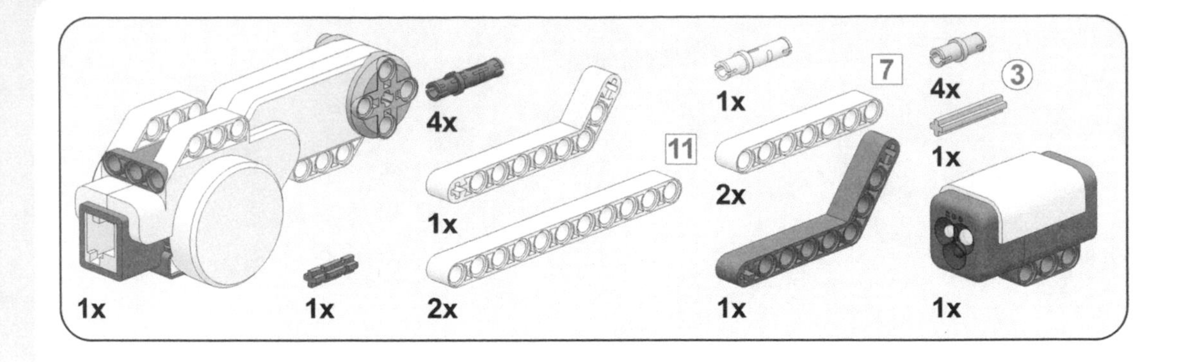
1x
4x
1x
2x
1x
1x
7
11
2x
1x
4x
3
1x
1x
1x

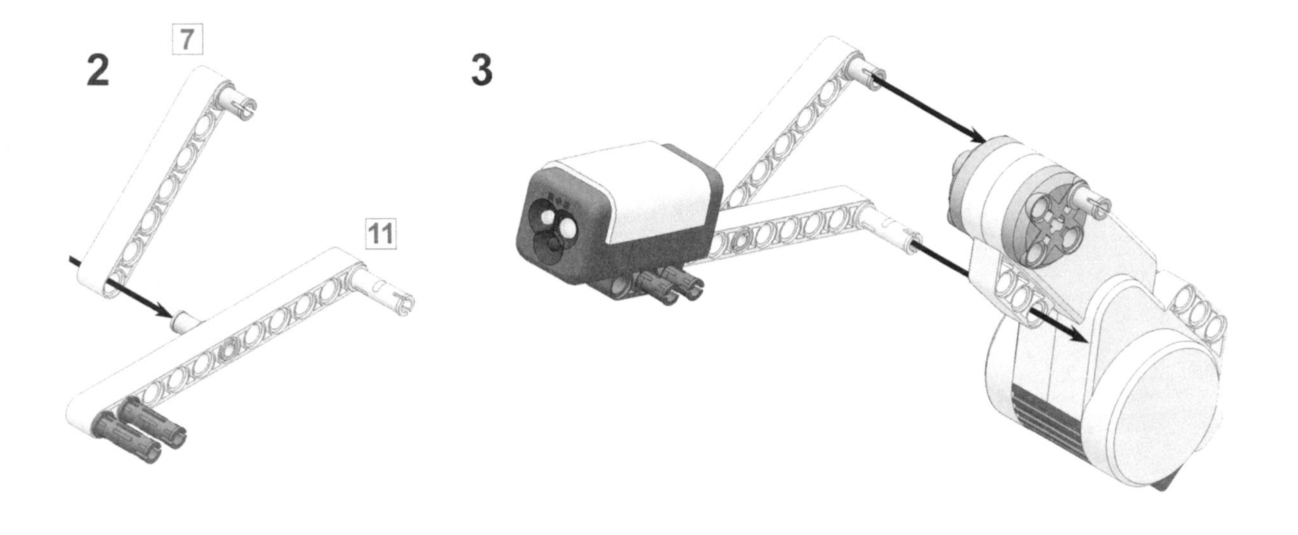
2
7
11
3

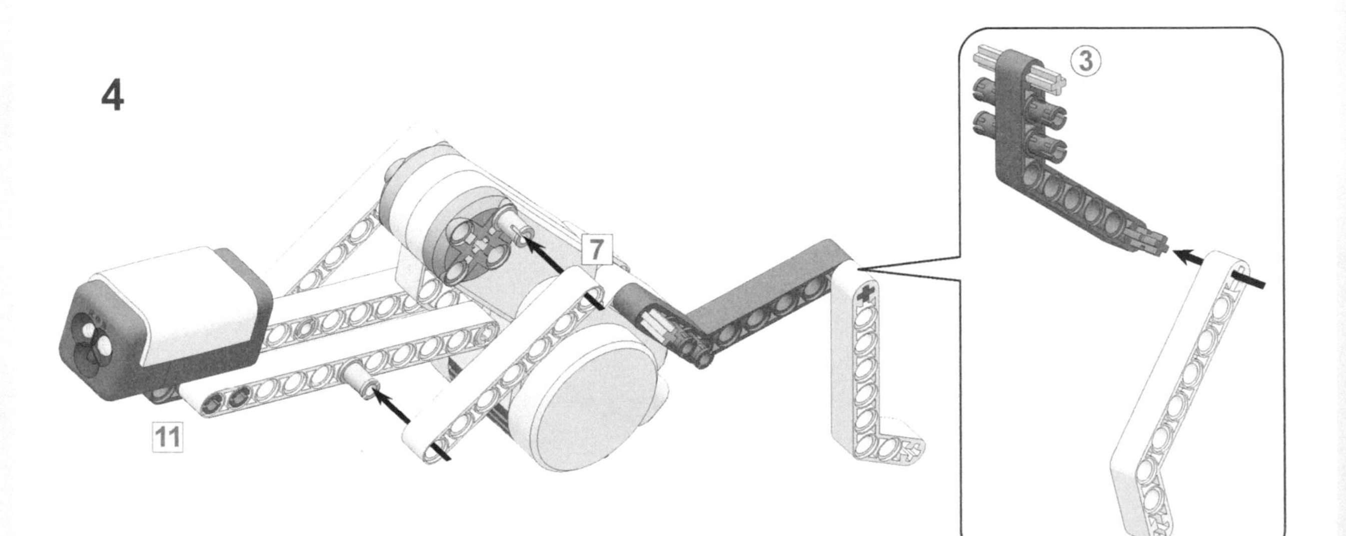
4
7
11
3

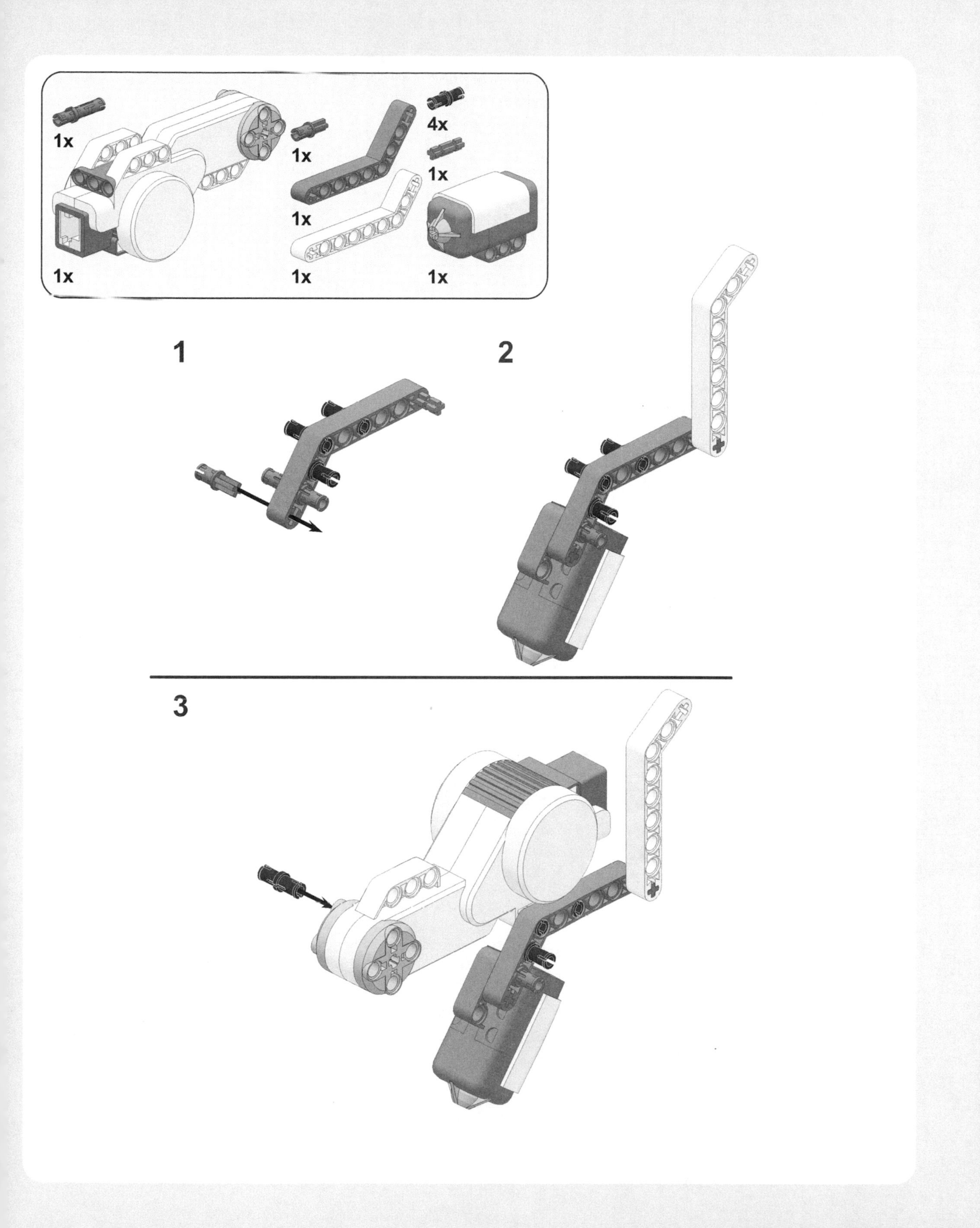
1x
1x
4x
1x
1x
1x
1x
1x
1x
1
2
3

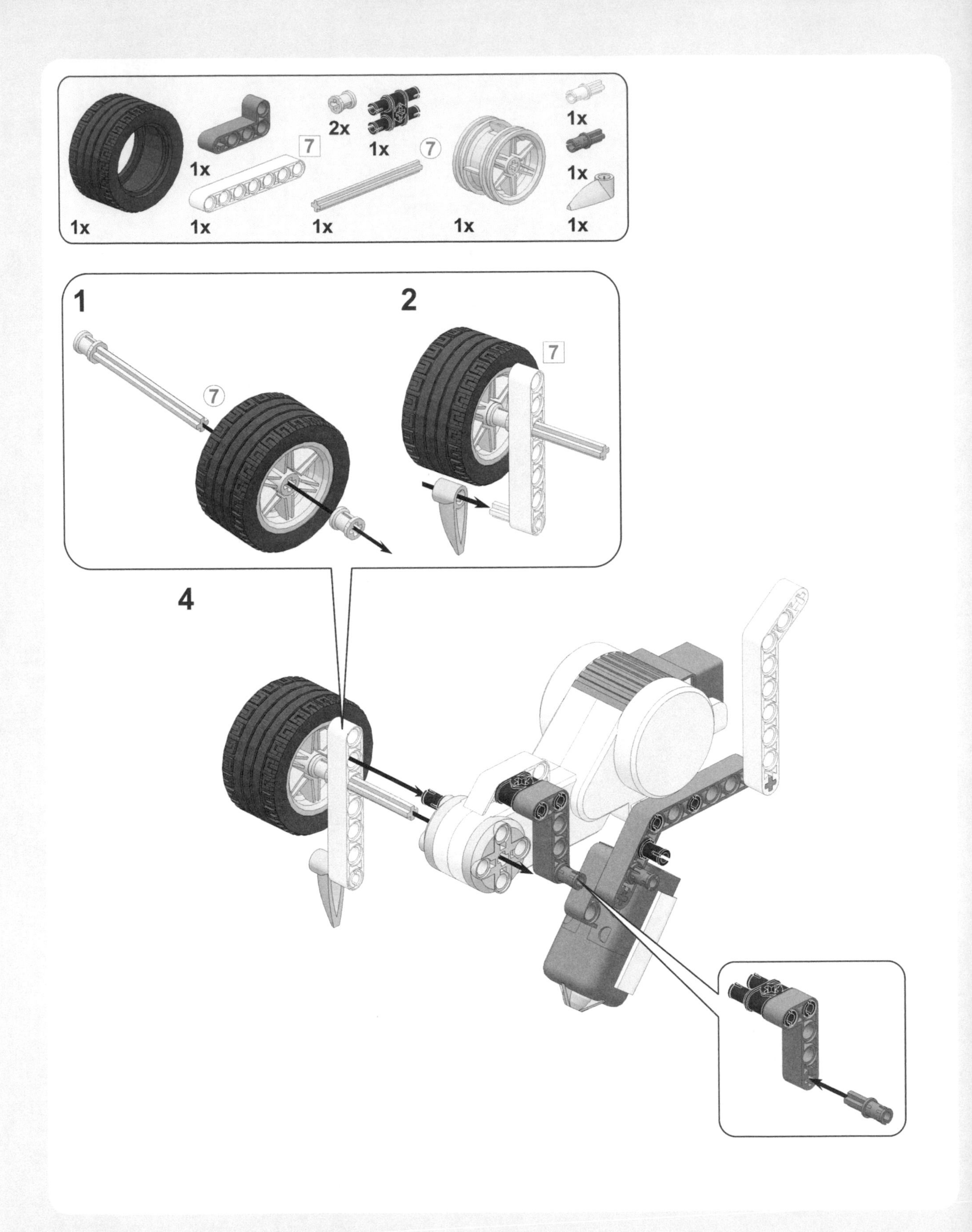
1x
1x
2x
1x
7
1x
7
1x
1x
1x
1x
1x
1x
1
2
7
7
4

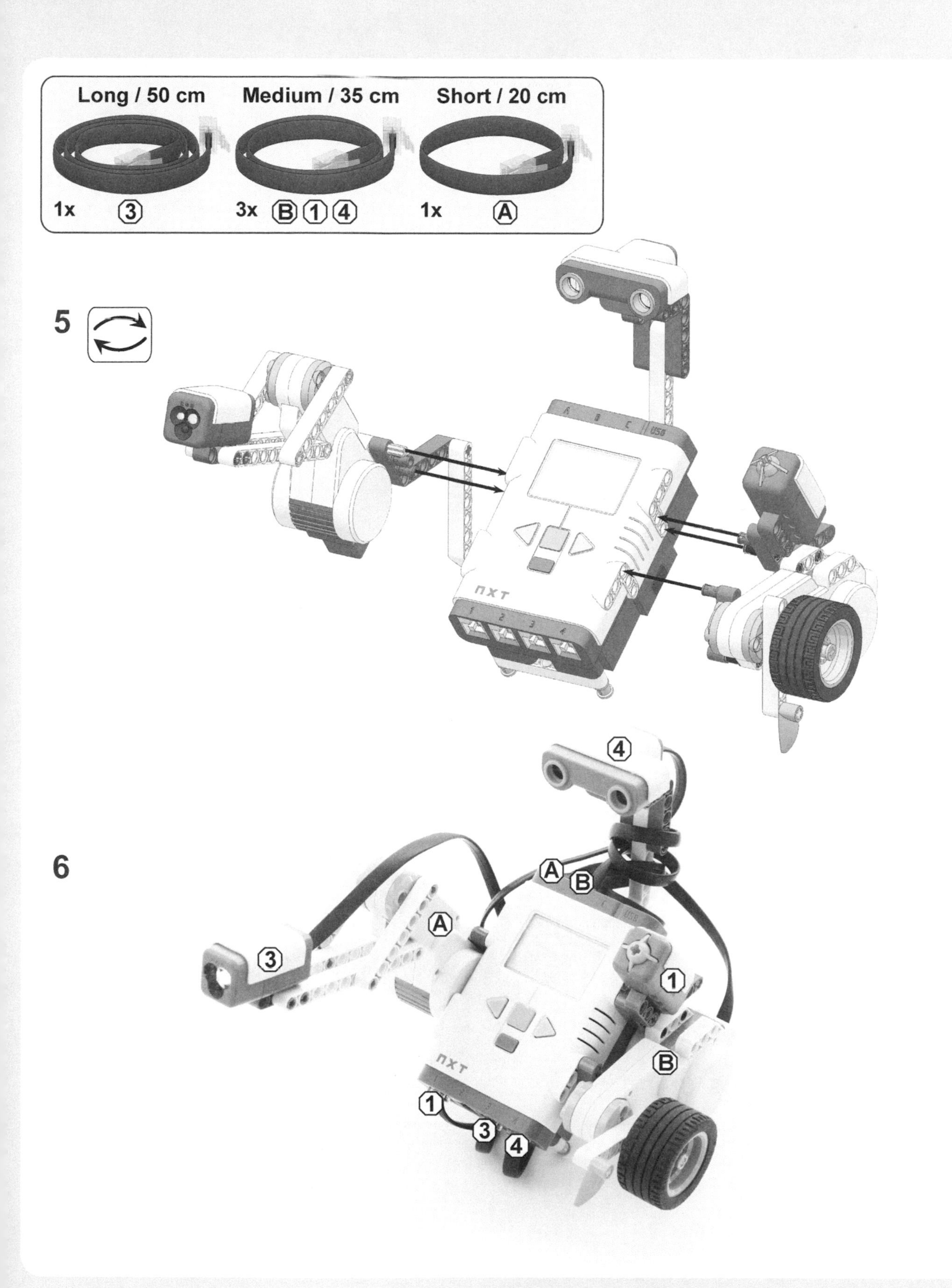
Long / 50 cm
Medium / 35 cm
Short / 20 cm
1x 3
3x B 1 4
1x A
5
6
4
A
B
A
3
1
B
1
3
4

a program to get started with data wires

To see how data hubs and data wires work, you'll create a small program that will make the SmartBot play a sound and then make the Hand motor rotate for three seconds. The motor's power level (and thus its speed) will respond to what the Ultrasonic Sensor sees: If the sensor sees something 43 cm away, the motor's power level will be 43 for three seconds; if the sensor reads 15 cm, the motor power will be 15; and so on. For example, if you keep a book close to the sensor and run the program, the hand should go up and down slowly, but move faster if you run the program again with the book farther away.

To accomplish these actions, you'll use the Ultrasonic Sensor block. You'll learn more about this block later, but for now just know that you use it to poll the sensor. Create the **Smart-Intro** program as shown in Figures 10-4 through 10-6, and then run the program to see how it works.

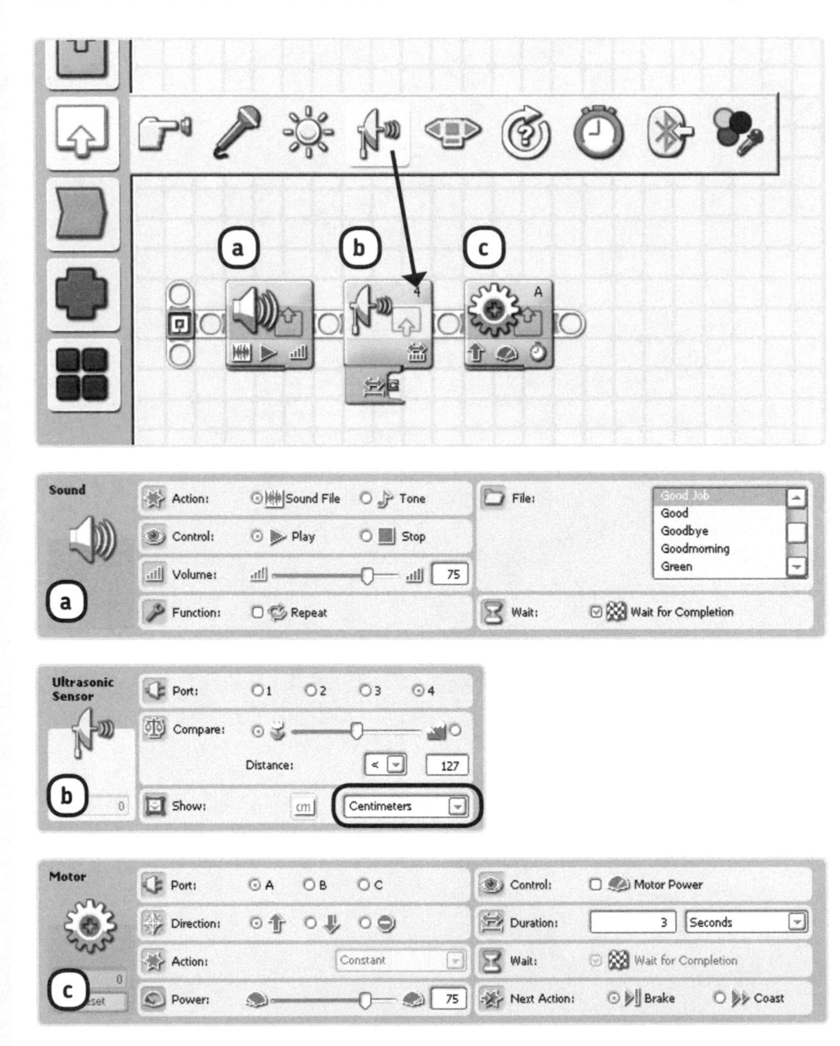

Figure 10-4: ***Step 1:*** *Place all the necessary blocks for the Smart-Intro program on the Work Area, and configure them as shown here.*

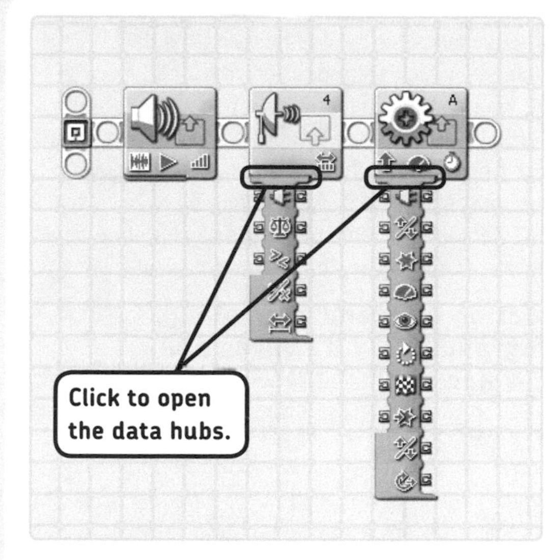

Figure 10-5: ***Step 2:*** *Open the blocks' data hubs by clicking the tabs at the lower-left edge of the block. You should already see a small data hub just below the Ultrasonic Sensor block, but you can open the complete data hub by clicking the same tab.*

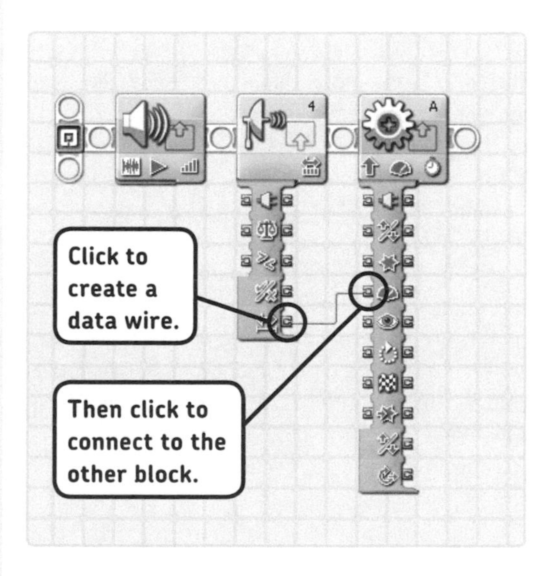

Figure 10-6: ***Step 3:*** *Create and connect the yellow wire as shown here. The yellow line is the data wire.*

Download the program to SmartBot, and run it while keeping a book about 20 cm (8 inches) away from the Ultrasonic Sensor. Next, run the program again with the book about twice as far from the robot. You should notice that each time you run the program, the Hand motor turns at a different speed.

understanding the sample program

Congratulations! You've just created your first program with data wires. Now you'll learn exactly how your program works by analyzing the function of each block.

The first Sound block simply plays a sound. Once the sound finishes playing, the Ultrasonic Sensor block polls the sensor once, resulting in a reading of, say, 35 cm. The yellow data wire then *carries* the sensor measurement to the other end of the wire to the Motor block, which then makes the Hand motor turn for three seconds, as specified in its Configuration Panel. The motor power (and thus its speed) depends on the value carried by the data wire; it's 35 in this case. Figure 10-7 shows an overview of what happens.

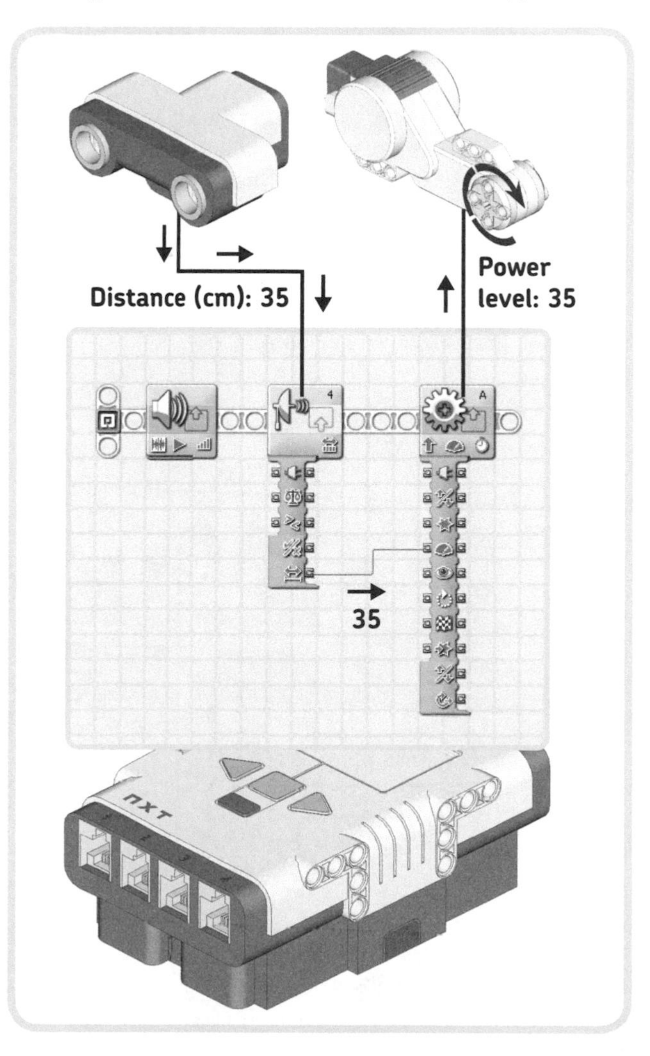

Figure 10-7: An overview of the Smart-Intro program. The Ultrasonic Sensor block polls the sensor and sends the sensor value through a data wire to the Motor block, which uses this value to set the motor speed.

how do data hubs and data wires work?

You'll now analyze some of the new features that you used in the program you just made. As you can see, you use a data wire to carry information between blocks. In the example program, the yellow data wire carries the sensor value to the Motor block to set the motor's power level.

You created the data wire by first clicking one of the data plugs on the data hub of the Ultrasonic Sensor block. Each data plug carries out a different value, but you connected your data wire to the Distance data plug, because you wanted to know the measured distance. Next, you connected the other end of the wire to the Power data plug on the hub of the Motor block. As a result, you actually reconfigured the Power setting of this block, and the motor moved at a speed based on the sensor reading.

You'll now take a more detailed look at the data hub. Open a new program, pick a Motor block from the Programming Palette, and place it on the Work Area. Then, open the block's data hub, as shown in Figure 10-8.

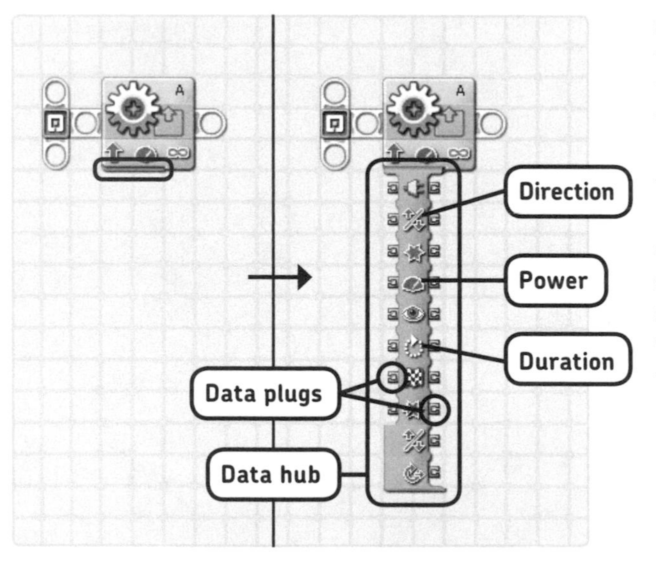

Figure 10-8: To open a block's data hub, click the tab at the lower-left edge of the block. To close the hub, click the tab again. Move your mouse pointer over the data plugs to figure out what each plug stands for. (I've shown only a few examples here.)

As you mouse over the various plugs on the data hub, you should see which setting is reconfigured when you connect a wire to this plug. For example, as shown in Figure 10-8, you'll see plugs for Direction, Power, and Duration. You'll also find these terms on the block's Configuration Panel. In other words, when wiring data wires into a block's data hub, you're actually reconfiguring the block's settings as if modifying the settings in the Configuration Panel.

creating a second example program with data wires and data hubs

Now you'll create a program to make the SmartBot's motor accelerate. It will start out slowly and increase its speed until it reaches the maximum power level.

To make this program, you place a Motor block that makes the motor turn for half a second inside a Loop block. Because the Motor block is inside the loop, the motor keeps spinning. You'll also use a new feature of the Loop block that makes it count the number of times it has repeated the Motor block inside it, called the *loop count*.

Because the Motor block keeps looping, the loop count increases over time, and you use this value as an input for the motor power. When you start the program, the loop count is 0, which makes the motor run at zero speed (it stands still) for half a second (as specified by the Duration setting). When the Loop block returns to the beginning, it repeats the Motor block; the loop count increases to 1, and therefore the motor's power level is 1. When it repeats, the motor speed is 2, and so on. Follow the instructions in Figures 10-9 and 10-10 to create the **Smart-Accelerate** program.

Download this program to your robot, and run it. If you configured everything correctly, the SmartBot's hand should start moving slowly, and as time passes, the Hand motor's speed should increase. The hand should stop accelerating when the loop count exceeds 100, because 100 is the motor's maximum power level.

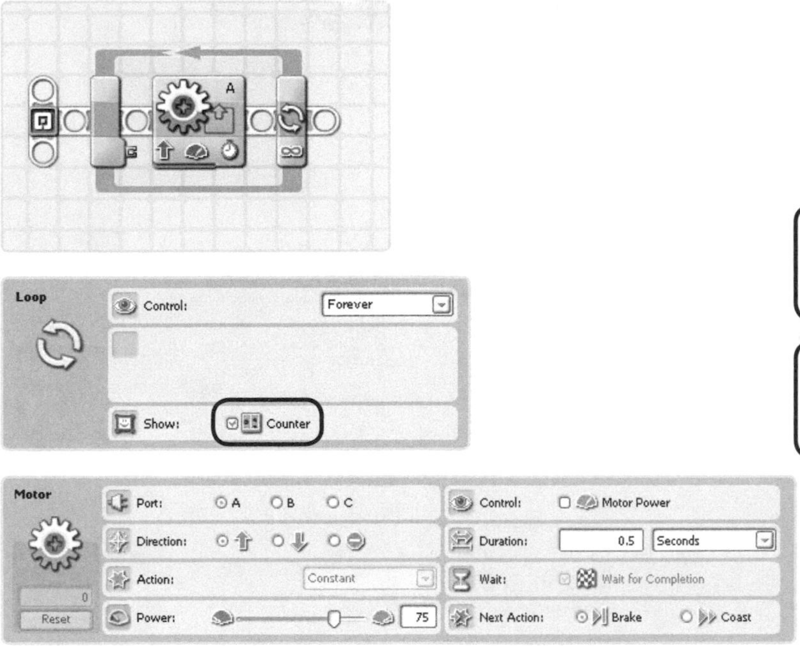

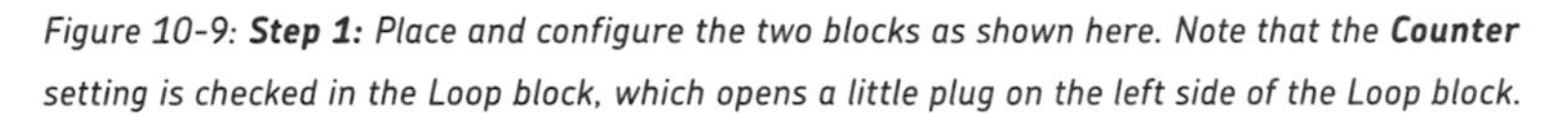

Figure 10-9: ***Step 1:*** *Place and configure the two blocks as shown here. Note that the* ***Counter*** *setting is checked in the Loop block, which opens a little plug on the left side of the Loop block.*

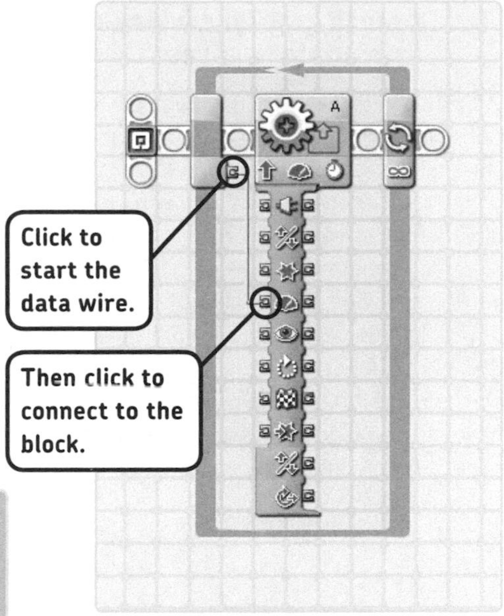

Click to start the data wire.

Then click to connect to the block.

Figure 10-10: ***Step 2:*** *Open the Motor block's data hub, and connect the data wire to the plugs as shown here.*

using data plugs: input and output

As you saw earlier, you connect data wires to data plugs on data hubs. The hubs contain two types of plugs, as shown in Figure 10-11: output plugs (on the right side) and input plugs (on the left side). Output plugs carry out a value and pass it to a data wire. For example, the Distance plug on the Ultrasonic Sensor block carries out the measured distance value. Input plugs retrieve the value from the data wire and pass it to the block it connects to so that the value can be used to reconfigure one of the block's settings. For example, you used the Power plug as an input plug.

Generally, a data wire carries information from an output plug of one block to an input plug of another block.

NOTE The Compare setting in the Ultrasonic Sensor block's Configuration Panel controls how the Yes/No output plug works. However, when using the Ultrasonic Sensor block just to poll the sensor as you did previously, you're only using the Distance plug of this block, meaning that you don't have to configure the Compare setting. (You'll learn how to use this setting in "The Logic Data Wire" on page 163.)

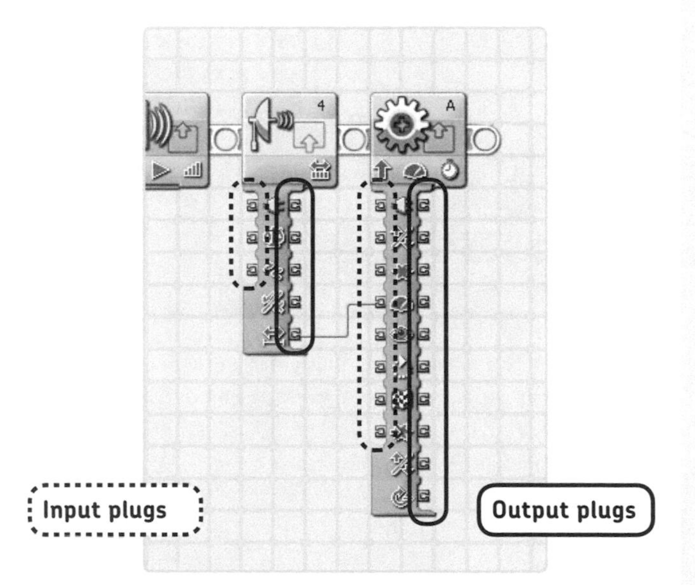

Figure 10-11: Input and output plugs on a data hub. The data wire carries information from an output plug to an input plug of another block.

block configurations when using data wires

When using blocks with data wires, what happens to the settings you configured in the Configuration Panel? In the two sample programs, you had the Motor block run a motor at a power level specified by a data wire, even though the Motor block's Power was set to 75 (Figure 10-12).

As a general rule, the data wire input overrides the setting specified in the Configuration Panel, and the setting in the Configuration Panel is ignored, as illustrated in Figure 10-12. All other settings that do not conflict with a data wire are in effect, as specified in the Configuration Panel. For example, the block here will make the motor turn forward because the Configuration Panel says so.

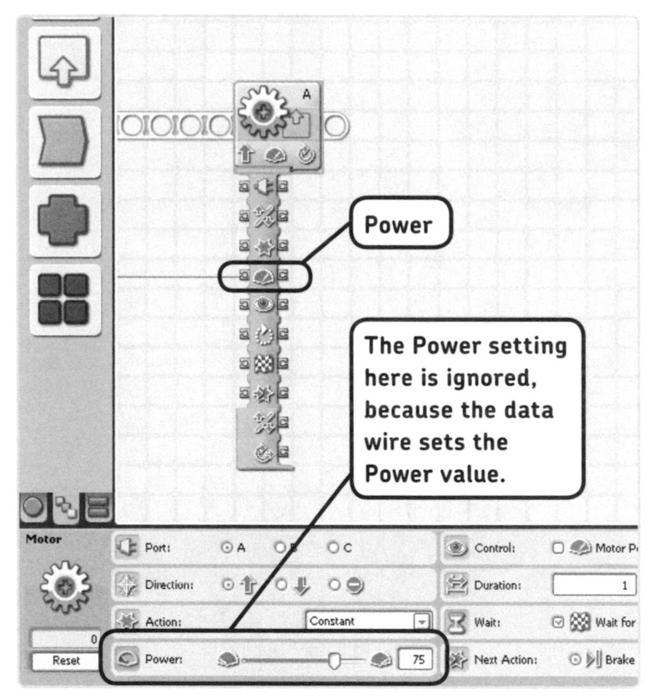

Figure 10-12: A Configuration block's setting is ignored when a data wire defines the same setting; the block will use the data wire's value.

deleting data wires

To delete a data wire, follow the instructions shown in Figure 10-13.

Figure 10-13: To delete a data wire that connects two blocks, click the data plug on the right end of the wire. If this doesn't work, you should be able to remove it by clicking the wire and then pressing the DELETE *key. Before you press this key, though, make sure that no block is selected, or you'll delete it with the data wire.*

DISCOVERY #47: GROWING CIRCLES!

Difficulty: Easy
Create the program shown in Figure 10-14. The Display block shown in the figure will display a circle in the middle of the NXT screen. You want the circle to grow bigger as you wait, so the Radius setting should increase over time. The program here is incomplete, because a data hub and a data wire are missing. Can you open the data hub and connect the data wire to finish this program?

HINT **Look for the Radius input plug.**

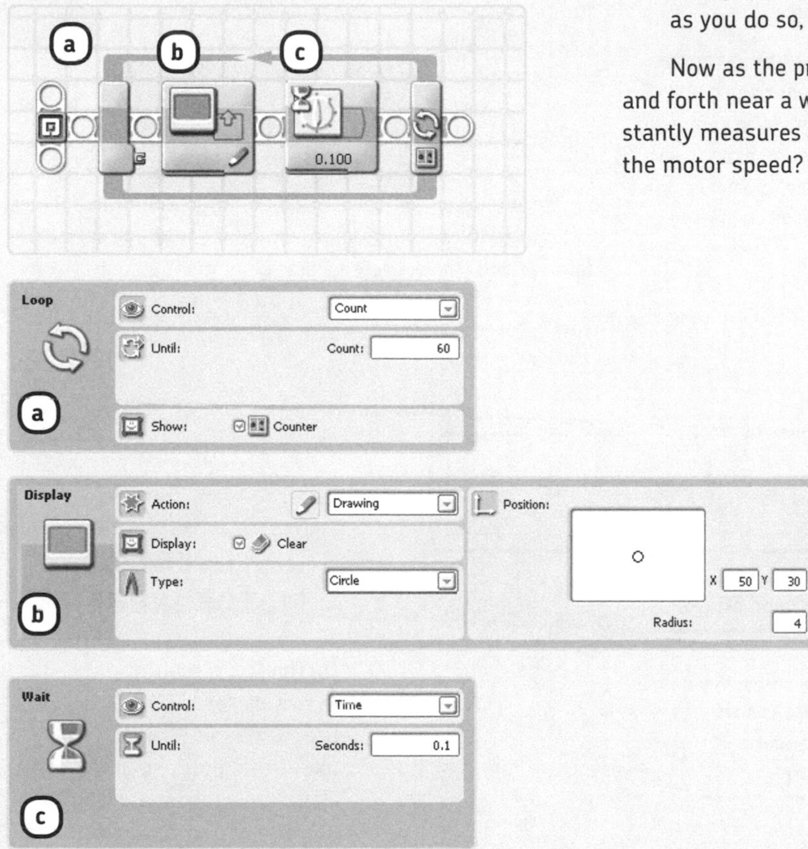

Figure 10-14: The incomplete program for Discovery #47. Can you complete it by connecting a data wire appropriately?

DISCOVERY #48: DYNAMIC SPEED!

Difficulty: Medium
The Smart-Intro program polled the Ultrasonic Sensor once and used the sensor value to set the speed of the Hand motor. You can modify this program so that the motor speed is continuously updated with a new sensor reading as follows:

1. Remove the Sound block.
2. Set the Motor block's Duration setting to **Unlimited**.
3. Place the two remaining blocks in a Loop block configured to loop forever. If the data wire breaks as you do so, reconnect it.

Now as the program runs, move the SmartBot back and forth near a wall so that the Ultrasonic Sensor constantly measures a different distance. What happens to the motor speed? Can you figure out why this happens?

sensor blocks

In Chapters 6 and 7 you learned to work with sensors by creating programs with the Wait, Loop, and Switch blocks. The final way to poll sensors is with *Sensor blocks*, as shown in Figure 10-15. These blocks are useful if you want to retrieve a sensor value and transfer it to another block with a data wire, as you saw with the Smart-Intro program.

There's a Sensor block for each sensor, including some sensors not found in the LEGO MINDSTORMS NXT 2.0 robotics kit. In the example programs so far, you've used the Ultrasonic Sensor block. Sensor blocks are of no use on their own; they must be connected to another block with a data wire in order to function.

configuring a sensor block

Configuring a Sensor block is very much like configuring a Wait, Loop, or Switch block, except that the Sensor blocks do nothing with sensor measurements themselves. Sensor blocks simply transmit sensor readings to other blocks with data wires. In addition to polling sensors, Sensor blocks can also compare a measured sensor value with a trigger value. Generally, these blocks have two output plugs. One plug outputs the sensor measurement (like the Distance plug in the Ultrasonic Sensor block); the other, the Yes/No plug, outputs the result of the comparison, as you'll learn in "Seeing a Logic Data Wire in Action" on page 164.

configuring a touch sensor block

The Touch Sensor block outputs its sensor value with the Logical Number plug, where a value of 1 represents pressed, and 0 means released. However, when you use this block, you'll more frequently use its Yes/No output plug, discussed in "Seeing a Logic Data Wire in Action" on page 164.

configuring a color sensor block

As mentioned previously, the Color Sensor can identify six different colors. The *Detected Color* plug on the Color Sensor block outputs a number between 1 and 6, with each number representing a particular color: black = 1, blue = 2, green = 3, yellow = 4, red = 5, and white = 6. (In Chapter 11 you'll create some programs that use this feature.)

Recall that the Color Sensor can function as a Light Sensor. Not surprisingly, you can use the Light Sensor's value in a program by selecting **Light Sensor** in the Action box on the Color Sensor block's Configuration Panel. Once configured in this way, when you subsequently connect a data wire to the Detected Color plug, it should carry out a value from 0 to 100, based on the brightness of the detected light (100 is brightest).

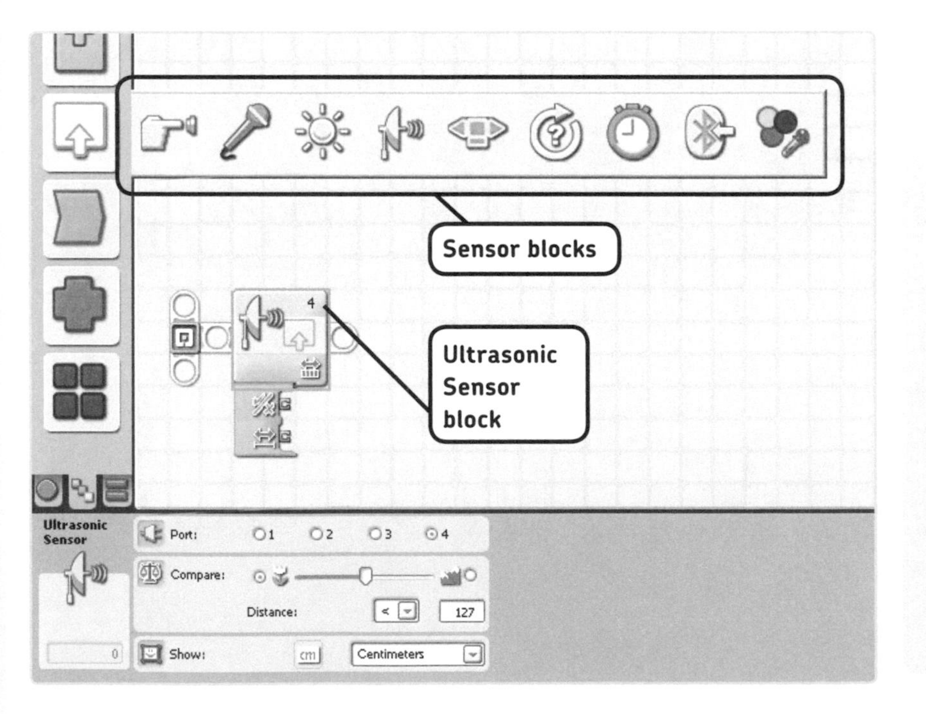

Figure 10-15: You can find a Sensor block for each sensor on the Complete Palette.

DISCOVERY #49: MOTOR INPUT!

Difficulty: Medium
In this discovery, you'll edit the program from Discovery #48 to set the motor speed of the Hand motor, based on the number of degrees you turned the Wheel motor. When you're ready, edit the program so that the motor speed depends on the amount of light detected by the Light Sensor.

configuring a rotation sensor block

In Chapter 7 you learned that each motor contains a Rotation Sensor, which tells you how many degrees a motor has turned since you started the program. The Degrees plug on the Rotation Sensor block's data hub outputs this number of degrees. If you rotate the motor backward (see Figure 7-20), the output value will be negative.

data wire types

Data wires carry information between blocks. So far you've used data wires to transfer numerical values only, but there are three types of data wires: Number, Logic, and Text data wires. Each type of data wire carries a specific type of information (numerical, logic, or text values), and each type has its own color, as shown in Figure 10-16.

the number data wire

The *Number data wire* (yellow) carries numeric information that may include whole numbers (such as 0, 15, or 1427), numbers with decimals (such as 0.1 or 73.14), and negative numbers (such as -14 or -31.47).

Examples of information carried by Number data wires are Ultrasonic Sensor readings and loop counts.

the logic data wire

The *Logic data wire* (green) can carry only two values: true or false. These wires are often used to define settings of a block that can have only two values, such as the turn direction of an NXT motor. For example, a motor will spin forward when a Logic data wire with the value true is connected to the Direction plug of a Motor block. Consequently, it spins backward when the Logic data wire value is false.

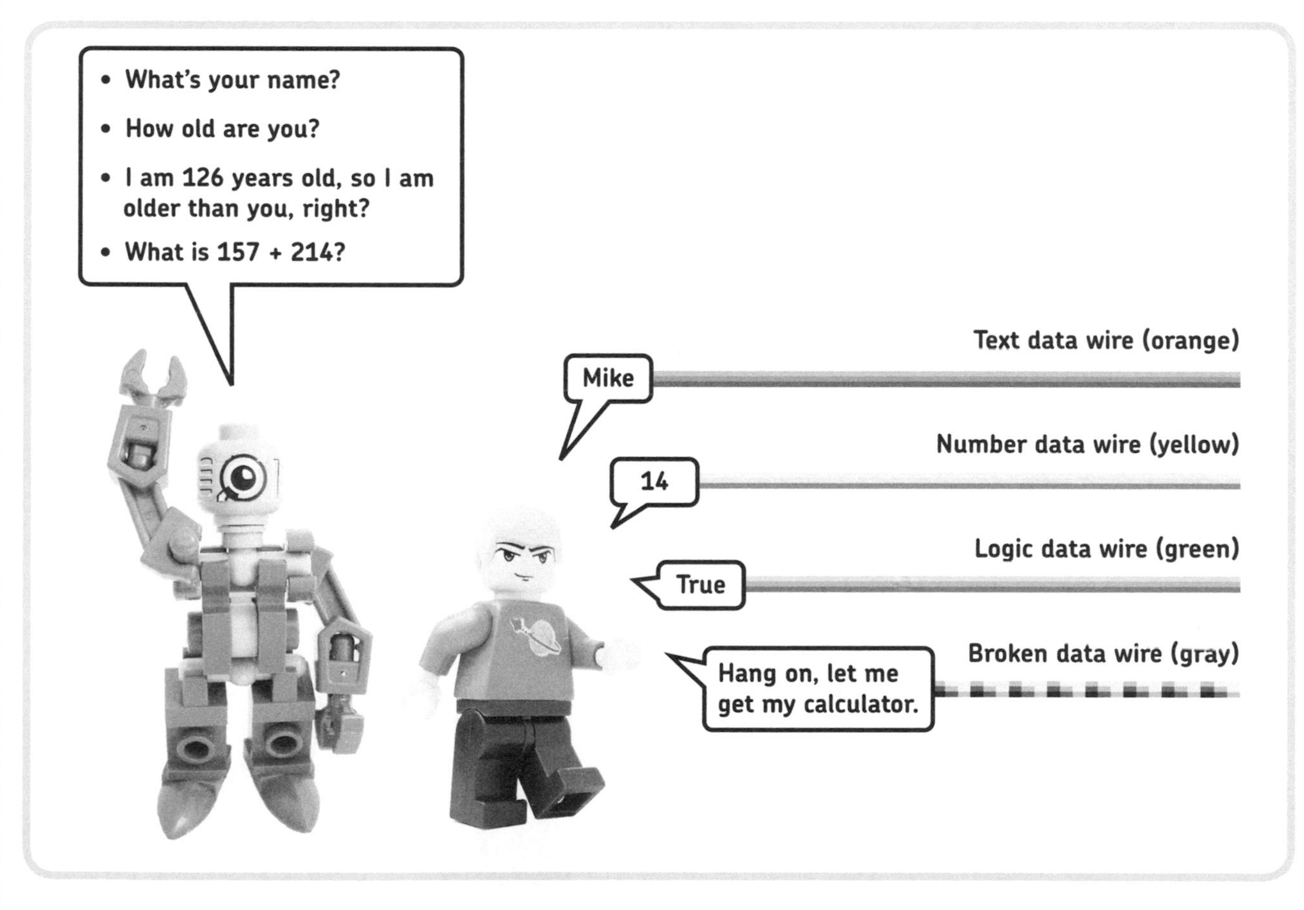

Figure 10-16: Examples of three types of values. Each value type is carried by its own data wire, as you'll learn in a moment. The last data wire here isn't functional, and it carries no information: The robot's last question requires a numerical answer, but Mike replied with a sentence, which would require a Text data wire. This mismatch results in a broken data wire, because the given information cannot be used.

seeing a logic data wire in action

You'll now create a program to see the functionality of the Logic data wire and the Compare function of the Ultrasonic Sensor block in action. This program will make SmartBot's Wheel motor spin forward as long as the Ultrasonic Sensor sees something closer than 40 cm and spin backward when this is not the case. The required input for the motor's Direction setting is a Logic data wire (true is forward, and false is backward). Therefore, you'll make a connection to a data plug on the Ultrasonic Sensor block that sends out a logic value using the Yes/No output plug. This plug will output a logic value (either true or false), based on the result of the Compare box in the Configuration panel. The Ultrasonic Sensor block will check to see whether the sensor reading is smaller than the trigger value (40 cm). If it is, the result is true; if not, the output value is false.

Now create the **Smart-LogicWire** program as shown in Figure 10-17.

NOTE Because you're not specifying the motor speed with a data wire, the motor power is 75 no matter which direction the motor spins, as set in the Configuration Panel.

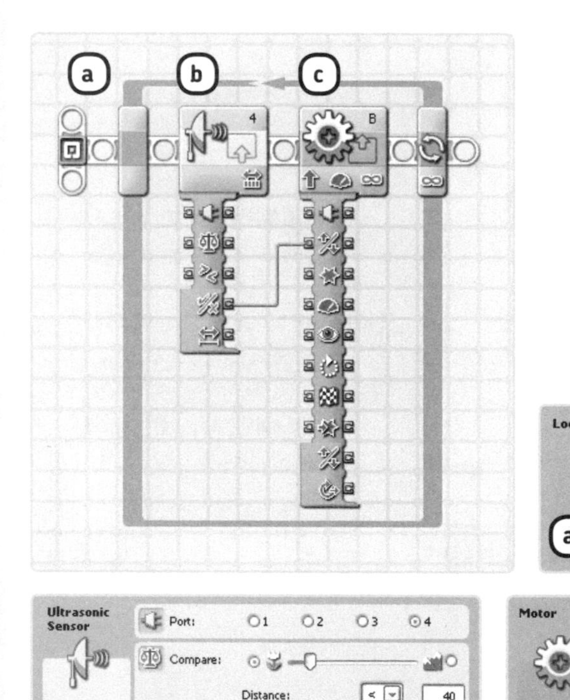

Figure 10-17: The configuration of the Smart-LogicWire program. The green Logic data wire connects the result of the comparison of the Ultrasonic Sensor block to the Direction plug on the Motor block. In this particular setup, the Sensor block checks to see whether the sensor reading is smaller than the trigger value, but you could also make it look for a value bigger than the trigger point. If you do, the result will be true if the sensor reading is greater than 40.

Once you've finished creating this program, download it to your robot, and notice how the motor direction flips as you move your hand closer to or farther from the sensor.

the text data wire

The *Text data wire* (orange) carries text between blocks to, for example, the data hub of a Display block so that it appears on the NXT screen. This text can be a word, like `Hello`, as well as sentences like `My name is Mike`.

displaying values with the number to text block

By entering a text line in the Configuration Panel, you can use Display blocks to show a text line on the NXT screen. You can also use a Text data wire to submit text to a block.

DISCOVERY #50: TOUCH SENSOR WIRES!

Difficulty: Medium

In this discovery you'll see the Touch Sensor block in action. The Yes/No plug on this block outputs a Logic data wire. Its value is true if the action specified in the Action setting (such as **Pressed**) takes place, and is false otherwise. Create a program that switches on the Color Lamp if the Touch Sensor is pressed and switches it off when it is not pressed. To do so, wire the Logic data wire from the Touch Sensor block to the Action plug of the Color Lamp block. What value should the Color Lamp block's Action plug receive for the lamp to be switched on?

You'll now make a program that continuously displays the Ultrasonic Sensor reading on the NXT screen. To do so, you use the Ultrasonic Sensor block to find the sensor value. You have a problem now, though, because you cannot use the Display block to display values from Number data wires. Therefore, you'll have to *convert* the numeric sensor reading into something that the Display block does accept: text.

The *Number to Text block* can do this for you. On the data hub's left side, it takes input from a Number data wire, such as a wire carrying a sensor value. On the right side, it outputs a Text data wire, which contains the same number, except in a format that the Display block can handle. You can see the Number to Text block in Figure 10-18, which also shows you how to make the **Smart-TextWire** program.

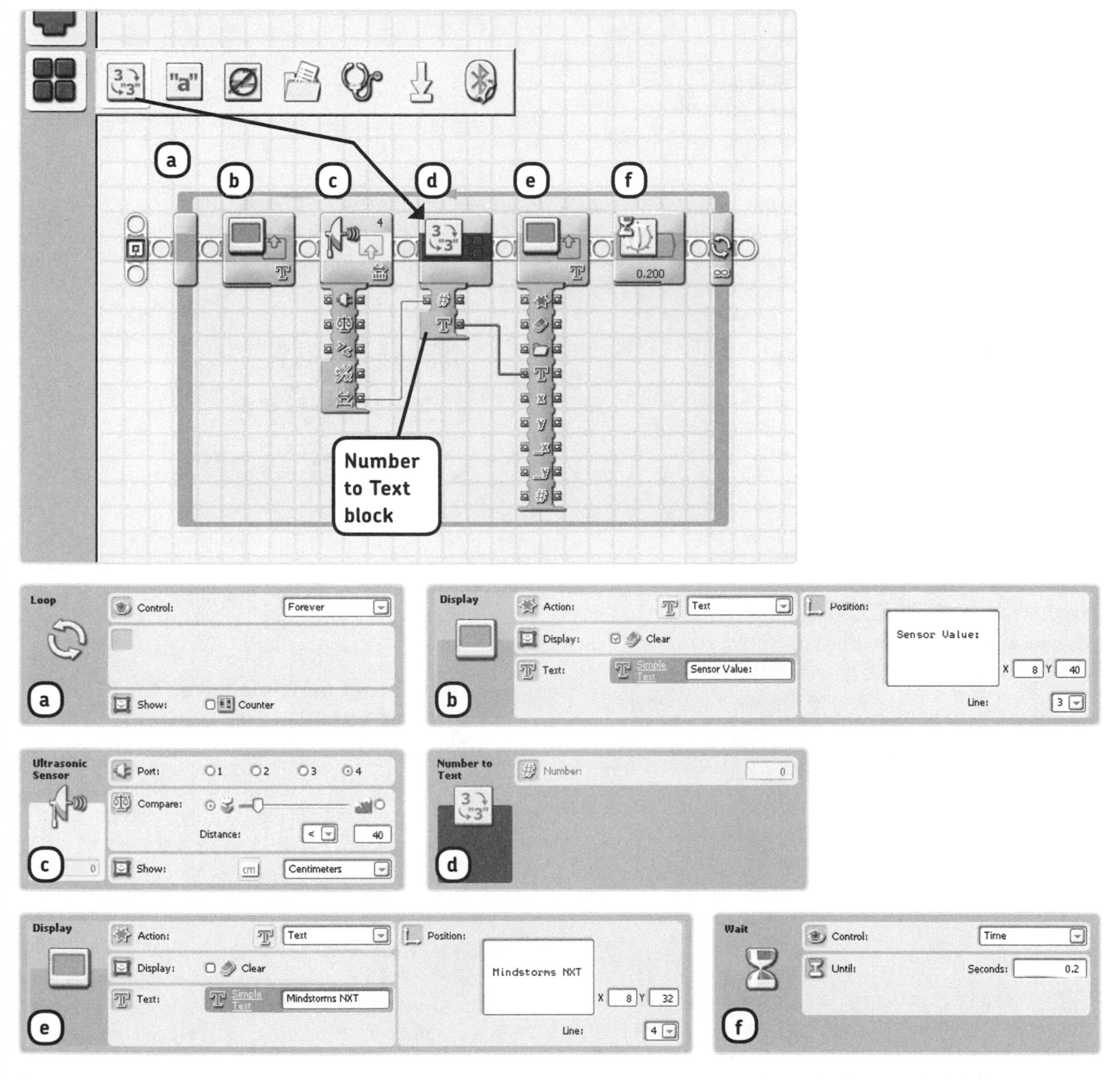

Figure 10-18: The configurations of the blocks in the Smart-TextWire program. You can find the Number to Text block between the Advanced programming blocks. Notice that the Display block (e) is configured to display the `Mindstorms NXT` *text line. However, the program ignores this setting, because you also use a Text data wire to specify what should be displayed.*

DISCOVERY #51: READ THE SENSOR READINGS!

Difficulty: Hard
Expand the Smart-TextWire program to make it display the Rotation Sensor reading by adding extra blocks inside the Loop block.

HINT Pay attention to the line numbers in the Display blocks, as well as the Clear checkbox.

NOTE When using a Sensor block only to perform a sensor measurement, you don't need to configure the block's Compare setting since you won't be using the result of this comparison.

the broken data wire

The *broken* data wire (gray) carries no information. When such a wire shows up, you know that you've made a mistake when making the data wire connection. The mistake can have several causes, as you'll learn in this and in the "Multiple Data Wire Connections" section. You'll have to delete this broken wire and reconnect it appropriately, or you won't be able to transfer the program to the NXT.

When you begin creating a data wire on a block's output plug, the NXT-G software automatically chooses a color for the wire, depending on the plug you clicked. For example, if you clicked the Ultrasonic Sensor block's Distance plug, you'd see a yellow wire because the distance value is a number. Because this is a number, you have to connect it to an input plug on another block that accepts Number data wires, like the Power plug on a Motor block.

If you connect the yellow data wire (from the Distance plug) to a plug that cannot handle Number data wires, a broken wire shows up, indicating that the connection isn't right, as shown in Figure 10-19. (In "Using Help for Data Plugs" on page 168, you'll learn how to find the right data wire for each plug on a data hub.)

NOTE When you try to send a program with a broken data wire in it to the NXT, you'll get an error message. The solution is to delete any broken data wire and reconnect it properly, as discussed in "The Broken Data Wire" and "Multiple Data Wire Connections."

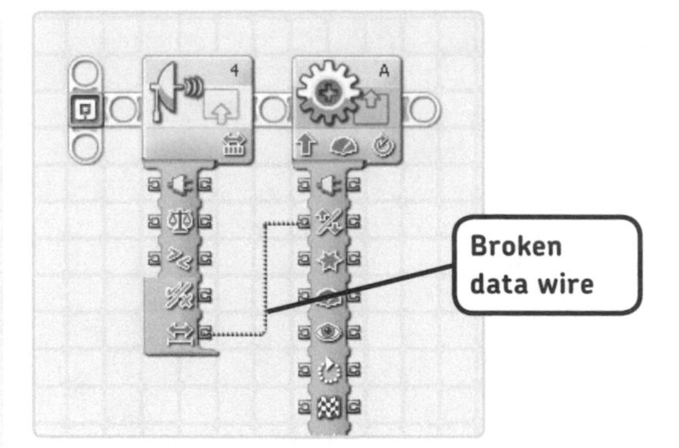

Figure 10-19: If you try to connect a data wire of a certain type (like a Number data wire) to an input plug on another block that doesn't accept this type, a broken data wire is displayed. The example shown here tells you that you can't connect the Ultrasonic Sensor reading (a number value) to the Direction plug (a logic value) of a Motor block.

multiple data wire connections

So far, you've used only one data wire connection per block, but you can use more data plugs on a single block. You can connect more than one data wire to a single block in several ways, but not every way to connect them results in a working program, as you'll learn in this section.

connecting multiple wires to different plugs

You can use multiple input plugs on a single block to reconfigure more than just one setting of a block with a data wire. For instance, you can control both the Power and Direction settings of a Motor block with data wires (using one wire for each setting).

In the same way, you can use multiple output plugs on a single block. For example, you can use the Ultrasonic Sensor block to poll a sensor (a data wire carries the sensor value) *and* compare it to the trigger value (another data wire carries a true or false value).

To illustrate this functionality, you'll expand the Smart-LogicWire program (Figure 10-17) by adding a Number data wire, as shown in Figure 10-20.

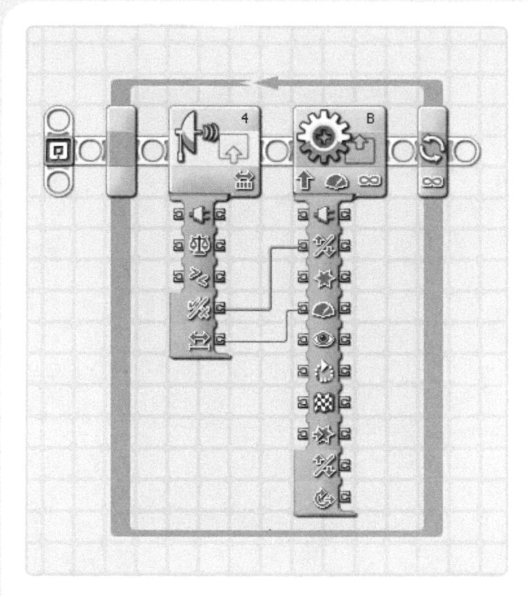

Figure 10-20: You can connect multiple data wires to a block. Here, SmartBot's Wheel motor rotates forward when the sensor reading is less than 40 cm, and backward otherwise. The actual sensor measurement defines the speed at which the motor turns.

DISCOVERY #52: MULTIFUNCTIONAL WIRES!

Difficulty: Medium
Create a program like the Smart-Accelerate program you made earlier in which the loop count controls the motor speed. Have the value that controls the speed passed on to a Number to Text block to enable the value to be displayed on the NXT screen.

connecting multiple wires to one data plug

You can use a single output plug to send information to more than one block, as shown on the left of Figure 10-21, but not the reverse: You cannot connect more than one wire to a single input plug since then the block would not know which wire to take as the input. If you try to make such a connection, a broken data wire will show up, as shown on the right of Figure 10-21. You can delete broken wires just like other data wires (Figure 10-13).

using settings with both input and output plugs

Some items on a data hub have both input and output plugs, as shown on the left of Figure 10-22. As you've learned, the input plug uses a data wire's value to reconfigure one of the block's settings. For example, you can use a data wire to set the Action setting of the Color Lamp block.

This Action setting also has an output plug on the right side, which will output the same value as the input value on the left side. Basically, the Color Lamp block on the left of Figure 10-22 takes input from the wire on the left side and passes that value to the next block so that the information

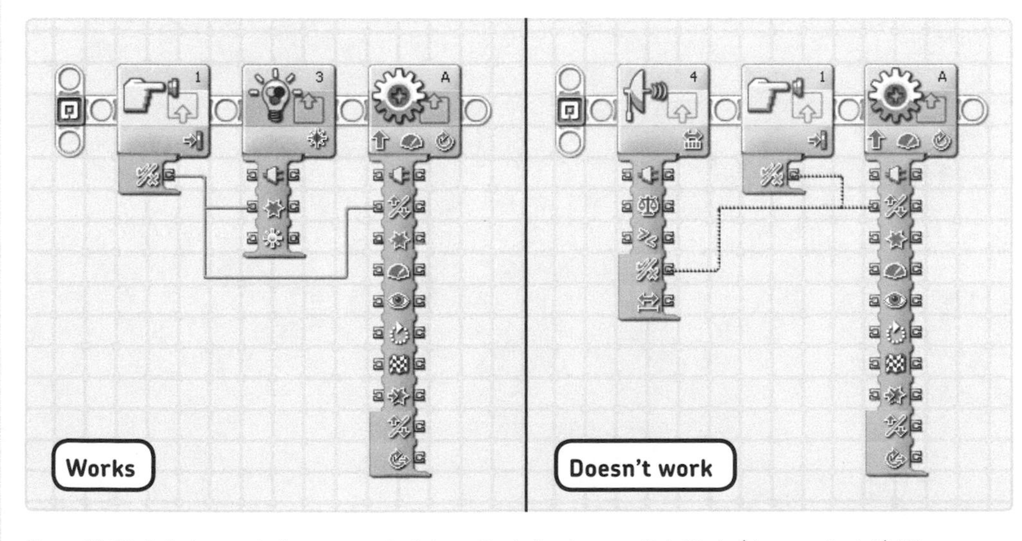

Figure 10-21: A single output plug can transfer information to inputs on multiple blocks (shown on the left). When you run this program while the Touch Sensor is pressed, the Color Lamp is switched on, and the motor rotates forward. It is impossible to connect multiple data wires to one input plug (shown on the right).

Figure 10-22: When an input plug has a corresponding output plug, the value received by the input is passed on to the next block with no change (shown on the left). If there is no value to pass on, a broken wire is displayed (shown on the right).

can be used again. If, however, there is no input on the left, there is no information to pass through, so the wire will show up as a broken data wire (on the right of Figure 10-22).

using help for data plugs

When creating programs, you may want to use a block's data plug that isn't discussed in this book. If you do, you can find information about data plugs specific to each programming block in the help section of the software. (To access it, click **more help** in the Little Help Window at the bottom right of the software screen, and after it opens, use the menu on the left to find information on a programming block.)

For example, when you open the help file for the Motor block by clicking the block's name in the menu on the left, you'll first see some general information about the block and its functions. As you scroll down on the page, you'll see a table with the data hub characteristics, a portion of which is shown in Figure 10-23. By looking at this table, you'll learn how to properly use the functions of each programming block and its data plugs.

	Plug	Data Type	Possible Range	What the Values Mean	This Plug is Ignored When...
	Port	Number	1 - 3	1 = A, 2 = B, 3 = C	
	Direction	Logic	True/False	True = Forwards False = Backwards	
	Action	Number	0 - 2	0 = Constant 1 = Ramp Up 2 = Ramp Down	Duration Type = Unlimited or Seconds
	Power	Number	0 - 100		
	Control Motor Power	Logic	True/False		

Figure 10-23: A few data hub characteristics of the Motor block

The table shown in Figure 10-23 contains specific information about each data plug, including the data wire type ("Data Type") that should be connected to the plug, the range of possible values for the plug for the block to function as expected ("Possible Range"), what each value means ("What the Values Mean"), and when a data wire is ignored by a block ("This Plug Is Ignored When . . .").

You'll now look at the Direction plug. As you've learned (and as you can see in Figure 10-23), you can only connect a green Logic data wire to the Direction plug. You also see that a value of true will make the motor spin forward, while false will make the motor rotate backward. Finally, you can see that when a data wire is connected to this plug, it is never ignored. Using this method, you'll be able to find information about each block's data hub.

DISCOVERY #53: LOOKING FOR HELP!

Difficulty: Easy

Create the program shown in Figure 10-24. Your goal is to finish the program to display an image somewhere on the screen (aim in the File list), depending on the reading from the Wheel motor's Rotation Sensor. The more the motor has turned, the farther the image should move to the right. Use the Display block's help file to find out which data plug you must use in order to set the horizontal position of the image. Now connect a data wire that will make the program work as described.

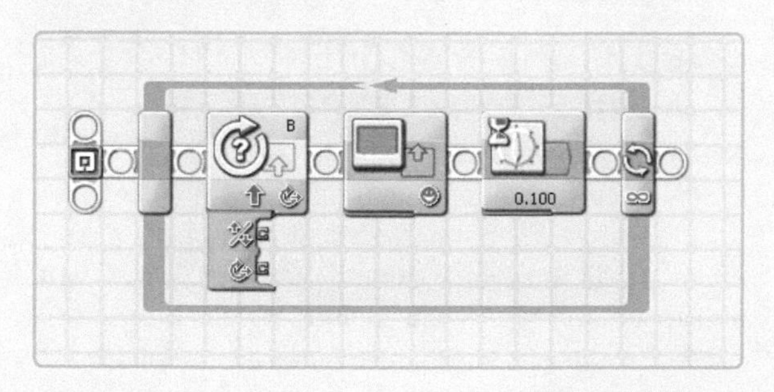

Figure 10-24: A starting point for the program in Discovery #53

tips for managing data wires

Now that you've learned how data hubs and data wires work, you can use them in your programs. But, as you make larger programs, it can get difficult to keep your programs organized and understandable, especially when you use a lot of data wires. The following tips will help you to keep your programs clean and readable.

hiding unused data plugs

You can hide unused data plugs on the data hub by clicking the tab at the lower-left edge of a block, as shown in Figure 10-25. To redisplay them, click the tab again. When creating programs, you'll first open the complete data hub, and then once you've connected the desired data wires, you can hide the other plugs.

NOTE From now on in this book, I'll hide unused data plugs in programs to make the programs more readable.

Figure 10-25: Hiding unused data plugs will make programs more compact and understandable.

using data wires across your program

When configuring programs with data wires, you don't have to connect a block to the one right next to it; you can connect blocks when there are other blocks in between, as shown in Figure 10-26.

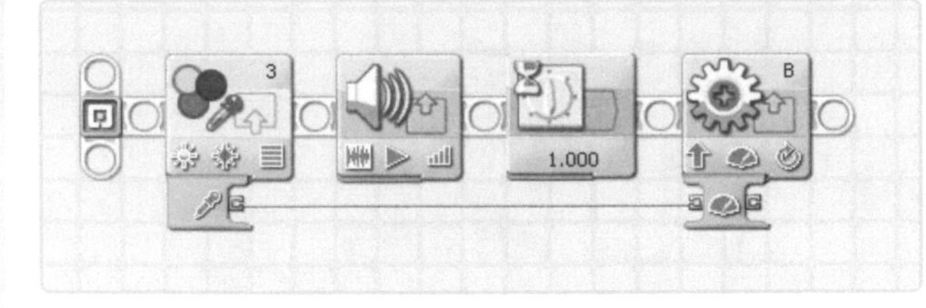

Figure 10-26: Data wires allow you to connect blocks spread over your program, as shown here. The connected blocks do not have to be right next to each other.

further exploration

In this chapter, you learned how to create programs using the features of data wires. Most of the programs you've made so far with data wires are fairly small, and data wires may not seem very useful yet. However, data wires are essential to making advanced programs for your NXT robots. In Chapter 11, you'll learn a few new ways to use data wires beyond simply controlling motor speeds and displaying sensor readings on the robot's screen.

The following discoveries will let you practice the programming skills you gained in this chapter. Post your solution to these discoveries to the companion website at *http://www.discovery.laurensvalk.com/* so that your fellow book readers can see what you've made!

DISCOVERY #54: TURN UP THE VOLUME!

Difficulty: Easy
Create a program that repeatedly says "Good morning" at a volume based on the Ultrasonic Sensor reading. Use the Sound block's data hub to wire in a value to change the volume. Look at the Sound block's help section to determine which data plug to use, and then put the blocks in a Loop block configured to loop forever.

DISCOVERY #55: CONTROL THE THROTTLE!

Difficulty: Medium
Create a program with only three programming blocks and two data wires that will make the Hand motor turn. The speed and direction of the Hand Motor should be controlled by how much and in which direction you turned the Wheel motor, meaning that you'll use the Wheel motor as a Rotation Sensor. Which data plugs do you need to use in order to create this program?

DISCOVERY #56: SMARTBOT IS WATCHING YOU!

Difficulty: Expert
Create a program that counts the number of people walking by your robot. Place your robot in such way that the Ultrasonic Sensor can sense people in front of it. Place two Wait blocks inside a Loop block, and configure the first block to wait until the sensor sees someone passing by; use the second one to wait until the person is out of sight. Ultimately, the loop should go around once each time it senses someone passing by. Now, when you display the loop count on the NXT screen, you're displaying how many people have walked by! To ensure that your program works properly, place a Sound block inside the loop so that you'll hear a sound each time someone passes by.

BUILDING DISCOVERY #10: A POLITE SMARTBOT!

In this discovery, you'll give the SmartBot a hat that it can take off when it notices that someone is passing by. To do this, you'll expand the robot design with an extra motor which you'll control with a Motor block. For the arm mechanism, look at the robot's Hand motor. When you finish creating the new robot design, create a program that makes the robot interact with people by making it take its hat off and talk if it sees someone. Can you turn the SmartBot into a polite robot?

using data blocks and using data wires with loops and switches

Now that you know how data wires work, you can do some really interesting things with more of the programming blocks. You can use some of these new blocks to make the NXT combine and process sensor values so that they can be used as input values for other actions. For example, with the tricks learned in this chapter, you'll be able to program your robot to do something only when two sensors are triggered at the same time or to do things randomly, rather than performing a series of preprogrammed actions.

In this chapter I'll show you how to do math on the NXT with the Math block so that you'll be able to have your robots calculate, for example, the distance to travel based on a sensor reading. I'll also introduce you to a few new data programming blocks such as the Random, Compare, and Logic blocks as well as a few new techniques to use with the Switch and Loop blocks, by providing sample programs and new discoveries.

These techniques and programming blocks are essential to advanced robot programs like the ones you'll create in Part IV, as well as for advanced programs that you'll create for your own robots. As in Chapter 10, you'll use the SmartBot to test your new programs.

The discoveries in this chapter might not seem so interesting at first, but once you've mastered many of the essential programming skills by solving them, you'll be able to create much more interesting programs for your robots and build some really smart robots!

data blocks

The Complete Palette contains a series of blocks that you have not used yet: *data blocks*. Data blocks (Figure 11-1) include the Math block, the Random block, the Compare block, and the Logic block. Each block has its own function, but they all process values carried by data wires and generate new values based on the input values. This section will explain how to use these blocks in your programs.

the math block

The Math block allows the NXT to do arithmetic operations such as addition, subtraction, multiplication, and division. In its Configuration Panel, you can fill in placeholders for two numbers (A and B), and choose which *operation* should be applied to them, such as multiplication (A will be multiplied by B). The result is output with a data wire. When you use the Math block, you'll often transfer one or both numbers to the Math block with a data wire instead of entering them yourself, as you'll see in the second sample program in this chapter.

The **Smart-Math** program in Figure 11-1 shows the Math block performing a multiplication. The Number to Text and Display blocks are used to display the result on the NXT screen.

Now you'll learn how to use the Math block in programs for your robots to do more than just display a value.

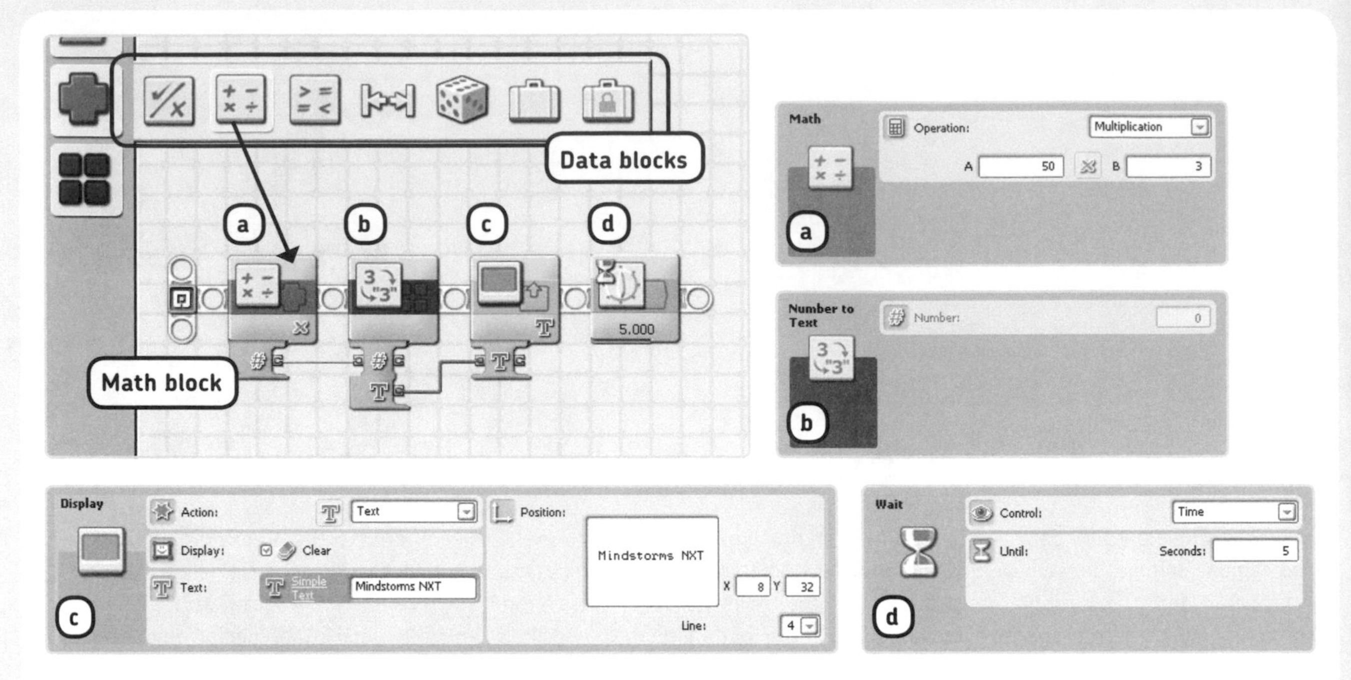

Figure 11-1: The Math block in the Smart-Math program multiplies 50 by 3 and displays the result (150) on the NXT screen. (I've hidden the unused data plugs to make the program easier to understand.)

using the math block in a more advanced program

The Sound block can play tones at specific frequencies: The higher the frequency, the higher the pitch you'll hear. You'll use the Sound block to create a program that plays tones that change based on readings from the Color Sensor. Your robot will repeatedly play short beeps, depending on which colored ball (from your robotics kit) is held in front of the sensor.

To make this program work, you'll have the Color Sensor block send sensor values to the Sound block with a data wire connected to the Detected Color plug (which outputs a value between 1 and 6 depending on the detected color). To have the Sound block play tones at a frequency that you specify, you transfer the sensor values to the block's *Tone Frequency plug*. If the block receives small values (around 250) from the data wire, it plays low buzzes. Large numbers (around 4,000) make the block play high pitches.

As you can see, this data plug requires fairly large input values. However, since the Detected Color value ranges between 1 and 6 (not 250 and 4,000), you need to process the reading somehow so it can be used as input to the Sound block. To accomplish this, you'll use the Math block to multiply the color value by 250. For example, if the Color Sensor sees blue, the sensor value is 2, and the Math block will multiply it by 250. The result is a frequency of 500, which will make the Sound block play a low tone. You'll hear a higher sound when the sensor sees a red ball (frequency = 1250). Create the **Smart-Sound** program shown in Figure 11-2 now.

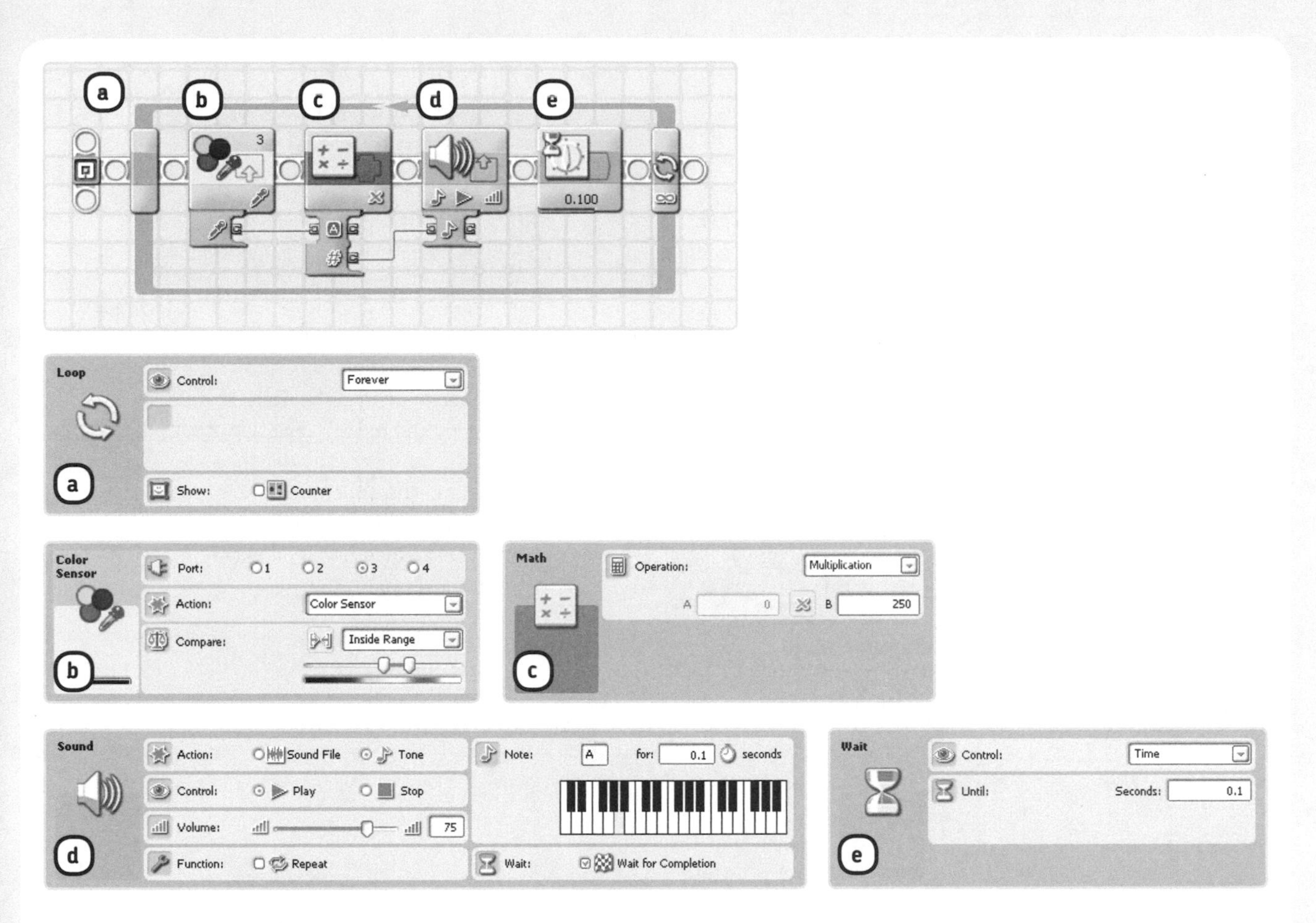

Figure 11-2: The Smart-Sound program. Instead of filling in two numbers in the Math block's Configuration Panel as in the previous program, you feed one value to the block with a data wire.

DISCOVERY #57: MATH PRACTICE!

Difficulty: Medium

Create a program that continuously displays the Light Sensor and Ultrasonic Sensor readings, as well as the sum of the two, on the NXT screen, as shown in Figure 11-3. Use a Math block to add the two readings, and add a Wait block configured to wait for 0.5 seconds (in the Loop block) so that you'll have enough time to read the display.

Ultrasonic:
68
Light:
27
Total:
95

Figure 11-3: The NXT screen in Discovery #57

the random block

You use the *Random block* to generate a random number for use in your program. Use the block's Configuration Panel to select a range for the random number, such as 35 to 47. The Random block will output the random value through a Number data wire, which in this case will be 35, 47, or some value within this range.

The Random block is useful when you want your robot to do something unexpected. For example, you could use a Random block to make the SmartBot's Wheel motor turn at a random motor speed, as shown in the **Smart-Random** program in Figure 11-4.

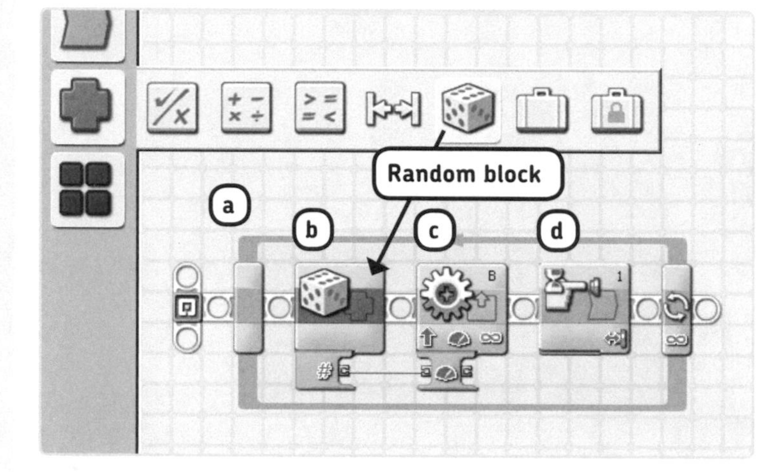

Figure 11-4: The configuration of the Smart-Random program. The Random block generates a random value between 0 and 100, which is carried to the Power plug on the Motor block, switching on a motor. Next the program waits until you bump the Touch Sensor, at which point the loop returns to the beginning, and the motor speed changes to a new random speed.

DISCOVERY #58: RANDOM SOUND!

Difficulty: Hard
Combine the Smart-Sound program with the Smart-Random program to make the NXT play a random tone. Use a Math block to increase the size of the random value generated by the Random block, and wire the new value into the Tone Frequency plug of the Sound block. Have the robot play tones for one second, and then wait until you bump the Touch Sensor. Repeat this sequence continuously. Finally, expand the program to also display the frequency on the NXT screen.

the compare block

The *Compare block* checks to see whether a value is greater than (>), less than (<), or equal to (=) another value. You can enter the values you'd like to compare in the Configuration Panel, or you can supply the values to the block with data wires (plug *A* and plug *B*).

The Compare block outputs one Logic data wire (true or false), based on the result of comparing value A to value B. For instance, if you set Operation to **Equals**, the block outputs true if value A is equal to B.

The **Smart-Compare** program in Figure 11-5 shows the Compare block in action. In this program, the Random block sends a random value (0, 100, or somewhere in between) to the Compare block, which checks to see whether this value is less than 50. If it is, the Wheel motor on Smart-Bot spins forward for one rotation; if not, it turns backward. This behavior repeats, and the motor continues to revolve in random directions as the random values change.

the logic block

In some of your programs you've used Sensor blocks to determine whether the Touch Sensor was pressed or whether the Ultrasonic Sensor value was bigger than the configured trigger value. The results were carried away from the blocks with Logic data wires.

You can use the *Logic block* to compare two Logic data wires. Depending on how you configure the block, it can see whether both wires carry the value true. If so, it will output a value true. If none or just one of the two input values are true, its output value will be false.

You can use a Logic block to create a program that turns on the Color Lamp if both the Touch Sensor is pressed *and* the Ultrasonic Sensor sees something farther away than 100 cm. If either or both conditions do not occur, the Color Lamp is switched off, because the Logic block's output is false. Figure 11-6 shows the **Smart-Logic** program.

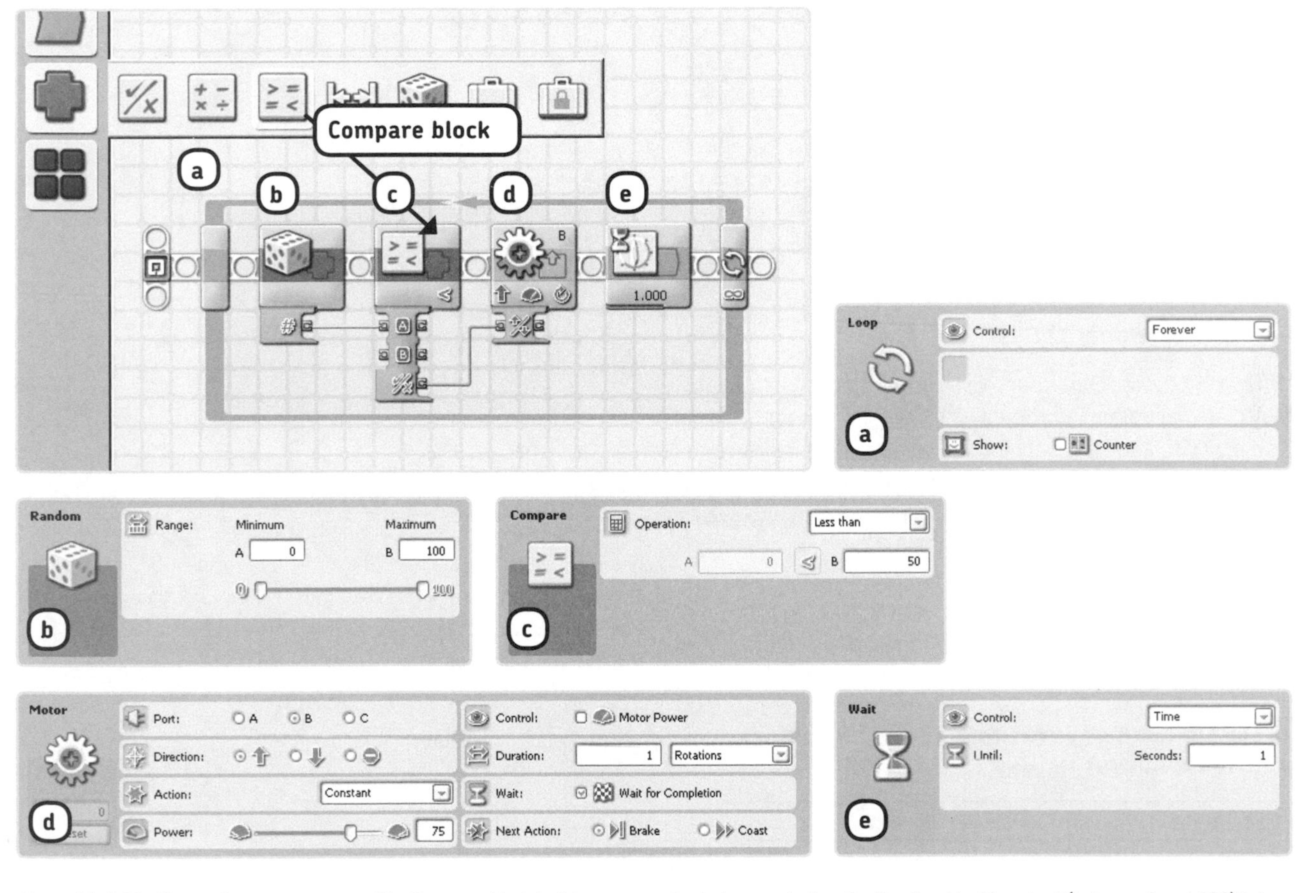

Figure 11-5: The Smart-Compare program. The Compare block in this program checks to see whether the Random block's output (between 0 and 100) is less than 50. If it is, the Compare block's output value is true, and the Wheel motor on the SmartBot turns forward. If not, the output value is false, and the motor turns backward.

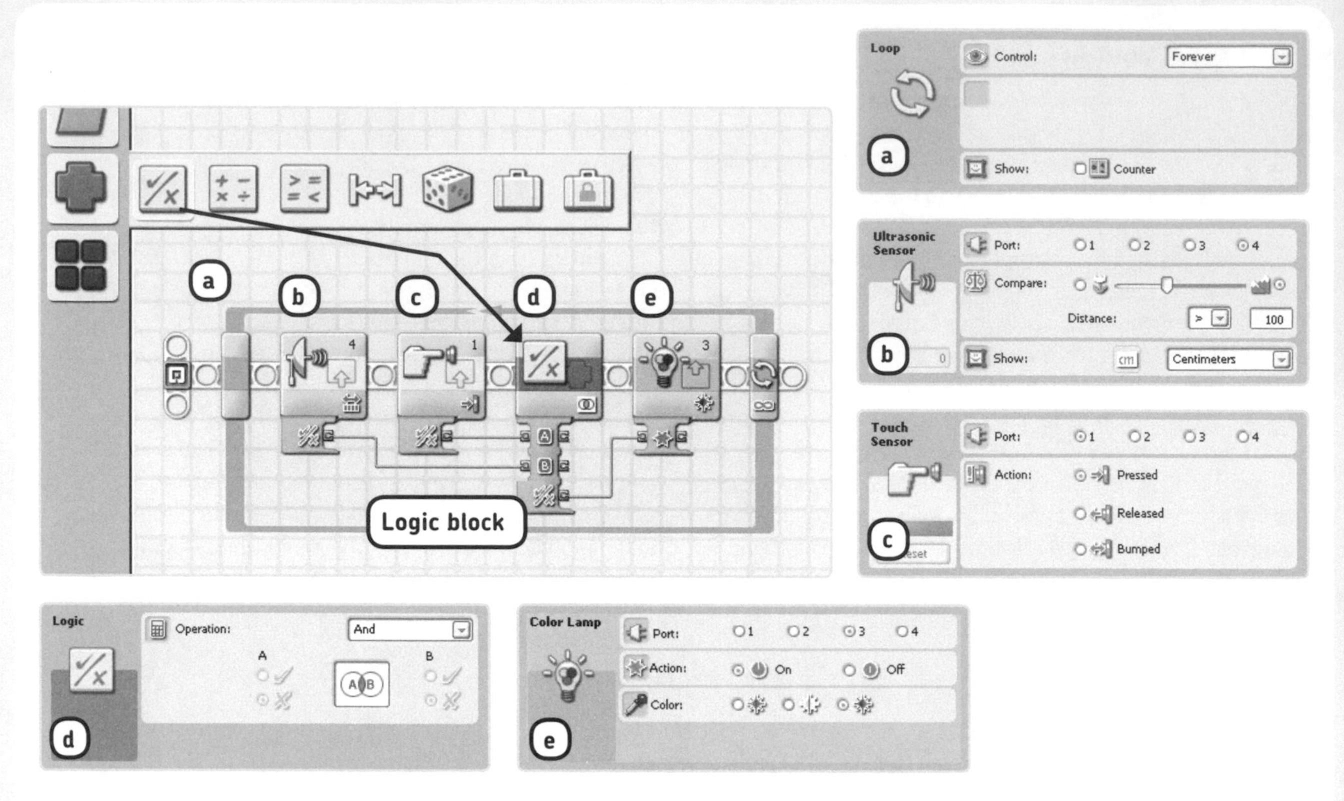

Figure 11-6: The Smart-Logic program. If the Touch Sensor is pressed and the Ultrasonic Sensor sees something that is farther away than 100 cm, the lamp on the Color Sensor is switched on. Because you use the results of the compare and action functions within the Sensor blocks, you connect the Logic data wires to the Yes/No output plugs on the data hubs. (I've hidden the other data plugs, such as the Distance plug on the Ultrasonic Sensor block.)

logic operations

When configuring the Smart-Logic program, you can select from four operations in the Logic block's Configuration Panel: And, Or, XOr, and Not. Each option will make the block compare the two logic input values differently, as you'll learn in a moment. The option you choose will depend on what you want your program to do.

Table 11-1 lists the available operations, as well as the input values that would lead to an output value of true with each of the operations. (The Smart-Logic program uses the And operation.)

table 11-1: the operations of the logic block and their output values

Operation	Output value is true when . . .
And	Both inputs are true
Or	One or two inputs are true
XOr	One input is true and the other is false

With these operations, you wire the two Logic data wires to be compared to the A and B input plugs of the block, but it doesn't matter which wire you connect to which plug.

not operation

When you select the *Not* operation, the Logic block doesn't compare two Logic data wires; it only does something with the logic value wired into plug A on the data hub. This operation just *inverts* the input signal: If the input A value is true, the output will be false; if the input A value is false, the output will be true.

DISCOVERY #59: AND, OR, XOR, OR NOT?

Difficulty: Medium
In Chapter 7, you learned that you could use the NXT buttons as Touch Sensors. Like the Touch Sensor block, the Yes/No data plug on the NXT Buttons block outputs a value that is either true or false, depending on whether the Action setting specified in the Configuration Panel takes place. Create a program where the Color Lamp is switched on when either the Enter button or the Touch Sensor is pressed (but not both). The lamp is switched on if a data wire with the value true is wired into the Action plug on the Color Lamp block. You'll have to use the Logic block to see whether only one of the sensors is pressed, but which operation do you use?

switch blocks and data wires

You've used the Switch block to enable a program to make decisions, such as to determine whether a Touch Sensor is pressed. As a quick reminder, see Figure 11-7, which shows a Switch block in action. Recall that a *condition* is a statement like "The Touch Sensor is pressed." When the Switch block determines that a Touch Sensor is being pressed, the condition is true, and the blocks on the upper half of the Switch block are run. If the condition is false, meaning the Touch Sensor is not pressed, the blocks on the lower half of the Switch block are run instead. (See Chapter 6 to learn more about the basics of Switch blocks.)

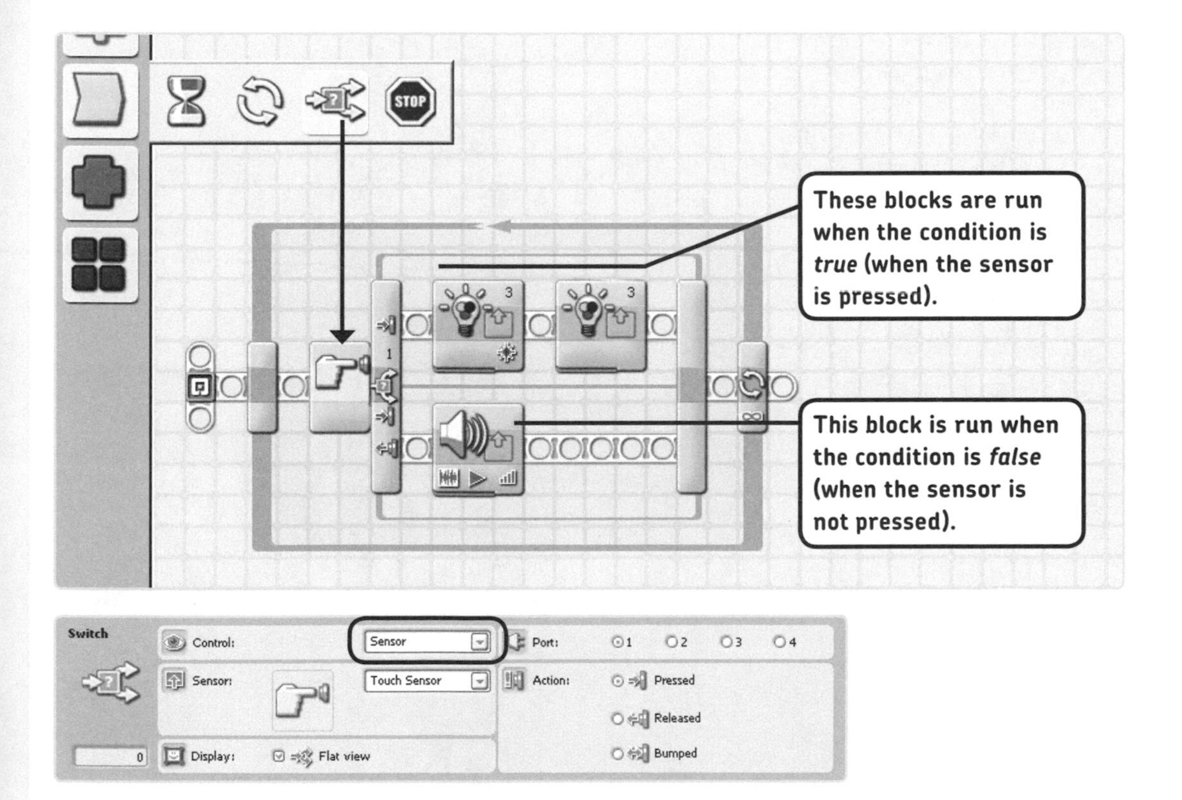

Figure 11-7: The Smart-Touch program continuously checks whether the Touch Sensor is being pressed. If it is, the Color Lamp blinks (it's switched on and off). If not, the robot says "No."

configuring switch blocks with data wires

In addition to working with sensor readings, a Switch block can also make decisions about *values* when a data wire is connected to the Switch block.

For example, you could configure a Switch block so that the blocks in its upper part run when the input value signal is true and so the blocks on the lower part run when the input signal is false. To do this, you connect a Logic data wire to the Switch block.

To be able to connect a data wire to a Switch block, you'll select a Switch block from the Programming Palette, place it on the Work Area, then set the Control parameter on its Configuration Panel to **Value**, as you'll see in the Smart-Switch sample program that follows.

Smart-Switch displays a random number between 0 and 100 on the NXT screen and also displays whether that number is greater or less than 40. The program first generates a random number between 0 and 100. Next, with the aid of a Compare block, it checks to see whether that number is greater or less than 40. If greater, the Compare block's output value will be true; otherwise, it will be false. A Logic data wire transfers this value to a Switch block. If the Switch block receives a value of true, the upper parts in the switch are run; if the value is false, the lower blocks are run.

You'll find the **Smart-Switch** program in Figure 11-8 and Figure 11-9.

DISCOVERY #60: TOTAL OR ELSE!

Difficulty: Medium

Create a program that adds together the Ultrasonic Sensor (measured in centimeters) and Light Sensor values. Use a Compare block to determine whether the total value is greater than 300, and wire the output value of this block into a Switch block. Add Sound blocks to the Switch block that will say "Yes" when the switch receives a value of true and will say "No" when the data wire carries a value of false. Put all the required blocks in a Loop block.

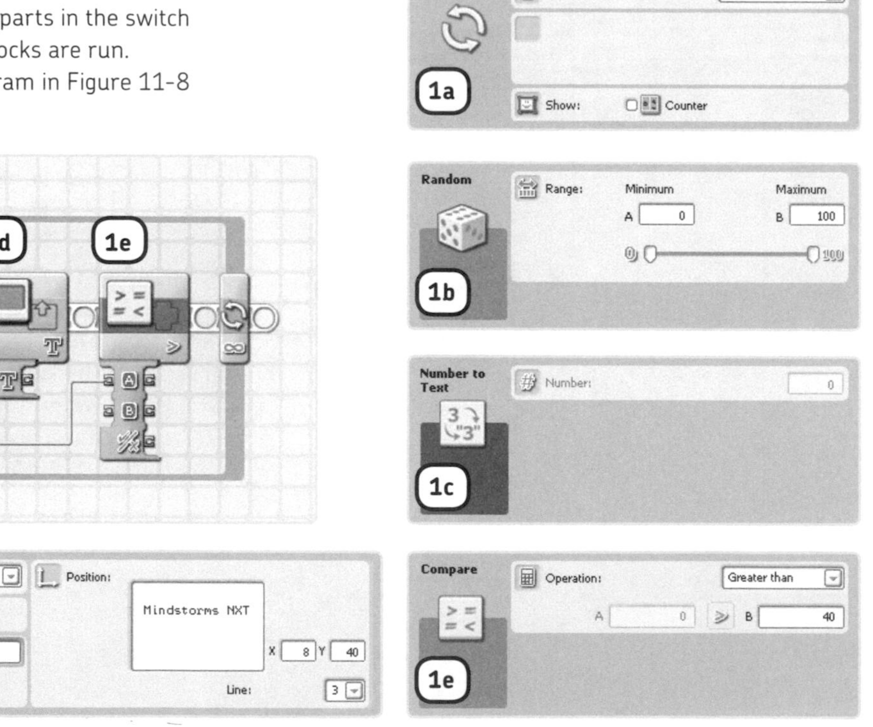

*Figure 11-8: **Step 1:** Because the complete program repeats continuously, you place all of the programming blocks inside a Loop block. The other blocks shown here allow the robot to display the random number on the NXT screen. That same random number is also sent to a Compare block, which checks to see whether it is greater than 40.*

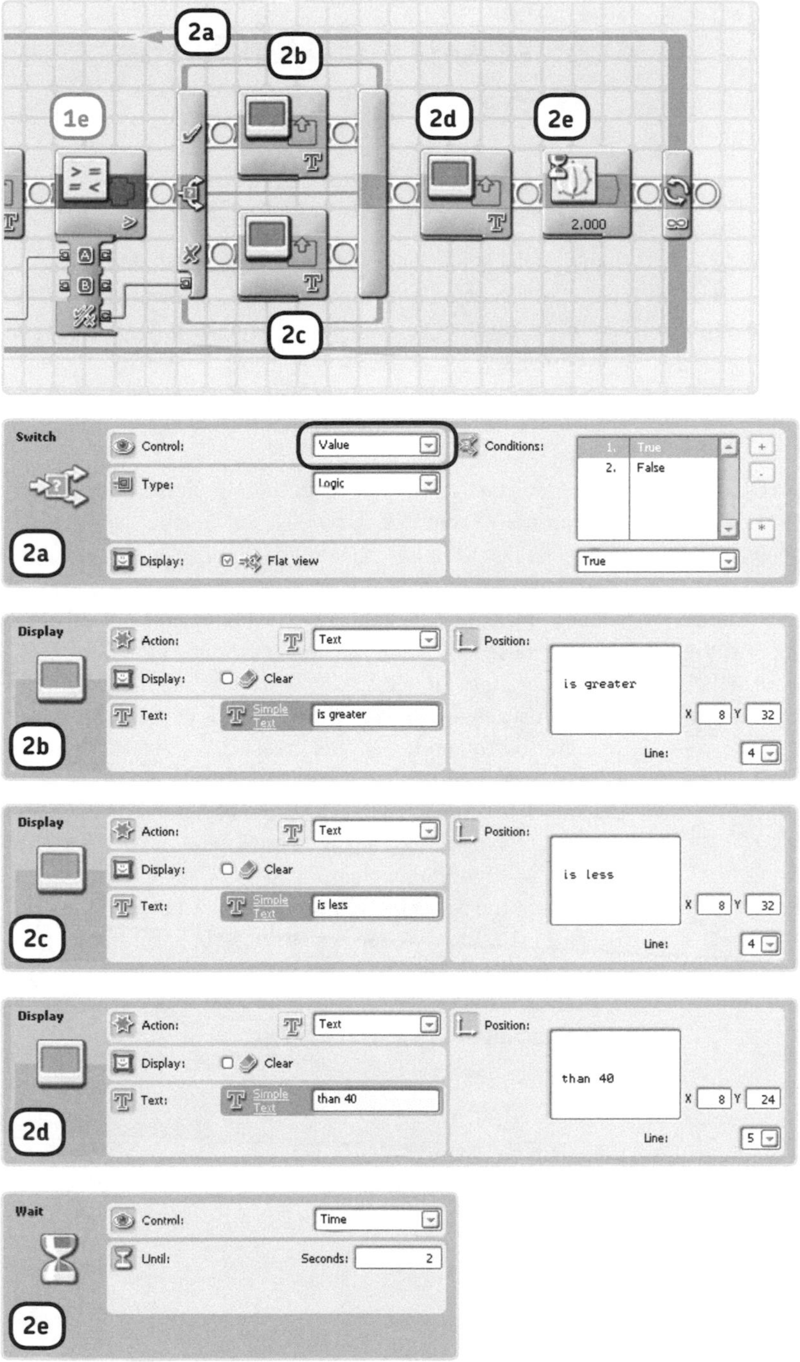

Figure 11-9: **Step 2:** If the value is greater than 40, a true signal is sent from the Compare block to the Switch block, causing the Display block in the upper part (in V) of the switch to be run. If the signal is false, the lower block (in X) is run. The next Display block (2d) in this step completes the line that is eventually displayed on the NXT screen. Every two seconds your screen will show a new line like `26 is less than 40`. The Wait block gives you enough time to read the display before the loop repeats.

using number and text data wires and switch blocks

Rather than wiring a Logic data wire into a Switch block, you can use Number or Text data wires. When using Number data wires, you could have blocks on the upper part of the switch run when the data wire value is, say, 5, and have the lower part run when the value is 3. The process is the same for Text data wires, except that when using them, you'll use text instead of numbers. You won't use this feature in this book, but if you want to learn more about it, you can refer to the NXT-G help section for the Switch block.

connecting data wires to inside switch blocks

In some cases it can be useful to connect data wires from outside a Switch block to blocks inside the switch, as shown in Figure 11-10. For example, you can do so when you want to use an Ultrasonic Sensor value outside the switch that you also want to use with a block inside the switch, without having to poll the sensor again. To do so, uncheck **Flat View**, and then connect the data wires. With Flat View turned off, click the condition icons to swap between the switch's two Sequence Beams.

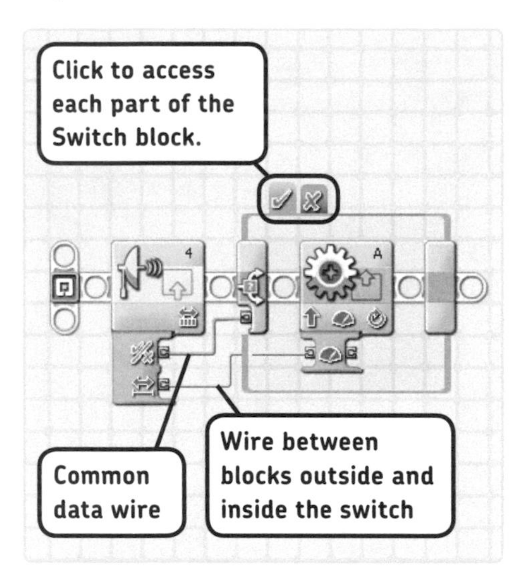

Figure 11-10: You can connect data wires from outside a Switch block to blocks inside it only when Flat View is turned off. The program shown here polls the Ultrasonic Sensor to compare its reading to a trigger value. If the result of the comparison is true, the motor is switched on, at a power level based on the Ultrasonic Sensor reading. Because there are no blocks in the other part of the switch, when the result of the comparison is false, nothing happens.

loop blocks and data wires

Previously when you've used Loop blocks, you've typically configured them to loop forever, causing the blocks inside the loop to repeat until you manually abort the program. You can also configure loop blocks to loop a certain number of times, to loop a certain number of seconds, or to repeat the blocks inside the loop until a sensor is triggered. You can also end a loop with a data wire.

When you set the Control parameter in the Loop block's Configuration Panel to **Logic**, you'll see an input plug on the right side of the Loop block, and you can connect a Logic data wire to this plug. Now you can specify whether the Loop block should stop looping when this plug receives a true or false logic value, as demonstrated in the Smart-Loop program in Figure 11-11. While looping, the program continuously checks to see whether the Touch Sensor or the Enter button on the NXT is pressed. If one or both are pressed, the Logic block outputs a true value. Because the Loop block is configured to repeat the blocks inside the loop until it receives the true signal, it ends when a sensor is pressed, and the Sound block has the NXT play a sound.

Create the **Smart-Loop** program now as shown in Figure 11-11. Smart-Loop is a very useful programming construction, because it has the program wait while checking two sensors. It actually functions like a Wait block, except that a Wait block can poll only one sensor.

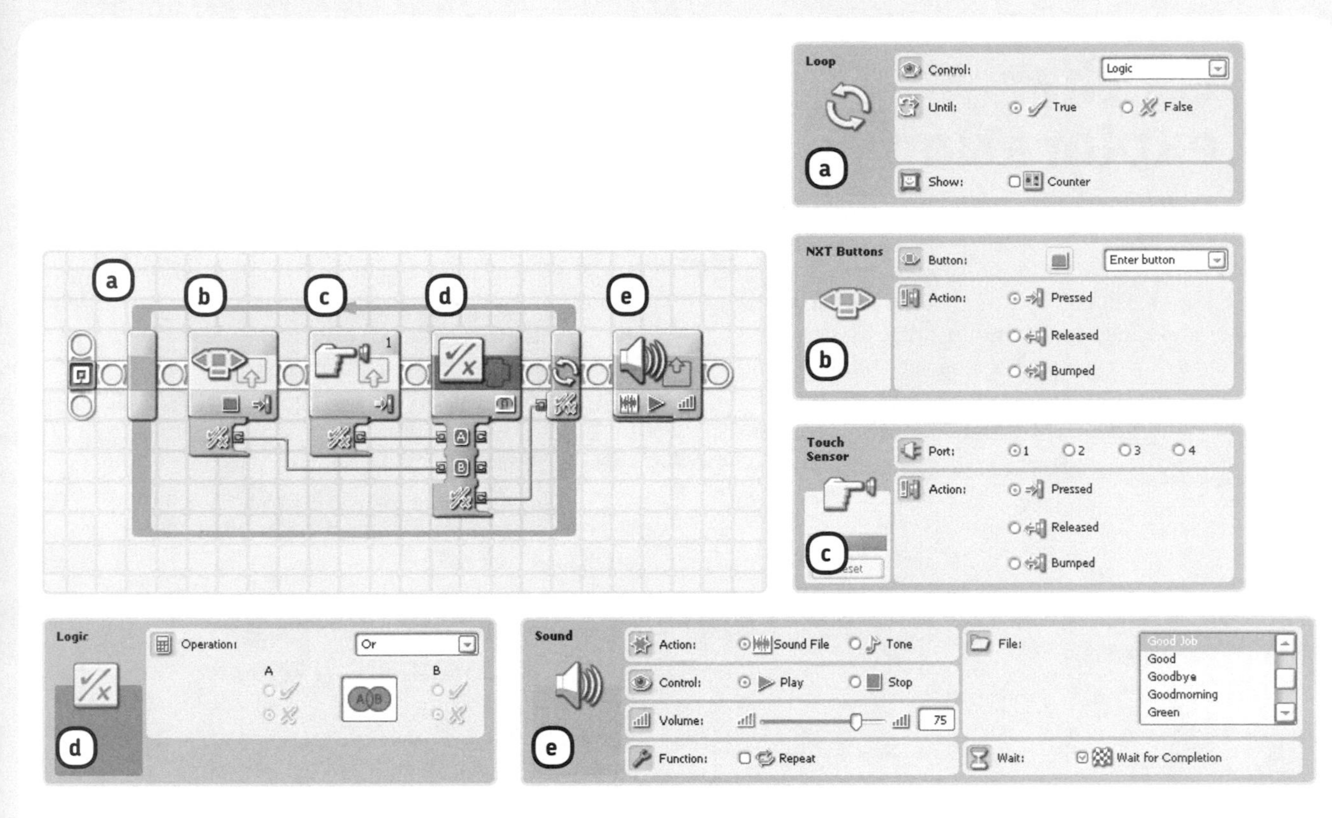

Figure 11-11: The configuration of the blocks in the Smart-Loop program

DISCOVERY #61: PRESS ANY KEY TO CONTINUE!

Difficulty: Easy

With the Smart-Loop program you learned how to create a program that pauses until one of two sensors is triggered. By adding some extra blocks, you can even poll three sensors (or more) at one time. Create a program that will make a sound when you press a button (either the Left Arrow, Right Arrow, or Enter button) on the NXT. You can see how to do this in Figure 11-12, but this program isn't finished yet. Can you connect the data wires and select the right operation in the Logic blocks to make it work?

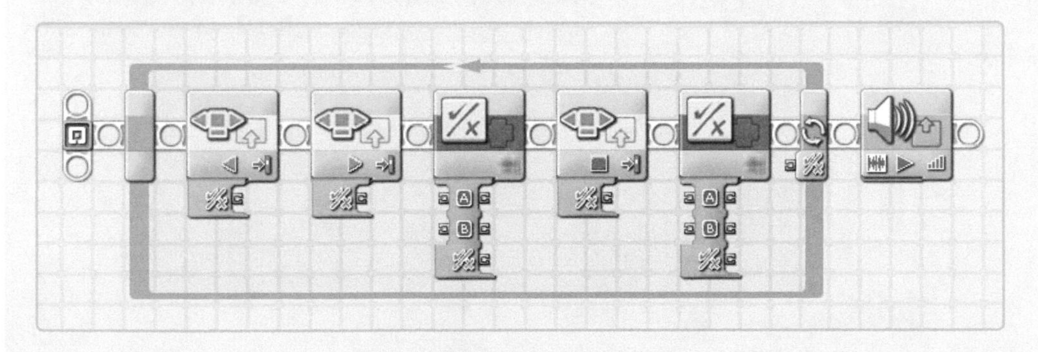

Figure 11-12: The unfinished program for Discovery #61

further exploration

In this chapter you learned about several data programming blocks, as well as some tricks for using data wires with the Loop and Switch blocks. These techniques will allow your robot to process and combine sensor values in order to use them as inputs for actions such as making sounds or moving. You'll see practical examples of this in the next three chapters of this book. For now, practice your newly acquired skills in the following discoveries.

DISCOVERY #62: ARITHMETIC ROTATIONS!

Difficulty: Hard
For this discovery, add a motor to SmartBot that you can rotate easily by hand (connect a wheel to it). Next, create a program to display the values of the wheel motor's Rotation Sensors on the NXT screen, and wire both values into a Math block. If the Touch Sensor is pressed, a Math block should multiply the two sensor values; otherwise, it should simply add them. Display the results on the NXT. As you rotate the motors, you should change sums or products on the NXT screen.

BUILDING DISCOVERY #11: BIONIC HAND!

The SmartBot has proved useful for experimenting with some advanced programs, but it can't really do much. In this discovery, you'll build a robotic claw, attach it to your hand, and use sensors or the NXT buttons to control it. Be sure to add the Ultrasonic Sensor and program the robot so that the arm will warn you when the sensor sees you approach a wall. Use the programming techniques you learned in this chapter to display several sensor values on the screen or to make sounds based on sensor measurements.

NOTE You may want to use SmartBot in the next chapter, but you don't have to do so. If you'd like to work on Building Discovery #11, take the SmartBot apart if you like.

using variables, constants, and playing games on the NXT

If you've made it this far, you're just a few steps away from mastering all of the programming skills in this book. By now you've learned how to use many different programming blocks, as well as how to work with essential tools such as data wires. This chapter completes the programming section of this book by teaching you how to use the NXT memory with variables as well as how to use constants.

In addition to learning about variables and constants, you'll create a program in this chapter that combines almost all of the programming skills you've learned so far. This somewhat advanced program will demonstrate yet another possibility of the NXT system: It will let you play a real game on the NXT!

using variables

Think of a *variable* as a kind of suitcase that can carry information. When a program needs to remember a value (such as a sensor reading) for later use, it puts that value in the suitcase and stows it away. When the program needs to use the value, it opens the suitcase and uses the stored value. The variable is stored in the NXT's memory until it's needed.

Once information is stored in a variable, you can access it from other parts of your program. For example, you could store an Ultrasonic Sensor value in the suitcase and compare it later to a new sensor reading.

The program can access this stored information at any time while running, but the data is lost once the program stops. To store and access variable information, you use a *Variable block*, which you'll recognize by a suitcase icon on the block. Figure 12-1 shows an overview of what happens when you use variables.

defining a variable

Each variable has a *name* and contains a value. For example, a variable might be called Reading with a value of 56. Like data wires, a variable can contain either a numeric value (such as 56), a logic value (such as true), or a text value (such as "hello").

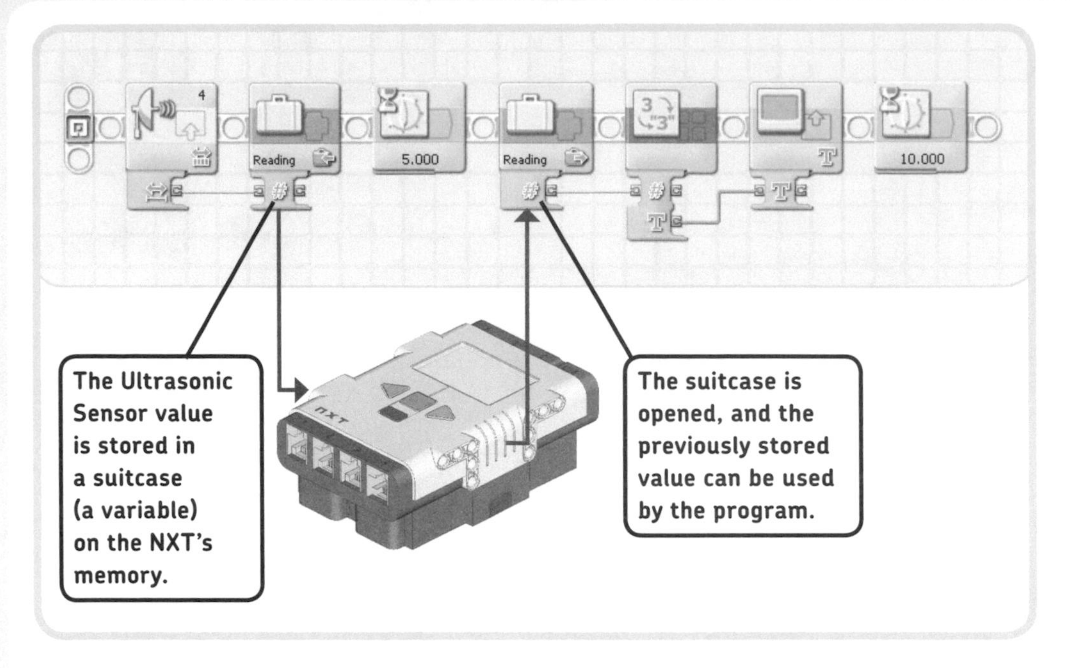

Figure 12-1: Values (like the Ultrasonic Sensor reading shown here) can be stored in variables in the NXT's memory. Once a value is stored, a program can retrieve it from memory in order to use it. You'll learn what this program does and how to create it later in this chapter.

But before you can use a variable in your program, you'll need to define it in the Edit Variables dialog box, as shown in Figure 12-2.

To delete a variable, open the Edit Variables dialog box, select the variable you want to delete, and click **Delete**.

NOTE You can use variables only in the program in which you've defined them.

using the variable block

Once you've defined a variable, you can use it in a program with the Variable block. The Variable block can *read* values from or *write* (store) values to a variable in the NXT memory. To configure a Variable block, first select the variable you want to read or write in the List box on the Configuration Panel. Next, in the Action box, specify whether you want to write a value to the variable or read a value from it. When you choose **Write**, the block stores the value that you enter in the Value box of its Configuration Panel.

If a data wire is connected to the block's Value data plug, the value carried by this wire is stored. If a value was stored in this variable previously, the old value is erased, and the new one is stored instead.

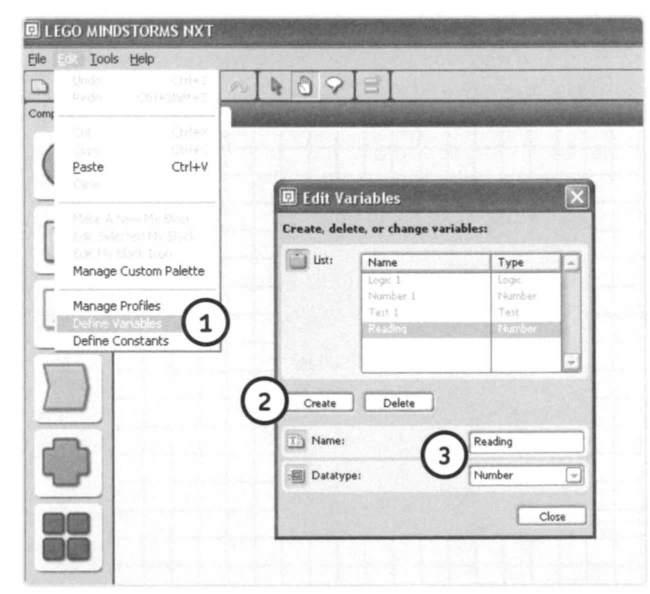

*Figure 12-2: Defining a variable in the variable editing window. **Step 1:** Select **Edit ▸ Define Variables**. **Step 2:** Click **Create** to make a new variable. **Step 3:** Enter a name for the variable (**Reading**), and select the data type (**Number**). Once you've done this, click **Close**.*

When configured to read a value, the Variable block retrieves the information from the NXT's memory and outputs it with a data wire so that the value can be used in the program to, for example, define a motor's Power setting.

When reading a variable's value, the value isn't changed, so if you read it again with another Variable block, you'll get the same value.

creating a program with a variable

The **Smart-Variable** program shown in Figure 12-3 stores the Ultrasonic Sensor value in a variable called Reading. After five seconds, it retrieves the value from the variable and displays it on the NXT screen, which means that the value you see on the NXT screen represents what the sensor measured five seconds ago. (Before configuring this program, define a variable called Reading to carry numeric information, as shown in Figure 12-2.)

Smart-Variable demonstrates the concept of using variables in a program, but it's a very basic program. Once you've created it, as shown in Figure 12-3, continue practicing with variables in Discovery #63.

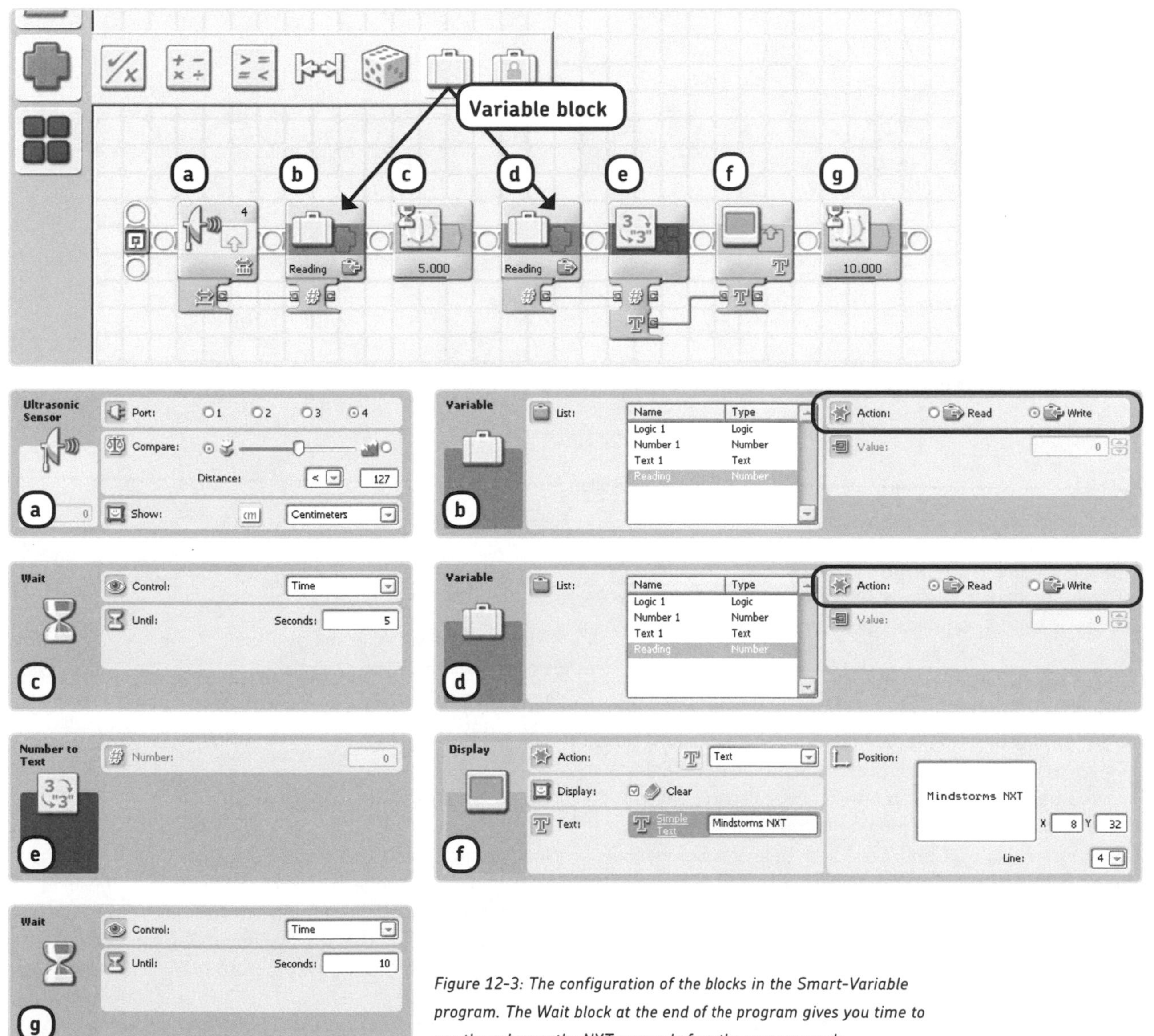

Figure 12-3: The configuration of the blocks in the Smart-Variable program. The Wait block at the end of the program gives you time to see the value on the NXT screen before the program ends.

DISCOVERY #63: OLD VS. NEW!

Difficulty: Medium

You'll create another program with variables that repeatedly compares new sensor readings to the sensor value stored in a variable at the beginning of the program. If the new reading is higher, the robot should say "Yes"; otherwise, it should say "No." Figure 12-4 shows part of the program.

HINT What is your first step when creating a program with data wires? Which Variable blocks must read or write a value? How do you connect the data wires?

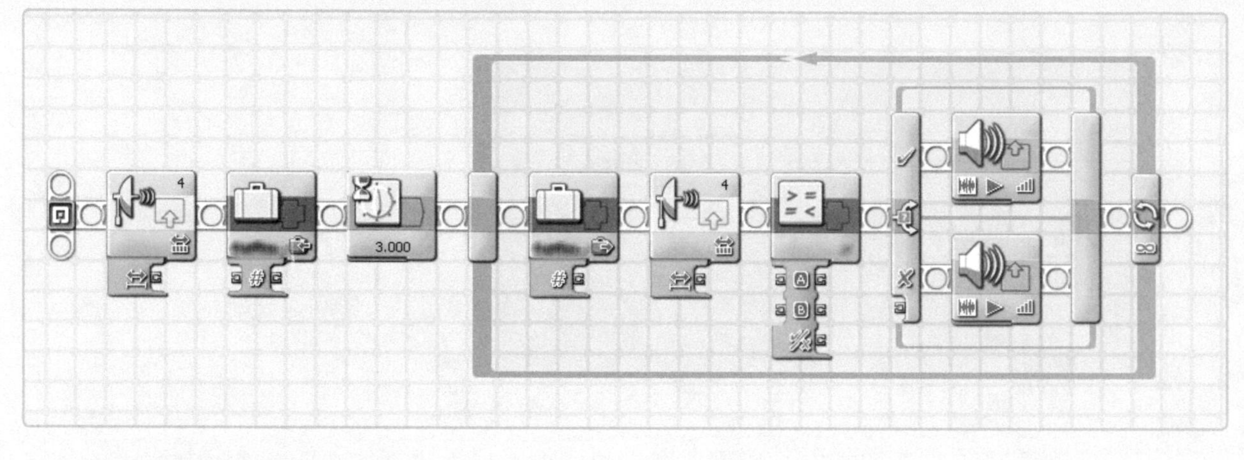

Figure 12-4: A starting point for Discovery #63

changing variable values

Previous sections have shown you how to write values to a variable and how to read values from one. Sometimes you'll want to change a variable's value, for example to increase its value by one, such as when you want to use a variable to track a high score or a total number of Touch Sensor presses. The **Smart-Count** program you'll now create demonstrates how to use a variable to track the number of times a Touch Sensor is bumped (pressed and then released).

You begin by defining a new number variable called **PressCount** to store the number of Touch Sensor bumps. The program will wait until the Touch Sensor is bumped, at which point the PressCount value will increase by one. To make the counting ongoing, you'll also use a Loop block in this program.

But how do you increase the variable's value by one? As shown in Figure 12-5, you use a Variable block to read the PressCount value. Then you transfer this value to a Math block, which adds one to the value. The result of this addition is wired into another Variable block configured to write (store) the new value in the PressCount variable, which is therefore now increased by one. The result of the addition is stored in the variable and displayed on the NXT screen. By using this method, you can change any variable's value. This sample has shown you how to add one to a value, but you can use the same method to subtract from a value.

Now create the program as shown in Figure 12-5.

initializing variables

When programming with variables, it's important to *initialize* them by giving them a starting value. You do this in the Smart-Count program in Figure 12-5 by setting the PressCount value to **0** at the beginning of the program. Initializing variables makes a program more reliable by making sure that each time you run the program, it will function the same way because it starts at the same place.

a b c d e f g h

PressC... 1 PressC... PressC...

A B

Variable

List:

Name	Type
Logic 1	Logic
Number 1	Number
Text 1	Text
PressCount	Number

Action: Read Write

Value: 0

a

Loop

Control: Forever

Show: Counter

b

Wait

Control: Sensor

Sensor: Touch Sensor

Port: 1 2 3 4

Action: Pressed Released Bumped

c 0

Variable

List:

Name	Type
Logic 1	Logic
Number 1	Number
Text 1	Text
PressCount	Number

Action: Read Write

Value: 0

d

Math

Operation: Addition

A 0 B 1

e

Variable

List:

Name	Type
Logic 1	Logic
Number 1	Number
Text 1	Text
PressCount	Number

Action: Read Write

Value: 0

f

Number to Text

Number: 0

3 "3"

g

Display

Action: Text

Display: Clear

Text: Simple Text Mindstorms NXT

Position:

Mindstorms NXT

X 8 Y 32

Line: 4

h

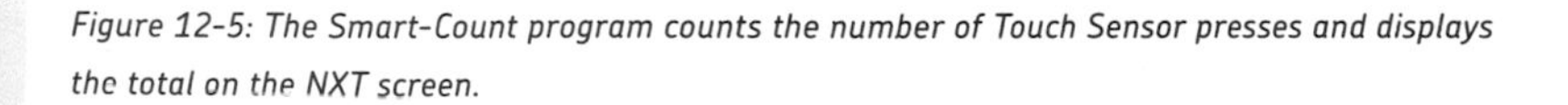

Figure 12-5: The Smart-Count program counts the number of Touch Sensor presses and displays the total on the NXT screen.

It's a good idea to initialize variables at the beginning of a program, especially when you're creating bigger programs. If you don't, you may find that the initial variable value is not equal to 0, which may cause the program to malfunction or to act unpredictably. Although usually not the case, your robot might, for example, say that you pressed the Touch Sensor five times, even when you haven't touched it all, because the variable count was left at 5.

DISCOVERY #64: SMARTER COUNT PROGRAM!

Difficulty: Hard
In this discovery, create a program based on the Smart-Count program that uses a variable to track the number of times the arrow buttons are pressed. If you press the Right Arrow button, the variable's value should increase by one; if you press the Left Arrow button, it should decrease by one. Display the variable's value on the NXT screen.

HINT Use a Loop block to wait until a button has been pressed (as in the Smart-Loop program, as shown in Figure 11-11), and then use a Switch block to determine which button was pressed. Into the Switch block add the blocks you're using to modify the variable's value and blocks that wait until the NXT button is no longer pressed.

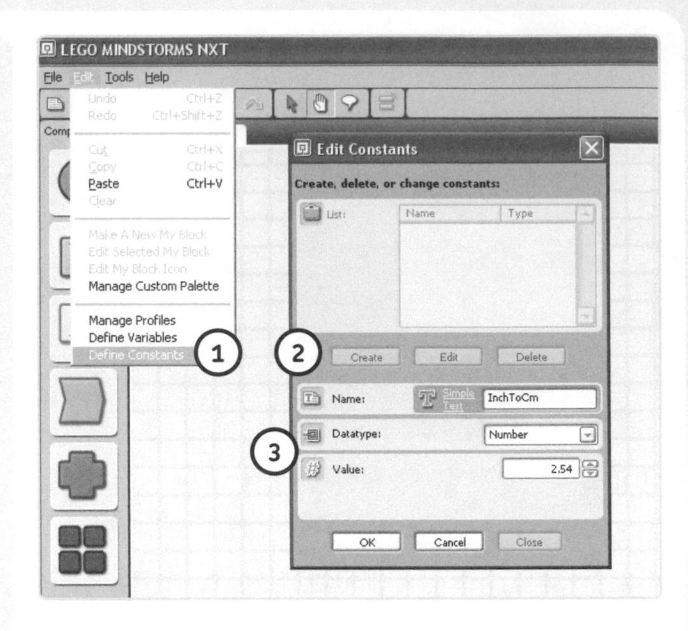

*Figure 12-6: To define a constant, select **Edit ▸ Define Constants** (1), and click **Create** (2). Give your constant a name, select the type of value it should contain, and enter that value (3).*

using constants

A *constant* is a value that cannot be changed while a program is running. For example, if your program needs to convert a distance measured in inches to one in centimeters, you would use a constant to multiply the inch value by 2.54. By creating a constant in an NXT program (let's call the constant *InchToCm*) with the value 2.54, each time you want to perform this conversion you would simply use the constant InchToCm.

You must define a constant before you can use it, as shown in Figure 12-6. Once you've defined one, you can use it in any of your programs.

using the constant block

You use Constant blocks to integrate defined constants into your program in one of two ways. In the Action box of the Constant block's Configuration Panel, you specify whether to select a constant from of a list of defined ones (Choose from list) or to use a custom constant (Custom). When you select **Choose from list**, the Constant block functions like a Variable block configured to read a value: The constant is read and output with a data wire, which then transfers the value to another block in the program. When you select **Custom**, you use the block to output a value that you enter in the Configuration Panel. (You don't need to define this constant in advance.) The Custom configuration is useful when a single value needs to be transferred to multiple blocks, as shown in Figure 12-7.

creating a program with constants

The **Smart-Constant** program (shown in Figure 12-7) uses both configurations of the Constant block as it converts inches to centimeters and displays both values on the NXT screen. You'll use a *custom* constant for the number of inches to convert so that when you run the program again to convert another value to centimeters, you'll need to adjust only the value in the Constant block, rather than in both the Math and Display blocks. You'll choose the constant for the conversion (2.54) from the list of constants as defined previously in Figure 12-6.

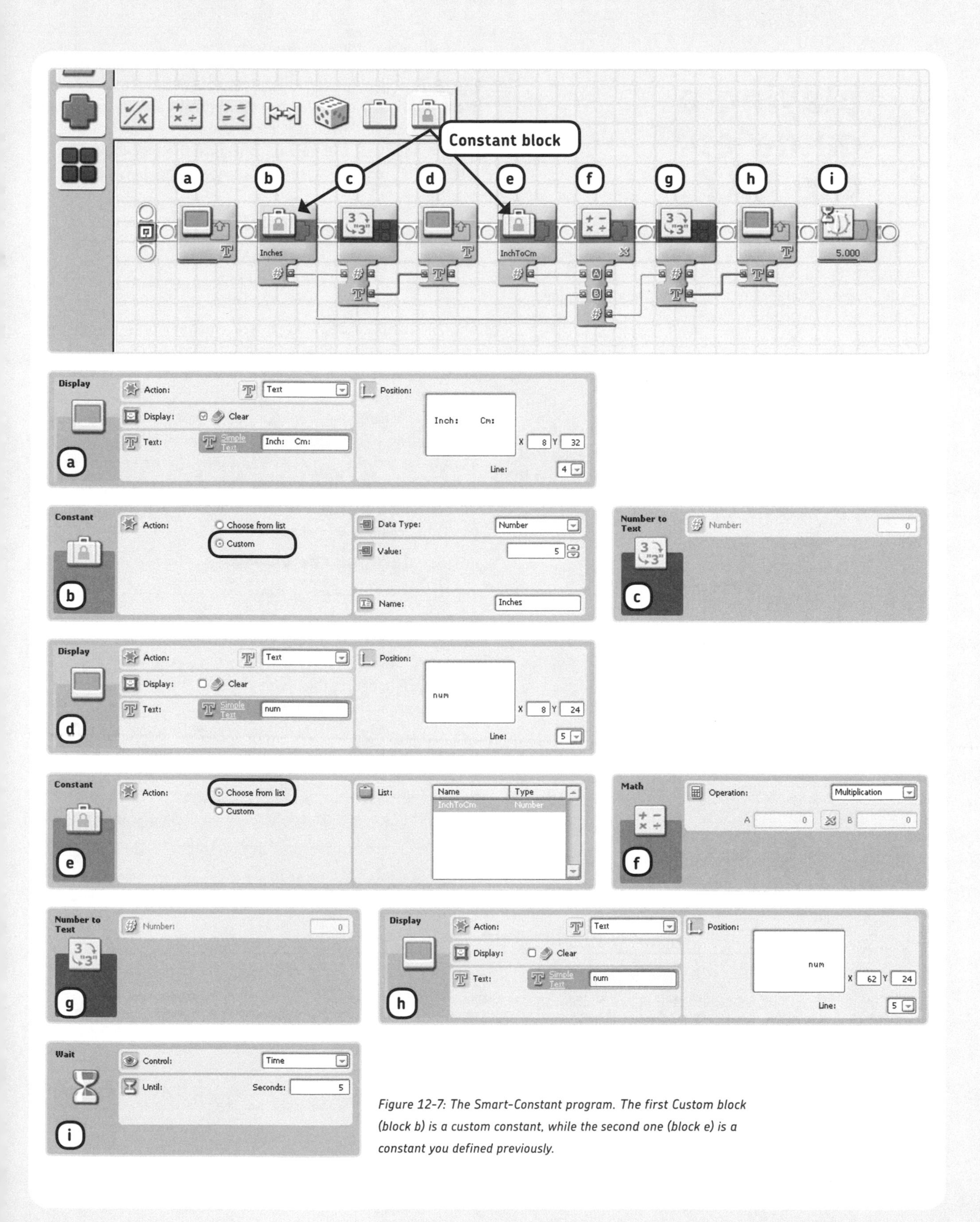

Figure 12-7: The Smart-Constant program. The first Custom block (block b) is a custom constant, while the second one (block e) is a constant you defined previously.

playing a game on the NXT

You'll now create one large program that combines many of the programming techniques that you've learned throughout this book. I've introduced each programming technique and block previously with short example programs, but now you'll see how to combine them for use in a larger program. Your goal will be to create a program that lets you play a game on the NXT. The NXT screen will display the game, and the NXT buttons will be the game controllers.

When playing the game, targets will appear on the NXT display randomly on the left and right. When a target appears, you'll press the left or right NXT button quickly to destroy it, and the program will pop up the next target, as shown in Figure 12-8. The more targets you hit within 30 seconds, the higher your score. If you miss a target, your score will decrease.

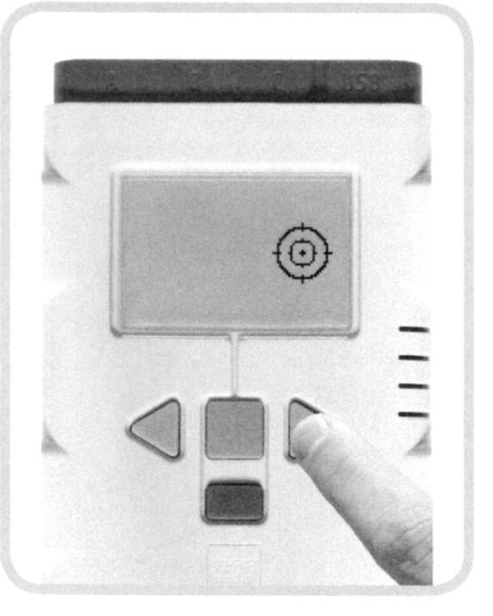

Figure 12-8: The Smart-Game program in action

NOTE At this point you should be able to re-create programs without viewing all the Configuration Panels, so from now on you'll see text comments on the programming images that tell how to configure the blocks. If the text doesn't tell how to set a certain setting, leave the default.

Because the **Smart-Game** program is much larger than any program you've created so far, take a good look at exactly what the program should do, as shown in Figure 12-9. The program consists of several sections, each of which has a specific function in the program. You'll create these sections step-by-step. As you create each section, you'll find descriptions of how each section works and why you need it in the program.

Steps 1 through 6 configure the blocks that display a target on the screen and adjust the score when you try to hit a target by pressing a button. You configure the Loop block in step 7 (repeating these actions for 30 seconds) and add a block to set the initial score to zero. Also in step 7, you add blocks to display the final score.

defining the variables

This program has three variables. The *Score* variable tracks the total score, and therefore it is defined to carry numerical values. The *Position* variable stores the target's position, and the *Button* variable stores which button was pressed.

Before updating the Score variable, the program checks to see whether Position and Button variables are equal. To make this comparison easier, you'll make Position and Button number variables. You'll indicate left with 1 and right with 2. For example, if the user presses the Left Arrow button, the program will set the Button variable to 1. Once you've defined the Score, Button, and Position variables as numerical, you're ready to begin creating the program.

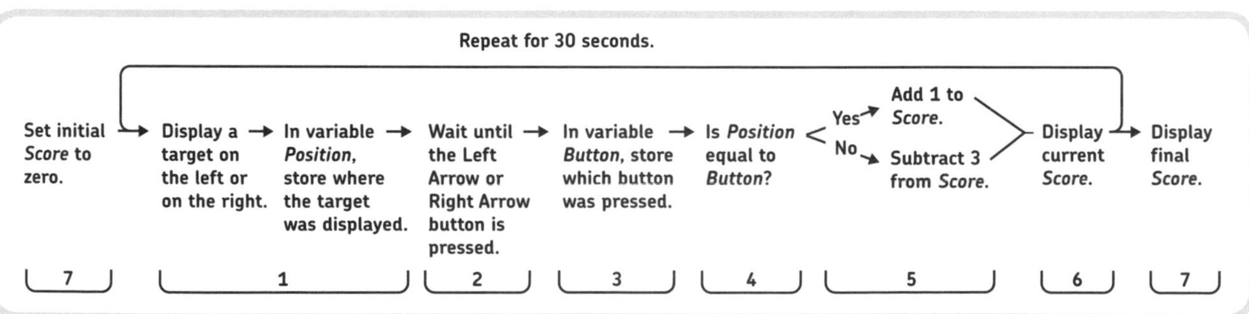

Figure 12-9: An overview of the Smart-Game program. The names of the variables used in this program are italicized. You'll re-create this program with programming blocks in seven steps.

step 1: displaying a target randomly

The program uses a Switch block to decide whether to display a target on the left or right side of the NXT screen, as shown in Figure 12-10. To make this a random decision, you use a Random block to generate a number between 0 and 99, and you use a Compare block to see whether the generated value is less than 50 (the block's output is then true). Because half of the numbers generated by the Random block will be less than 50, half the time that the program runs these blocks, the Compare block will output a true value, causing the target to be displayed on the left. The rest of the time the Compare block's output will be false, and the target will be displayed on the right. The target's position is stored in a variable, to be compared later with the user input. For the left position, 1 is written to the Position variable; for the right, 2 is written to the variable. Figure 12-10 shows the blocks that display the random target and set the Position variable.

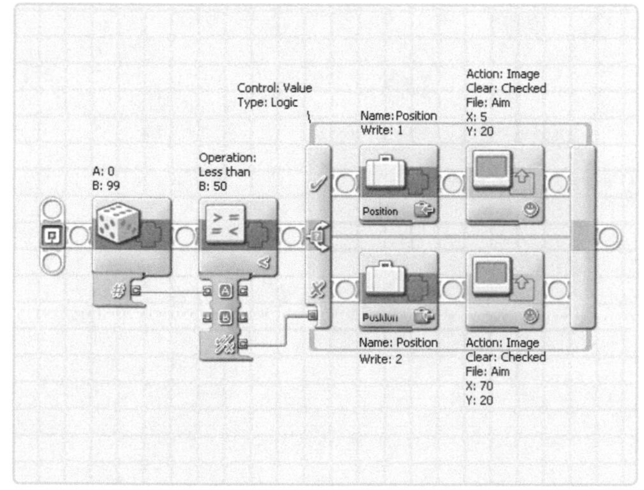

Figure 12-10: The configuration of the blocks placed in step 1

step 2: waiting until a button is pressed

Once the target has been displayed, the program waits until the user presses an arrow button on the NXT. You accomplish this with the technique of polling two sensors at the same time, as you used in Figure 11-11. Configure the blocks for step 2 as shown in Figure 12-11.

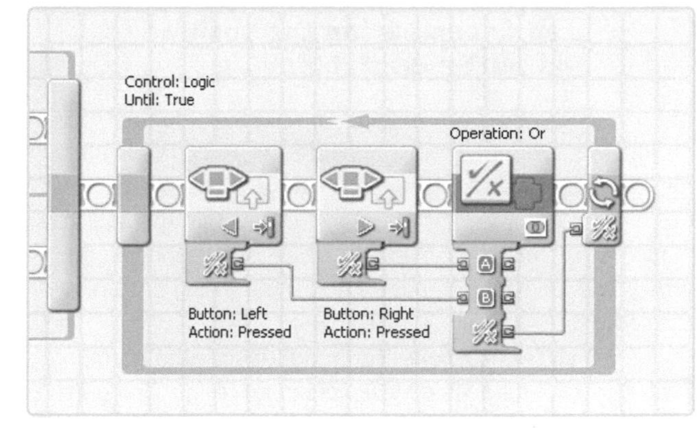

Figure 12-11: The configuration of the blocks placed in step 2

step 3: storing which button is pressed

When you press a button in the previous step, the Loop block stops looping, and the program jumps to the blocks you'll place in this step. Because programming blocks are run so quickly, the next block will run before you can release the pressed button, so you'll assume that the button is still pressed as the program runs through the blocks in this step. As a result, you'll simply use a Switch block to see whether the Left button is pressed. If so, you set the Button variable to 1, and you wait until the button is released. If the left button is not pressed, you know that the right button is, and you set the Button variable to 2 and wait until the right button is released, as shown in Figure 12-12.

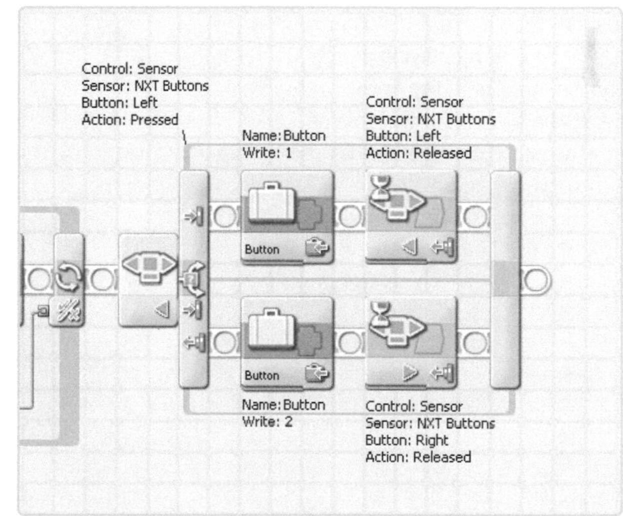

Figure 12-12: The configuration of the blocks placed in step 3

step 4: comparing the position and button variables

At this point in the program you know the position of the target (as stored in variable Position), as well as which button the user pressed (stored in Button). If these values are equal (both are 1 or both are 2), you know that the user has made the correct choice, and the score is increased by one. If they are not equal, the user clicked the wrong button, and the score is decreased by three.

You'll place the blocks required to modify the score in a Switch block. The Switch block gets input from a Compare block that determines whether Position is equal to Button. If the two values are equal, the Compare block outputs true, and the blocks in the upper part of the Switch are run.

You'll place the blocks to modify the score in the next step. In this step, configure the blocks to check whether Button equals Position, as shown in Figure 12-13.

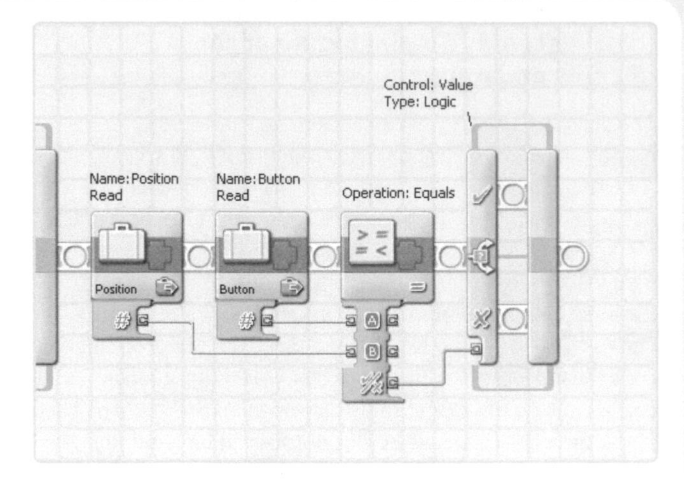

Figure 12-13: The configuration of the blocks placed in step 4

step 5: adjusting the score

If Button and Position were equal, the blocks in the upper part of the Switch shown in step 4 are run. In this step, you'll place the blocks that add one to the Score variable. The NXT will tell you that you've made the correct choice by displaying a *V* sign on the screen and by playing a high note. The blocks in the lower part of the switch subtract three from the score, display an *X* sign on the screen, and play a low note. Figure 12-14 shows how to configure the blocks.

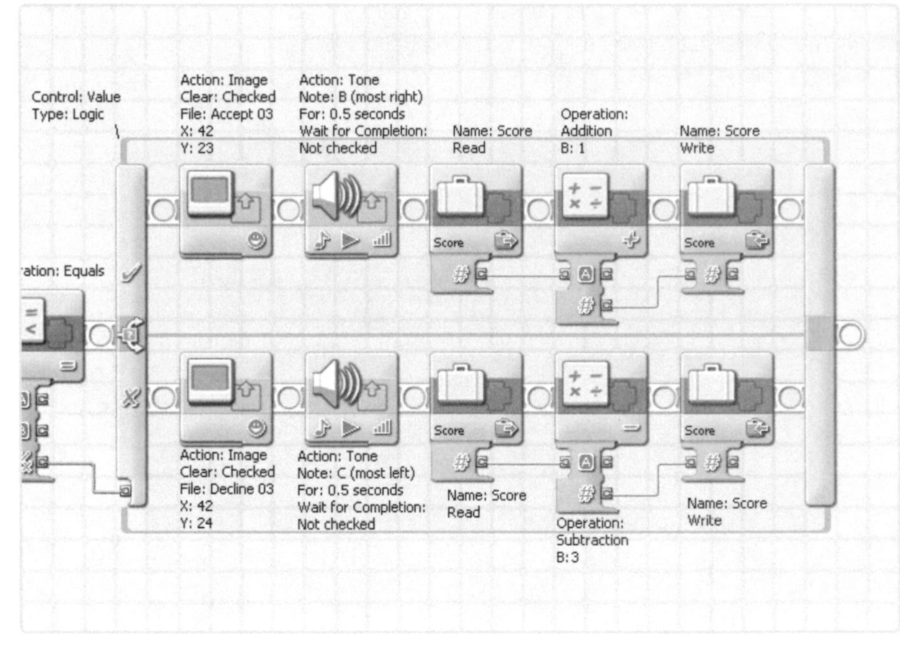

Figure 12-14: The configuration of the blocks placed in step 5

step 6: displaying the current score

The blocks in step 6 display the current score and give you time to see the score as well as the *V* or *X* sign on the NXT screen before the program goes back to display a new target. Figure 12-15 shows the blocks that display the score.

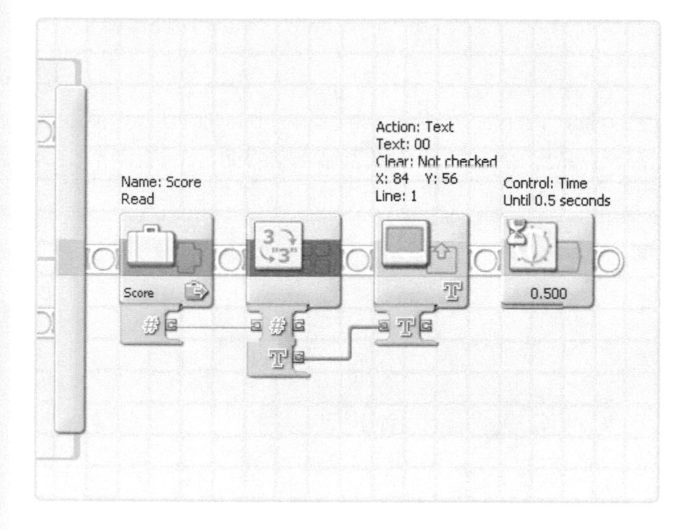

Figure 12-15: The configuration of the blocks placed in step 6

step 7: making the program repeat for 30 seconds

The blocks you've placed up to now display a target, wait for a button press, and modify the score based on user input, but you don't yet have a working game. To turn your steps into a working game, drag all the blocks you placed in steps 1 to 6 into a Loop block configured to loop for 30 seconds.

Finish the program by adding a block at the beginning to set the initial score to zero, and add blocks to display the final score. Figure 12-16 shows the finished program.

Congratulations! You've finished the Smart-Game program. Now run it and see which score you can get in 30 seconds.

NOTE If you're unable to get this program to work as described, download a working version from the companion website.

expanding the program

You've now seen how to combine your programming skills to create one large program. But the fun doesn't have to end here. Here are some suggestions for customizing this program that will make it truly your own:

* Add more sounds to the program: Play applause when the correct button is pressed and a scream if you press the wrong one. Also, make your program play a tune of your own creation once the game is over.

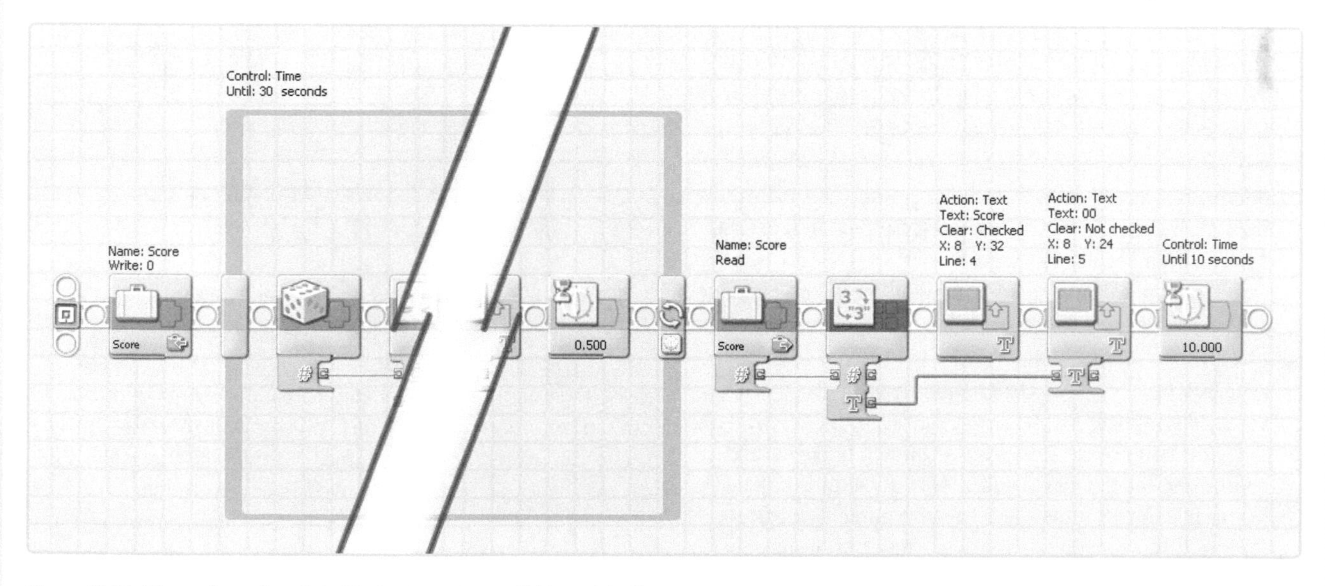

Figure 12-16: The configuration of the blocks placed in step 7. Most of the blocks in the Loop have been hidden here to better focus on the newly placed blocks in this step.

* Create a startup screen that is displayed on the NXT at the beginning of the program. Use images, lines, and drawings to customize your screen. Add extra blocks to have the NXT display `Press the enter button to start the game!` on its screen.
* Although the Smart-Game program uses only the NXT, that doesn't mean you can't use the motors and sensors connected to the robot. Think of a way to use the Color Lamp in your program, or have the Hand motor move faster and faster as you make more mistakes.

further exploration

Congratulations! You've completed all of the programming chapters in this book. Now you're ready to learn how to build and program three more cool robots in the last part of this book.

But before you do, take a look at the following discoveries. These are more difficult than the other ones you've done so far, but remember that there are multiple solutions to each. Give them a try. If you think you have a good solution, post it to the companion website (*http://www.discovery.laurensvalk.com/*) to see what other readers think!

DISCOVERY #65: SMART-GAME PROGRAM ADVANCED!

Difficulty: Hard

Expand the Smart-Game program so that it will display targets at three locations instead of only two. Display the third target in the middle of the screen, and use the Enter button to hit it.

DISCOVERY #66: BRAIN TRAINER!

Difficulty: Expert

Try this if you're up for a really complicated discovery. Create a program that displays random addition problems (for example, 7 + 6). The user decides whether the displayed answer is correct by pressing the Left Arrow button (incorrect) or the Right Arrow button (correct), as shown in Figure 12-17.

Half the time, a correct answer should be displayed. The game should display new sums for 30 seconds, after which it should display the total score. Each correct button click is worth one point, but each mistake costs five points.

You can create this program using only the programming blocks discussed in this book, but it's still quite difficult to make. See the companion website for the full program, as well as a description of how it works.

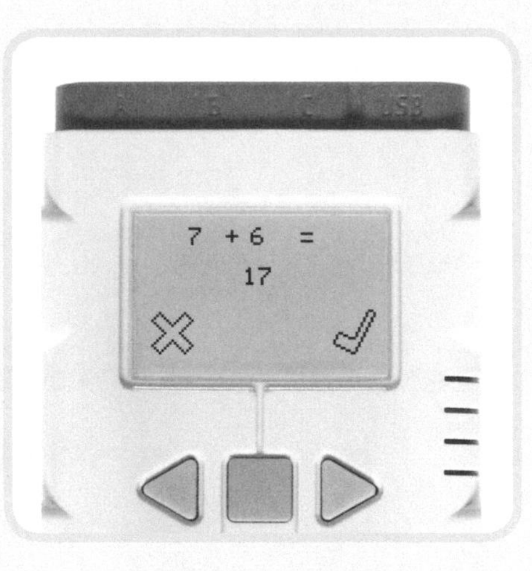

Figure 12-17: An example addition problem in NXT Brain Trainer game

BUILDING DISCOVERY #12: WHACK-A-MOLE!

The Smart-Game program is just one example of a game you can play on the NXT. This discovery challenges you to build a construction similar to the Whack-a-Mole game. Use NXT motors to randomly make moles pop up out of a frame of LEGO pieces, and then use the NXT buttons or Touch Sensors to whack the moles down. Create your own advanced program for this machine to turn it into a real Whack-a-Mole game. You can make the NXT display high scores, play sounds, or increase the difficulty level as time passes.

advanced robot projects

snatcher: the autonomous robotic arm

Previous chapters taught you a great deal about programming NXT robots. Now that you've reached an advanced level of programming, you're ready to build some more complicated robots in this and the next chapters. This chapter will teach you to build the Snatcher, an autonomous robotic arm that can find and pick up objects, as shown in Figure 13-1.

The Snatcher uses two NXT motors to control a set of treads, allowing the robot to move in any direction. You control its movement like you controlled the Explorer in Chapter 4 by adjusting the power and direction of the Driving motors to control the robot's speed and driving direction (as shown in Figure 4-4).

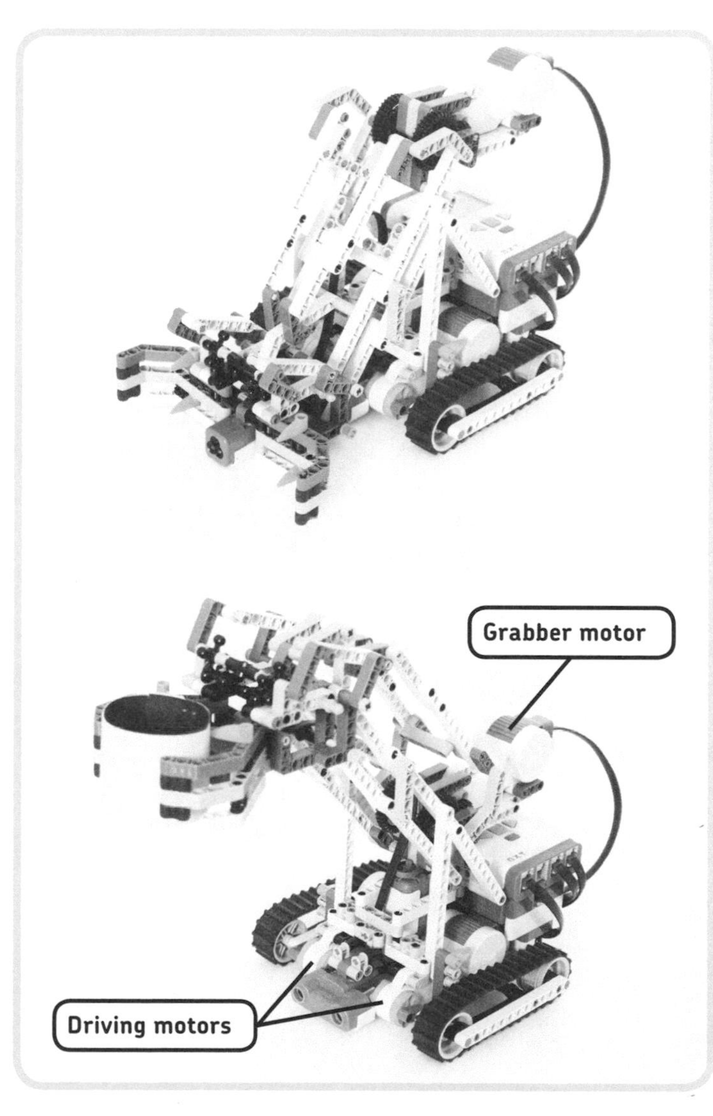

Figure 13-1: The Snatcher can find and pick up objects.

understanding the grabber

Driving around on treads may be interesting, but the really cool part of this robot is its multifunctional grabber. Normally grabbing and lifting objects requires two motors: one to grab the object and another to lift it. The Snatcher robot requires just one motor (which I call the *Grabber motor*) to accomplish both tasks, because of a unique construction of LEGO beams, axles, and gears. You'll take a look at this technique now, but you will really understand how it works when you build the robot.

the grabbing mechanism

Figure 13-2 shows how the Snatcher grabs objects. As the NXT motor spins forward, a small gear (indicated with a number *1*) makes a bigger gear (*2*) rotate in the direction shown by the arrow in the figure. This rotation starts a chain reaction of moving beams, which ultimately causes the grabber to grasp objects positioned between its fingers (*6*). When the motor spins backward, the reverse occurs, and the grabber opens. The beams marked *3*, *4*, and *5* simply transfer the rotational movement of the motor to the grabber so it can close its claws. Construction *4* connects the beams numbered *3* and *5* in order to allow smooth movement, even when the grabber arm is positioned as shown in the second image of Figure 13-1.

the lifting mechanism

Once the Snatcher has grabbed an object, it can lift it. But before you look at how the robot does this, you'll see a simplified version of the situation.

As shown on the top of Figure 13-3, as you move the big gear (*2*) with your hand, the beams numbered *7*, which represent the grabber and the motor, move as indicated by the gray arrows. No matter what the movement is, these parts remain parallel to the ground, and the parts labeled *8* remain perpendicular to the ground. To really understand how this works, build the structure shown with the parts in your NXT robotics kit.

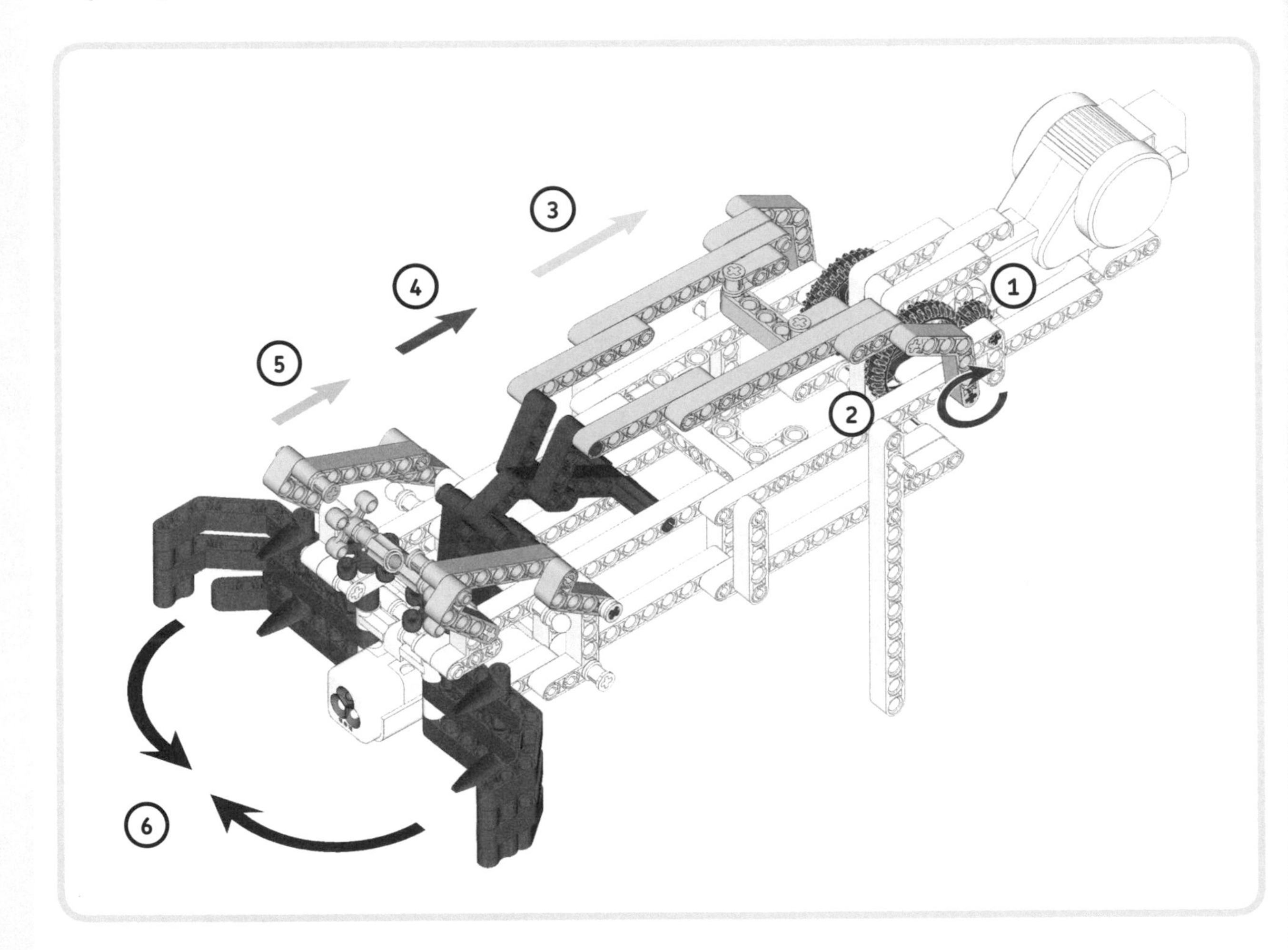

Figure 13-2: Grabbing objects by turning the motor forward

This mechanism works only because gear *2* does not move relative to the beam labeled *9*, since they're connected with a pin (*10*) as shown in the figure (it makes parts *2* and *9* form one fixed part). So, turning gear *2* directly makes the beam labeled *9* move.

The bottom of Figure 13-3 shows how the Snatcher lifts objects. The mechanism is actually quite similar to the one shown on the top, except that the real Snatcher doesn't have this pin (*10*), which would turn parts *2* and *9* into one fixed part, enabling the gear (*2*) to directly control the beams labeled *9*. This robot uses different parts to lock the beams to the gear.

Once the Snatcher has grabbed an object, the parts labeled *10* (shown on the bottom of Figure 13-3) no longer move as they did in Figure 13-2. Instead, they lock in position and move just like the beams numbered *9* shown in this image. Since the gear (*2*) is connected to the number *10* parts, it is now indirectly also connected to the number *9* parts (because the constructions are fixed), and the robot can lift objects.

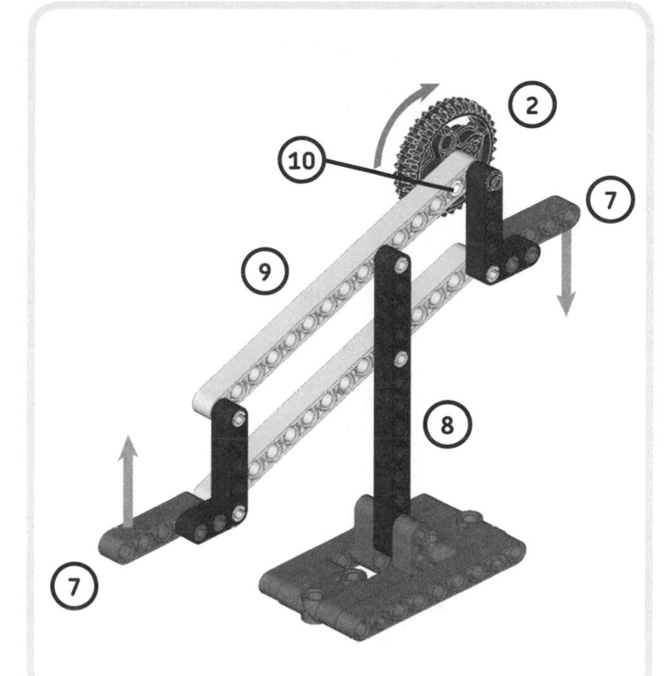

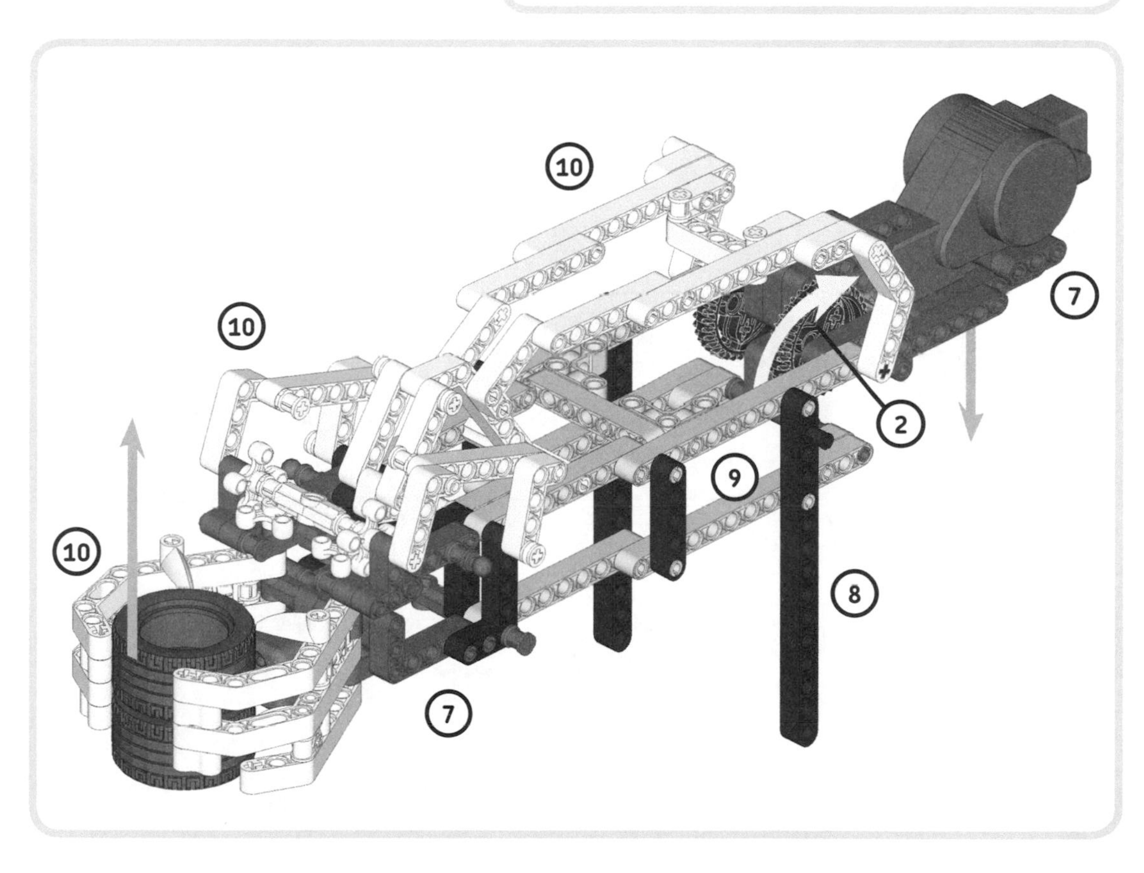

Figure 13-3: Once the Snatcher has grabbed an object, it can lift it. Here is a simplified overview of the lifting technique (top), with an illustration of the actual robot (bottom).

building the snatcher

Now that you've gotten a sense of how the Snatcher's grabber mechanism works, it's time to build the robot to learn how it really works. To do so, follow the directions on the next pages, but first select the pieces you'll need, as shown in Figure 13-4.

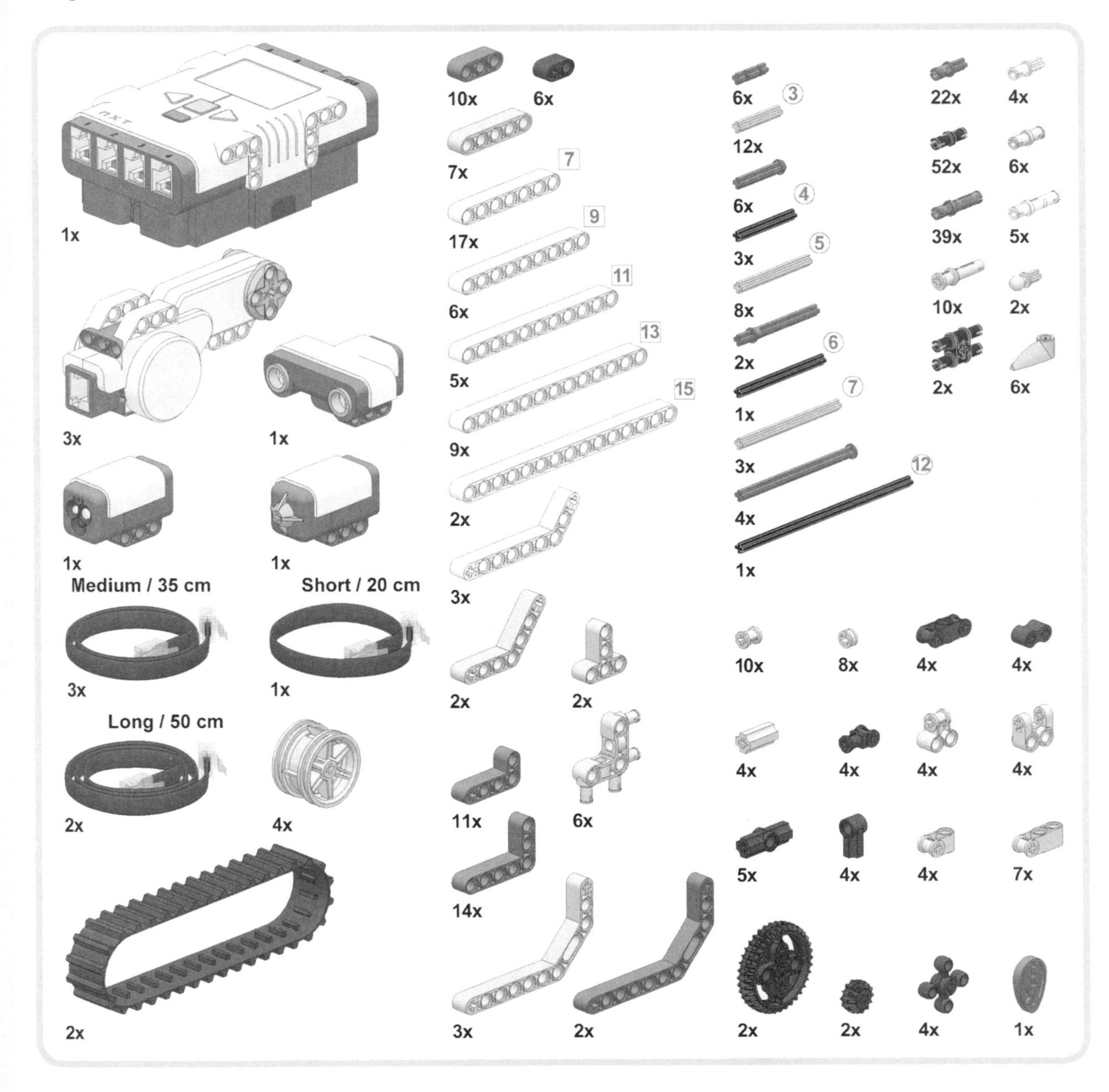

Figure 13-4: The required pieces to build the Snatcher

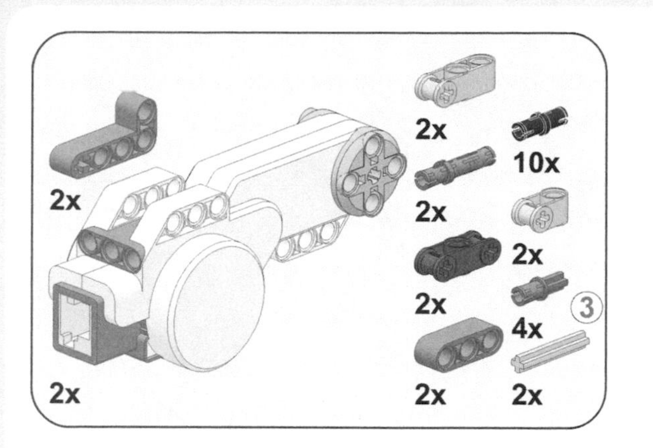

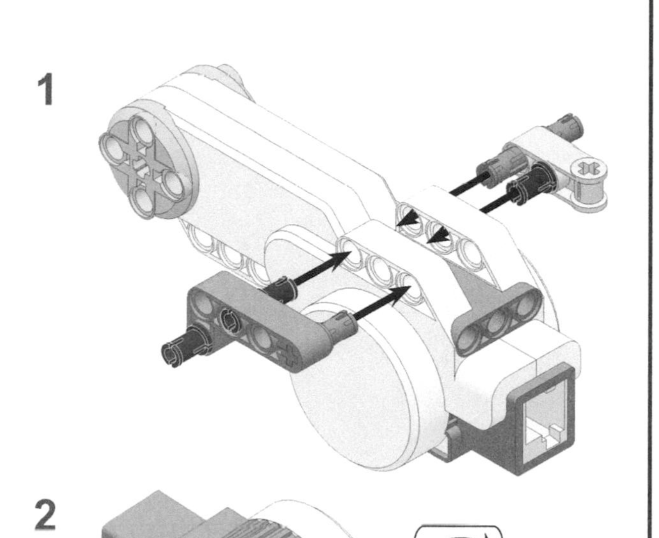

2

3

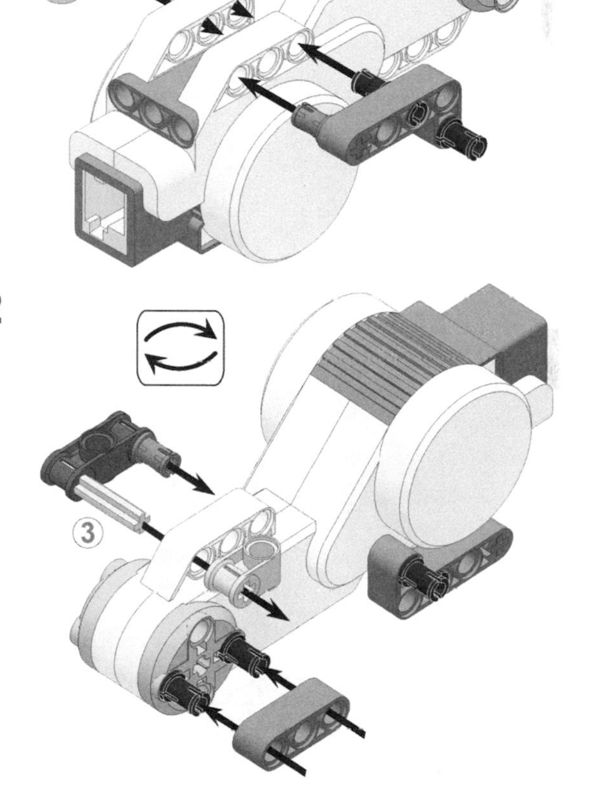

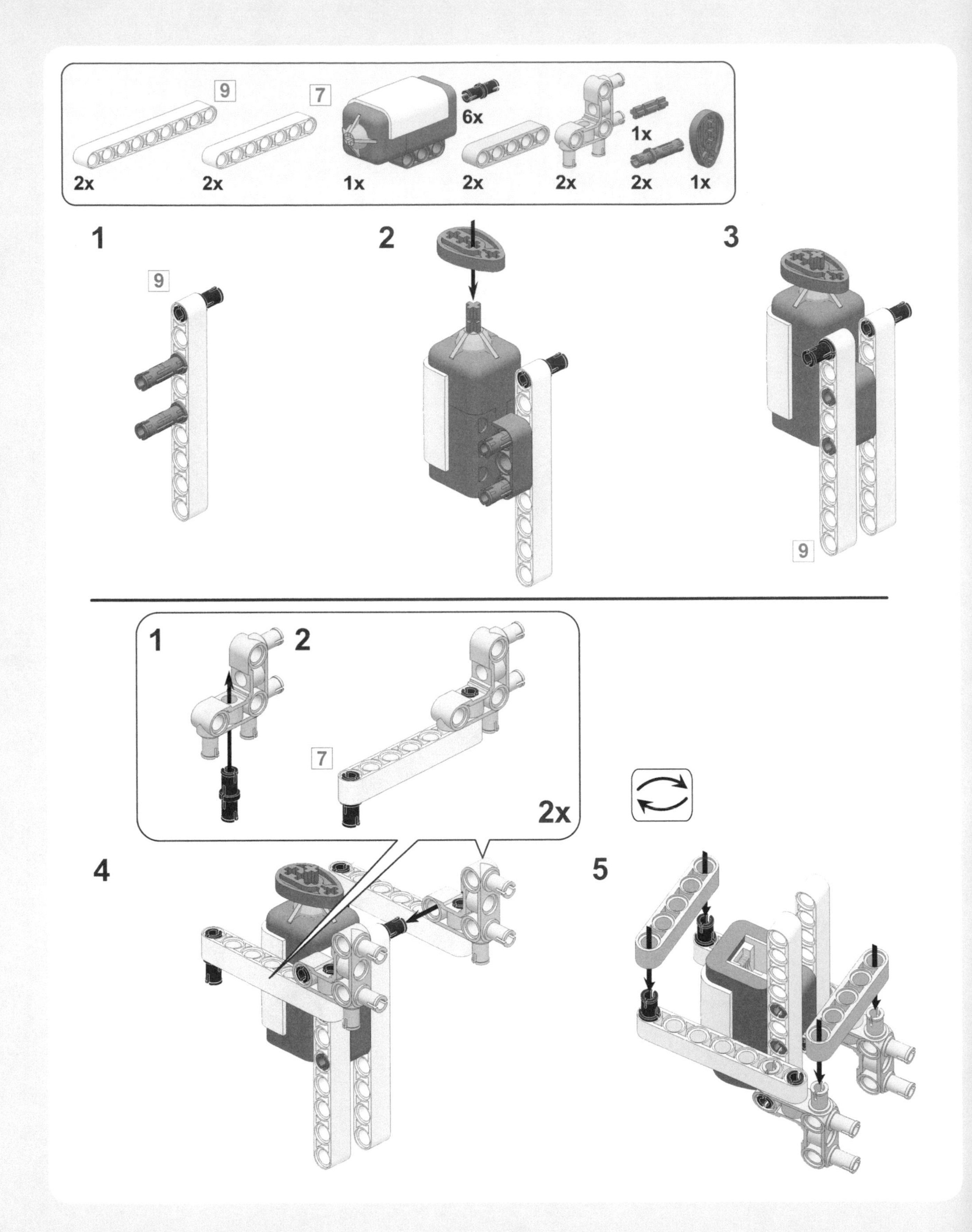
9
7
2x
2x
1x
6x
2x
2x
1x
2x
1x
1
2
3
9
9
1
2
7
2x
4
5

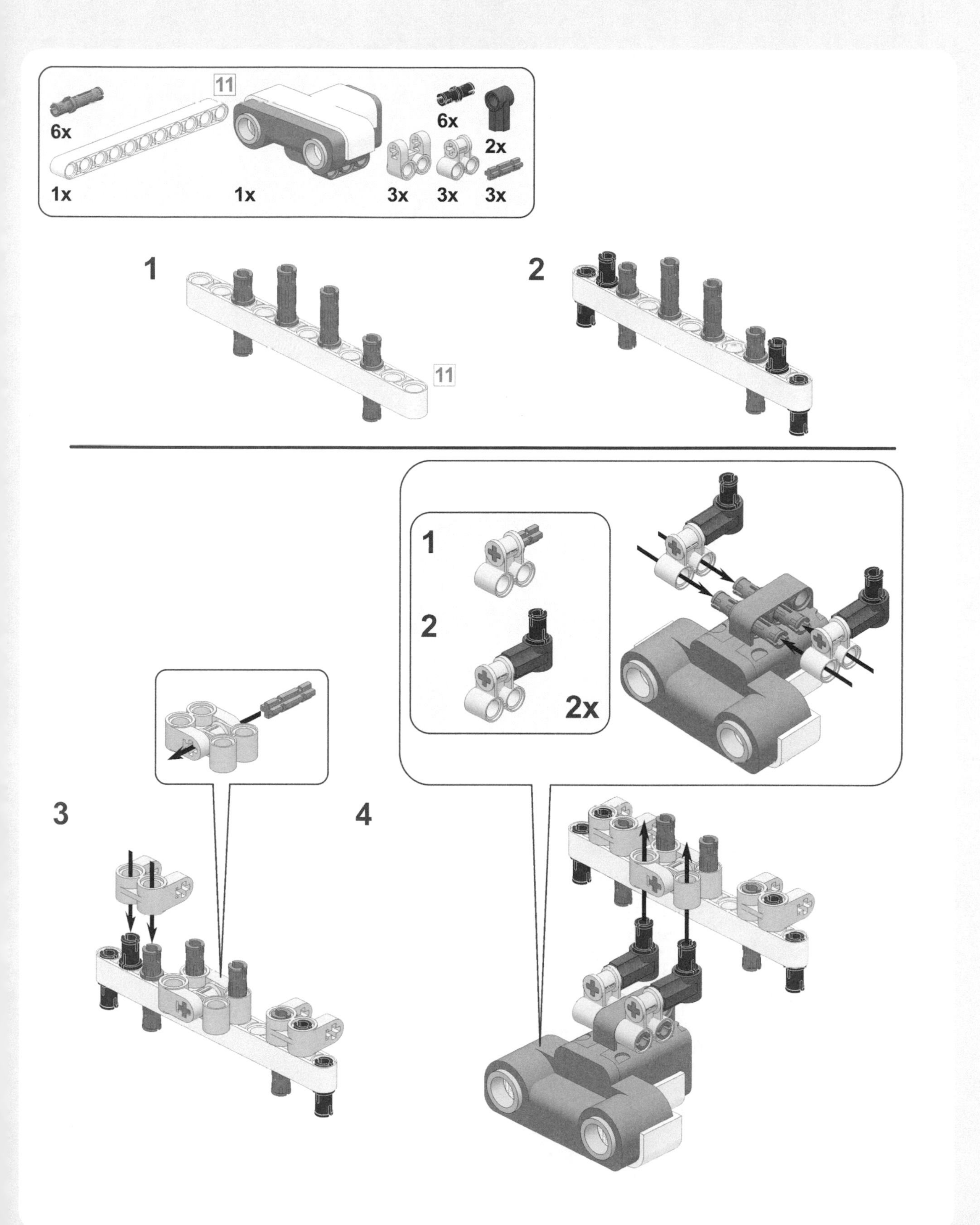
11
6x
6x
2x
1x
1x
3x
3x
3x
1
11
2
1
2
2x
3
4

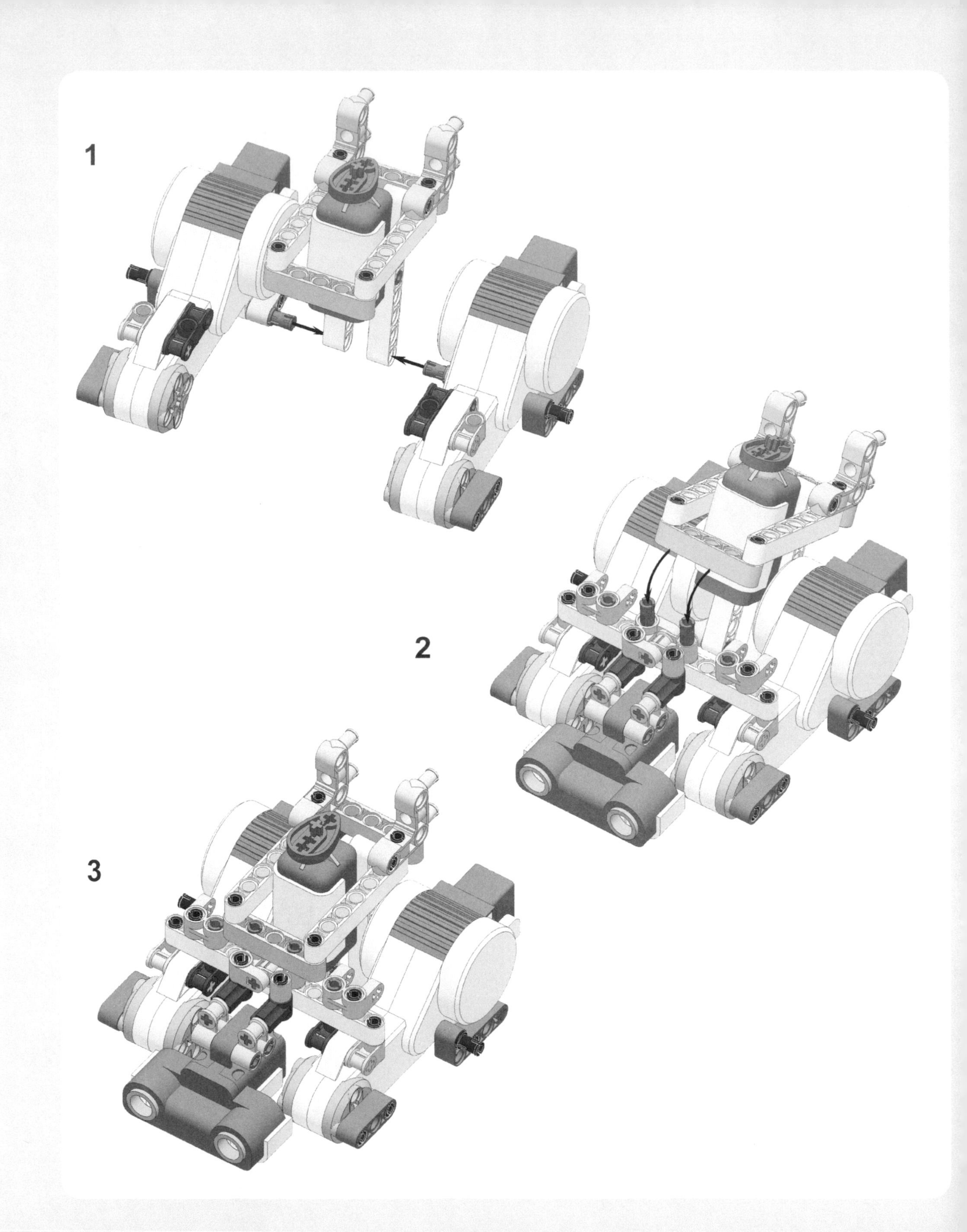
1
2
3

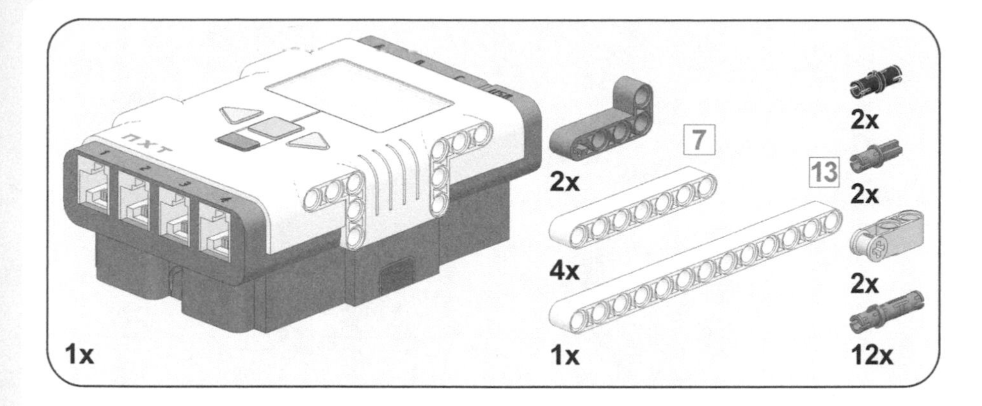

7
13
2x
2x
2x
4x
2x
1x
1x
12x

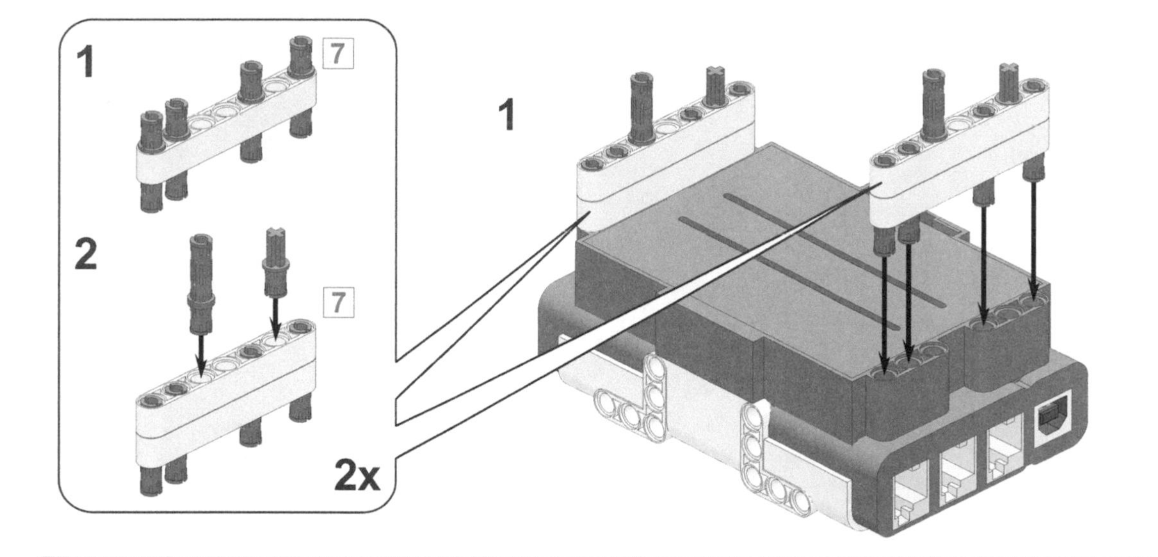

1
7
2
7
2x
1

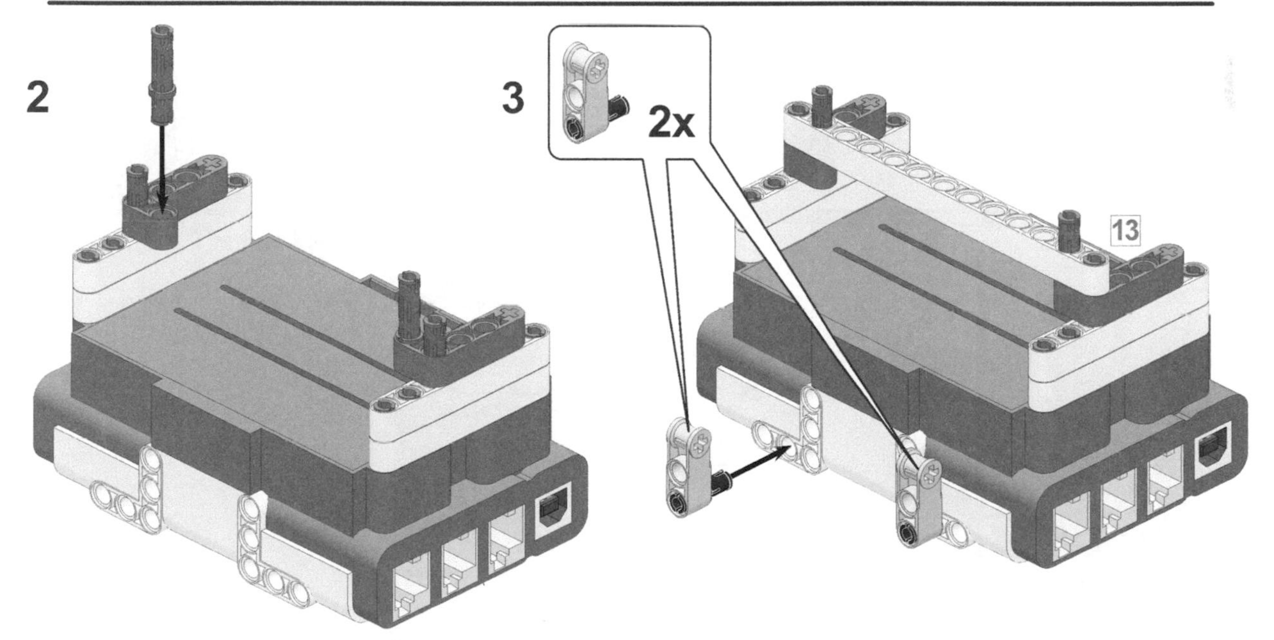

2
3
2x
13

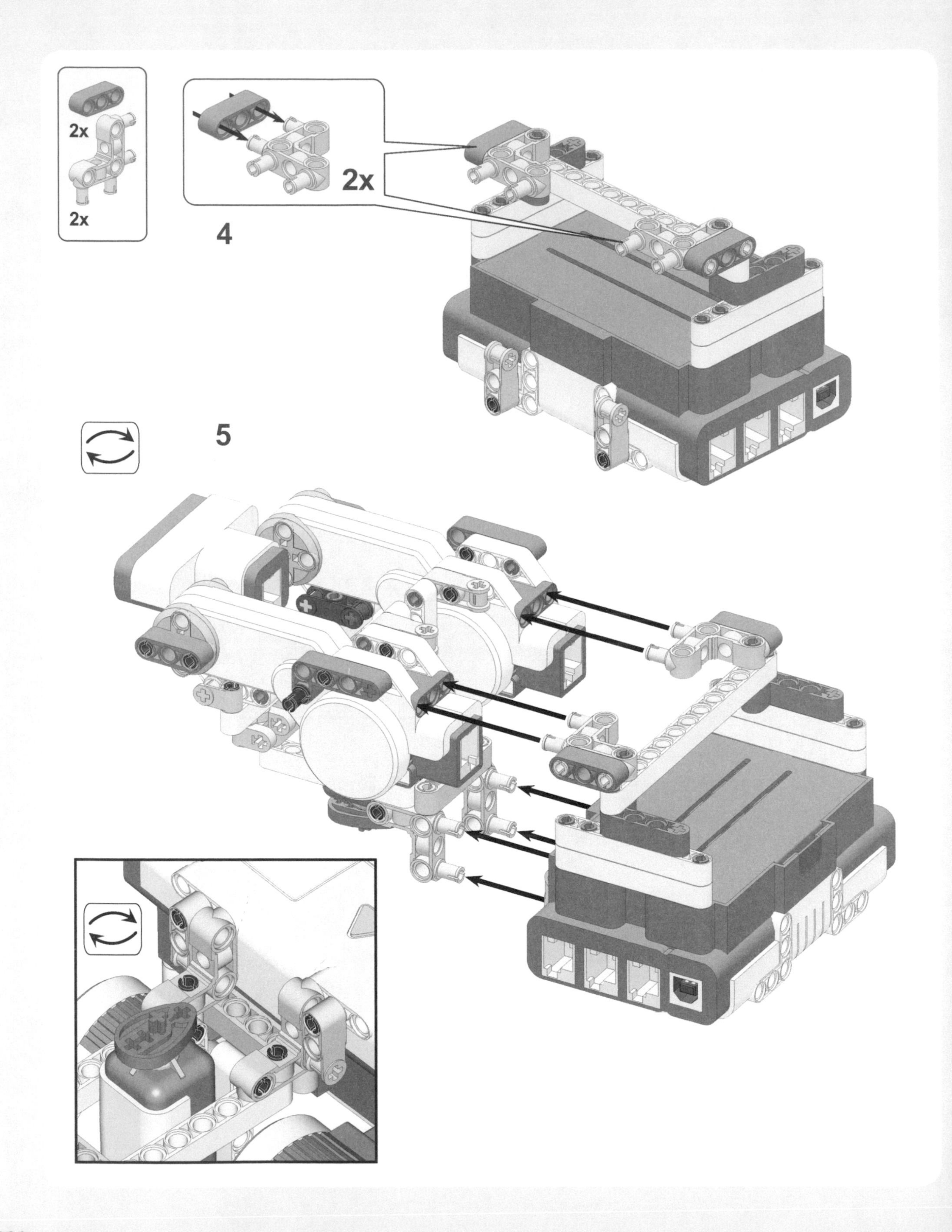
2x
2x
2x
4
5

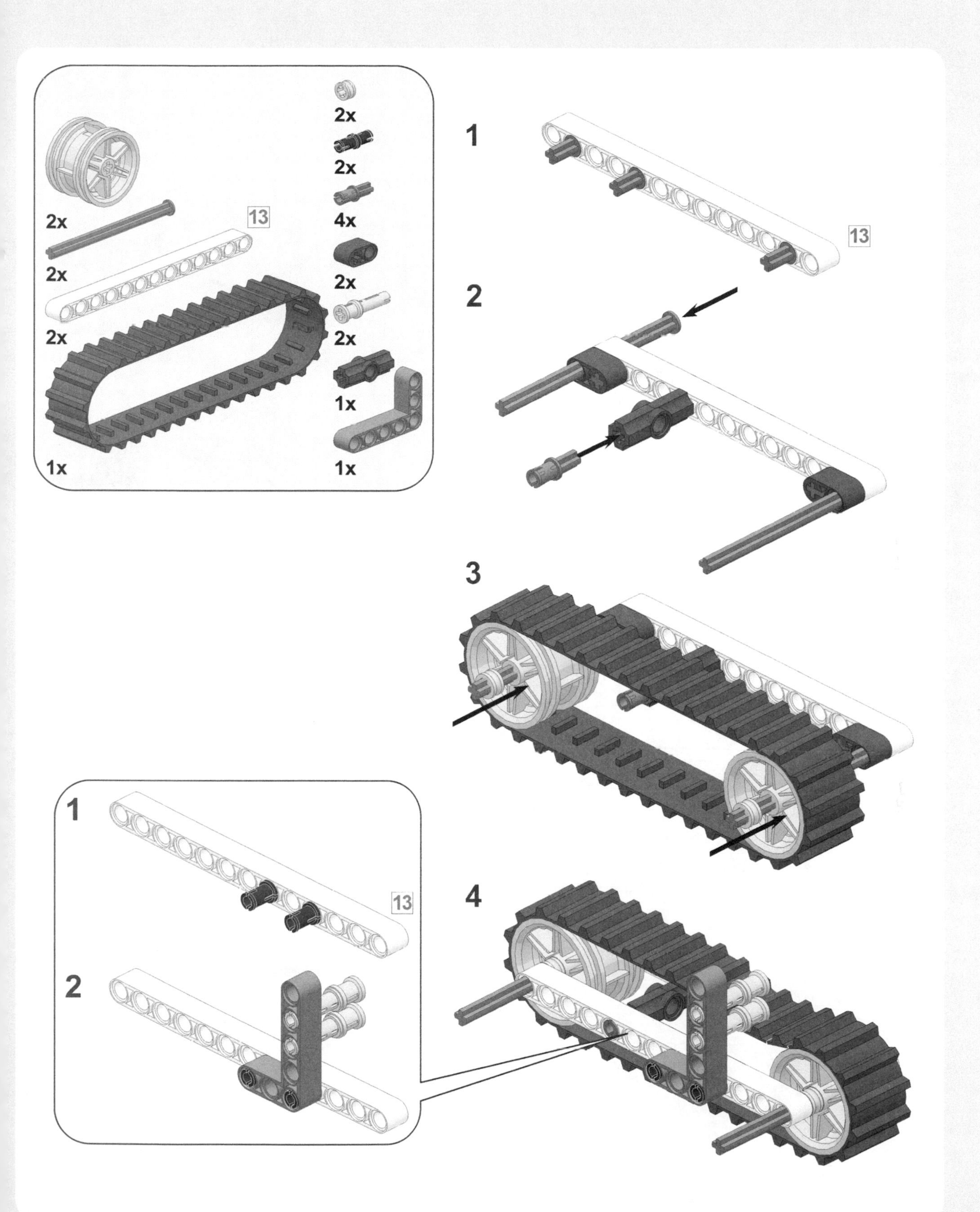
2x
2x
2x
4x
13
2x
2x
2x
2x
2x
1x
1x
1x
1
13
2
3
4
1
13
2

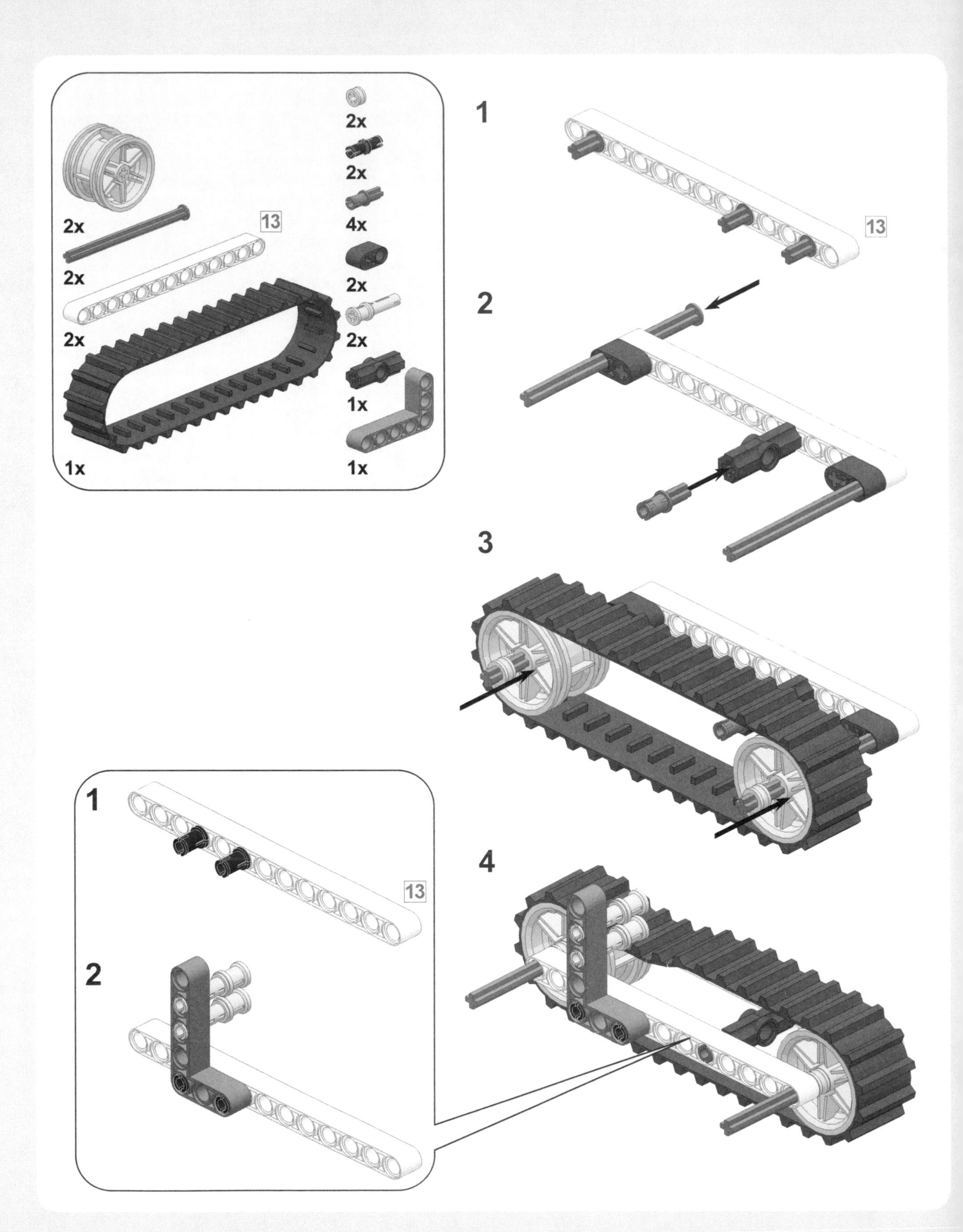
2x
2x
2x
13
4x
2x
2x
2x
2x
1x
1x
1x
1
13
2
3
4
1
13
2

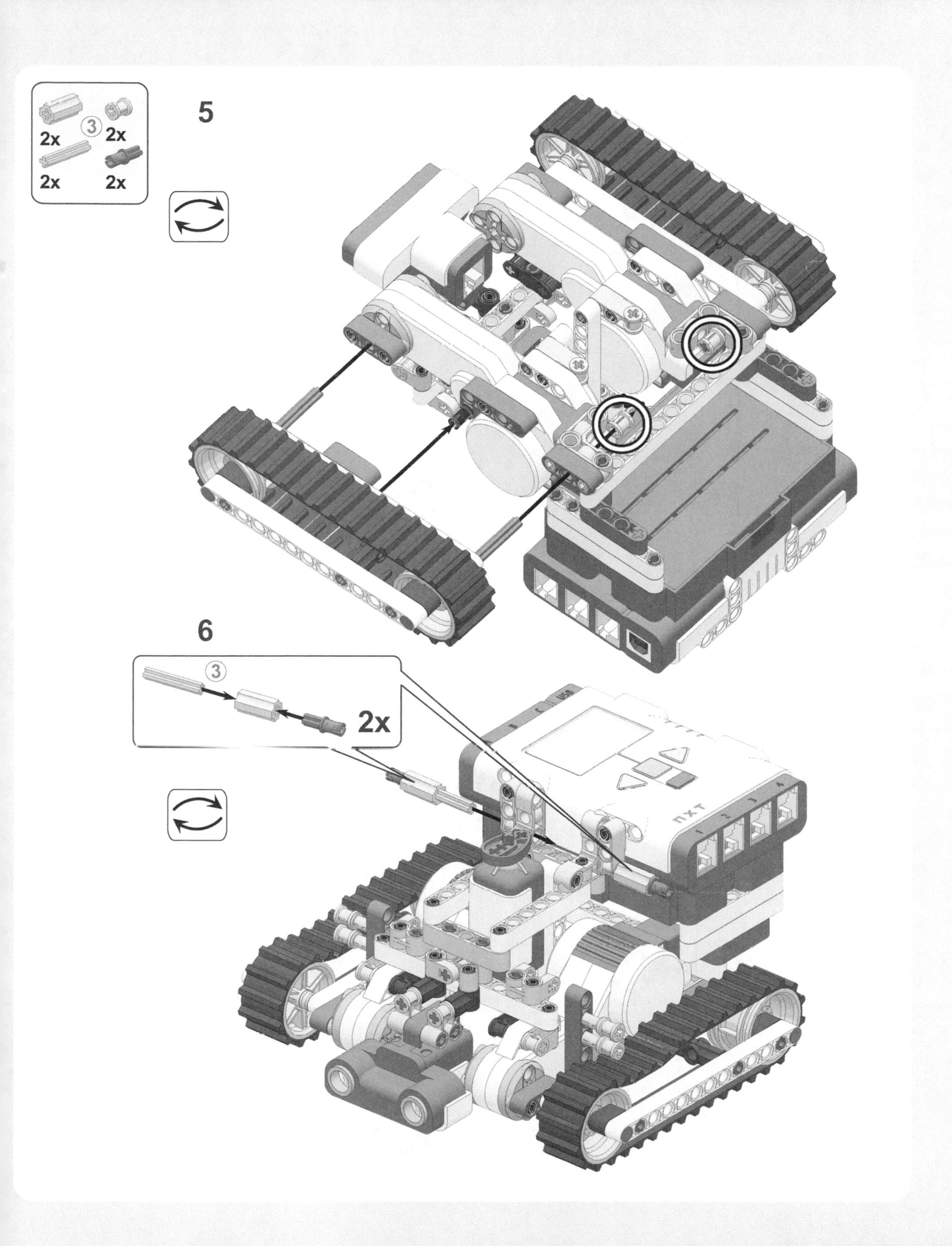
5
2x
3
2x
2x
2x
6
3
2x

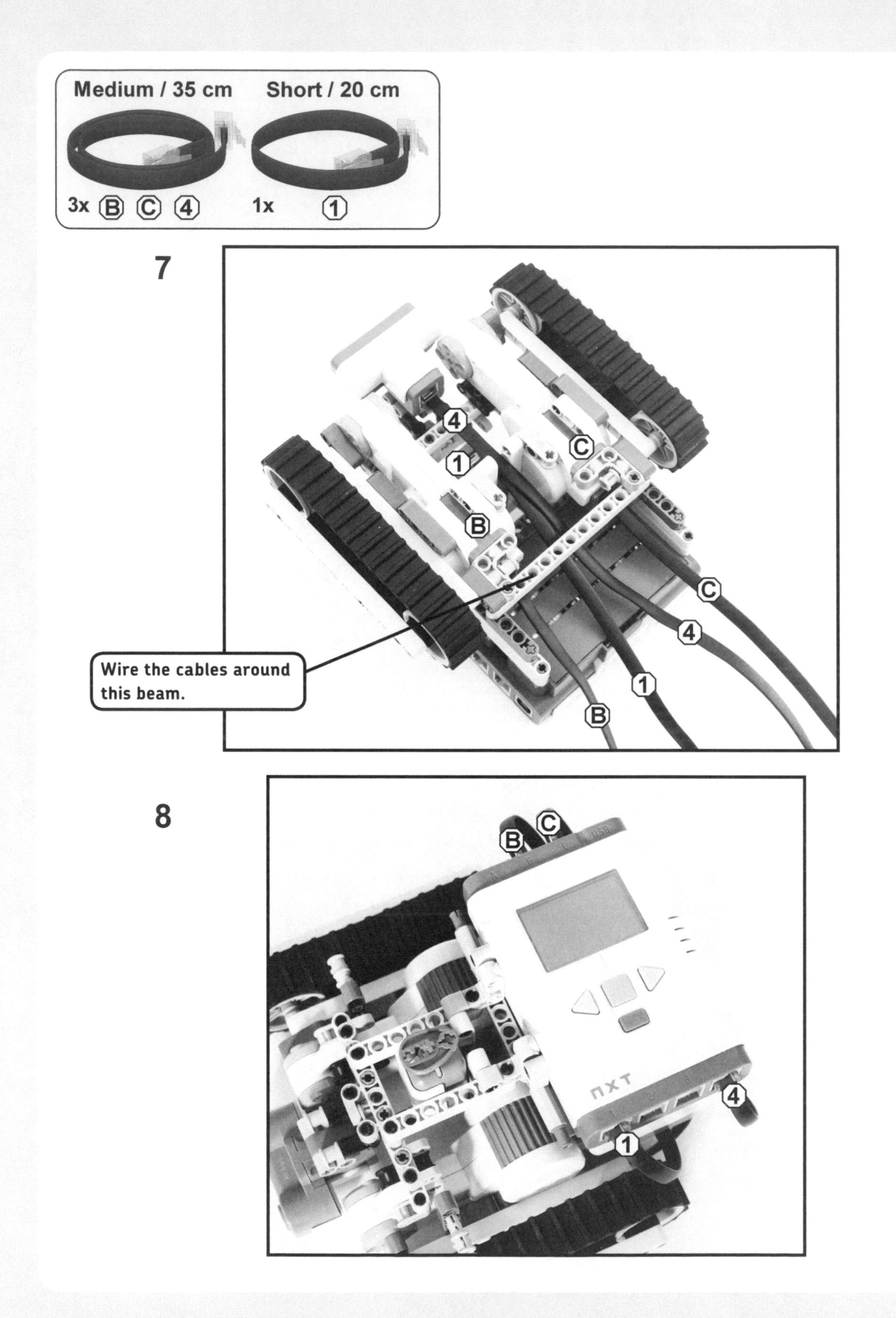
Medium / 35 cm
Short / 20 cm
3x B C 4
1x 1
7
4
C
1
B
C
4
1
B
Wire the cables around this beam.
8
B
C
NXT
4
1

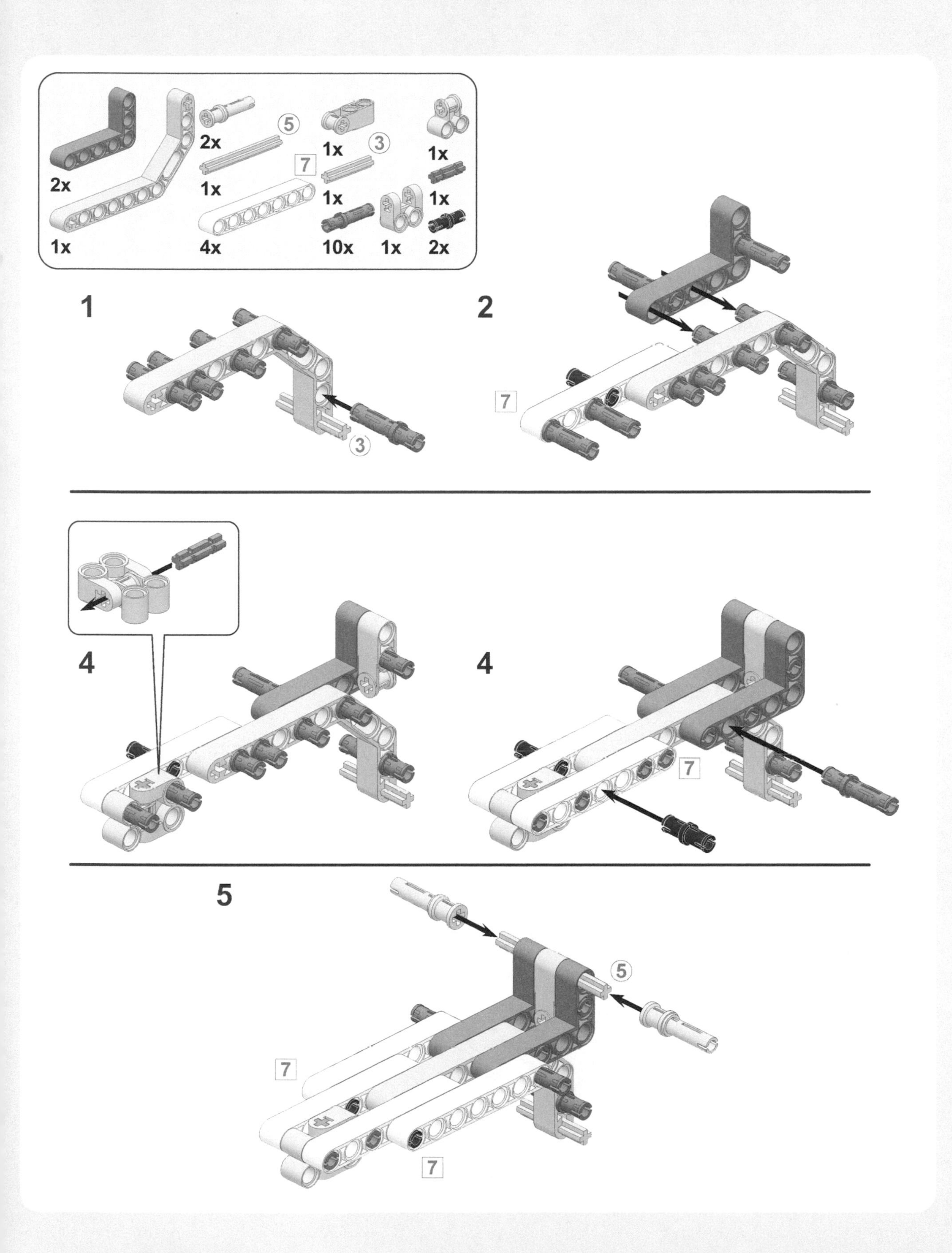
2x
1x
2x
1x
4x
1x
1x
10x
1x
1x
1x
2x
5
3
7
1
2
3
7
4
4
7
5
5
7
7

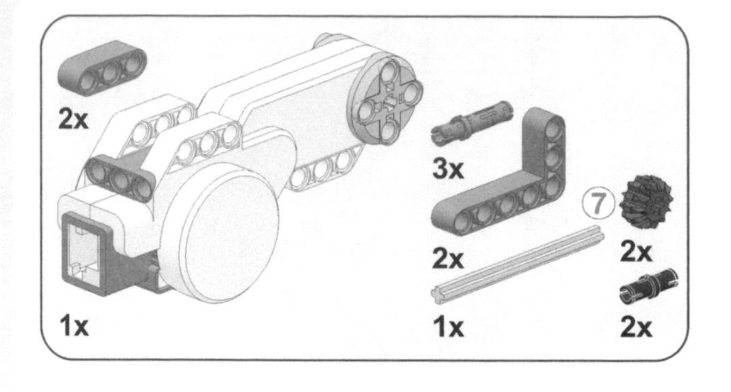

6

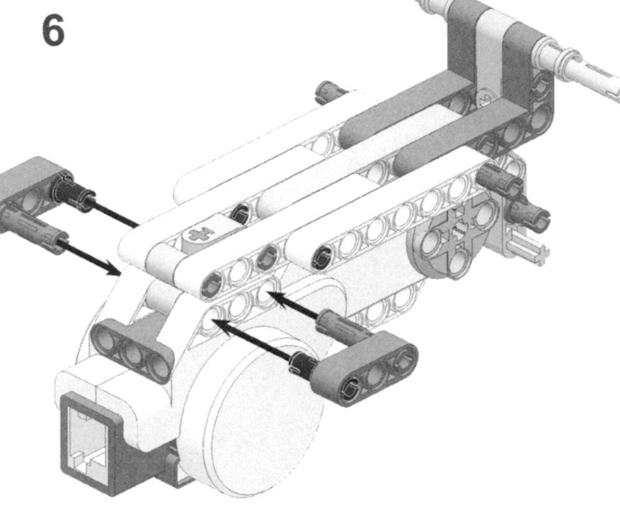

7

8

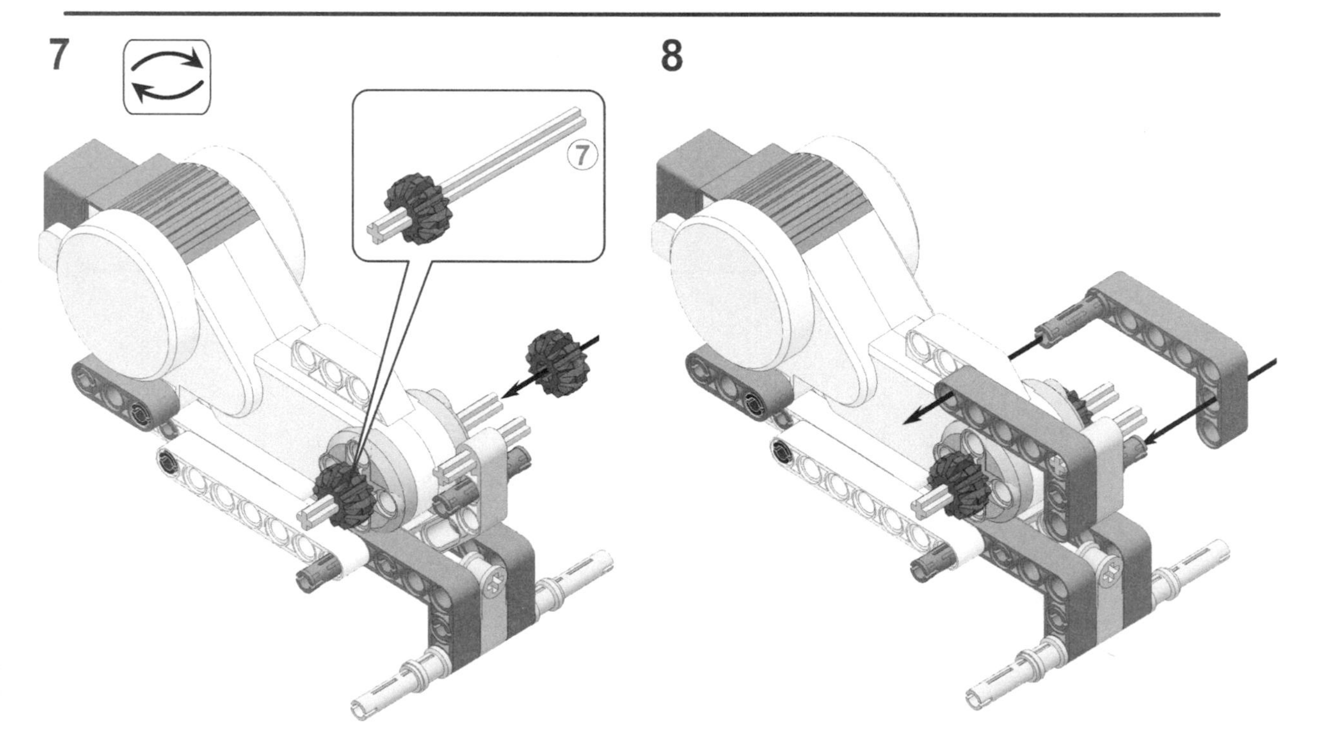

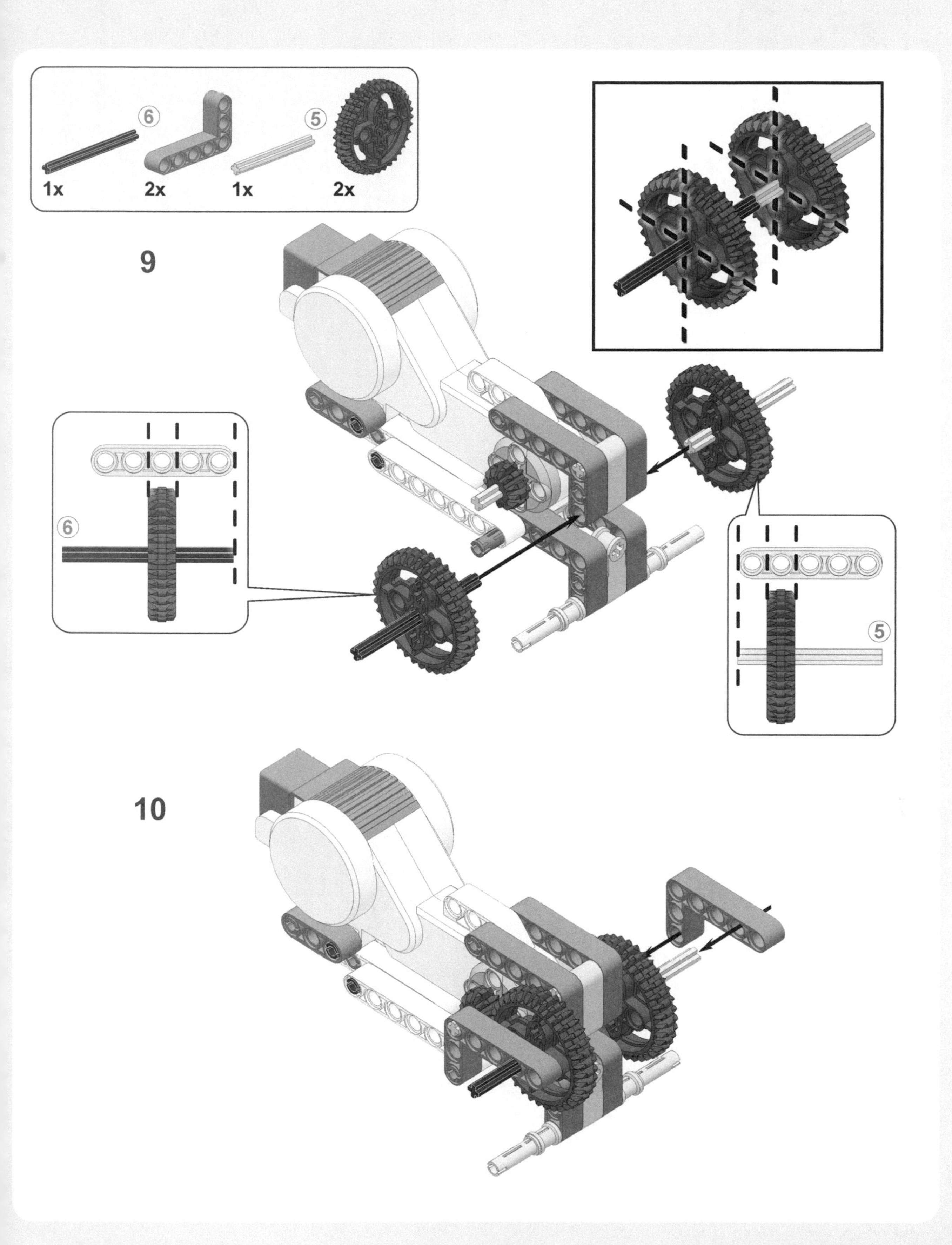
6
5
1x
2x
1x
2x
9
6
5
10

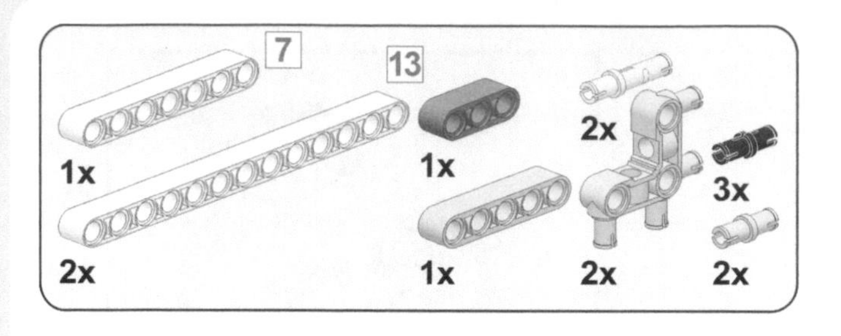
7
13
1x
1x
2x
3x
2x
1x
2x
2x

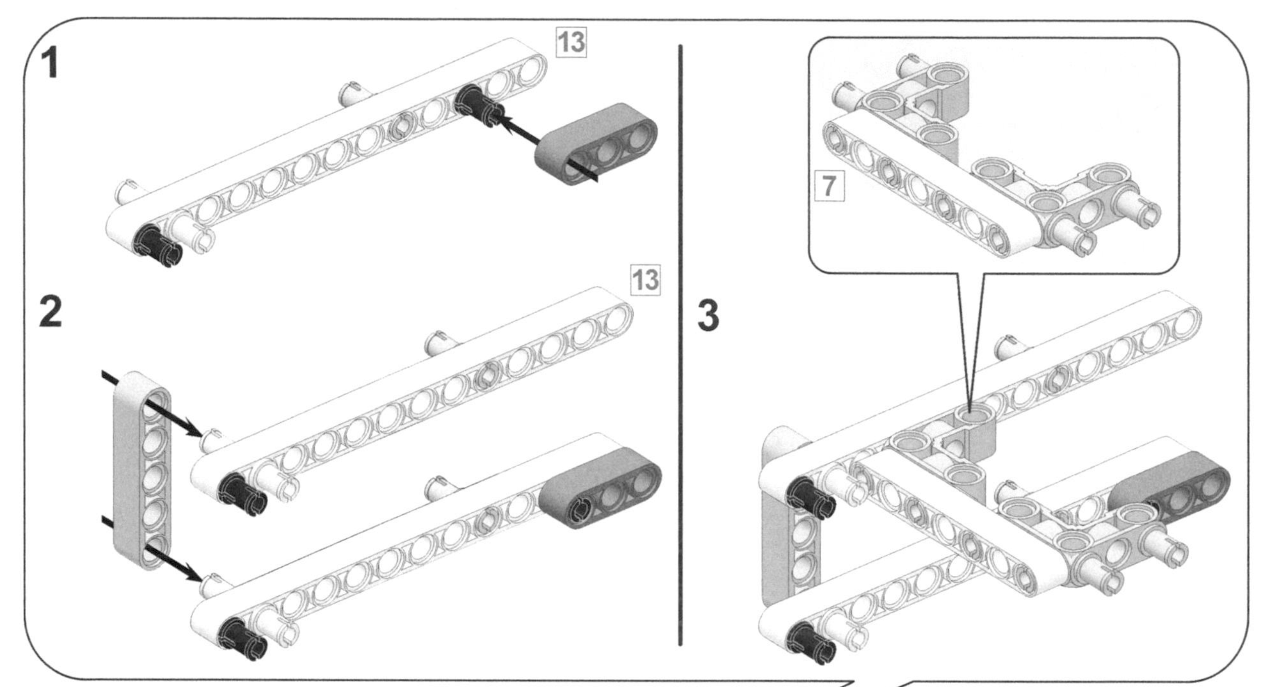
1
13
2
13
3
7

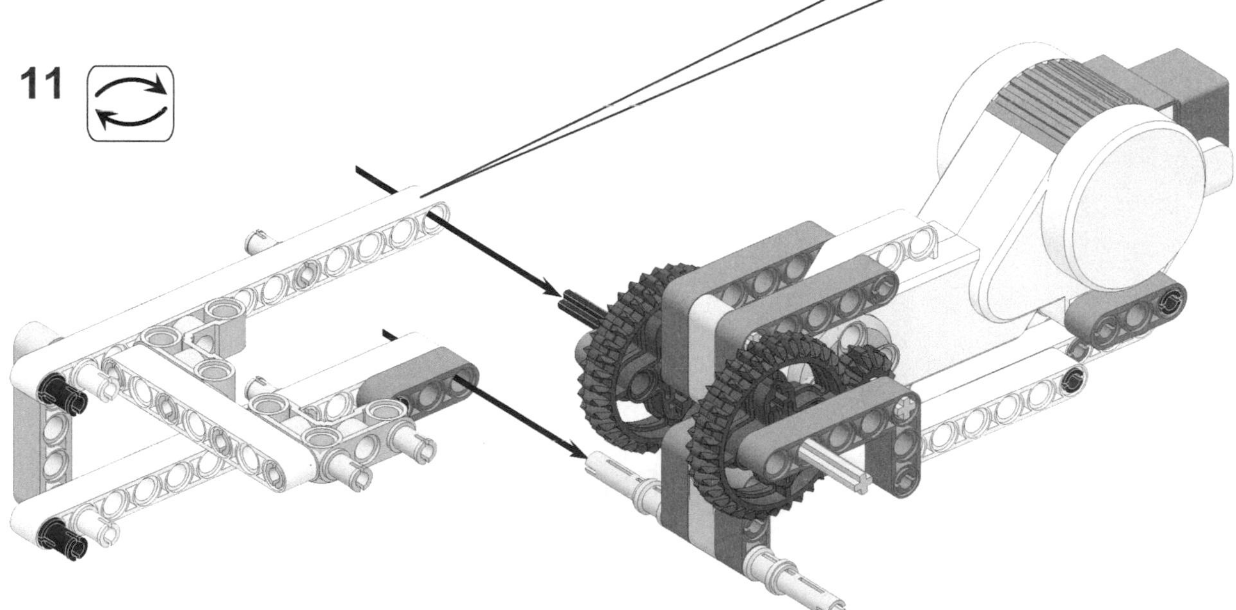
11

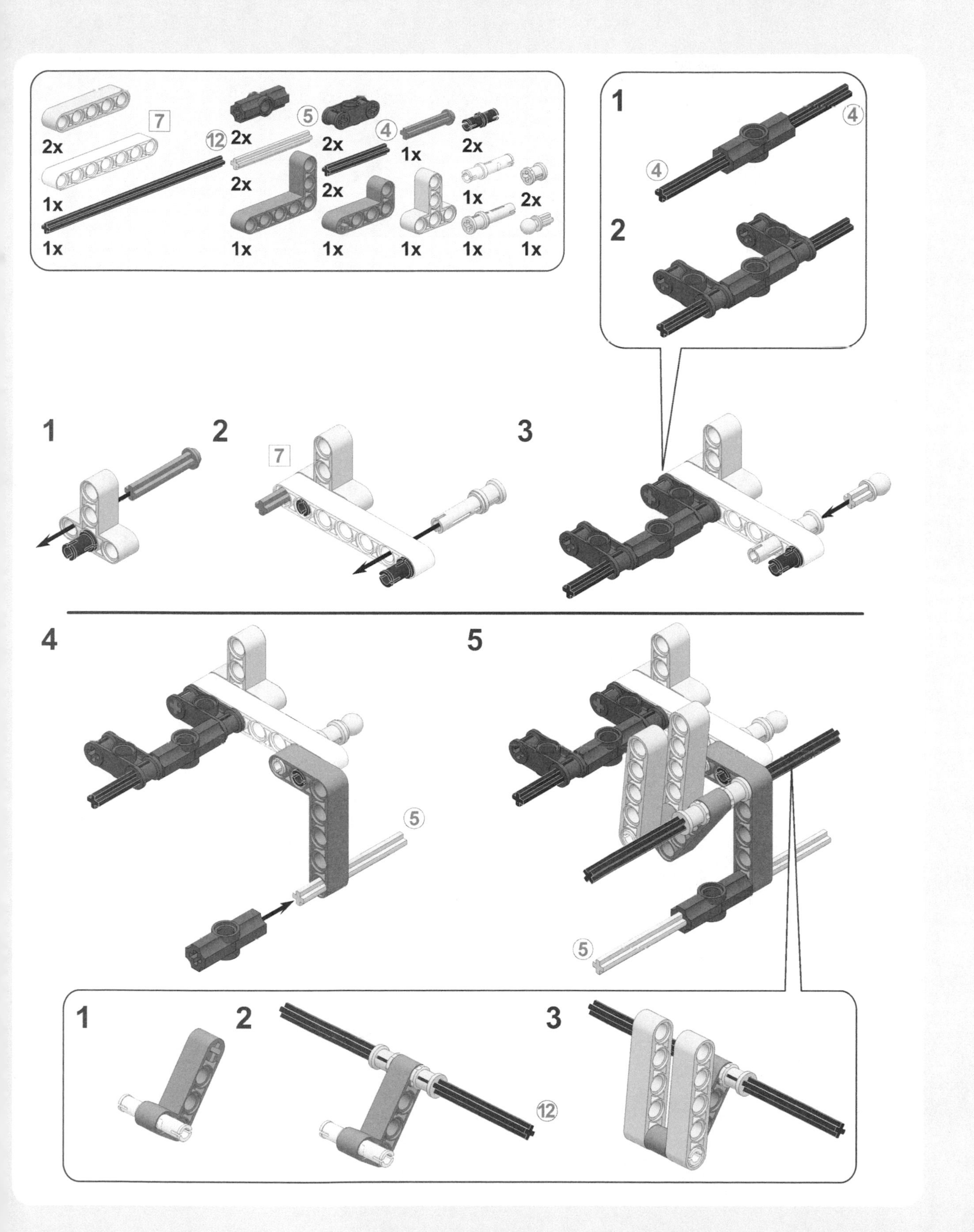

7
12
5
4
2x
2x
2x
1x
2x
1x
2x
2x
1x
2x
1x
1x
1x
1x
1x
1x
1
2
3
4
5

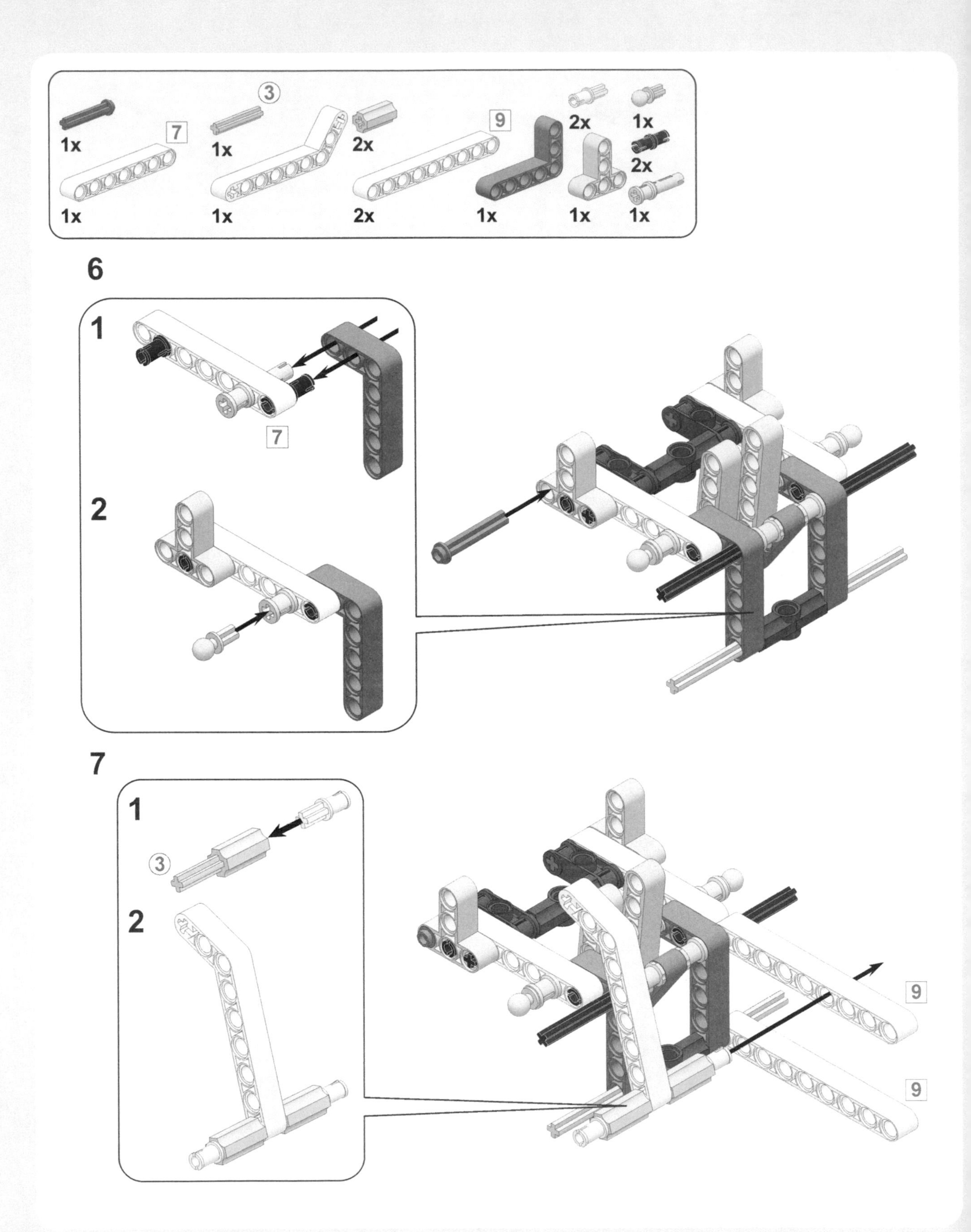

1x
1x
③
1x
1x
2x
7
9
2x
2x
1x
1x
1x
2x
1x
6
1
7
2
7
1
③
2
9
9

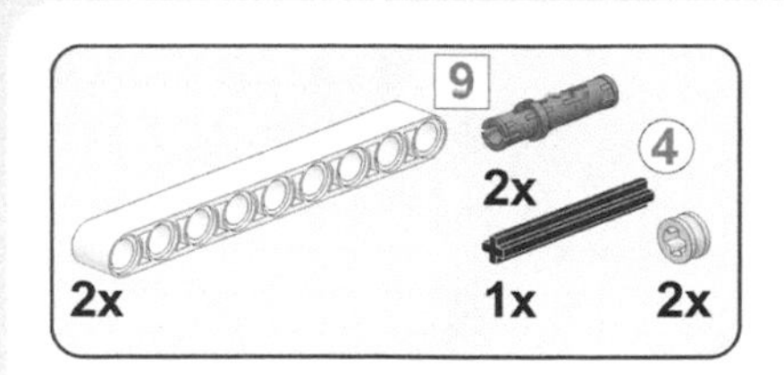
9
2x
4
2x
1x
2x

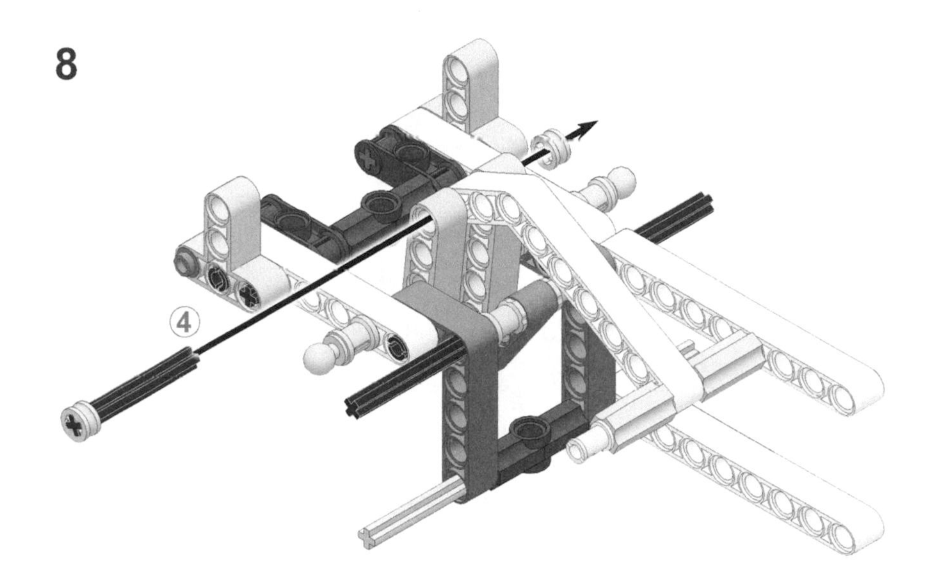
8
4

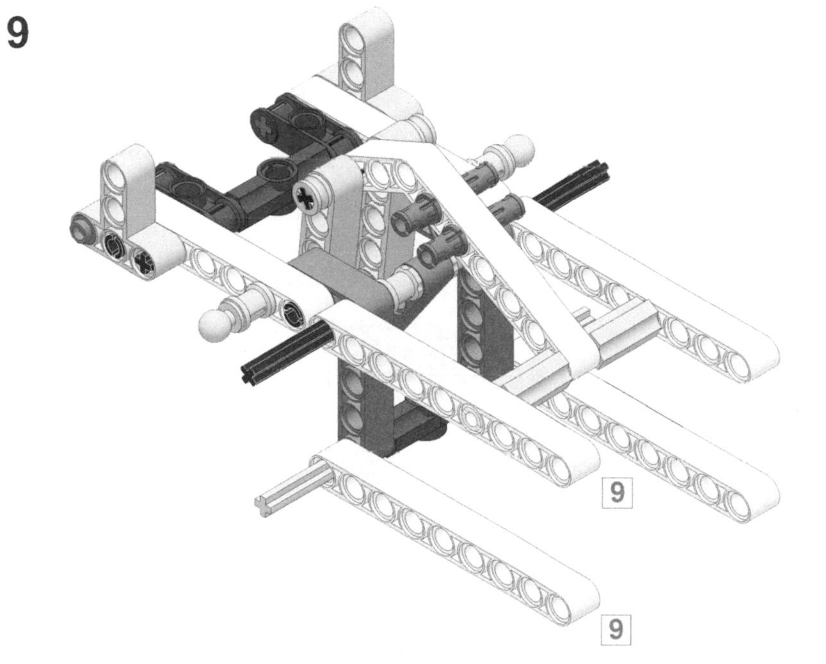
9
9
9

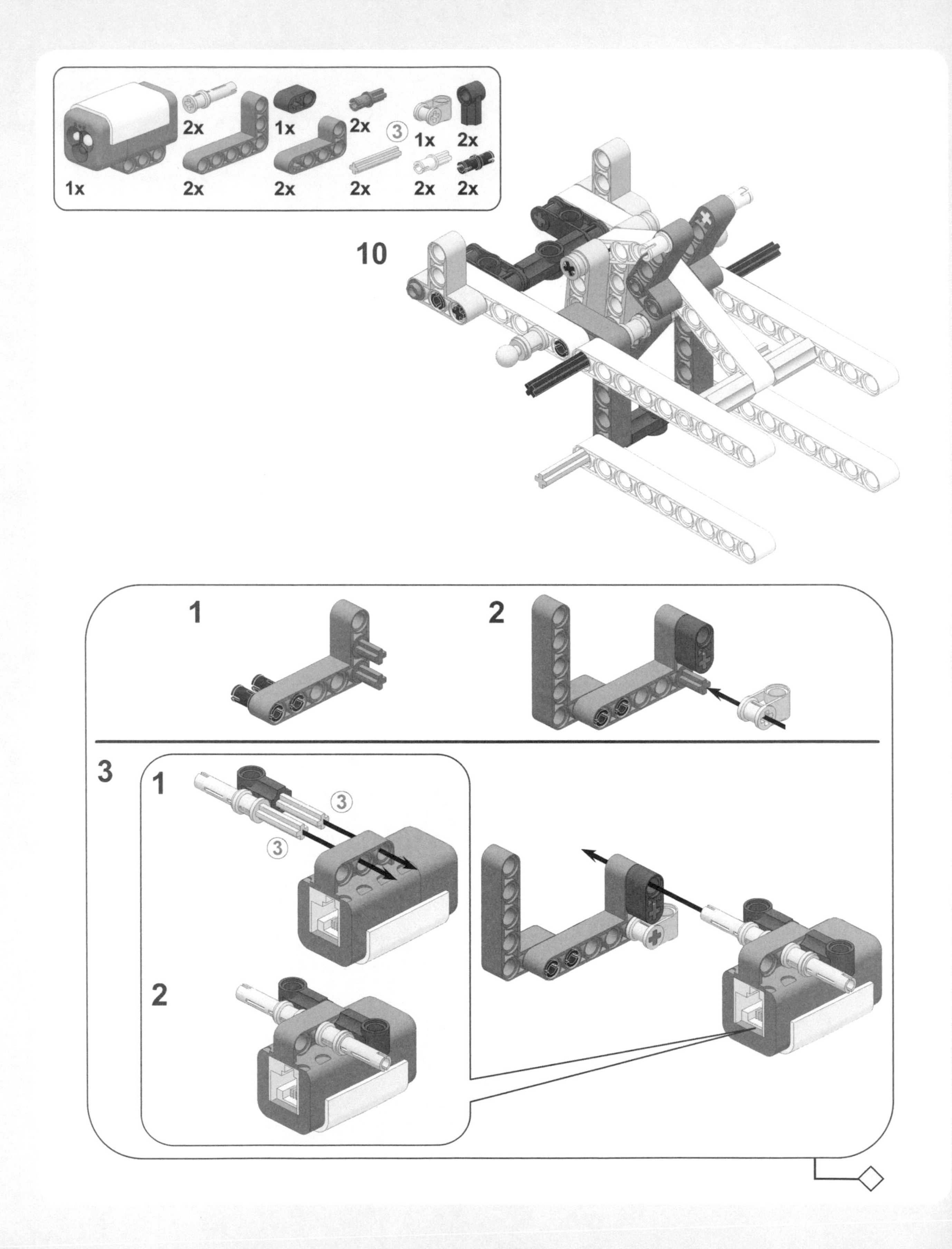
1x
2x
2x
1x
2x
2x
2x
3
1x
2x
2x
2x
10
1
2
3
1
3
3
2

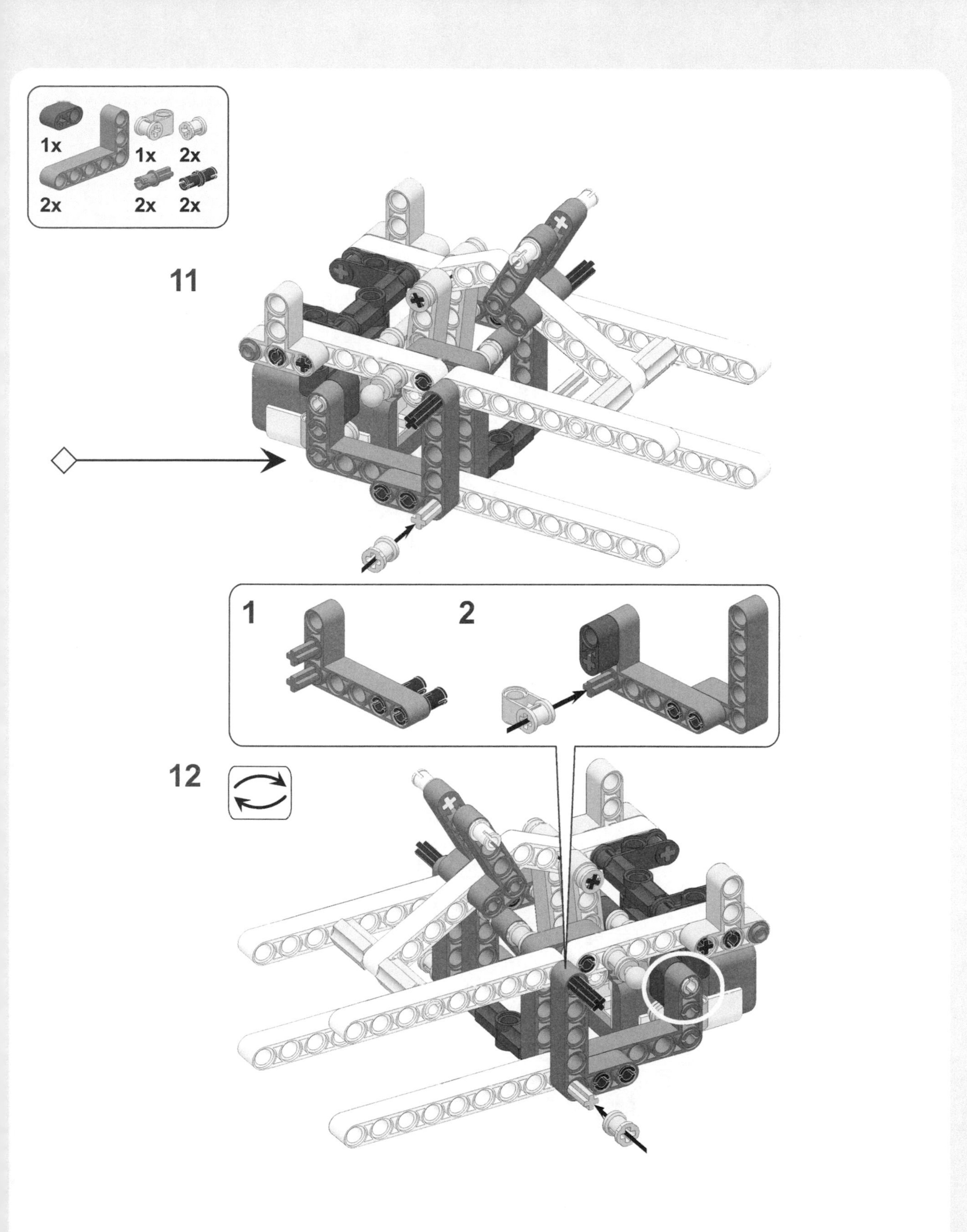
1x
1x
2x
2x
2x
2x
11
1
2
12

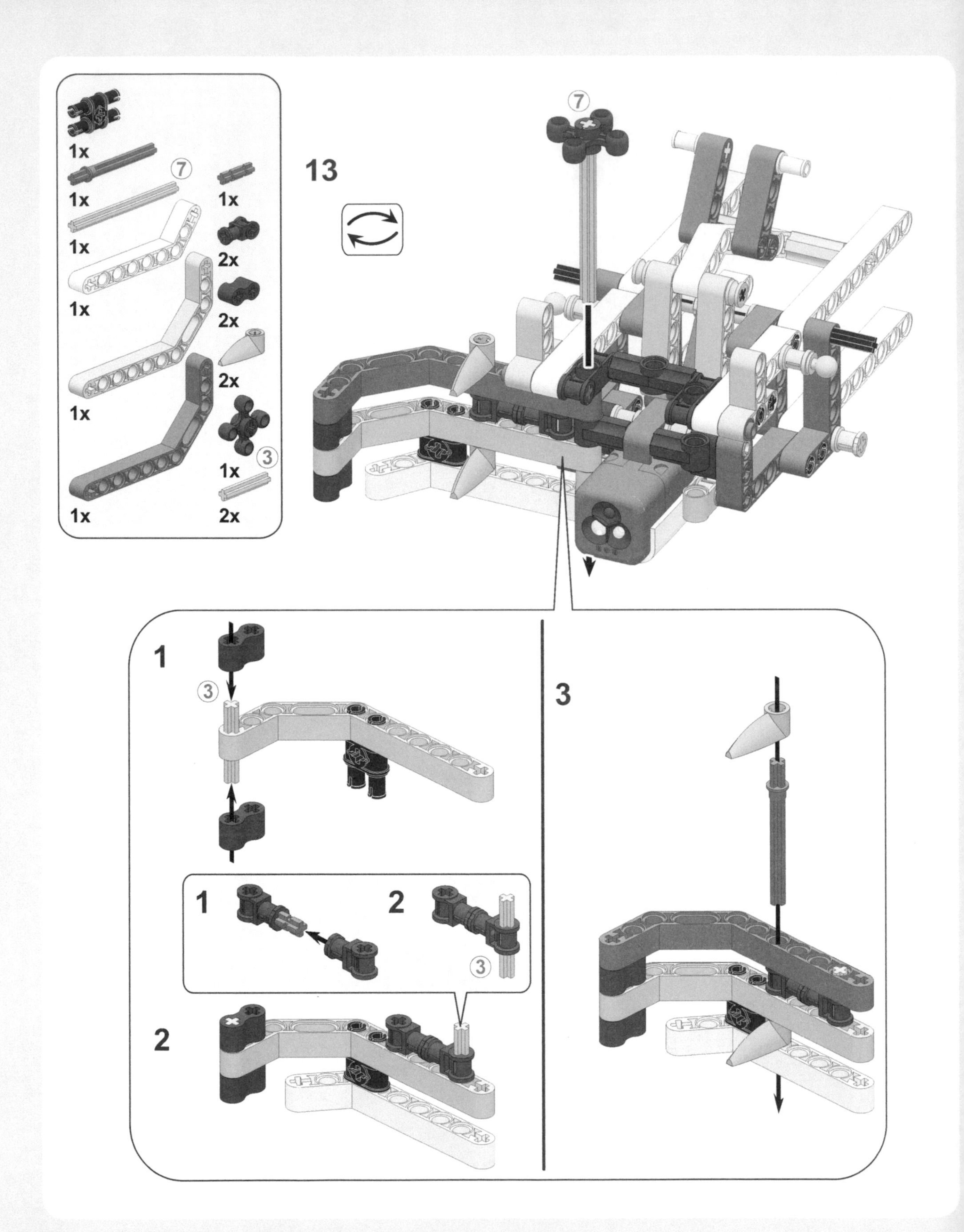

1x
1x
7
1x
1x
1x
1x
1x
1x
2x
2x
2x
1x
3
2x
13
7
1
3
1
2
3
2
3
3

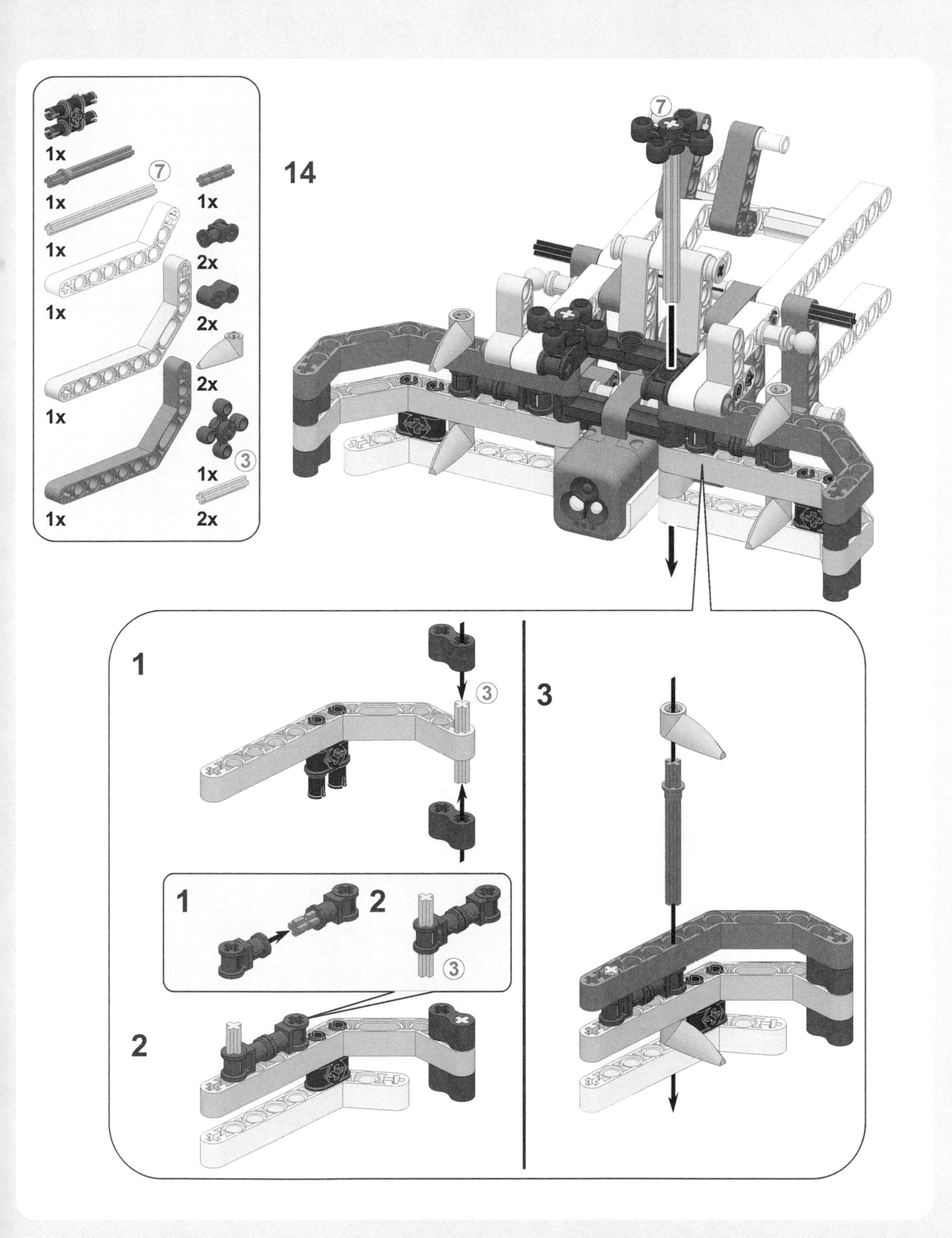
1x
1x
7
1x
1x
2x
1x
2x
2x
1x
1x
3
1x
2x
14
7
1
3
3
1
2
3
2

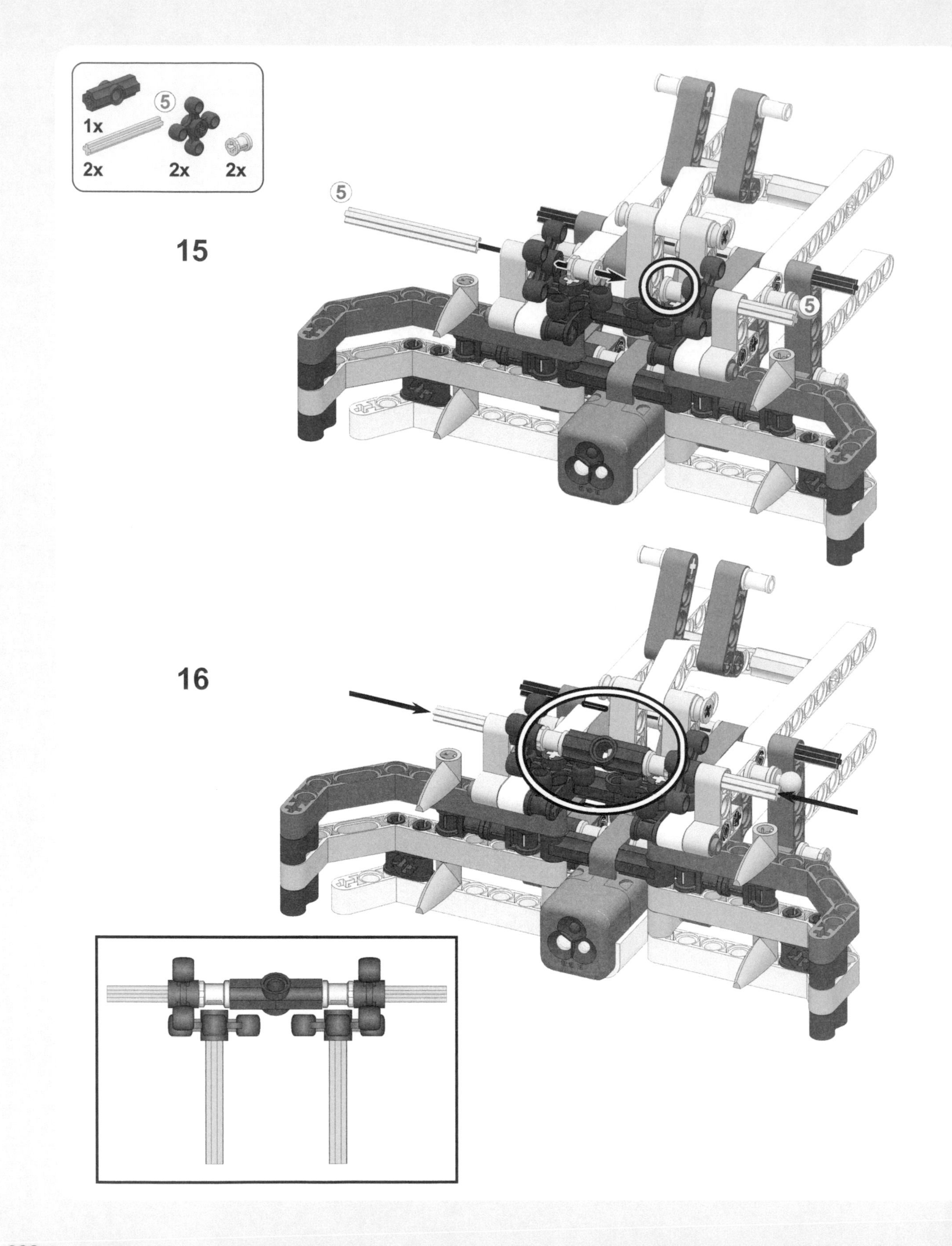
1x
5
2x
2x
2x
15
5
5
16

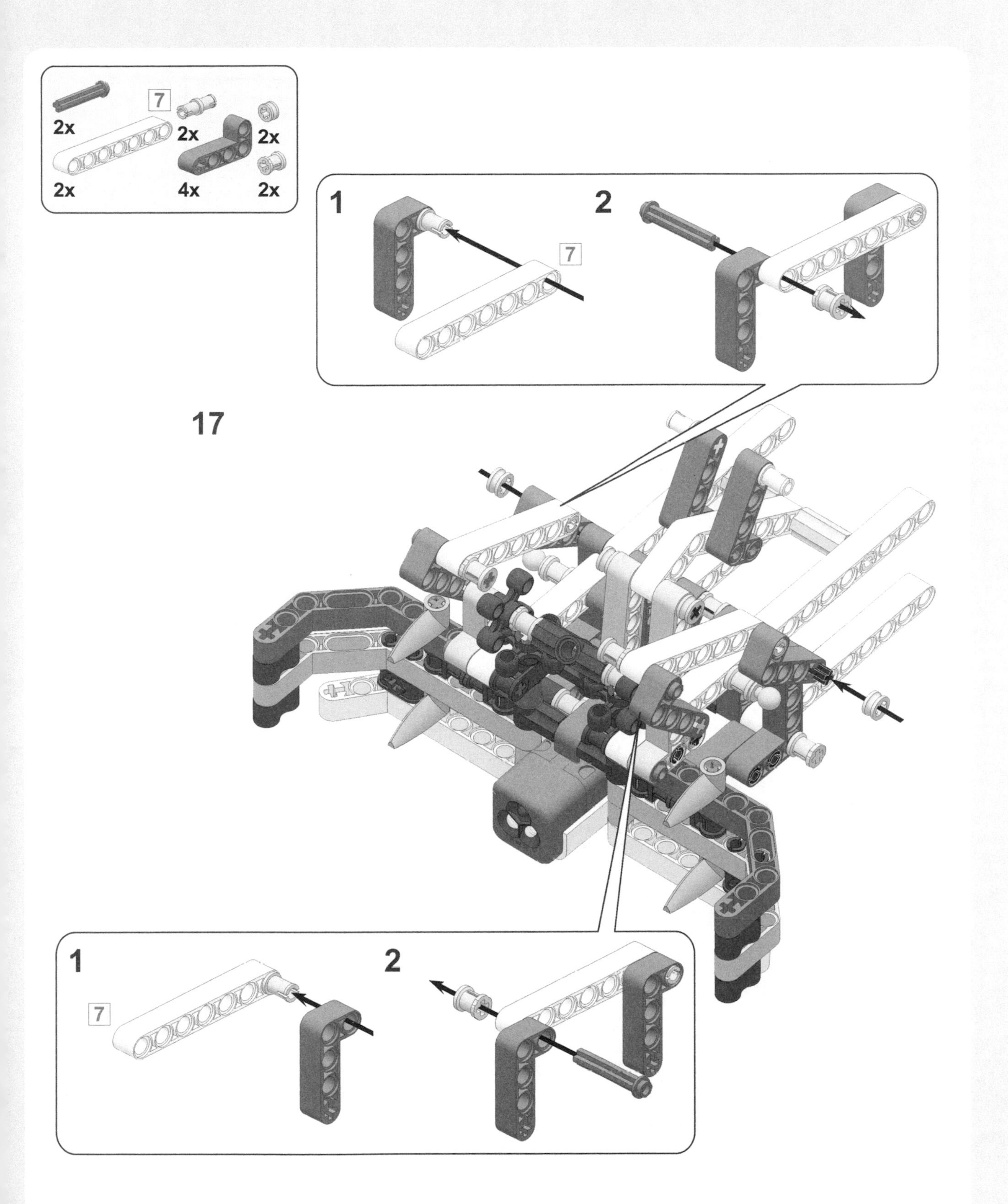
2x
7
2x
2x
2x
4x
2x
1
2
7
17
1
2
7
7

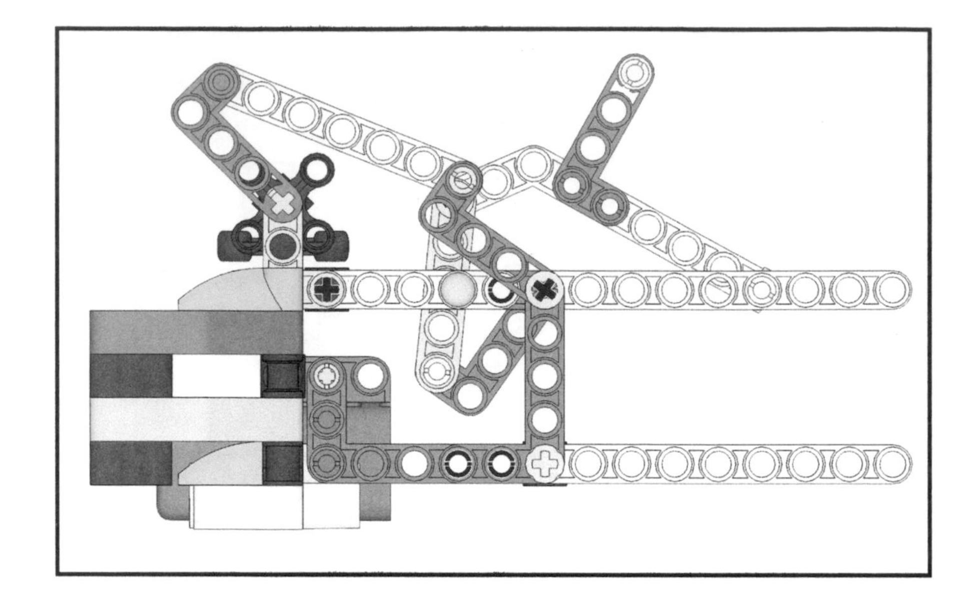

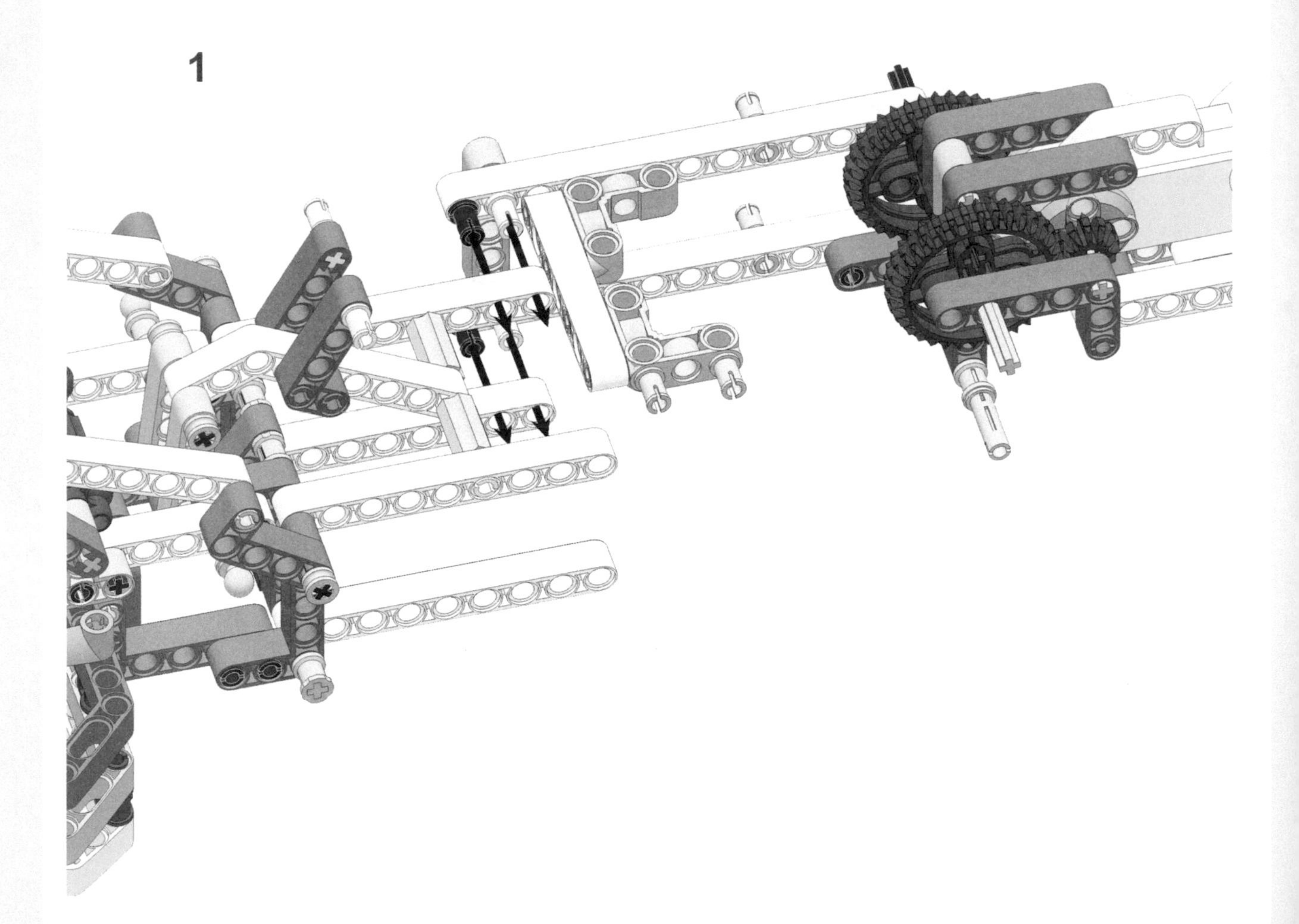
1

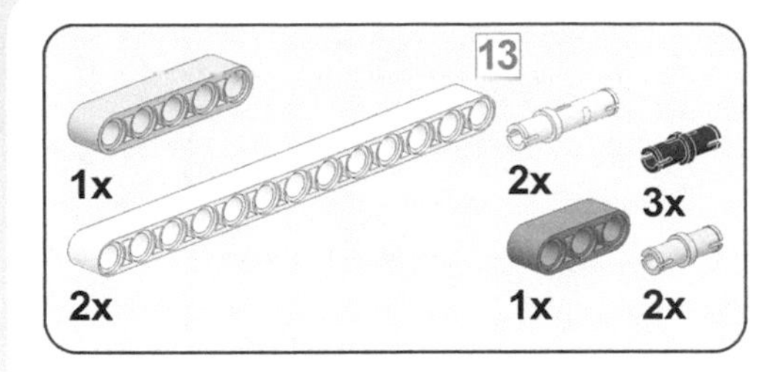
13
1x
2x
2x
3x
1x
2x

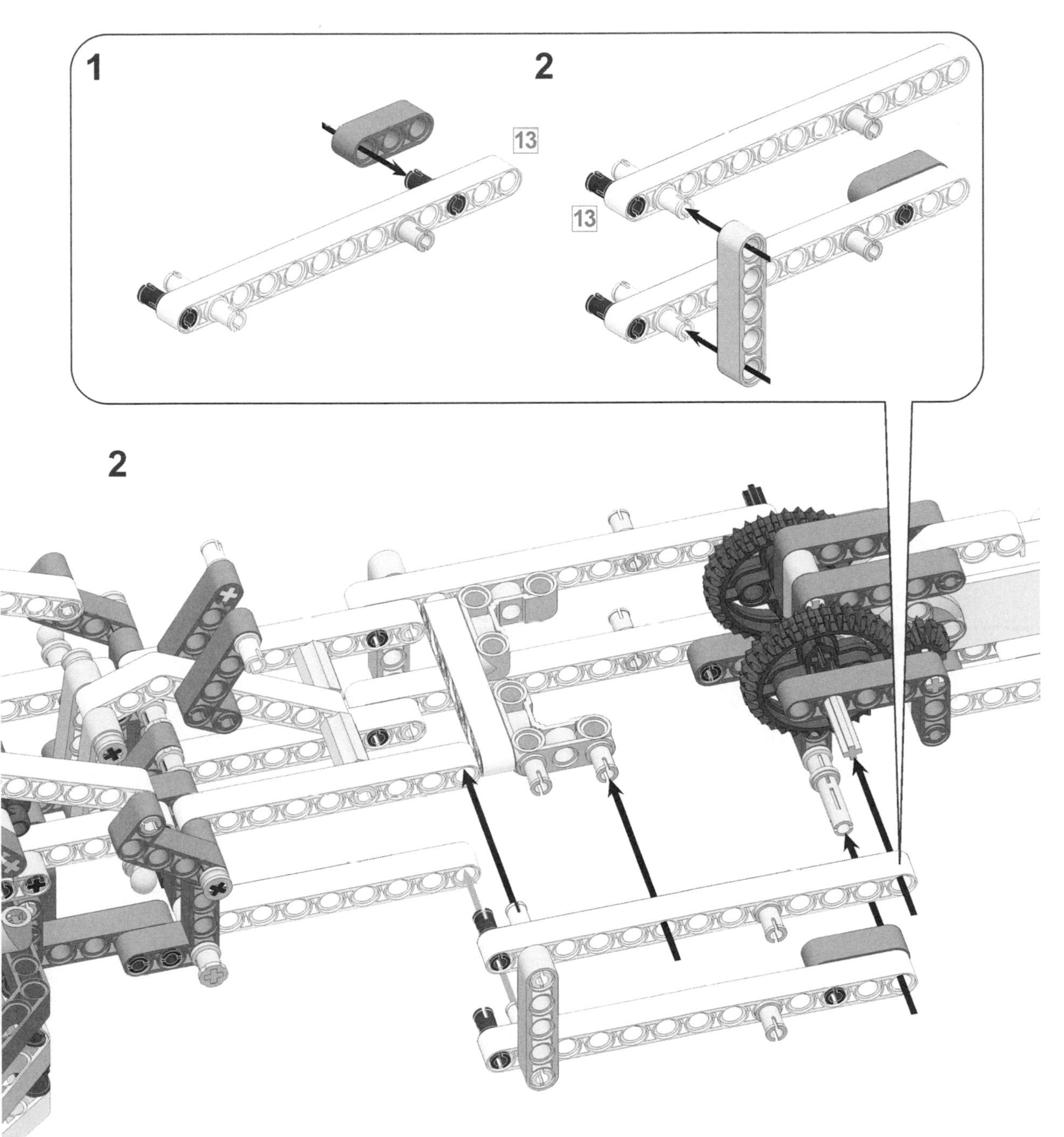
1
2
13
13
2

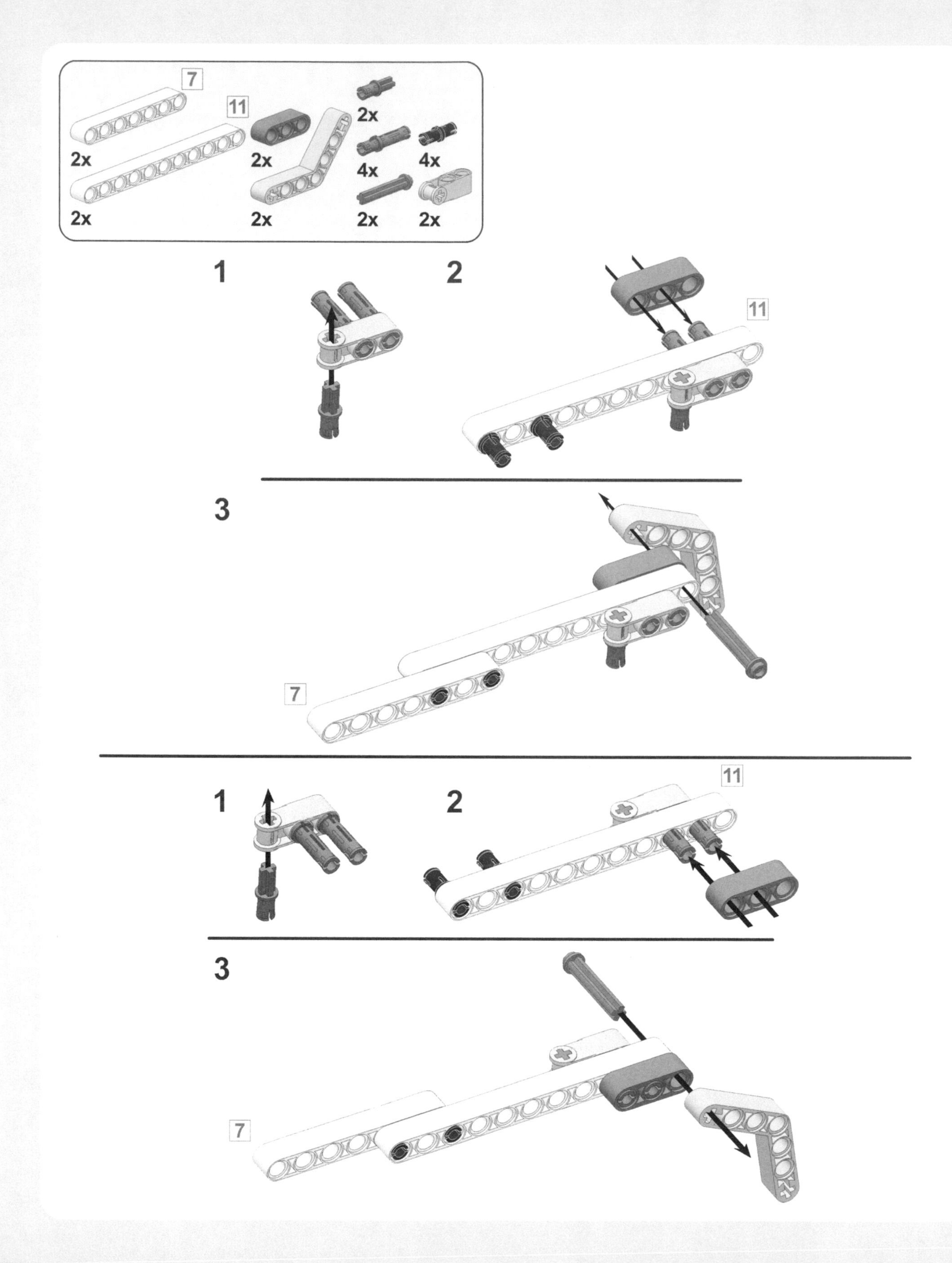
7
11
2x
2x
2x
2x
2x
4x
4x
2x
2x
2x
1
2
11
3
7
1
2
11
3
7

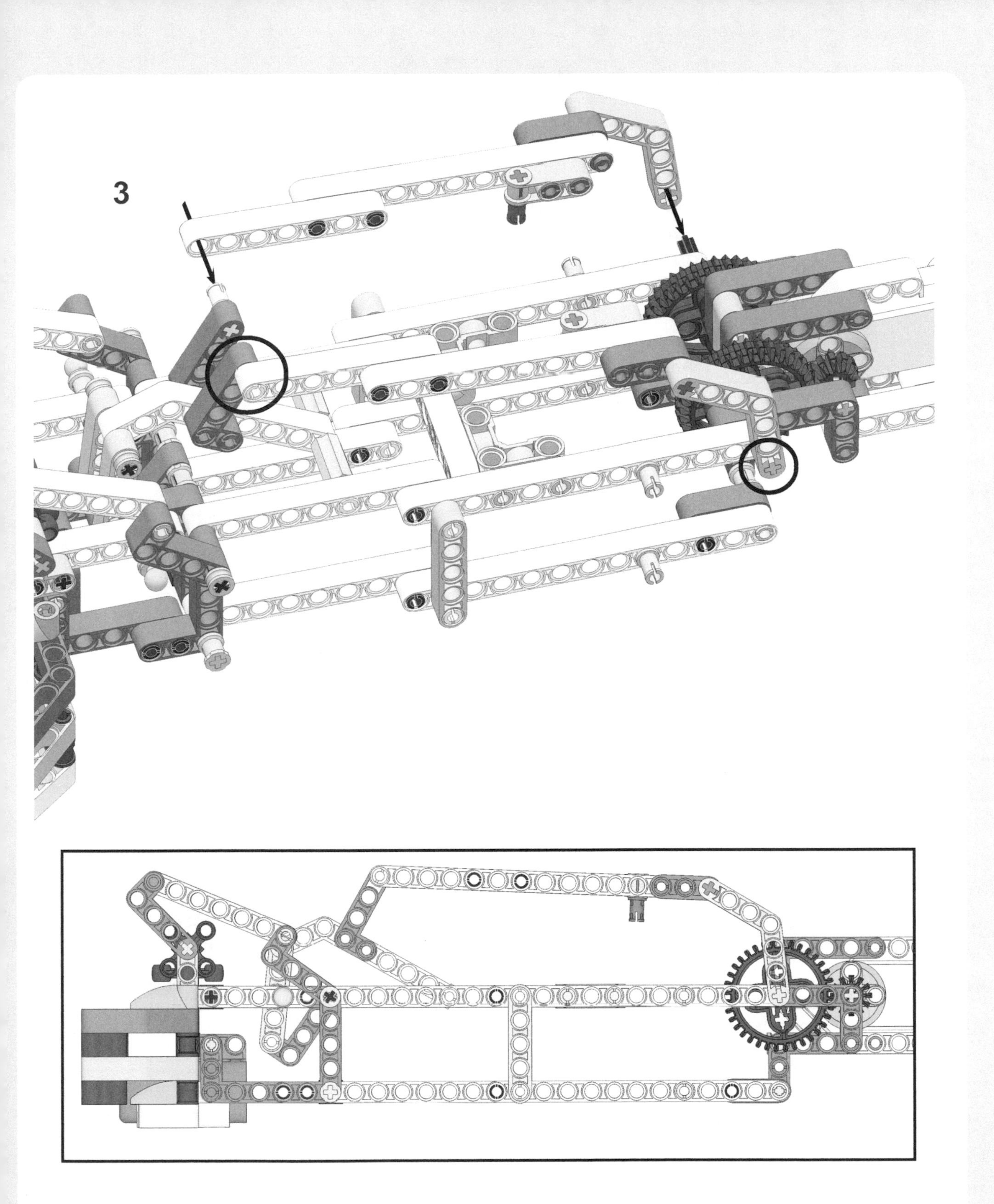
3

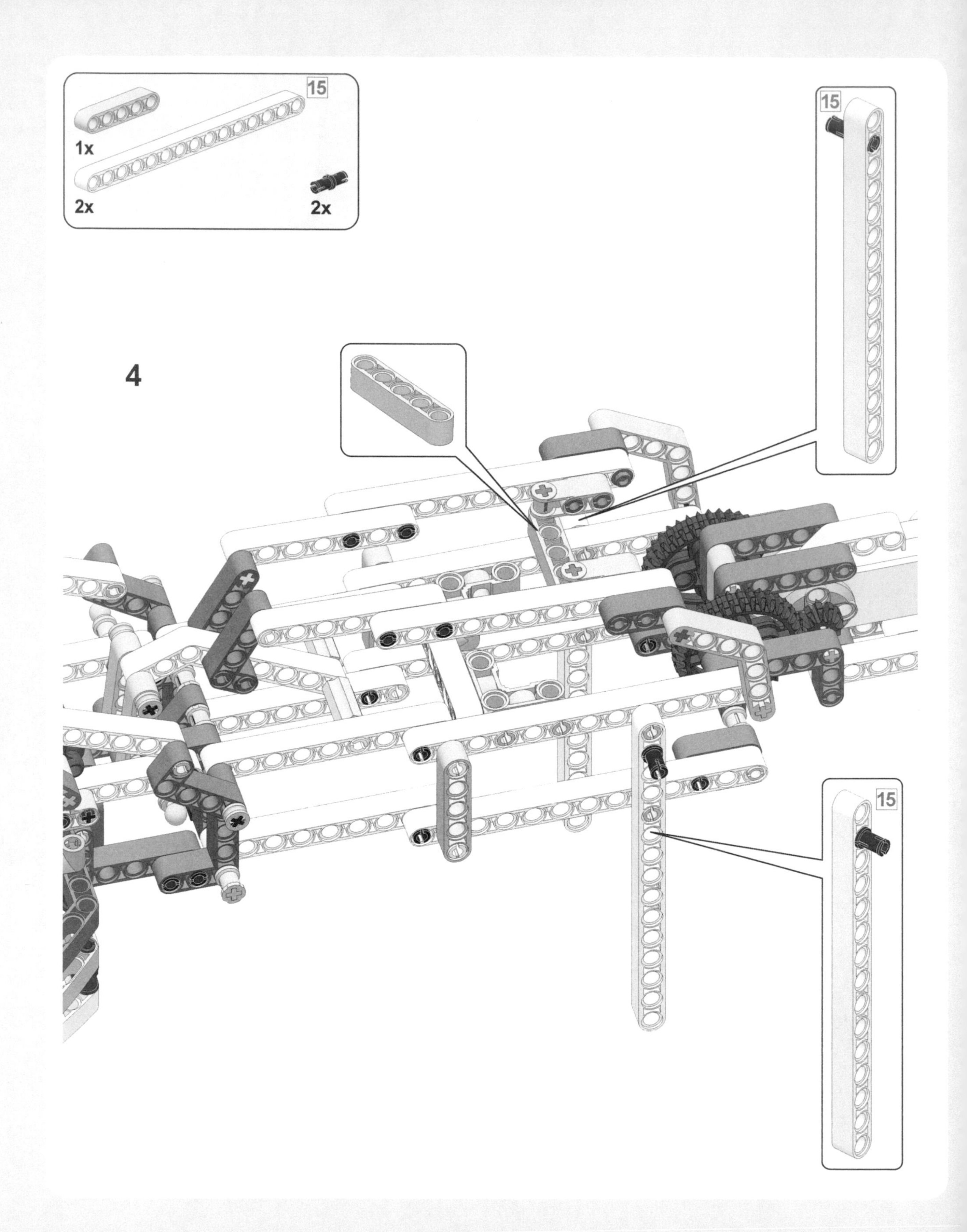
15
1x
2x
2x
4
15
15

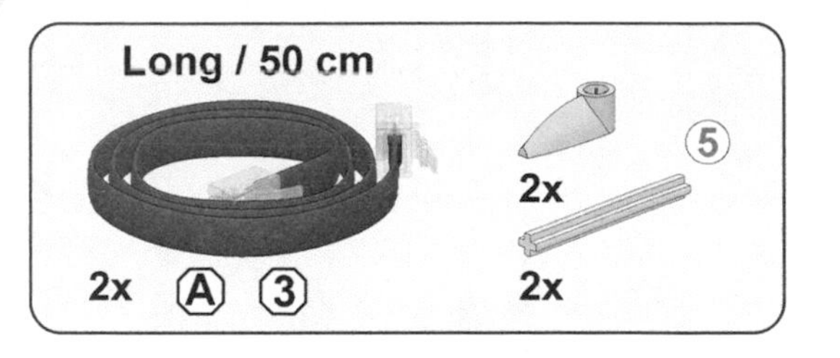

1

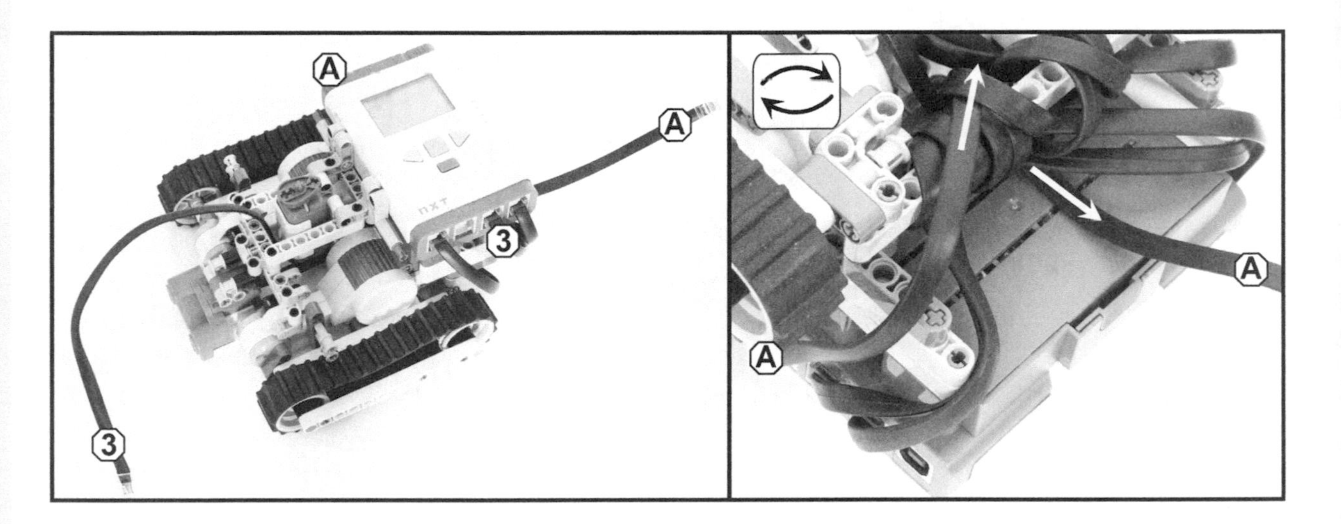

2

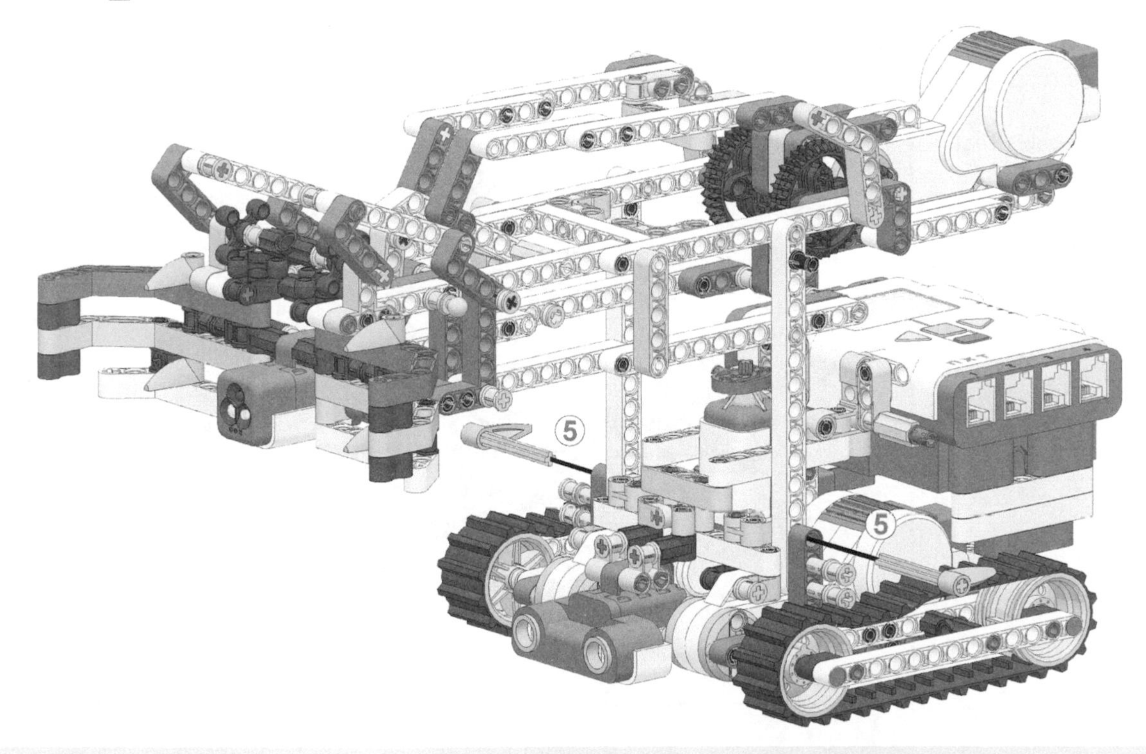

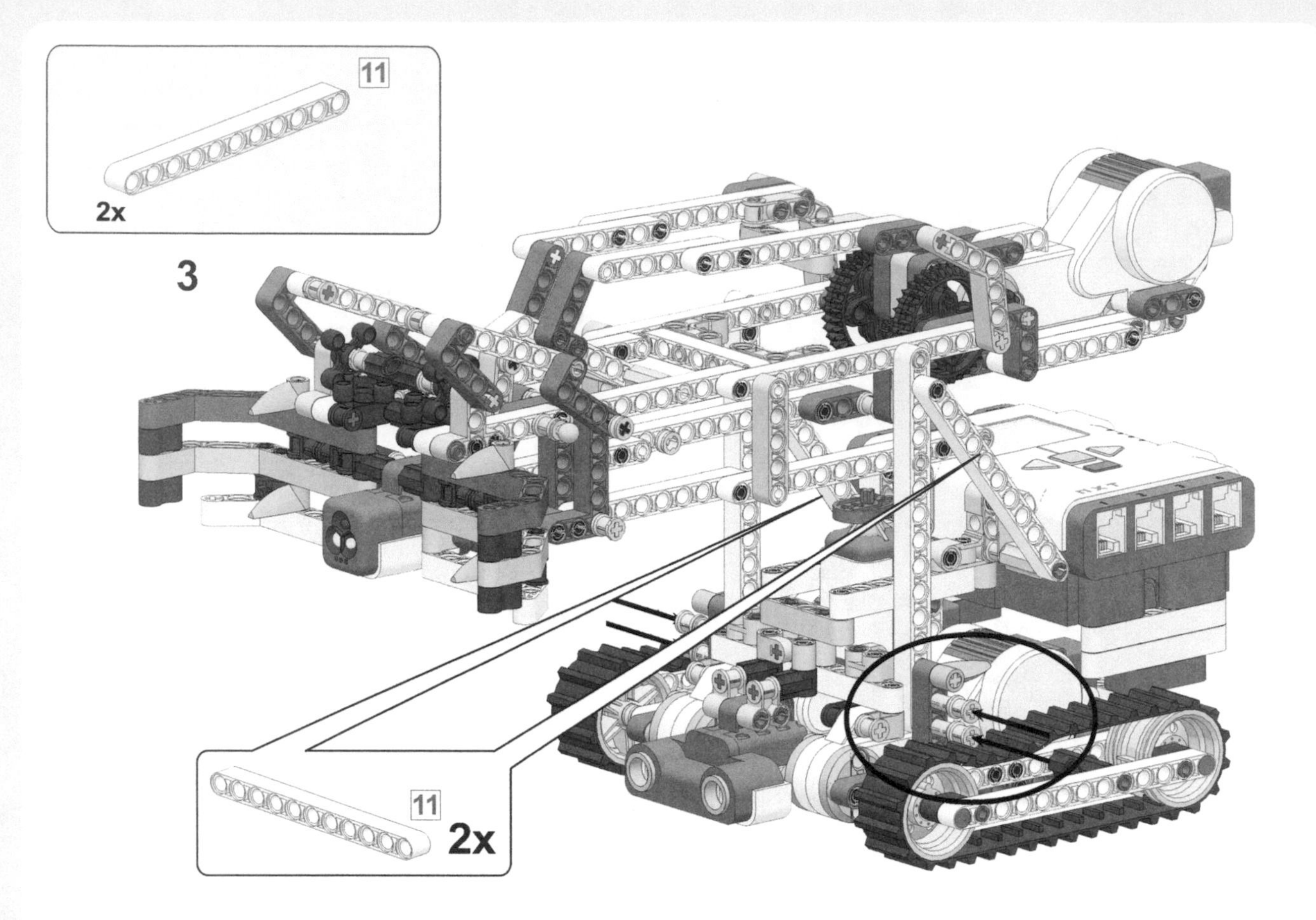

4

NOTE Guide the cable that you previously connected to input port 3 through the frame as shown, then plug it into the Color Sensor.

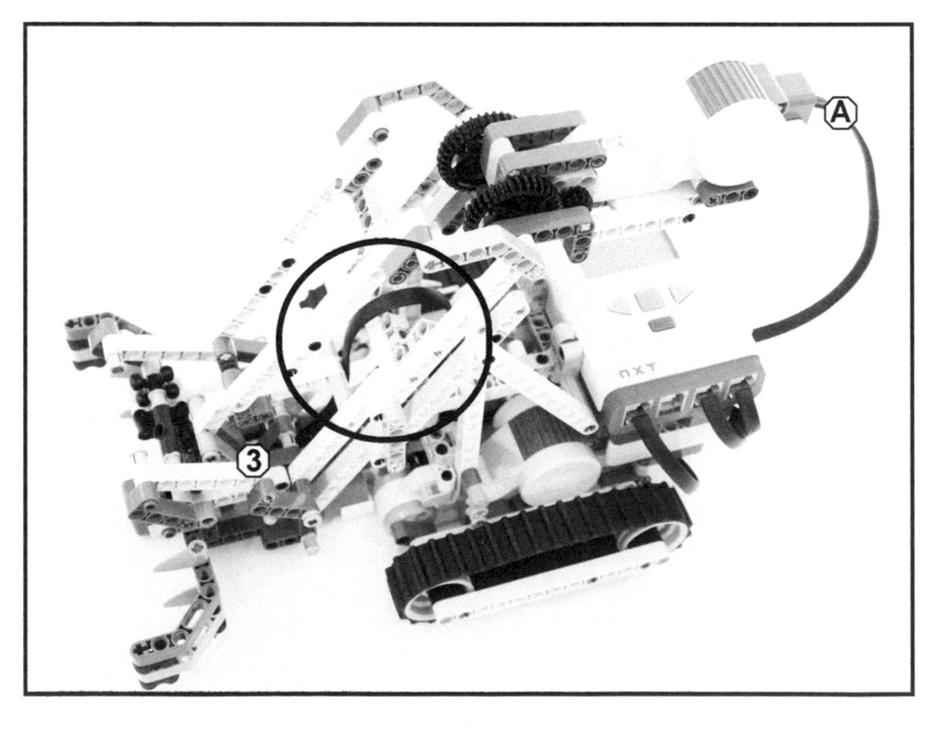

creating objects

You can modify the Snatcher's arm to grab almost anything as long as the object isn't too heavy, but the version you've just built is designed to pick up paper rings, like the ones shown in Figure 13-5. Use construction paper to make about four paper rings in different colors (yellow, blue, red, and green) before you program the robot.

programming the snatcher

Having built the Snatcher and created objects for it, you're ready to program it. You'll create a program that makes the Snatcher find, grab, lift, and move an object, as well as identify the object's color. Each task should run autonomously, which means that all tasks must be performed without human interaction.

You'll use the Ultrasonic Sensor to find the object and use the Driving motors to approach the object and position it between the robot's fingers. The Grabber motor will grab and lift the object, and the Color Sensor will identify the color of the object that was picked up. Finally, the robot will turn around and then drop the object elsewhere. Figure 13-6 shows an overview of the Snatcher program.

creating the my blocks

Because this program will require many programming blocks, you'll use five My Blocks to make the final program easier to understand and create, as shown in Figure 13-6.

NOTE If a programming figure in this chapter doesn't mention changing a particular setting on a block, just leave that setting unchanged.

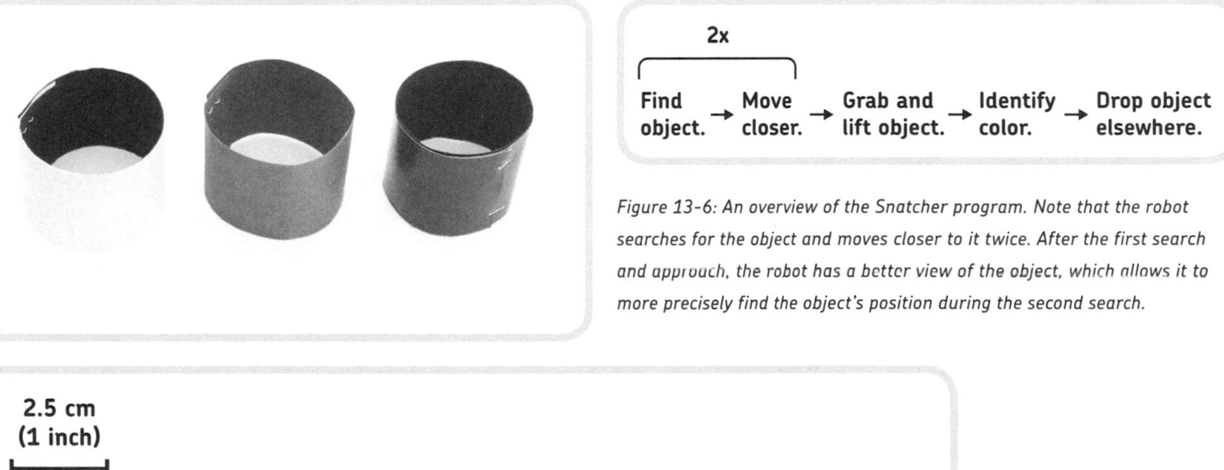

Figure 13-6: An overview of the Snatcher program. Note that the robot searches for the object and moves closer to it twice. After the first search and approach, the robot has a better view of the object, which allows it to more precisely find the object's position during the second search.

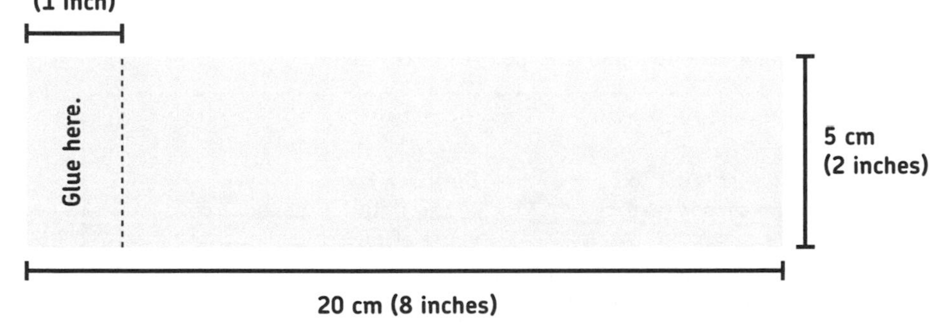

Figure 13-5: The Snatcher will grab these objects. Instead of gluing the objects, you may also staple them. If you don't have construction paper in different colors, pick thick paper of any color, such as black or brown, and staple or glue lighter-weight, colored paper around it.

my block #1: grab

This block will make the robot grab and lift the object positioned between the robot's fingers. If there's nothing to grab, this block simply closes and raises the grabber.

You configure the Grabber motor to turn forward until the Touch Sensor inside the Snatcher detects that the grabber is lifted all the way up. Configure the blocks that do this as shown in Figure 13-7, and then turn them into a My Block called **Grab**. (Also, select some appropriate icons to make it easier to remember what the blocks do.)

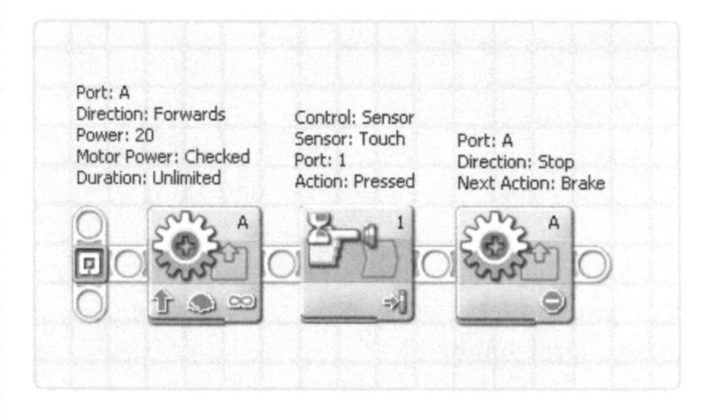

Figure 13-7: The configurations of the blocks in the Grab My Block

my block #2: release

This next My Block lowers the robot's grabber and opens its claws, releasing the object.

You should use this block only when the grabber is already lifted up (because of the Grab My Block). Although only one block is required to release an object, you'll create a My Block for this action to make the main program easier to understand. Configure the block as shown in Figure 13-8, and then turn it into a My Block called **Release**.

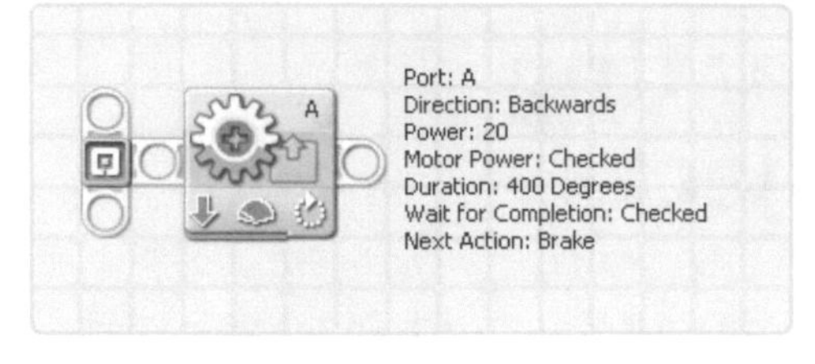

Figure 13-8: The configuration of the block in the Release My Block

my block #3: find object

Figure 13-9 shows how the Snatcher finds objects. When pinpointing objects, Snatcher looks for the closest object in range while turning to the right. After scanning, it turns in the opposite direction to the point where it saw the closest object. Once the Snatcher has run this block, it should be facing the object.

As the robot turns for 180 degrees, the Ultrasonic Sensor constantly measures distances. The lowest value it records is the distance to the object, and you store it in a variable called *Closest*. You set Closest's initial value to **256** (cm) because the sensor cannot measure anything farther away. Any time the sensor measures a distance closer than the value in Closest, the old value is thrown away, and the new sensor measurement is stored. This way, you have the robot forget all but the closest measurement.

The values of the motors' Rotation Sensors change as the robot turns. Because the robot needs to remember *where* it saw the closest object, you store the Rotation Sensor value of motor C in a variable called *Direction* whenever the Closest value is updated with a new value. In the end, the Direction variable should contain the value of the Rotation Sensor as measured when the robot saw the closest object. After turning for 180 degrees, the Snatcher turns left and

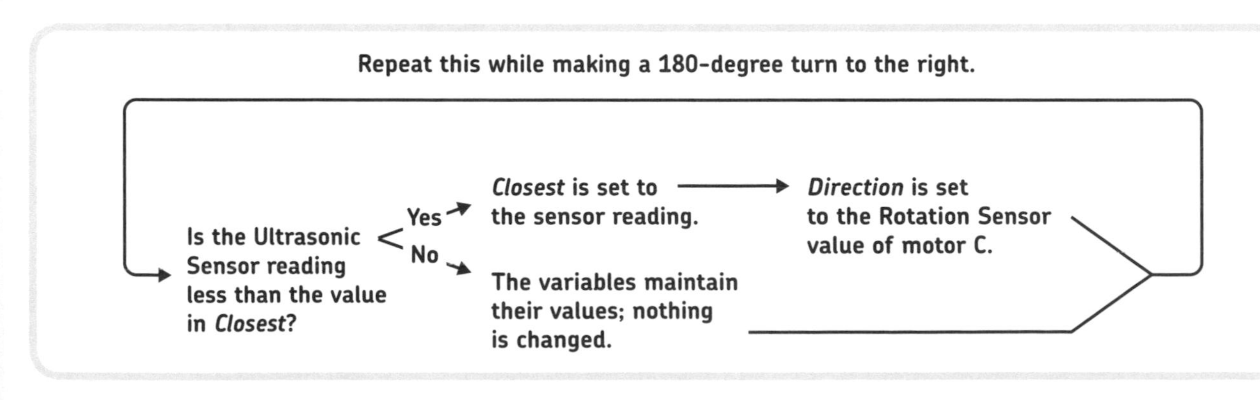

Figure 13-9: Finding the closest object in range

stops when the Rotation Sensor value equals that of the Direction variable. When it stops, the robot should point straight at the object and be ready to approach and grab it.

Define two variables called *Closest* and *Direction* (both numerical variables), configure the required blocks to find objects as shown in Figures 13-10 to 13-12, and then turn them into a My Block called **Find Object**.

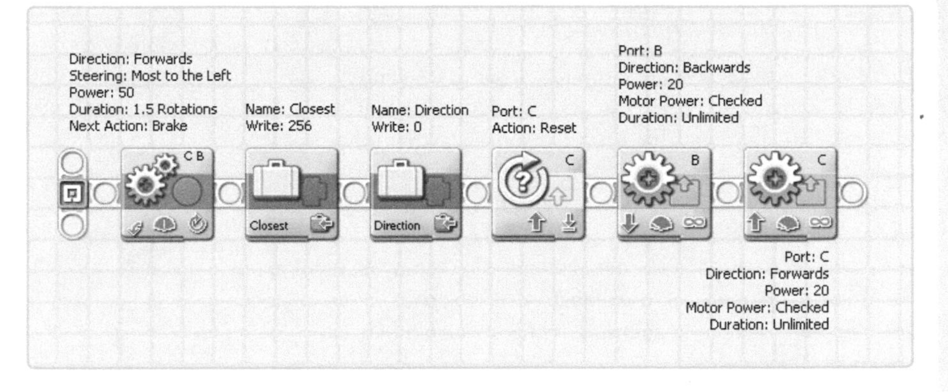

Figure 13-10: These blocks prepare the Snatcher to find objects: the robot turns left, initializes the variables, resets the rotation sensor in motor C, and starts turning right.

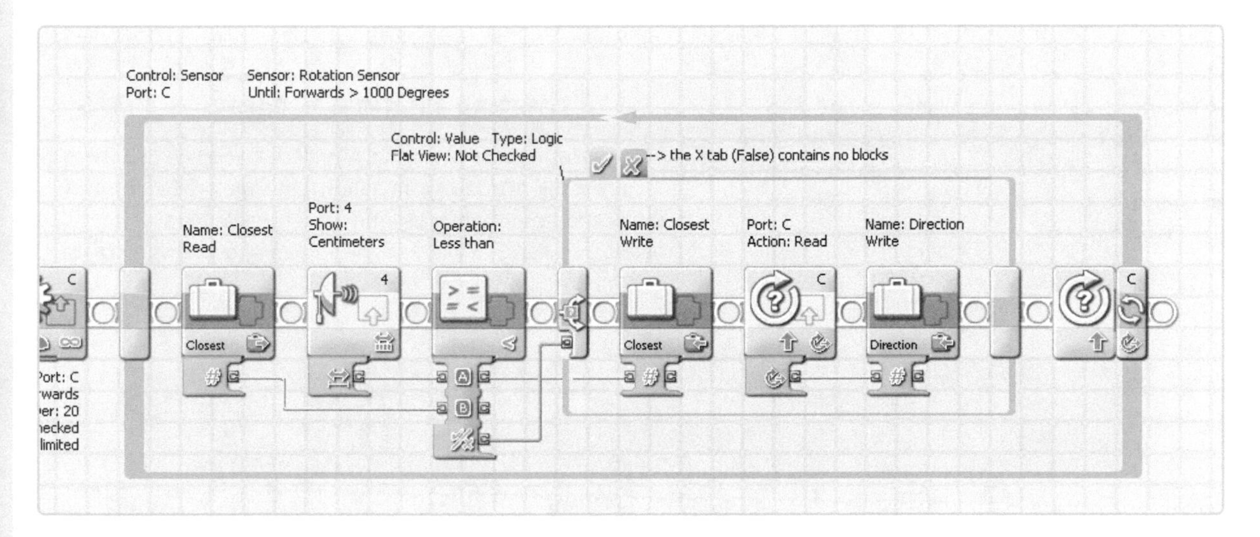

Figure 13-11: These blocks take care of searching for objects, as discussed in Figure 13-10.

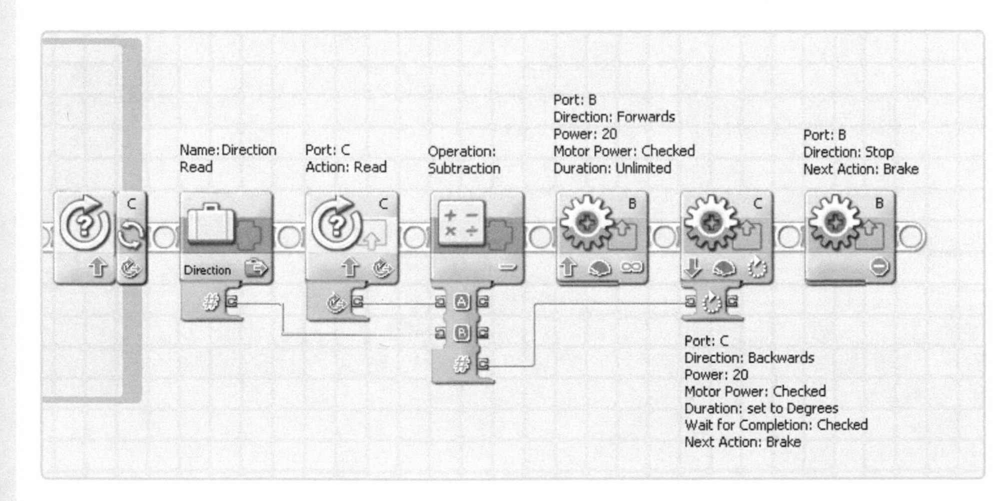

Figure 13-12: Once the search completes, these blocks make the robot return to the place where it spotted the closest object.

my block #4: move closer

This block makes the Snatcher move closer to the found object, based on the previously measured distance to it. If the object is far away, the Snatcher moves farther than when it is close already. To accomplish this, a Math Block multiplies the value of the Closest variable by 45 and transfers the result to the Duration setting of the Move Block. For example, a closest recorded distance of 10 cm would make the robot approach the object by turning the motors 450 degrees, while a measurement of 5 cm would make it turn only 225 degrees.

Configure the required blocks to make this movement as shown in Figure 13-13, and turn them into a My Block called **Move Closer**.

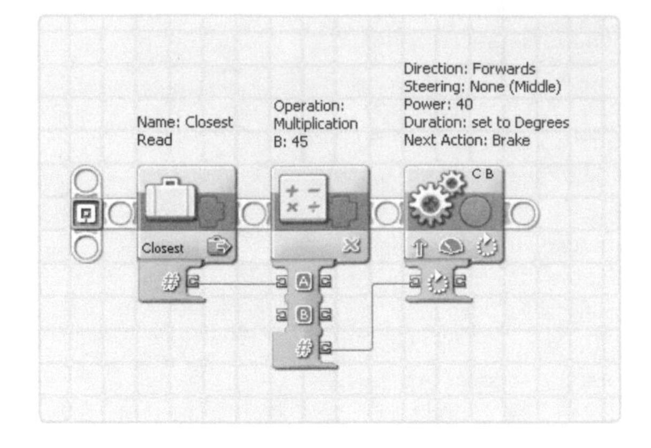

Figure 13-13: The configurations of the blocks in the Move Closer My Block. To be able to select ***Closest*** *in the Variable block, you may need to define this variable again.*

my block #5: say color

This My Block simply plays a sound file (such as "Red") based on the detected color of the object. You may have already created this block. If you haven't, do so by following the directions in Figure 7-16, or download it from the companion website.

creating the final program

Now that you've created your My Blocks, you can use them to create the main program (shown in Figure 13-6) to make the Snatcher autonomously find, grab, lift, and move an object. Figure 13-14 and Figure 13-15 show how to create the final program.

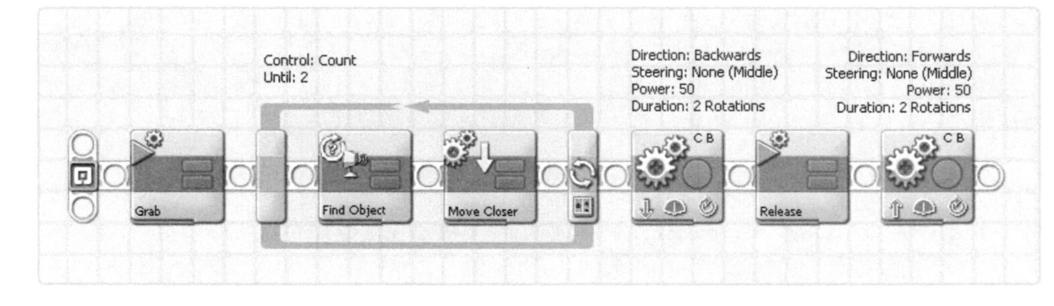

Figure 13-14: Before the robot starts looking for objects, it lifts up the grabber (using the Grab block) so that the Ultrasonic Sensor's sight isn't blocked by the grabber. After searching and before grabbing an object, it lowers the grabber (Release). The Move Blocks here prevent the grabber from crushing the object when it is being lowered.

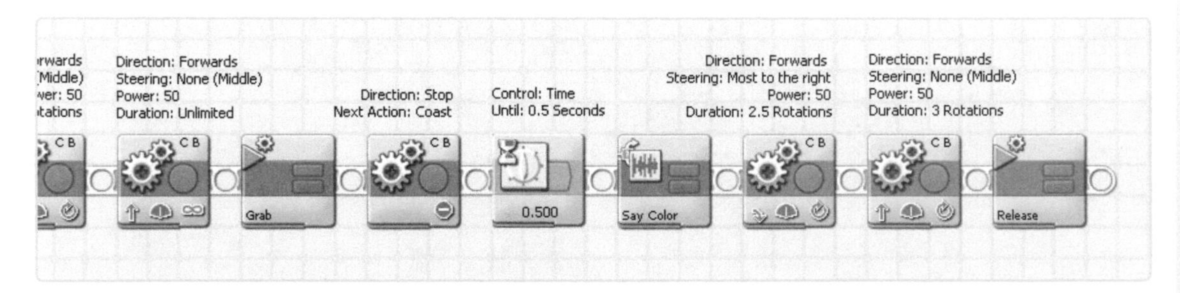

Figure 13-15: The next step is to grab the objects. To make sure that the object remains positioned between the robot's fingers, the Snatcher moves forward as it grabs. Next it identifies the object's color and drops it elsewhere.

troubleshooting the snatcher

If the Snatcher's program isn't functioning properly, carefully reconfigure the blocks or download (and examine) a functioning program from the companion website.

If the grabber arm on the Snatcher doesn't seem to work properly, you may have connected some LEGO parts too tightly. Try removing the small gray part as shown in Figure 13-16 to free up some space around the dark gray beam that connects to the axle.

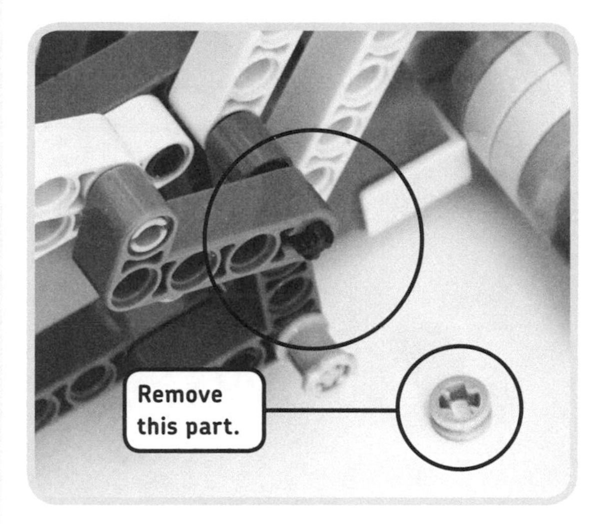

Figure 13-16: Remove the part shown here and then free up space around the highlighted dark gray beam if the Snatcher's grabber jams while trying to grab something.

further exploration

You've just completed one of the most complex robots in this book. Congratulations! Now that you've built and programmed it, you're probably looking for more things to do with the Snatcher. The following discoveries will help you further improve your programming and building skills.

NOTE The Snatcher's grabber was specifically designed to grab the objects you've used in this chapter, but you can easily modify it to grab anything you like, as long as it is not too heavy. To begin, remove the orange teeth-shaped parts from the Snatcher's grabber, and extend the fingers by adding extra LEGO beams.

DISCOVERY #67: I DON'T LIKE BLUE!

Difficulty: Medium
Modify the Snatcher program to make it picky about the objects it grabs. If it picks a blue object, it should drop it far away and then return to look for a new object. If it picks any other color, the program should say "Yes" and end instantly.

DISCOVERY #68: LIGHT IN A CORNER!

Difficulty: Hard
In this chapter you learned to program the Snatcher to look for the closest object. Now, using the Color Sensor in Light mode, create a program to look for the brightest light source in your room. Try making a program that finds lamps positioned on the ground. When it finds one, make the robot stop and play a sound.

BUILDING DISCOVERY #13: TABLETOP CLEANER!

Remove the entire grabber arm from the Snatcher so that only the driving base remains, and then position the robot on a table and create antennas for your robot to detect when the table edge is being approached. As the robot approaches the table edge, it should turn around and continue its path until it reaches another table edge. How do you create the antennas to detect a table edge? How do you build a robot so that it will never fall off the table? For more fun, build a sweeper module with the third NXT motor to wipe any LEGO part off the table!

14

hybrid brick sorter: sort bricks by color and size

The LEGO MINDSTORMS NXT 2.0 robotics kit comes with instructions to build a robot that sorts the colored balls in the set. This chapter goes a bit further by showing you how to build and program a sorter that sorts by both color and size. The Hybrid Brick Sorter robot separates regular LEGO bricks (2-by-4 studs) from smaller bricks (2-by-2 studs) and also separates red, yellow, green, and blue bricks, as shown in Figure 14-1. It drops each type of brick into different buckets or simply onto different places on the ground as shown in the figure.

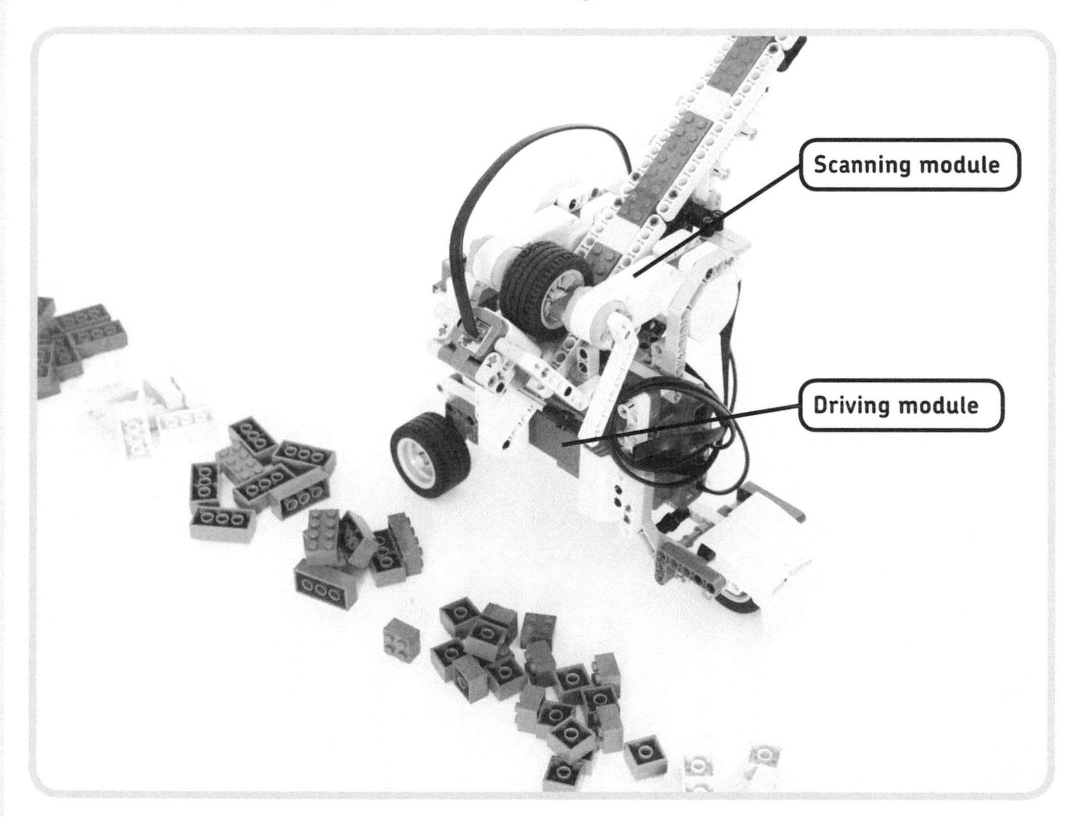

Figure 14-1: The Hybrid Brick Sorter sorts LEGO bricks by color and size.

understanding the sorting technique

The Brick Sorter consists of two modules, as shown in Figure 14-1: the *Scanning module*, which identifies the size and color of each brick, and the *Driving module*, which moves the robot toward the appropriate bucket to drop the brick.

the driving module

The Driving module navigates the robot to different buckets before it drops the bricks by turning the *Driving motor* as shown in Figure 14-2. The motor's Rotation Sensor tells the NXT where the robot is positioned.

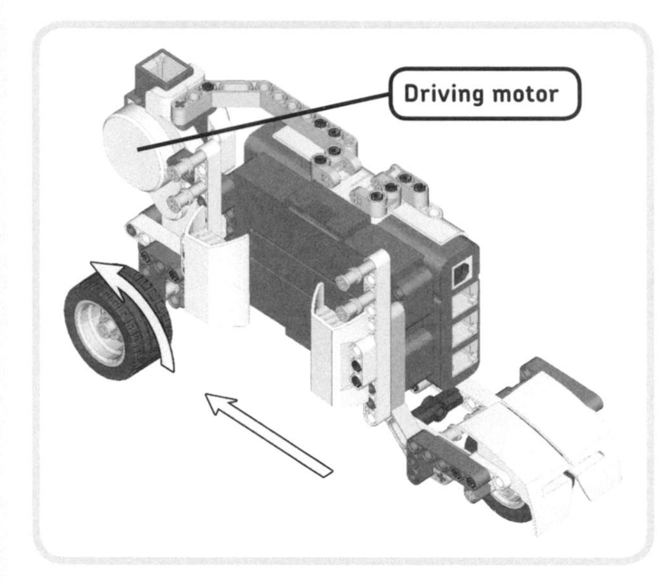

Figure 14-2: The Driving motor allows the robot to move back and forth in a straight line between different buckets. The arrows show what happens when the motor turns forward.

the scanning module

The Scanning module identifies the color and size of a LEGO brick and drops the brick once the robot has moved to the correct bucket. Before the bricks are sorted, they're stacked in the *chute*, and the *sorter wheel* prevents the bricks from sliding down, as shown in Figure 14-3. As this sorter wheel rotates, one brick slides down the chute until it is stopped by the *brick lock*. At this point, the Color Sensor identifies the brick's color, after which the brick lock is raised, and the brick falls out of the sorter into the appropriate bucket.

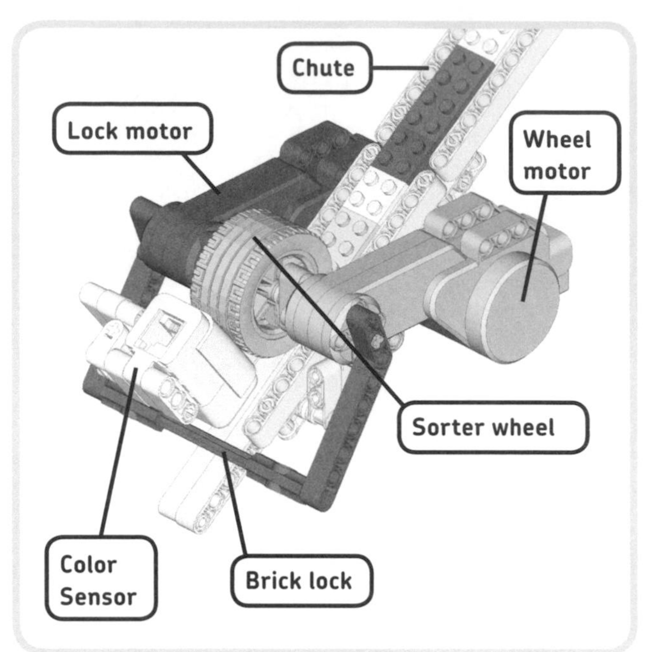

Figure 14-3: The Scanning module. Some parts have been removed here for clarity.

identifying a brick's size

You've just learned how the sorter reads the color of a brick, but how do you find its size? The answer lies in how much the sorter wheel should turn in order to let one brick slide down to the Color Sensor, where it is scanned. Tests with the sorter show that turning the *Wheel motor* (and thus the sorter wheel) by 46 degrees is just enough to let one small brick (2-by-2 studs) slide down toward the brick lock. A big brick (2-by-4 studs) requires a 92-degree turn in order to reach the sensor.

You identify the size of a brick by first turning the sorter wheel by 46 degrees. If you do this and the Color Sensor then sees a brick, you know that you have a small brick; if it doesn't, you have a big brick. (The sensor knows that it sees a brick when it doesn't see the white chute in the background.) If you have a big brick, the sorter wheel should turn another 46 degrees to allow the brick to slide down to the sensor, as shown in Figure 14-4. Once the sorter knows which brick it is dealing with, it moves to the correct bucket, and the brick lock is raised to release the brick.

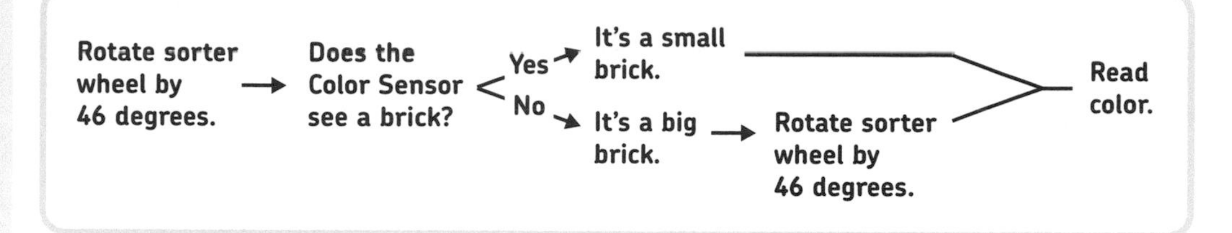

Figure 14-4: Finding the size and color of a brick

building the hybrid brick sorter

Before you build the sorting machine, select the required pieces, as listed in Figure 14-5. Once you have all your pieces, follow the building instructions to build it.

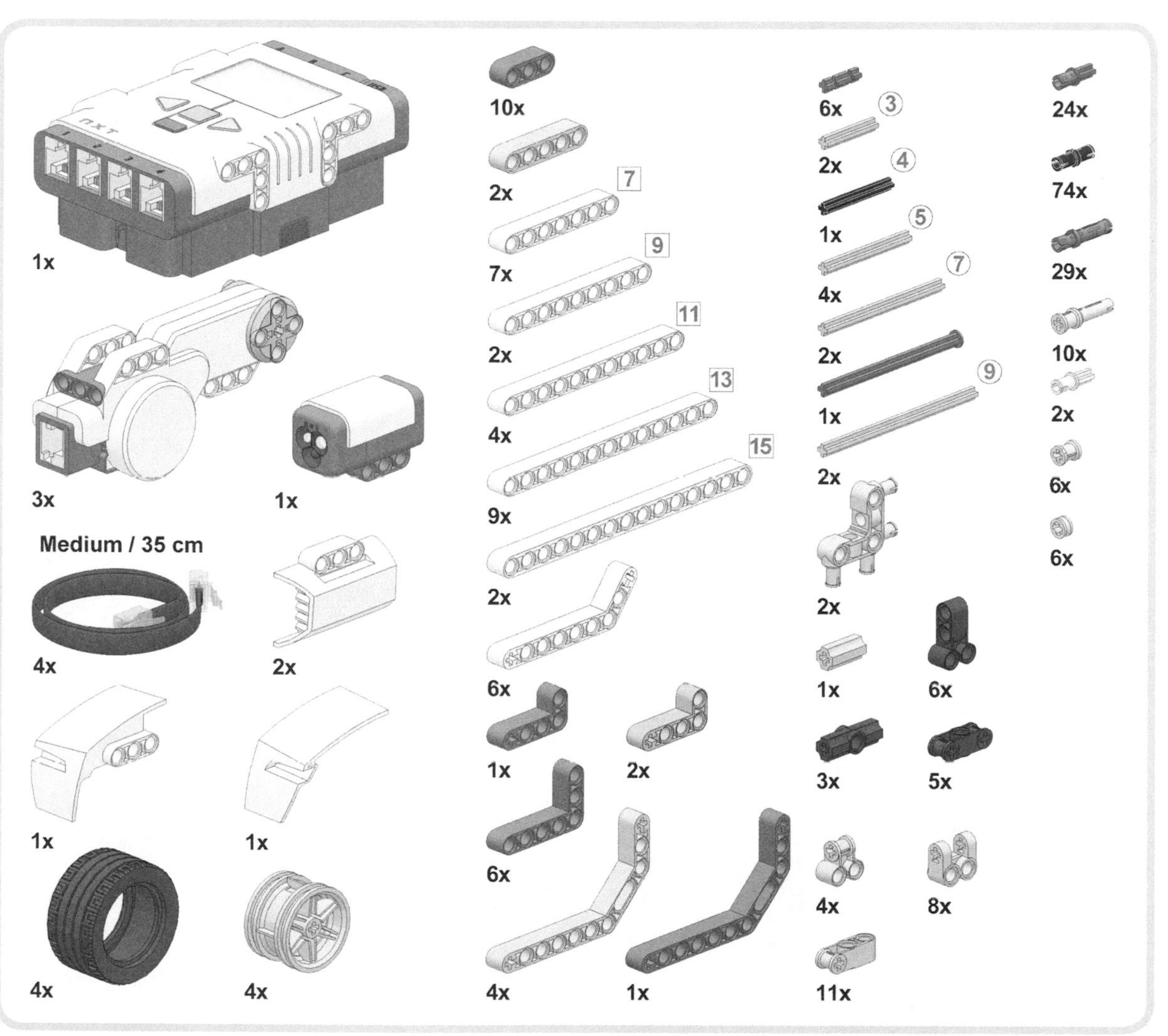

Figure 14-5: The required pieces for the Hybrid Brick Sorter

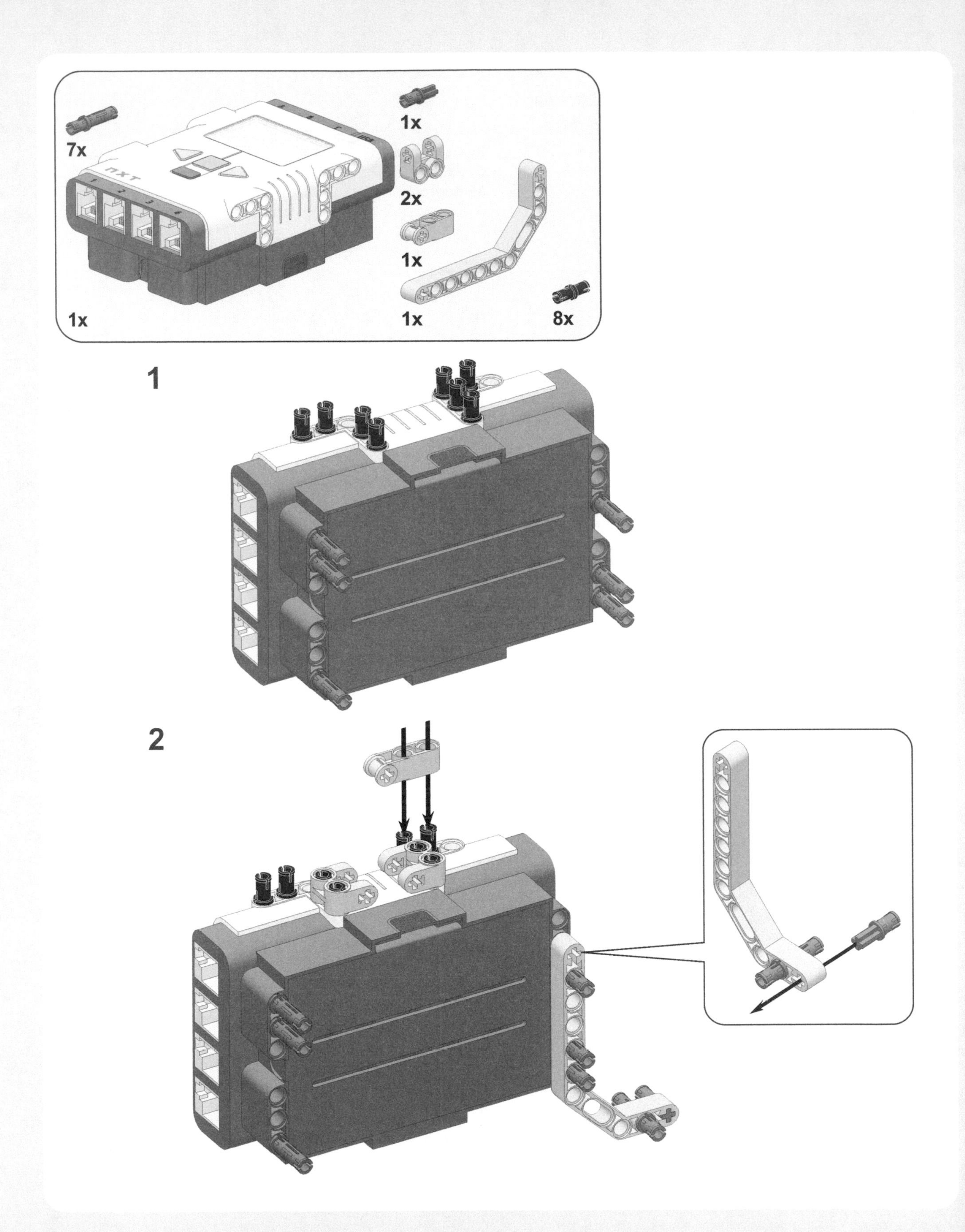
7x
1x
2x
1x
1x
1x
8x
1
2

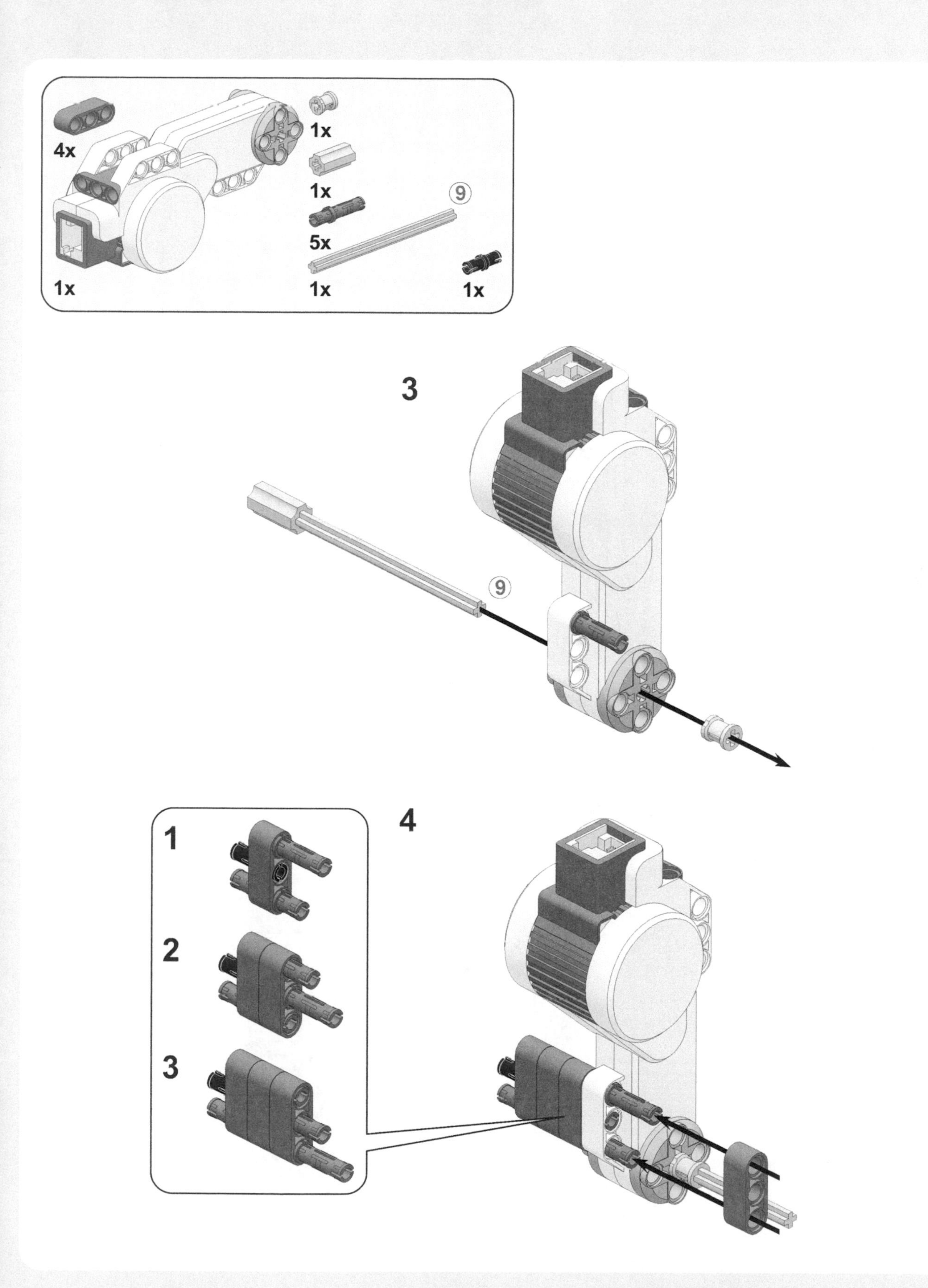
4x
1x
1x
5x
9
1x
1x
1x
3
9
4
1
2
3

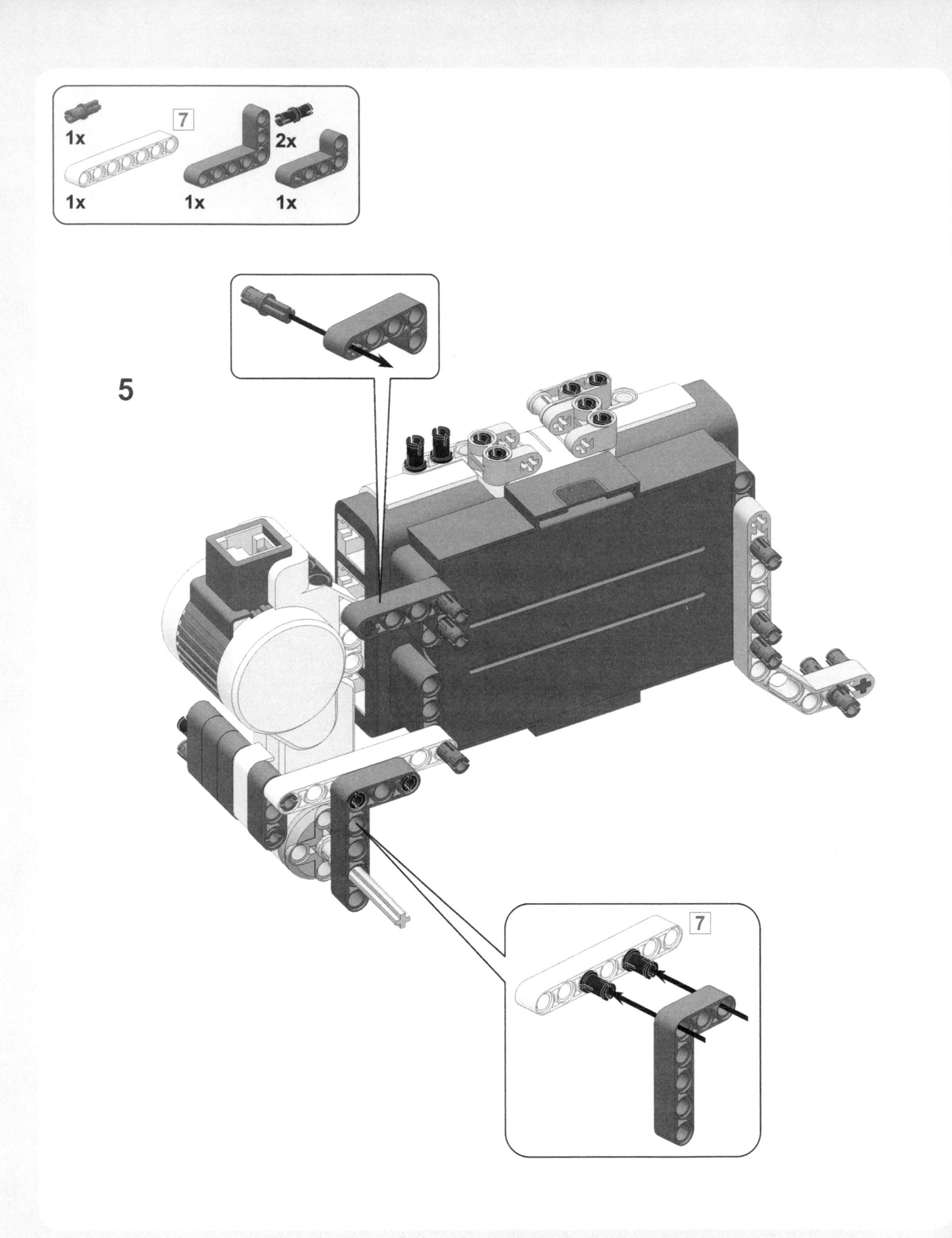
1x
7
2x
1x
1x
1x
5
7

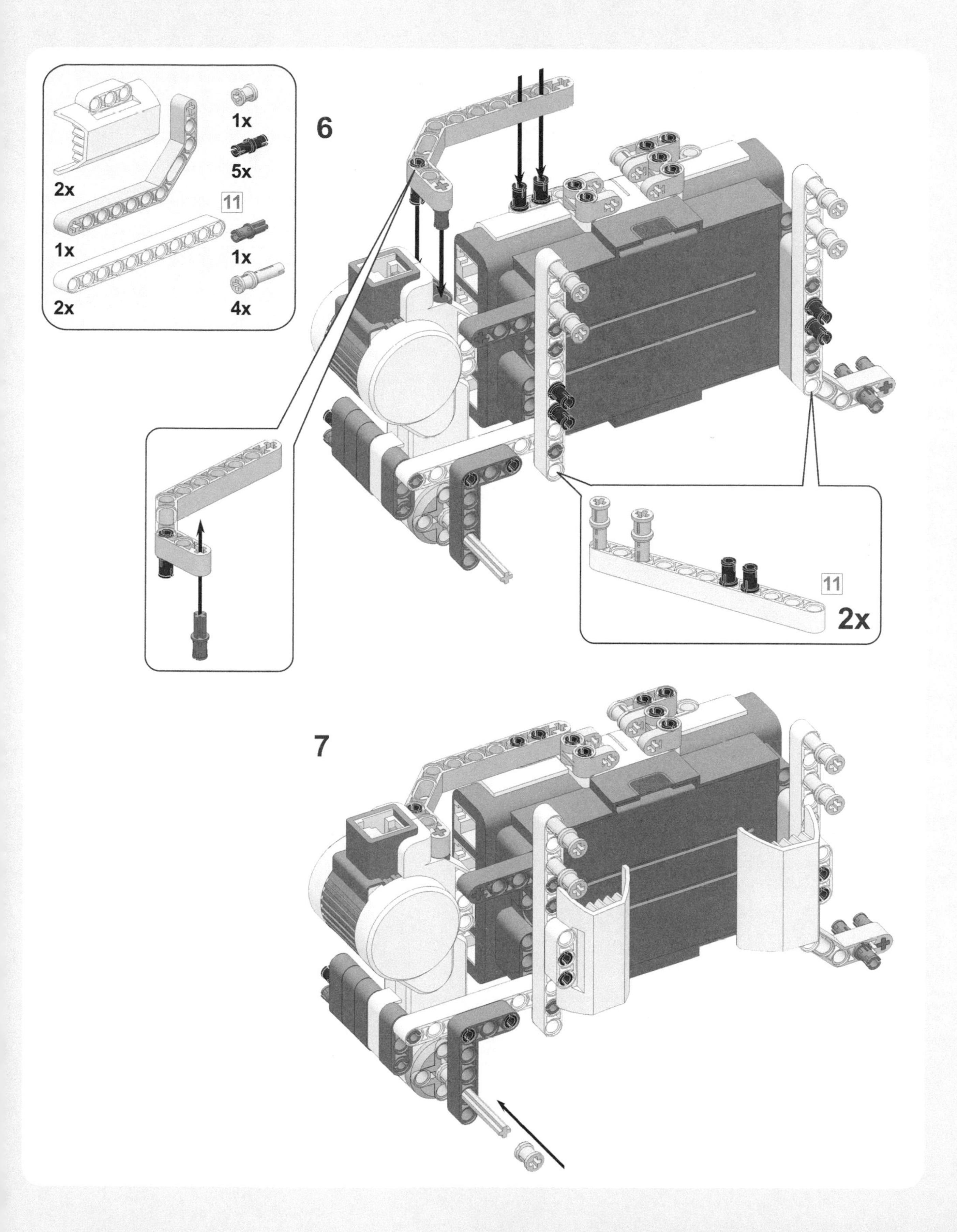

6
2x
1x
2x
1x
5x
11
1x
1x
4x
11
2x
7

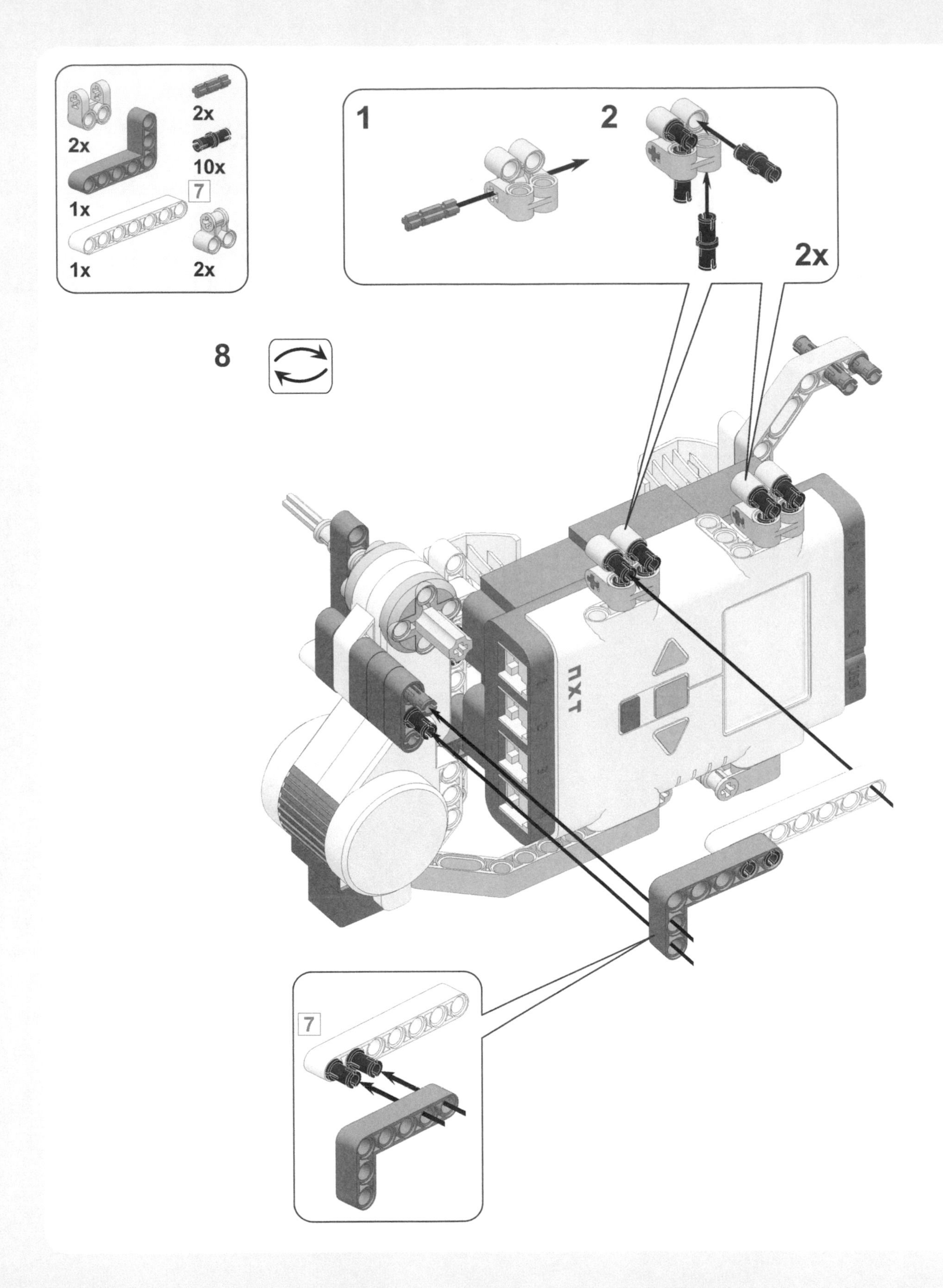
2x
2x
10x
7
1x
1x
2x
1
2
2x
8
7
nxt

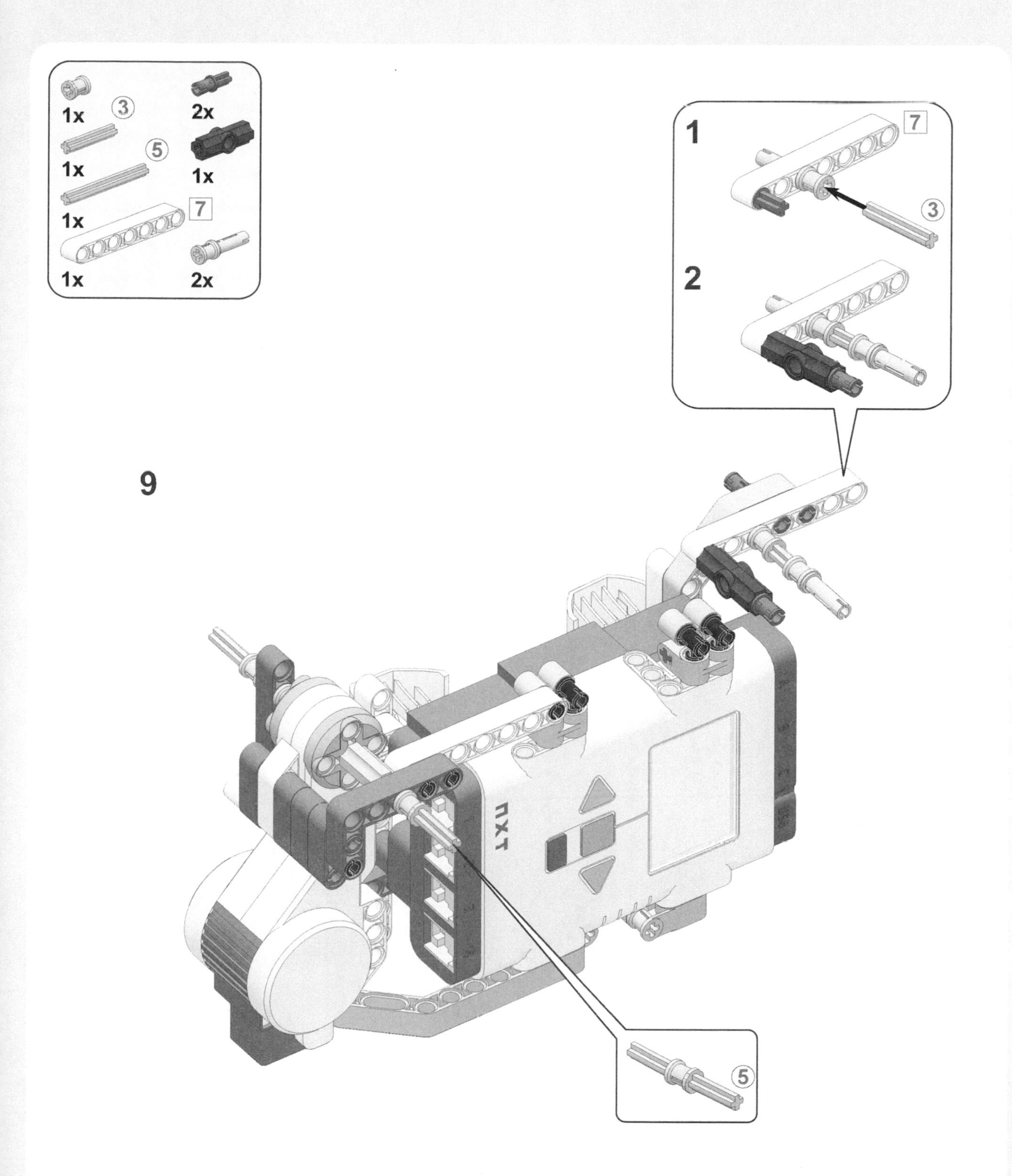
1x
3
2x
1x
5
1x
1x
7
1x
2x
1
7
3
2
9
NXT
5

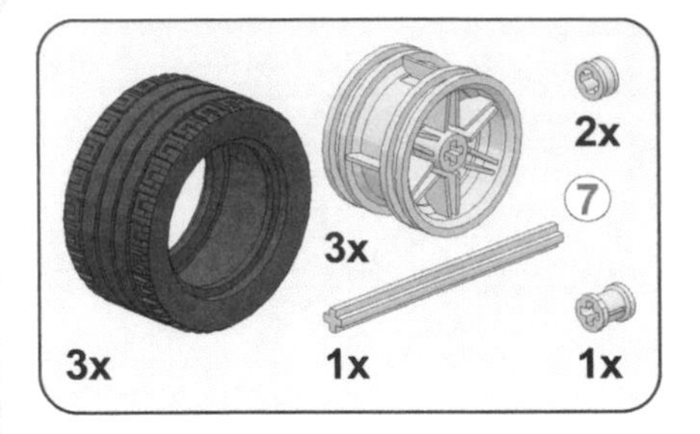
2x
7
3x
3x
1x
1x

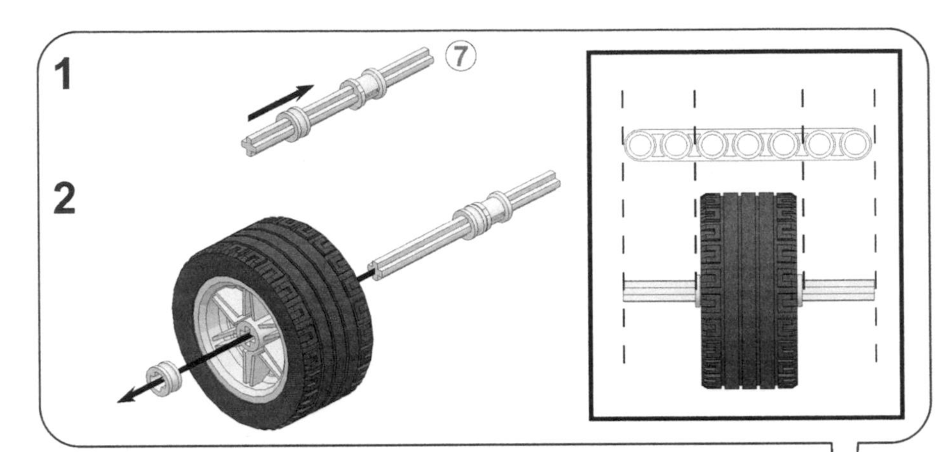
1
7
2

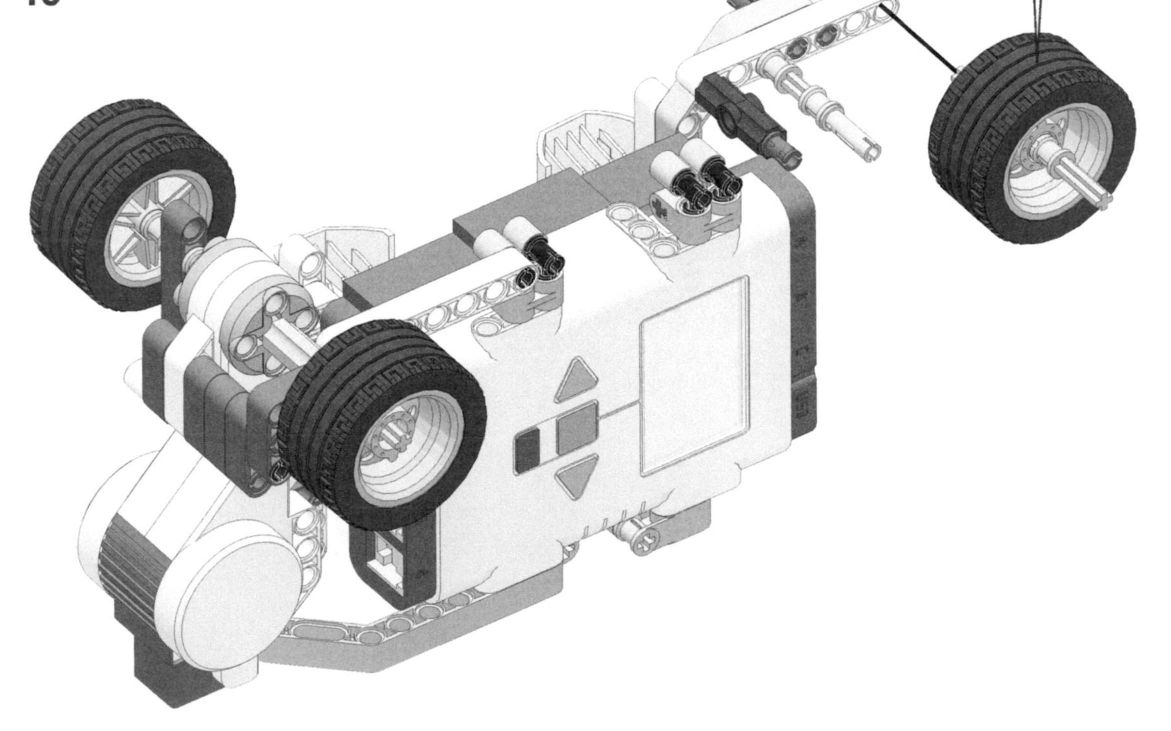
10

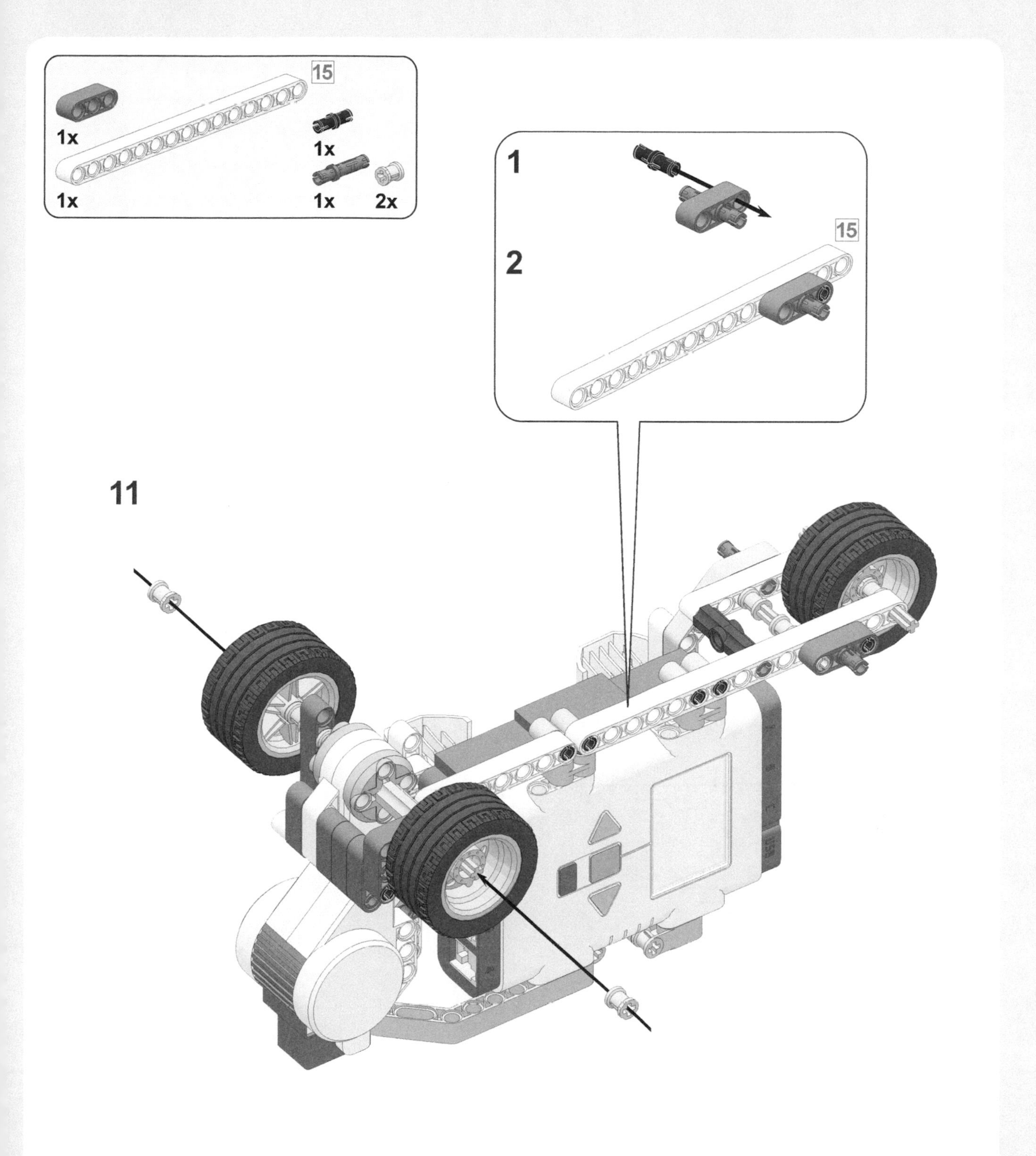
15
1x
1x
1x
1x
2x
1
2
15
11

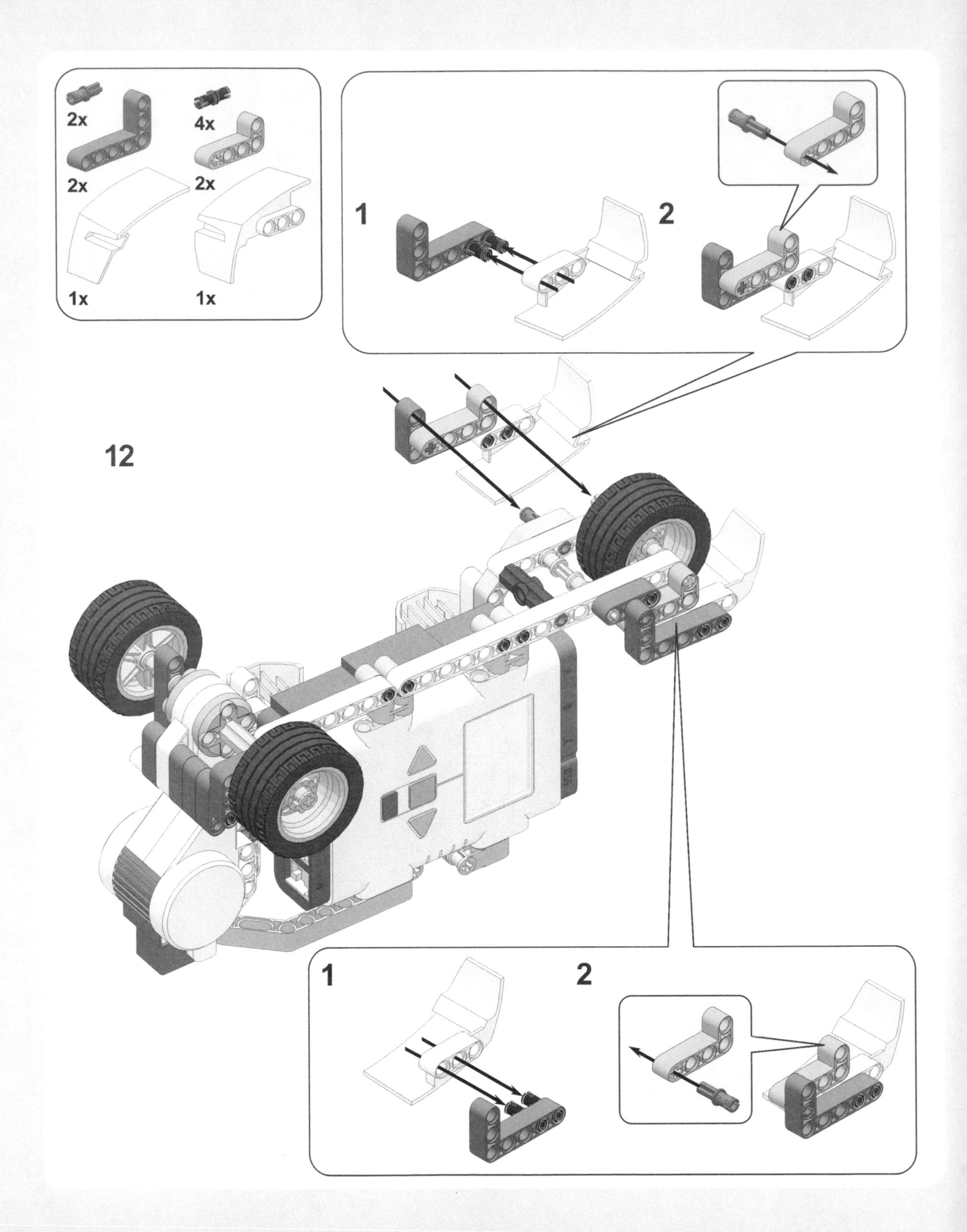
2x
4x
2x
2x
1x
1x
1
2
12
1
2

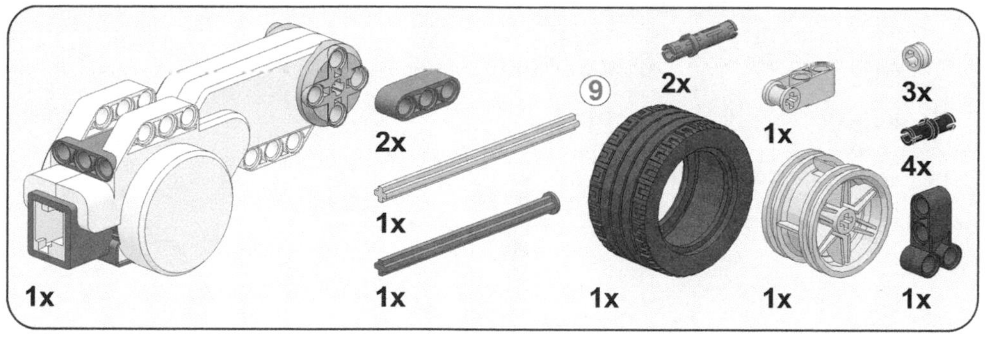
1x
2x
9
1x
1x
2x
1x
3x
4x
1x
1x
1x
1x

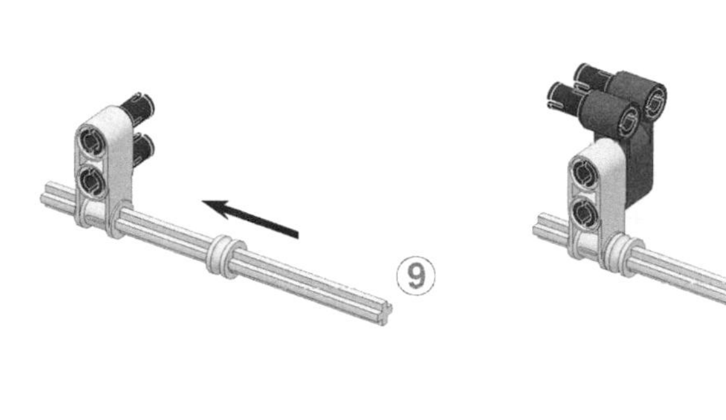
1
2
9

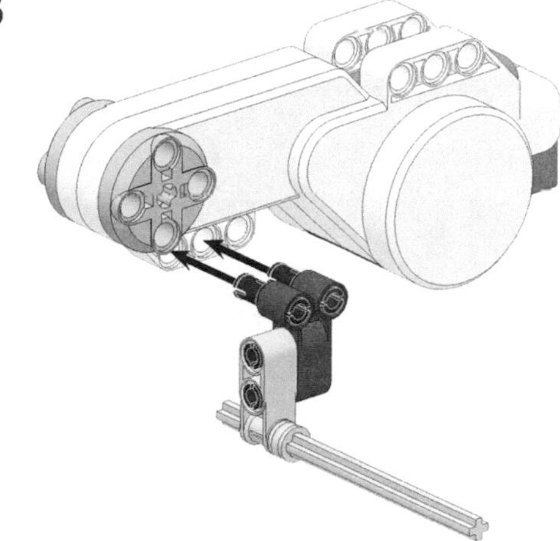
3

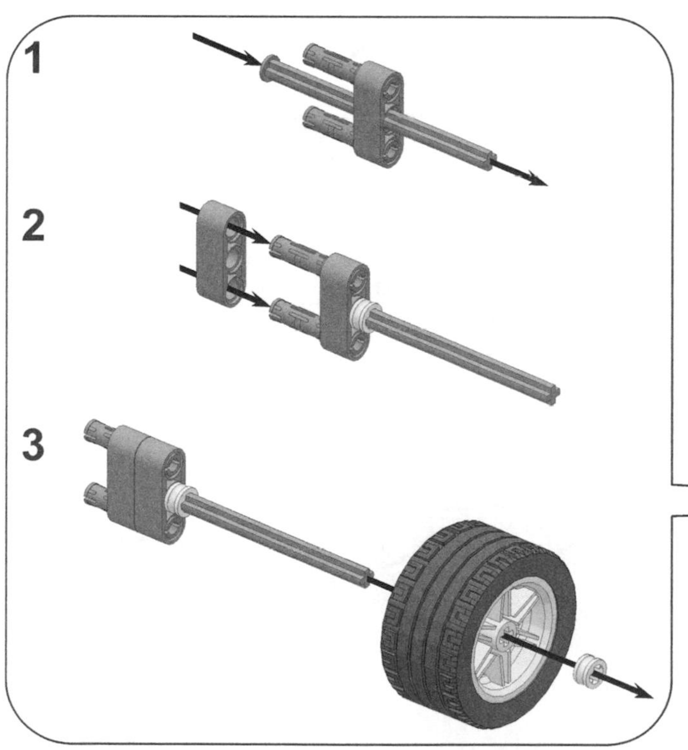
1
2
3

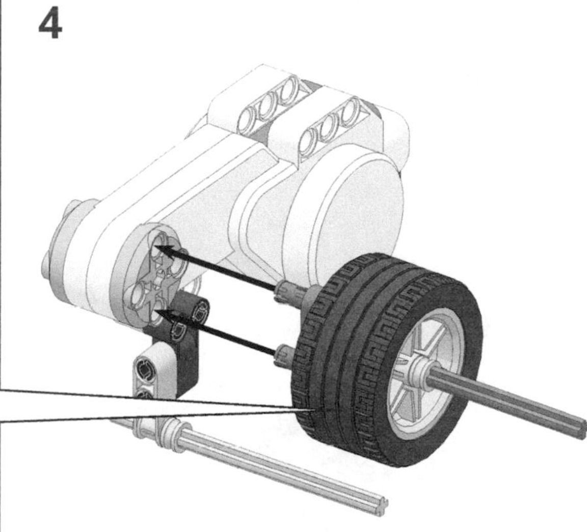
4

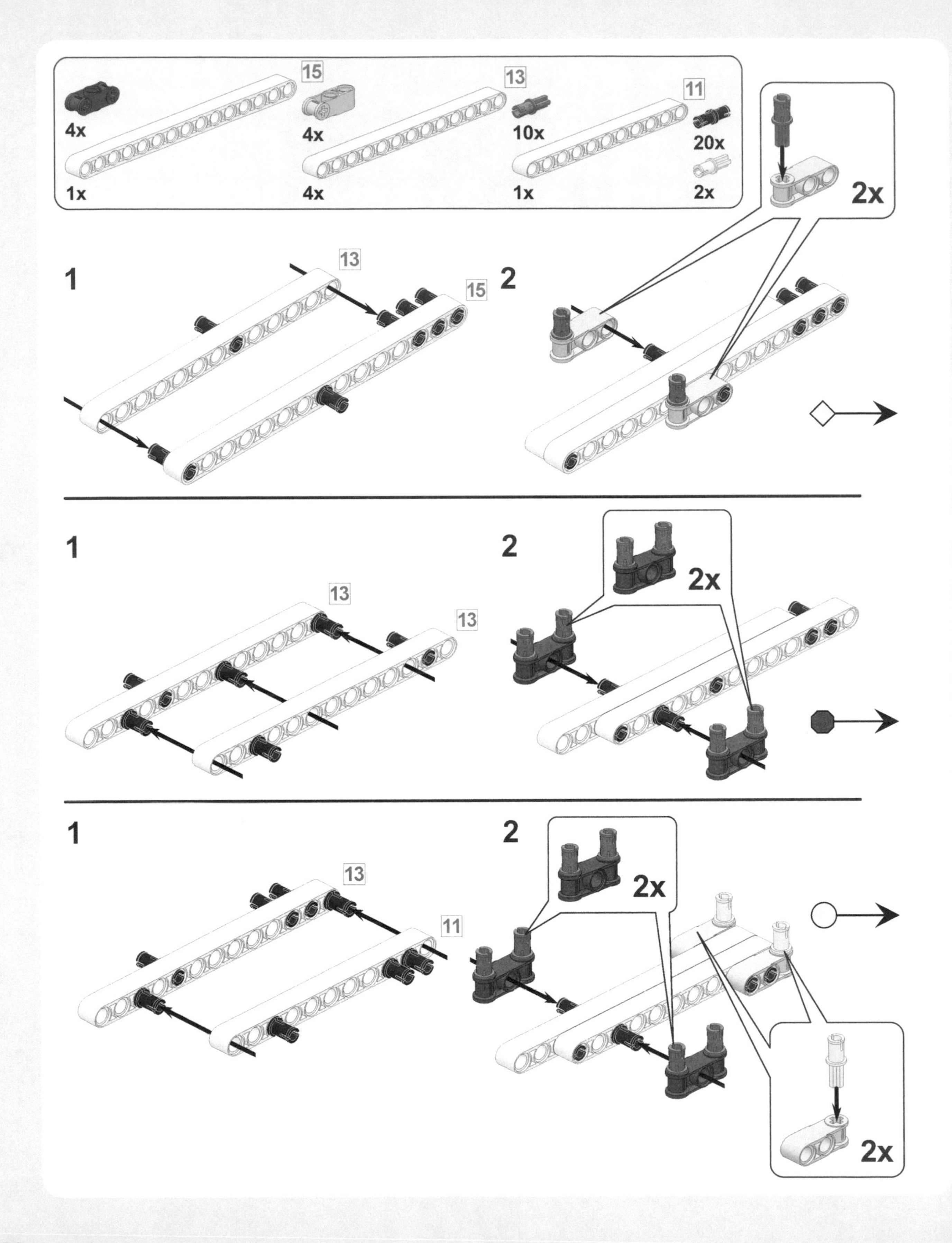
15
13
11
4x
4x
10x
20x
1x
4x
1x
2x
2x
1
13
15
2
1
13
13
2
2x
1
13
11
2
2x
2x

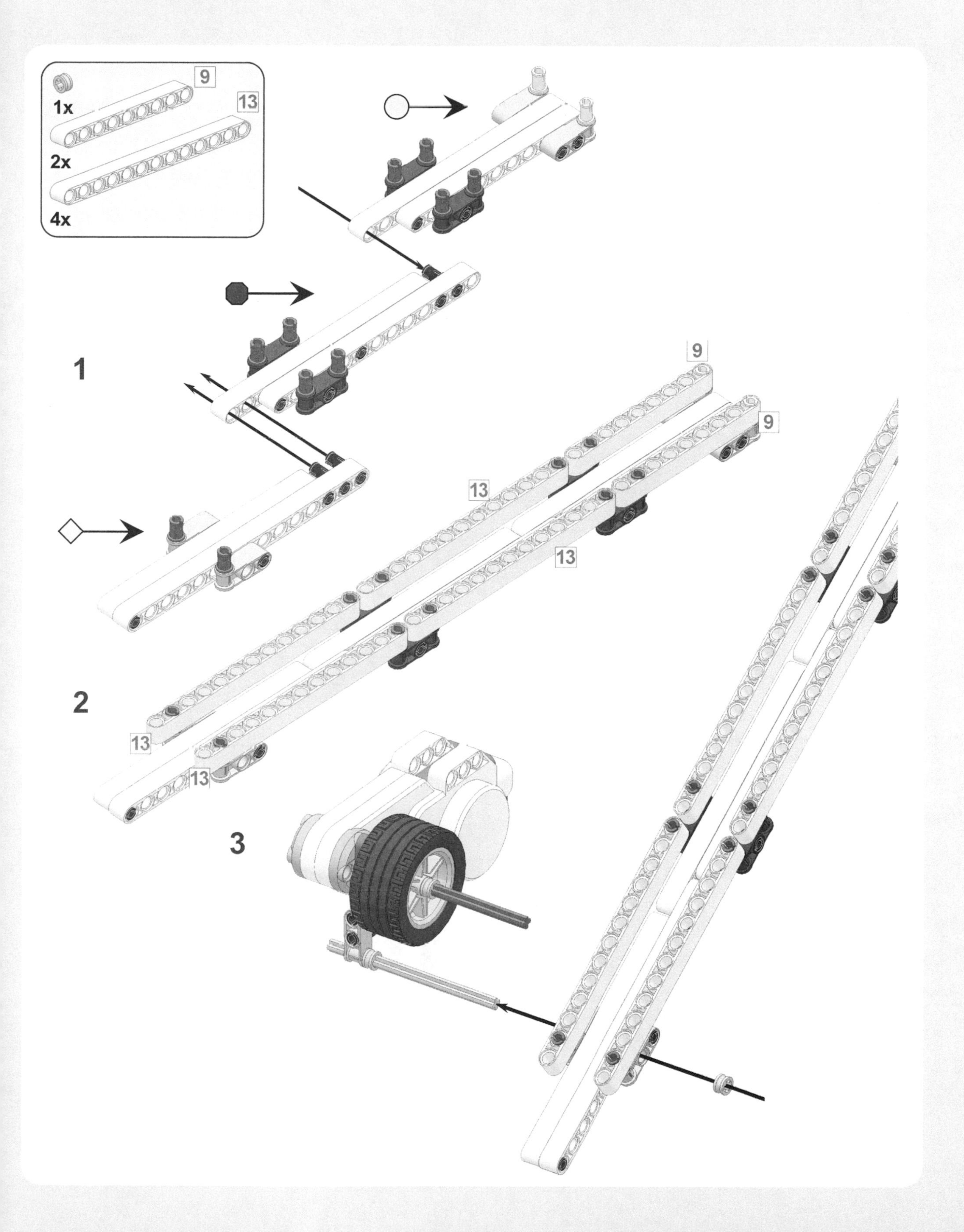

1x
2x
4x
9
13
1
2
3

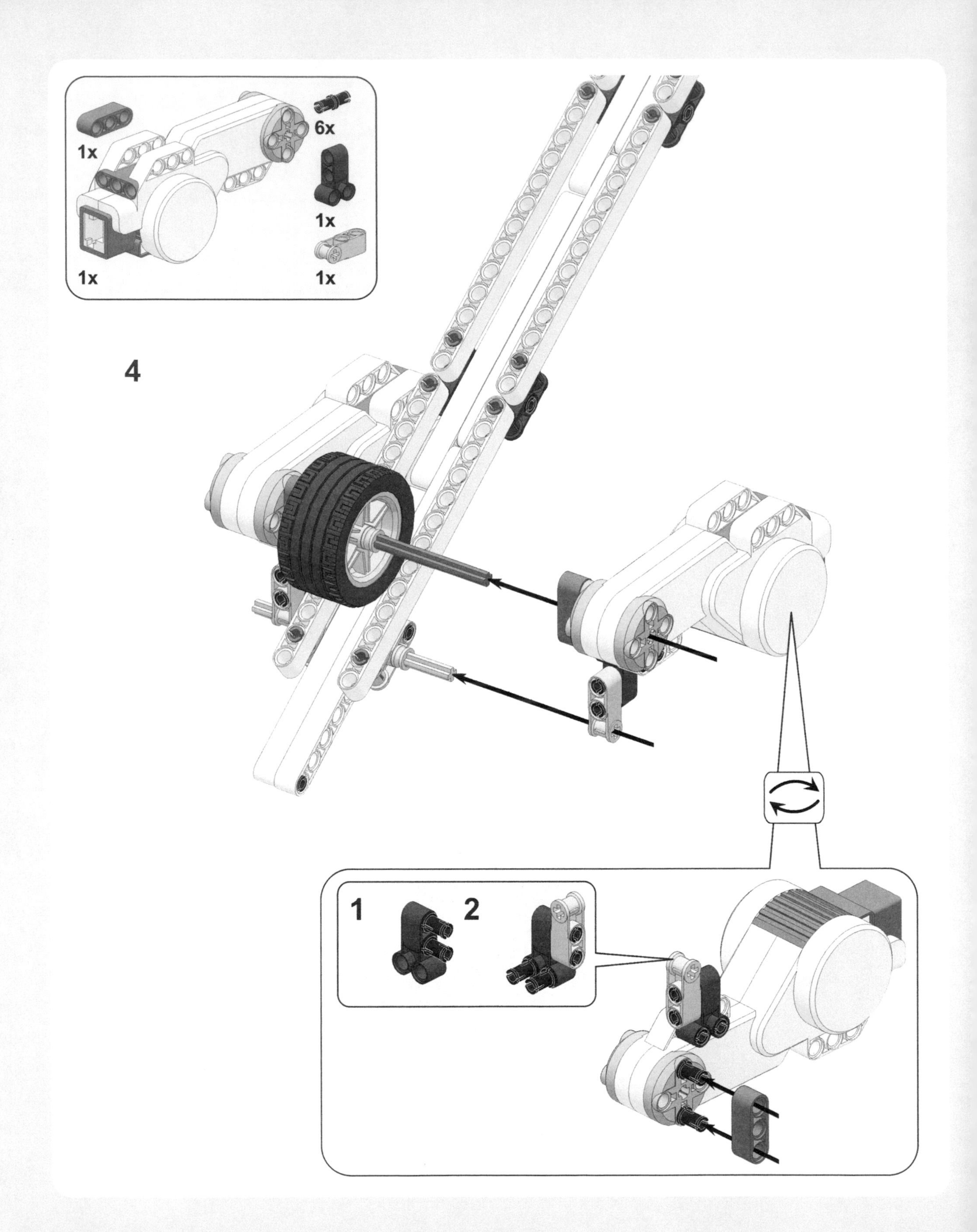

1x
6x
1x
1x
1x
1x
4
1
2

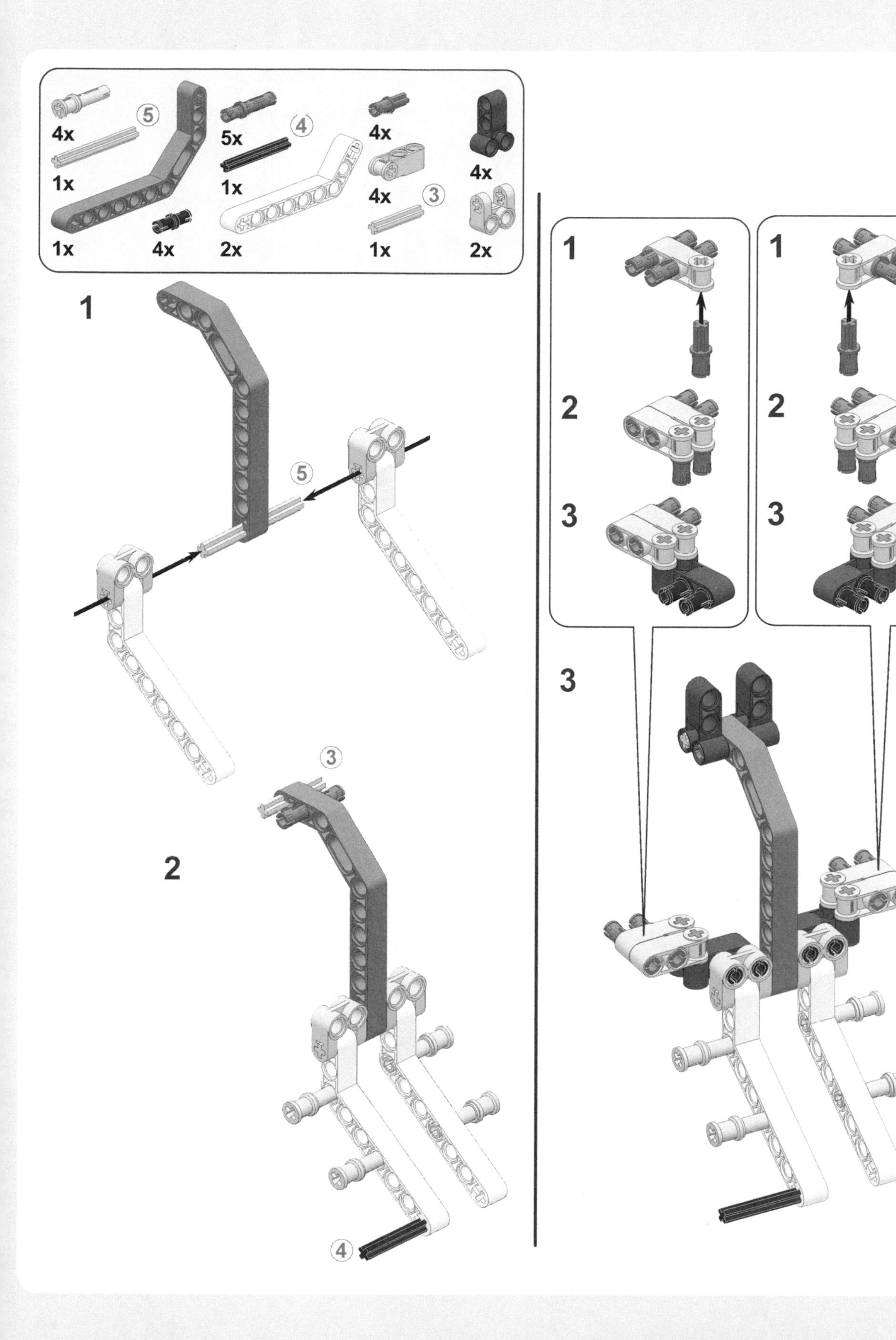
4x
1x
1x
4x
5x
1x
2x
4x
4x
1x
4x
2x
1
2
3

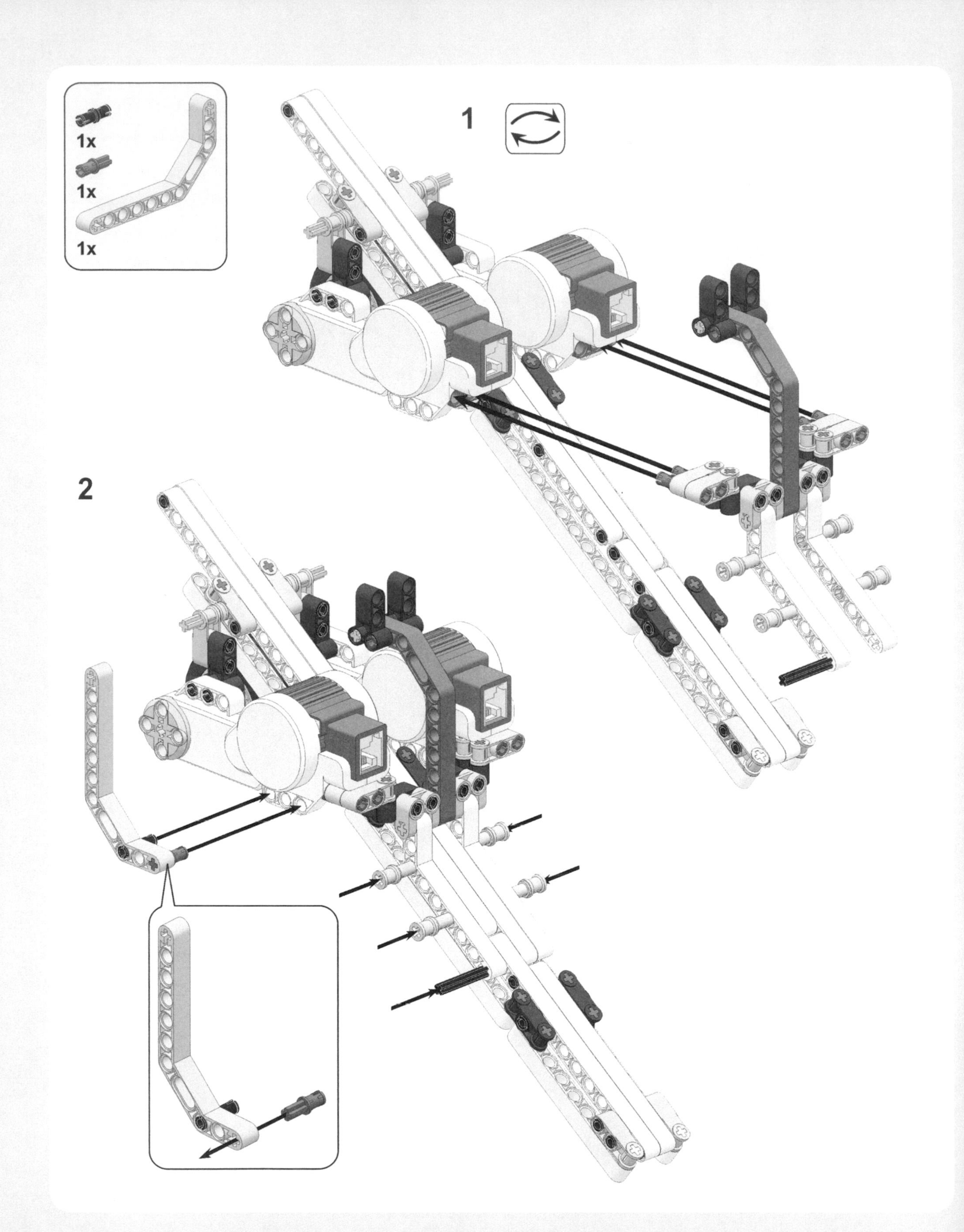
1x
1x
1x
1
2

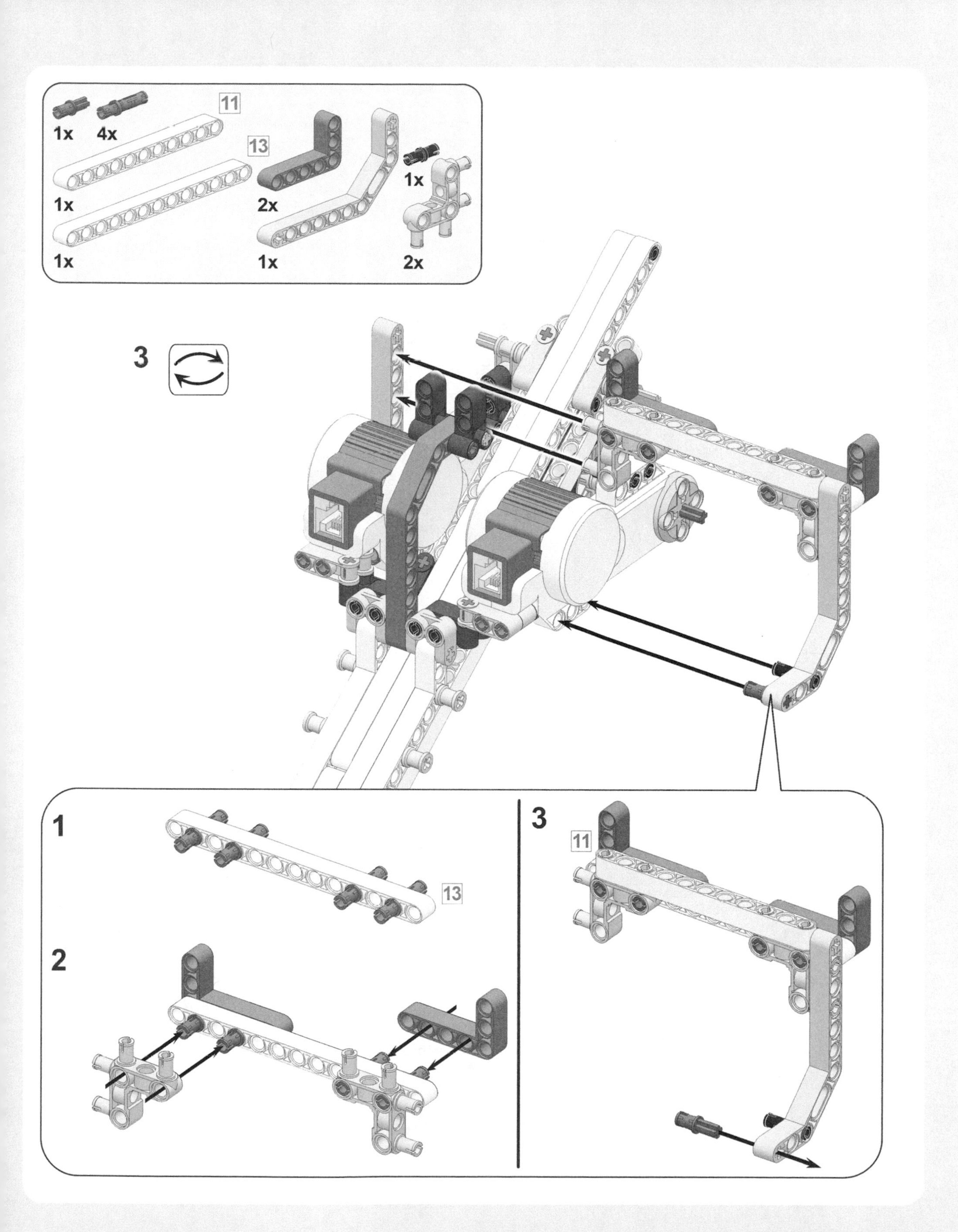

1x
4x
11
13
1x
2x
1x
1x
1x
1x
2x
3
1
13
2
3
11

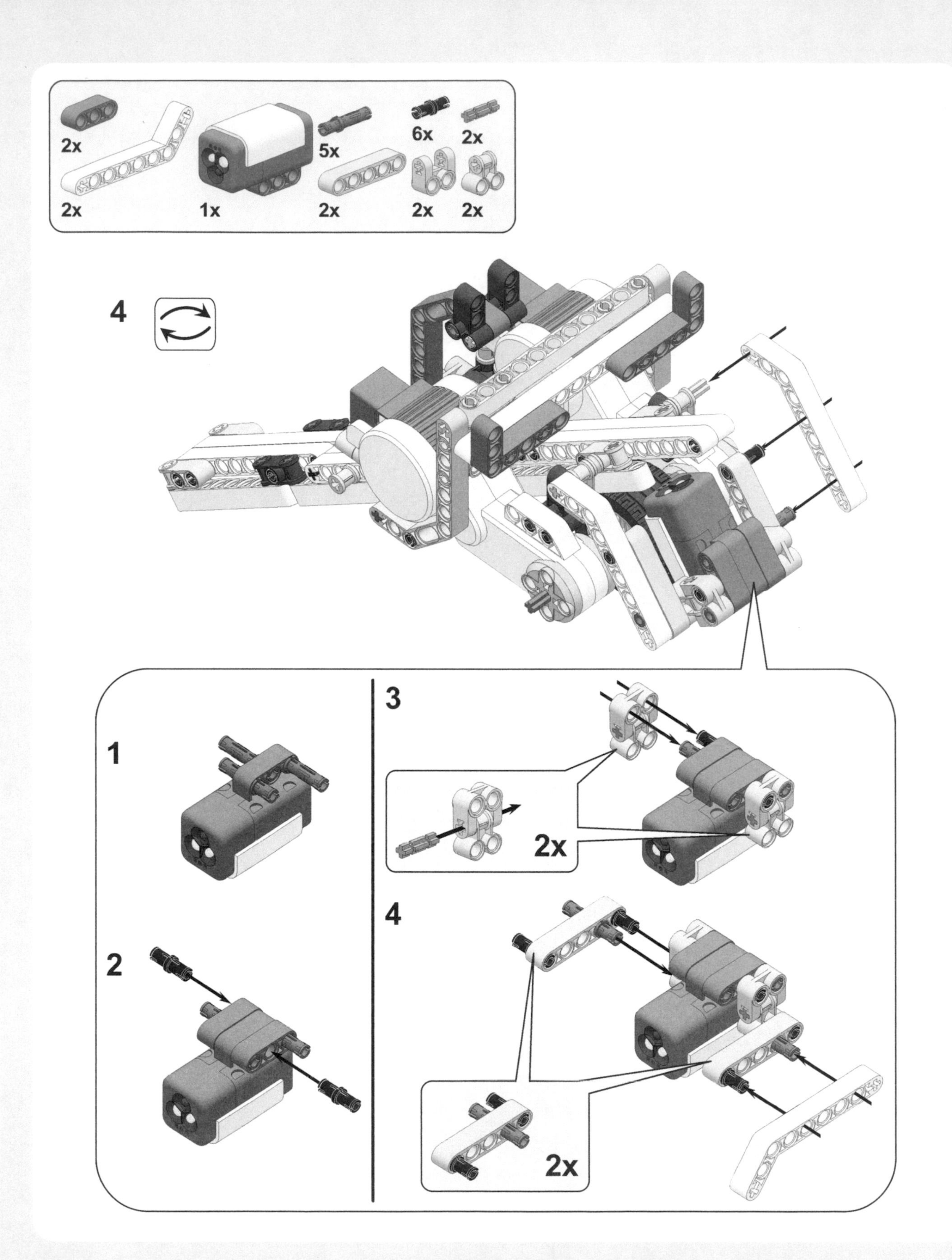
2x
2x
1x
5x
2x
6x
2x
2x
2x
4
1
2
3
2x
4
2x

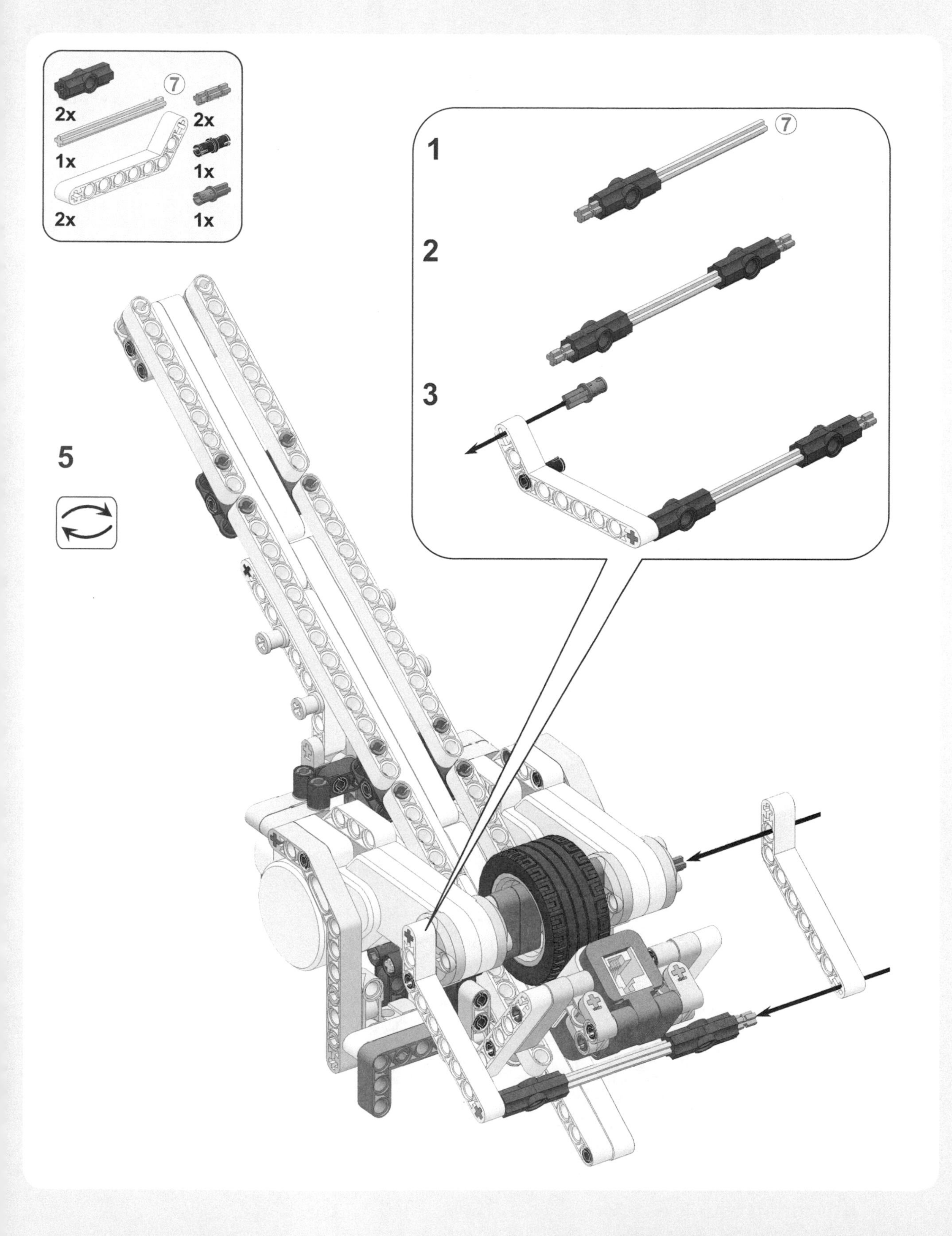
2x
⑦
2x
1x
1x
2x
1x
1
⑦
2
3
5

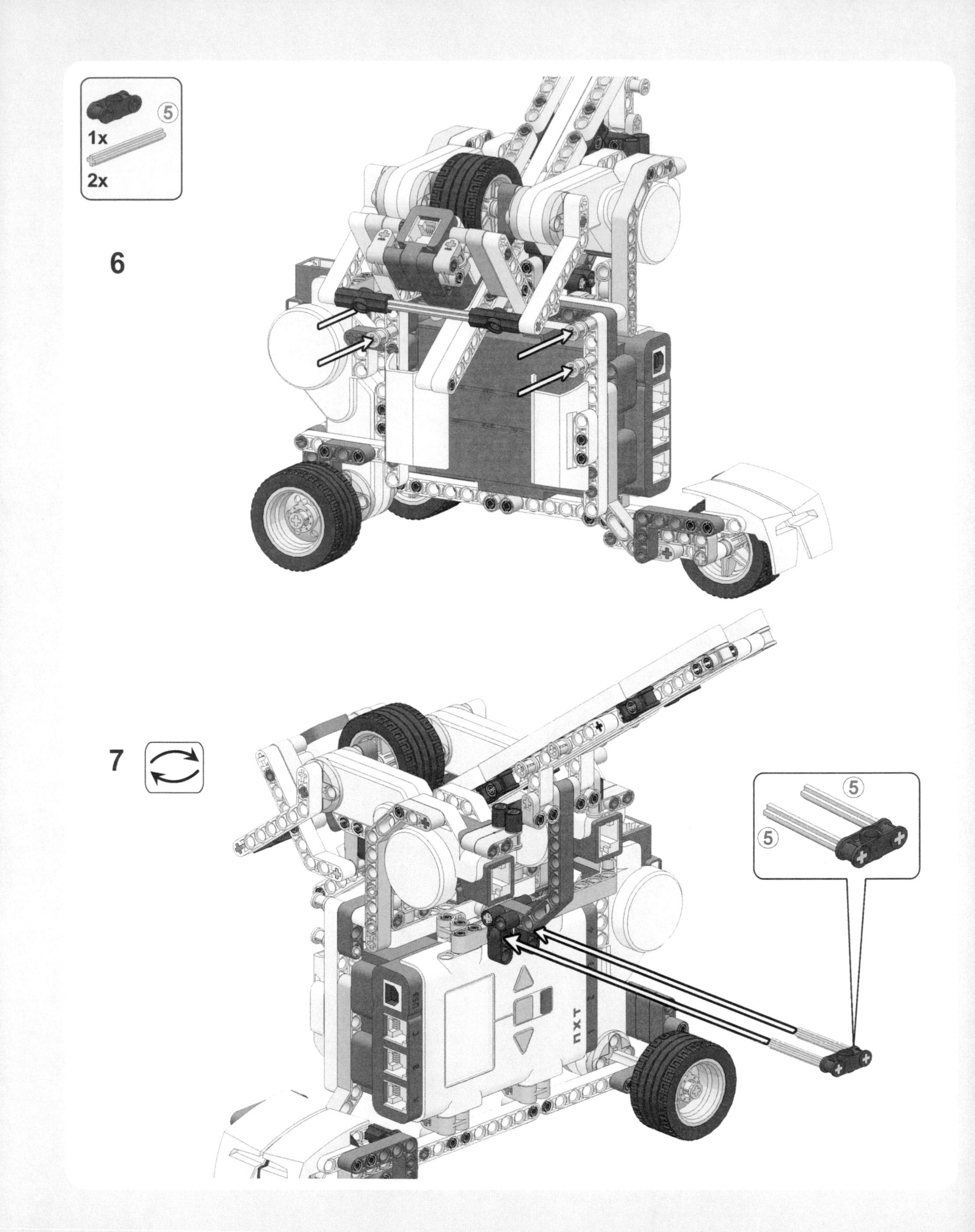
1x
5
2x
6
7
5
5

connecting the cables

Having built the sorting machine, it's time to connect the cables, as shown in Table 14-1. Make sure not to connect the Color Sensor cable too tightly, or it will make the Color Sensor interfere with the sorter wheel.

table 14-1: cable placement for the hybrid brick sorter

From motor/ sensor	To NXT brick port	Cable length
Wheel motor	Output port A	Medium (35 cm/15 inches)
Driving motor	Output port B	Medium
Lock motor	Output port C	Medium
Color Sensor	Input port 3	Medium

finding bricks to use with your sorter

Your Brick Sorter requires a handful of regular LEGO bricks in various colors and two sizes. The program you'll create is designed to sort green, blue, red, and yellow bricks, and to separate 2-by-4 studs from 2-by-2 studs, but if you modify it slightly, you can make it sort black and white bricks as well. (If you don't have these basic bricks in another LEGO set, visit the companion website for hints on finding them.)

finding buckets

Since your sorter will sort bricks in four colors and two sizes, you'll need eight buckets to store the bricks. Make your own buckets out of paper or other LEGO bricks or use ready-made buckets such as small cups, as shown in Figure 14-6.

Figure 14-6: An example of buckets the machine can store its bricks in

programming the hybrid brick sorter

You'll now program your machine to sort LEGO bricks. Figure 14-7 shows an overview of the complete **Hybrid Brick Sorter** program.

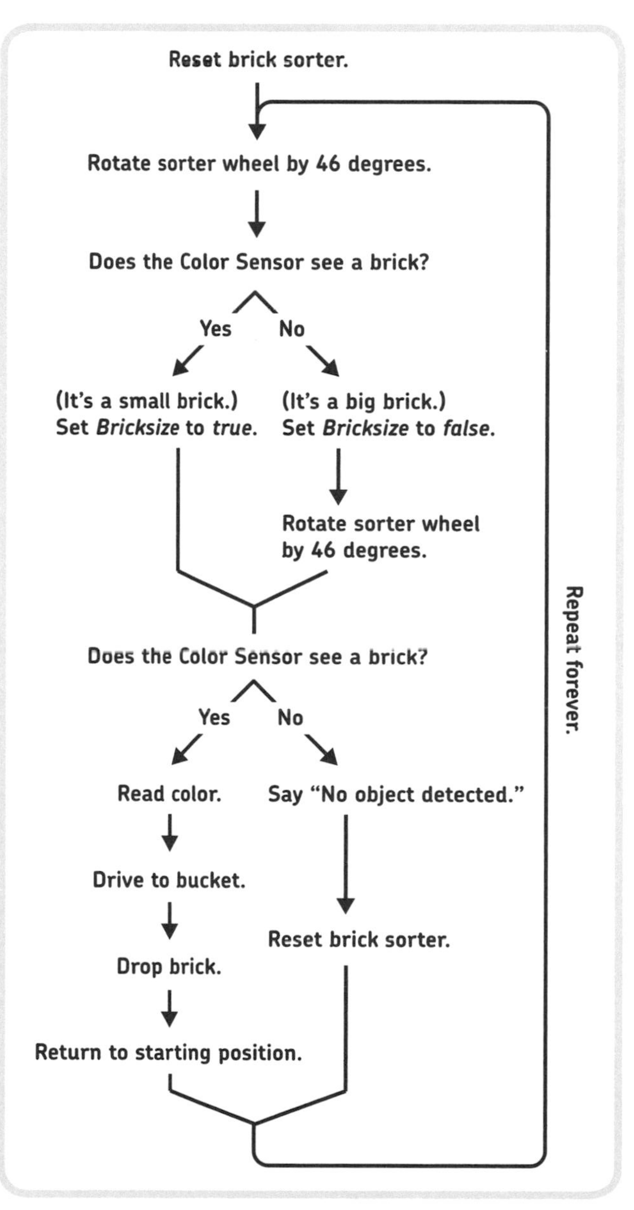

Figure 14-7: An overview of the program for the Hybrid Brick Sorter

Notice that this overview is an expanded version of the one in Figure 14-4. In addition to identifying the size of a LEGO brick, this program also checks to see whether there are any bricks left to sort. If the sorter wheel has turned 46 degrees twice and if the sensor still doesn't see a brick, there are no bricks left to sort, or there's been an error (that is, a brick got stuck). In this case, the sorter will report "No object detected!" and reset by turning the sorter wheel in the opposite direction so that bricks that got stuck can be sorted again. If there is no error and if the machine can identify the size and color of a brick, it proceeds to the appropriate bucket to drop the brick, after which it returns to its starting position to sort out the next brick.

creating the my blocks

You'll create two My Blocks for this program—Reset Sorter and Drop Brick—both of which will contain a selection of programming blocks that you'll use twice in the final program. You'll also use the Say Color block that you created in Discovery #30 on page 80.

my block #1: reset sorter

The Reset Sorter block will prepare the sorting machine by first turning the sorter wheel in the direction opposite from the way that it normally turns while sorting, moving any potential error-causing brick back up the chute. It will also lower the brick lock if it's raised and put the first brick from the chute directly under the sorter wheel to prepare it to be sorted. Configure the blocks that perform these actions as shown in Figure 14-8, and turn them into a My Block called **Reset Sorter**.

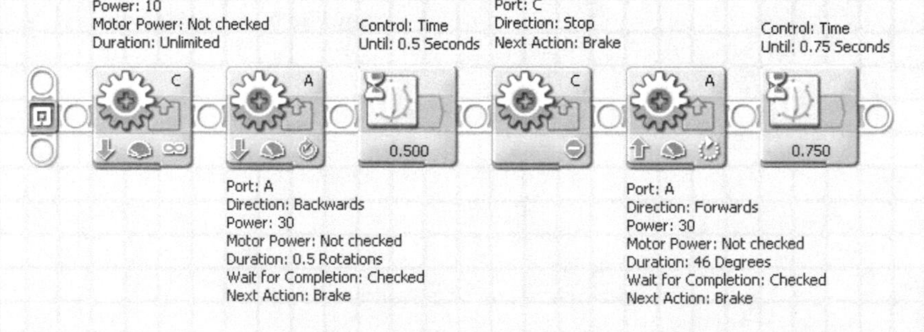

Figure 14-8: The configurations of the blocks in the Reset Sorter My Block

my block #2: drop brick

This simple block raises the brick lock, pauses the program to allow the brick to slide down to the bucket, and then lowers the brick lock again. Configure the blocks as shown in Figure 14-9, and then convert them into a My Block called **Drop Brick**.

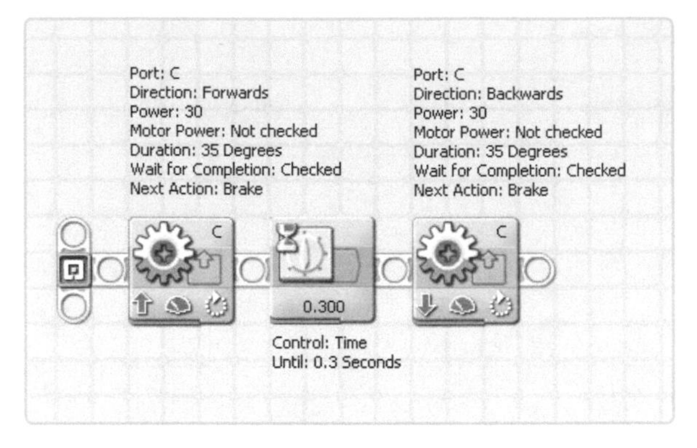

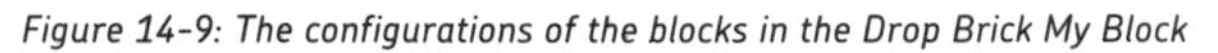

Figure 14-9: The configurations of the blocks in the Drop Brick My Block

creating the final program

Before you create your final program (see the overview in Figure 14-7), you'll define a logic variable called *Bricksize* and a numerical variable called *Bucket*.

step 1: getting the brick's size

You begin the program with the Reset Sorter My Block; you place all other blocks in a Loop Block configured to loop forever so that the robot will keep sorting bricks until you abort the program. You'll first place the blocks that identify the brick's size. If a small brick is detected, the robot says "Positive," and the *Bricksize* variable is set to *true*. If a big brick is detected, the robot says "Negative," and the variable's value is set to *false*. Configure this part of the program as shown in Figure 14-10.

step 2: checking where there is a brick to sort

Continue by configuring the blocks that check to see whether a brick is available to be sorted. (The robot will know when it sees a color other than the white of the chute.) If no brick is available, the robot says "No object detected!" and the Reset Sorter block runs.

As shown in Figure 14-10, the program first rotates the sorter wheel by 46 degrees. If the robot spots a brick at this point, it knows that it's a small one. If it doesn't spot a brick, there's either a big brick in the chute or none at all. To find out, the wheel rotates another 46 degrees, and the blocks that you add in this step check to see whether there is a brick to sort.

It might sound strange to first configure blocks to check the size of a brick (step 1) and to then check whether there's a brick to sort at all (step 2), rather than reverse. However, the reverse is impossible: If you were to check the availability of a brick first, that brick would already have moved all the way down to the brick lock, making it impossible to check its size.

Configure this part of the program as shown in Figure 14-11.

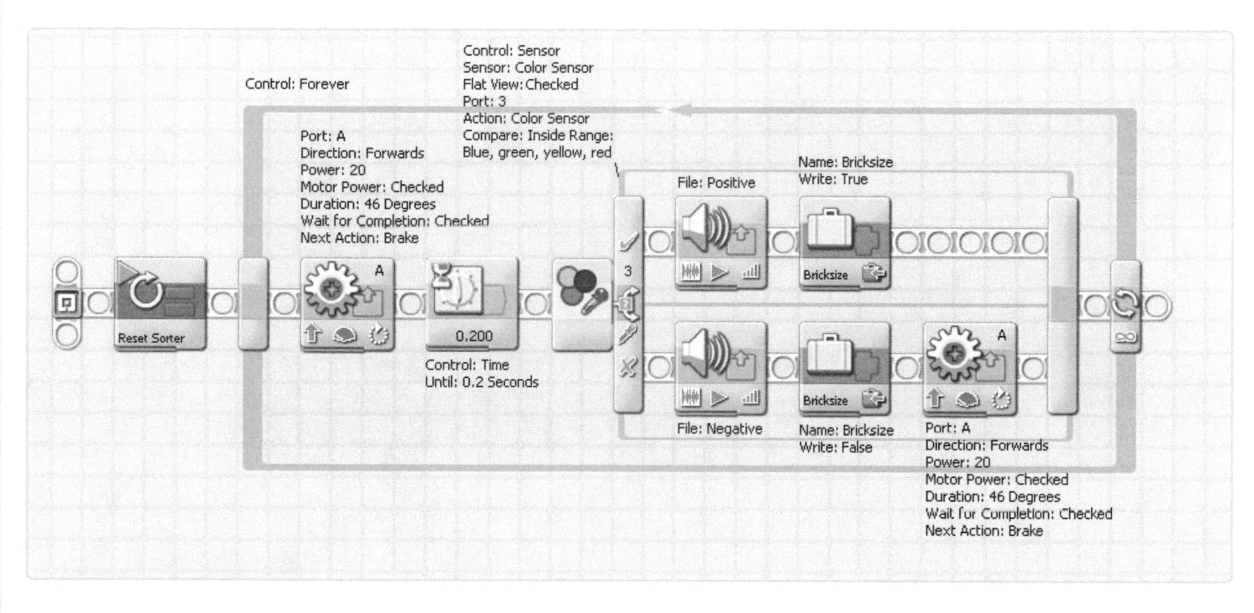

Figure 14-10: The configuration of the blocks that identify the brick's size. The Wait block shown here gives a small brick some time to slide down to the Color Sensor before the sensor detects it.

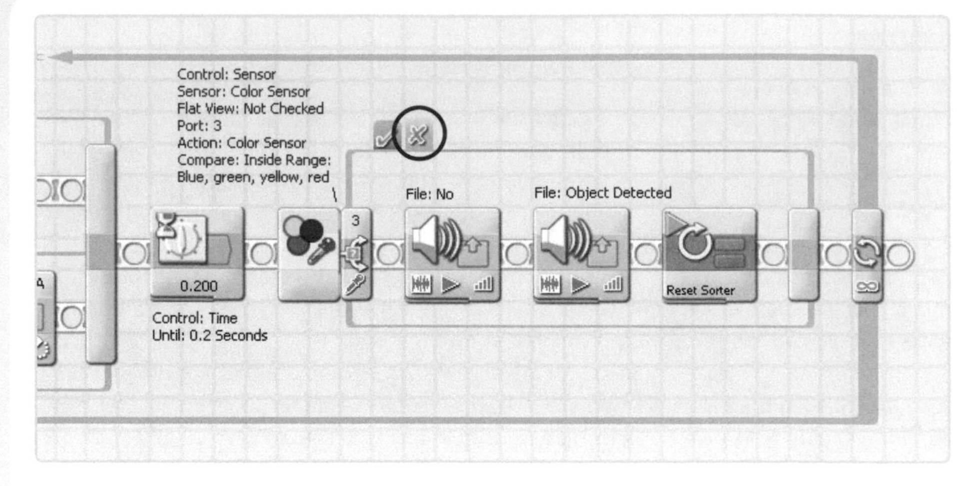

Figure 14-11: The configuration of both the blocks that check whether a brick is available to be sorted and the blocks that run when there isn't a brick left to sort

step 3: calculating the position of a bucket

Now you'll configure the blocks that run when the Color Sensor sees a brick. Figure 14-1 (page 237) shows how the bricks end up if you run this program. Notice that the sorter drops small bricks to the right and big bricks to the left of the starting position (in between the big and small blue bricks).

For example, when sorting large green bricks, the driving motor moves forward a certain number of degrees to reach the appropriate bucket. It moves the same number of degrees when sorting small green bricks, but in the opposite direction. The programming blocks in this step calculate this number of degrees based on the detected color and stores that number in the Bucket variable. Configure the blocks in this step as shown in Figure 14-12.

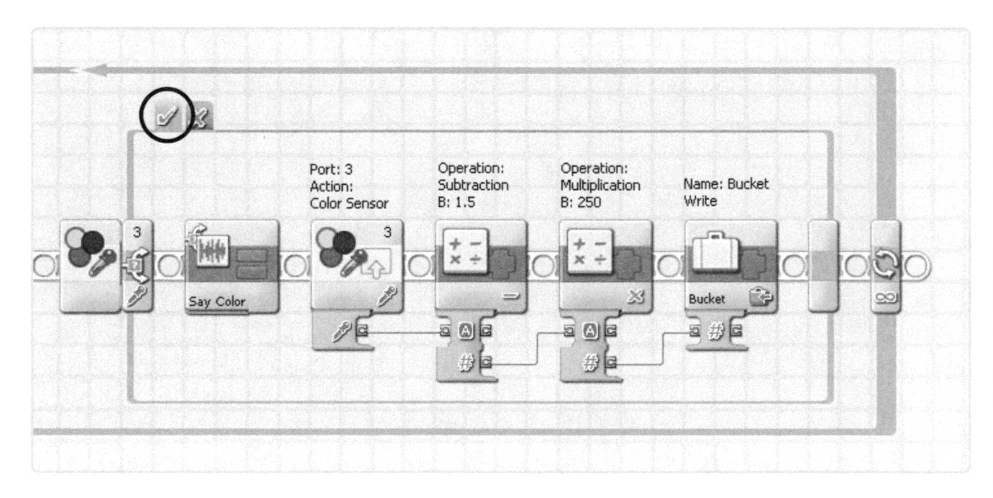

Figure 14-12: The configuration of the blocks that find the right bucket position based on the detected color

step 4: dropping a brick in the right bucket

Once you know how many degrees you need to move in order to reach the right bucket, you use this number together with the value in the Bricksize variable. You transfer the value in Bricksize to a Switch block specified to run the blocks in the upper part of the switch when the Bricksize variable is true (a small brick) and specified to run the blocks in the lower part of the switch when the value is false (a big brick).

The blocks in the switch make the robot drive toward the bucket (based on the value in the Bucket variable), drop the brick, and then return to the starting position. You use a Wait block in the loop to pause the program briefly each time it sorts a brick. Configure the blocks in this step as shown in Figure 14-13.

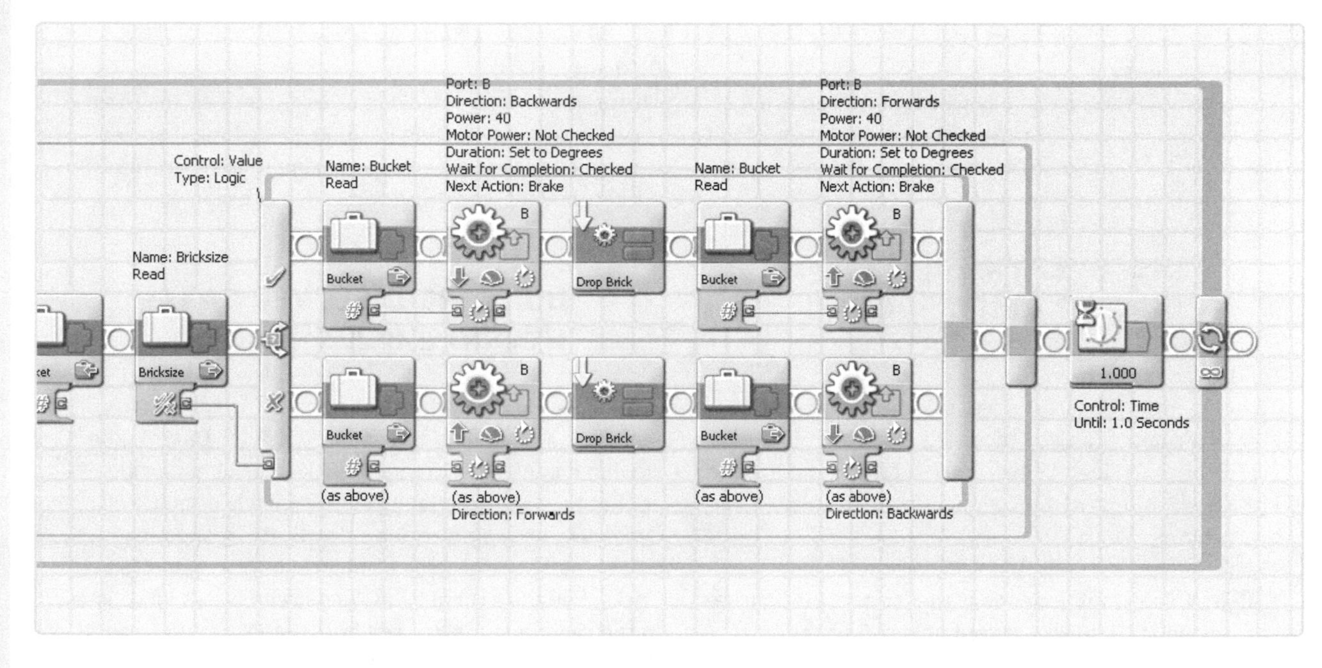

Figure 14-13: The configuration of the blocks that make the robot move to the right bucket, drop the brick, and return to the starting position

Now it's time to test everything to make sure that it works. Insert LEGO bricks into the brick chute with the studs facing upward, download this program to your robot, run it, and get ready to sort your bricks!

further exploration

You've just completed another advanced NXT 2.0 robot. Well done! Of course, you should enjoy this machine by running the sorting program a couple of times, but it's even more fun to make this robot a starting point for your own creations. The following discoveries will help you get started doing so. Let others know about your creations by sharing ideas on the companion website forums!

DISCOVERY #69: HIGH-SPEED SORTING!

Difficulty: Easy

The Hybrid Brick Sorter may work very reliably, but you can probably sort bricks faster on your own. Modify the program to make it challenging for anyone to stay ahead of the brick sorter when sorting bricks.

DISCOVERY #70: QUATRO BRICK SORTER!

Difficulty: Medium

The current brick sorter sorts bricks in only two sizes and four colors. Can you modify the program to also sort 2-by-3 studded bricks, 2-by-6 bricks, and even black and white bricks? You might need to enlarge the brick chute!

HINT **If you want your robot to also sort black bricks, look at the Math blocks placed in step 3 of the final program. How does the calculation work, and how should you modify the settings in these Math blocks to make black bricks part of the sorted bricks?**

DISCOVERY #71: INTELLIGENT SORTING!

Difficulty: Hard

The program you've made in this chapter had the robot move back to its starting position every time it sorted a brick. You can avoid this by making it go to the next bucket straightaway. How would you program this behavior?

BUILDING DISCOVERY #14: HYBRID BRICK SHOOTER!

The sorter currently drops bricks in different buckets, but how about adding a little more action? Destroy the Driving module, and create a brick-launching module so that the robot can launch bricks slowly or quickly depending on their size and color. Start by shooting just two types of bricks, such as only small green and red bricks. Separate them by making green bricks land close to the robot and making red bricks land farther away. Once you have a design that works reliably, expand your shooting program to separate more types of bricks.

15

CCC: the compact chimney climber

Many NXT robots move on wheels, some move on legs, and still others do something while remaining in place. But the Compact Chimney Climber (CCC) that you'll build in this chapter is different: It moves vertically. This robot climbs between two walls, as if up a chimney, as shown in Figure 15-1. When it reaches the top of the "chimney," it safely returns to the ground.

NOTE This is the last robot you'll build—not because it's the most difficult but because the CCC is the trickiest one to get to work. Beware! If done incorrectly, this robot can damage your LEGO.

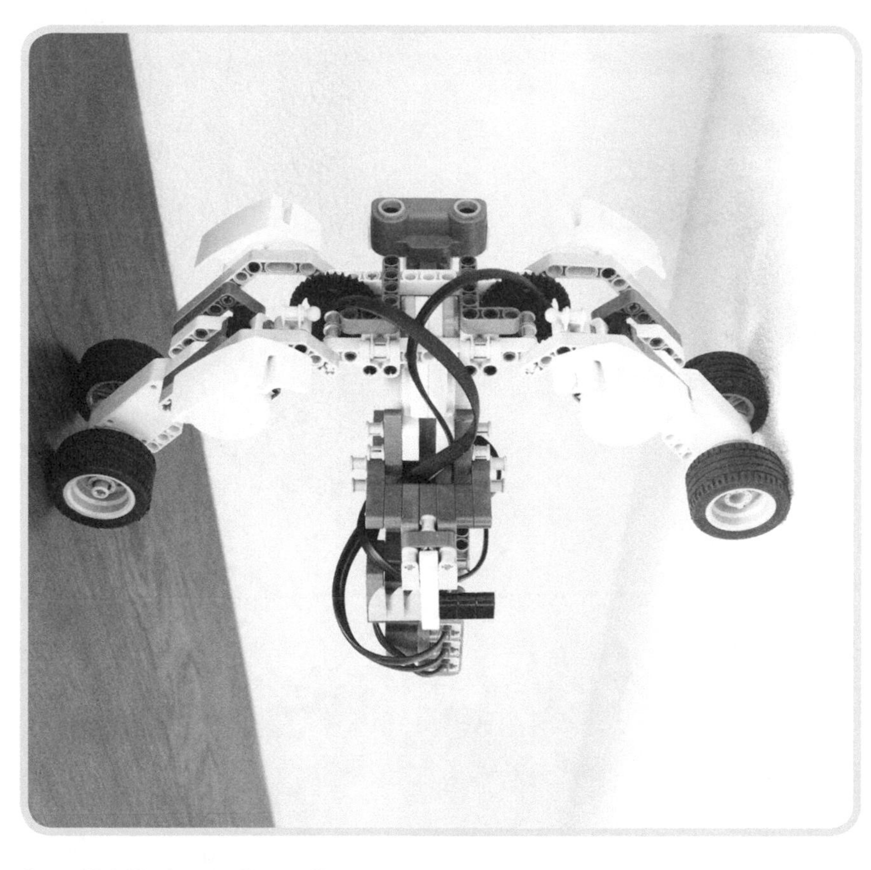

Figure 15-1: The Compact Chimney Climber moves vertically between two solid objects, such as a bookcase and a wall.

understanding the climbing technique

The CCC consists of two arms that extend to touch the walls it will climb so that the robot can move up and down by turning the wheels on its arms (see Figure 15-1). This climbing technique isn't very complicated and it works fairly well, but a simple robot with this technique will eventually lose contact with the wall and fall down. To prevent the robot from crashing to the ground, you'll need to keep it balanced as it moves, on both the x- and y-axes, as shown in Figure 15-2.

staying balanced on the x-axis

The CCC robot balances on the x-axis automatically by design. The main weight of the robot, the NXT with its batteries, is positioned below the axis, causing the robot to behave like the swing shown in Figure 15-3. If you were to try to turn the axis in this figure with your hands, you'd notice that gravity tries to stop you. Something similar happens when the climber tries to rotate along the x-axis: Gravity keeps it upright.

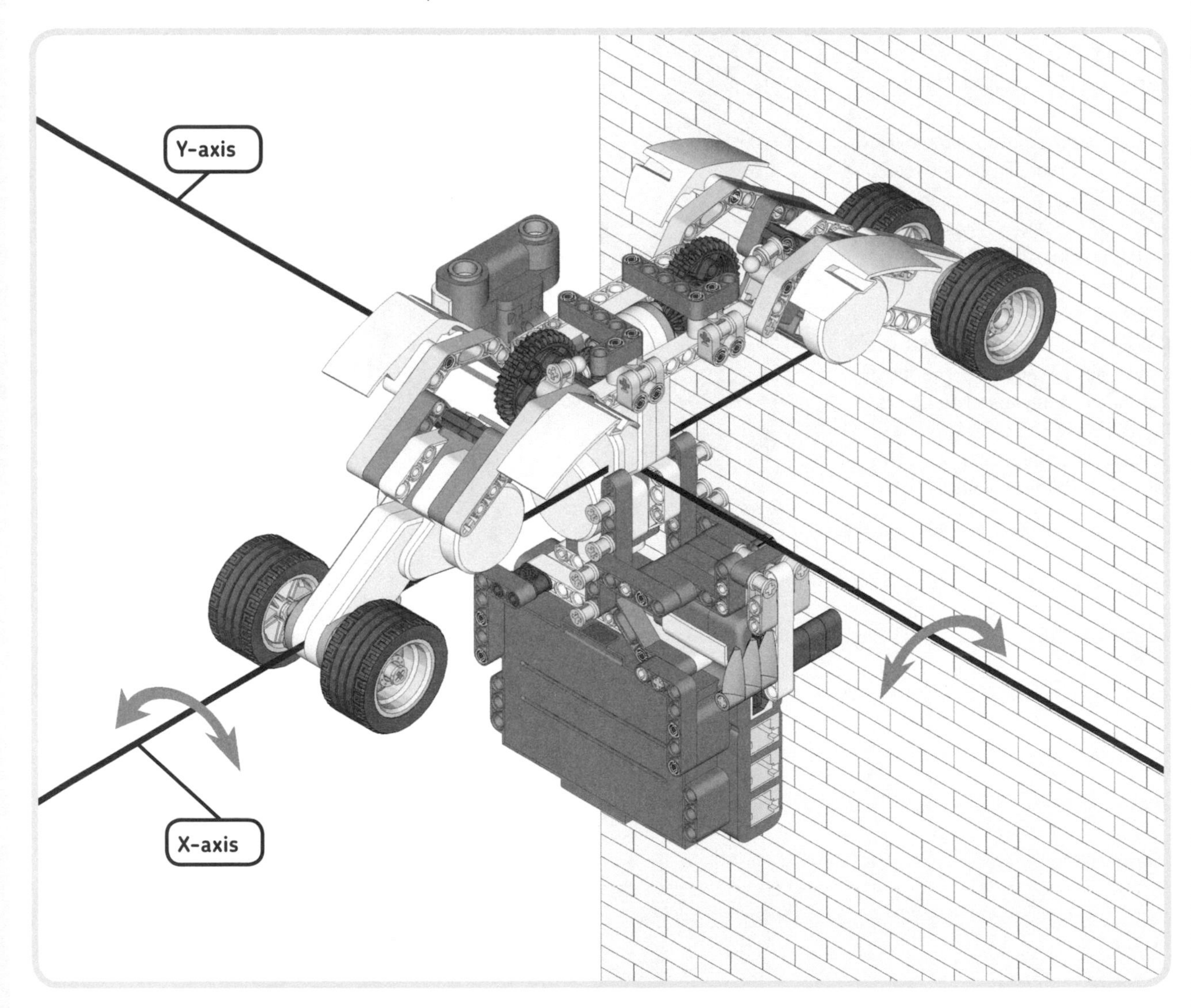

Figure 15-2: To be able to climb, the CCC has to balance on two axes. (Only one wall is shown here for better visibility.) If balance isn't maintained, the robot will eventually tilt in one of the directions indicated by the gray arrows and lose contact with the walls.

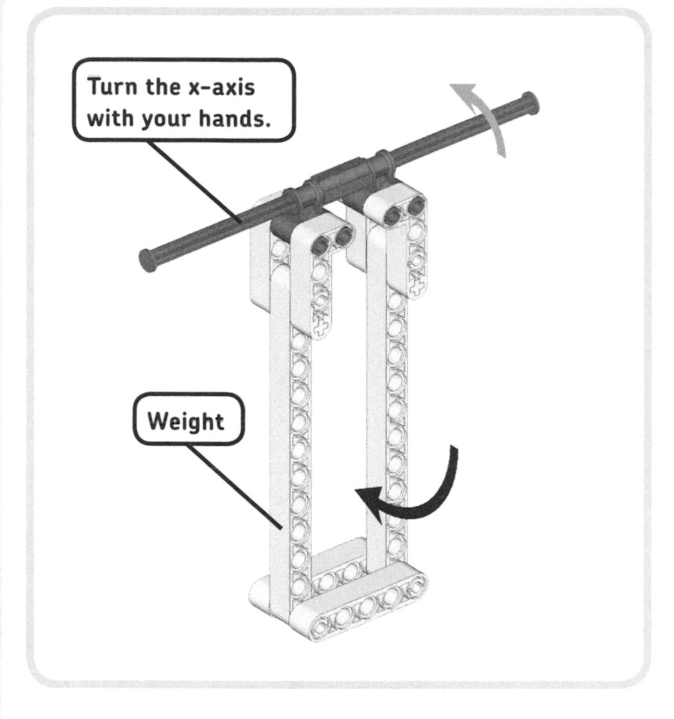

Figure 15-3: A simplified version of the climber shows how balance is maintained on the x-axis. Build this or a similar construction with the pieces in your NXT set. If you try to turn the axis (gray arrow), gravity will try to keep the weight below the axis (black arrow).

balancing on the y-axis

The CCC robot climbs by turning the wheels on both of its arms (the Wheel motors) at the same speed, as if the robot were driving up the walls. While climbing, the wheels on one side of the robot may slip, or one may grip the wall more than the other. When this happens, one side of the robot climbs faster than the other, which causes the robot to tilt along the y-axis, as shown on the left of Figure 15-4.

detecting balance errors

It is easy for us humans to see when the robot starts to become unstable on the y-axis, but the robot can also catch this error by using the Color Sensor as a *balance detector*, as shown in Figure 15-4. The sensor is tightly connected to the robot, but a swing in front of it with different colored parts moves freely; because of gravity, it always points downward. Because the sensor tilts along with the robot and the swing does not, the sensor sees different colors depending on how the robot is titled.

solving balance errors

The robot easily corrects balance errors by temporarily stopping the Wheel motor of the arm positioned higher than the other one, which allows the lower arm to keep up. For instance, if the robot sees black, the wheels on the right side should stop, as shown on the left of Figure 15-4. Once the wheels on both arms are at the same altitude (the sensor sees white), they should move at the same speed, allowing the robot to continue up the chimney (shown on the right of Figure 15-4). This same control technique is used when the robot goes down, except that the opposite wheel is stopped when a balance error is detected.

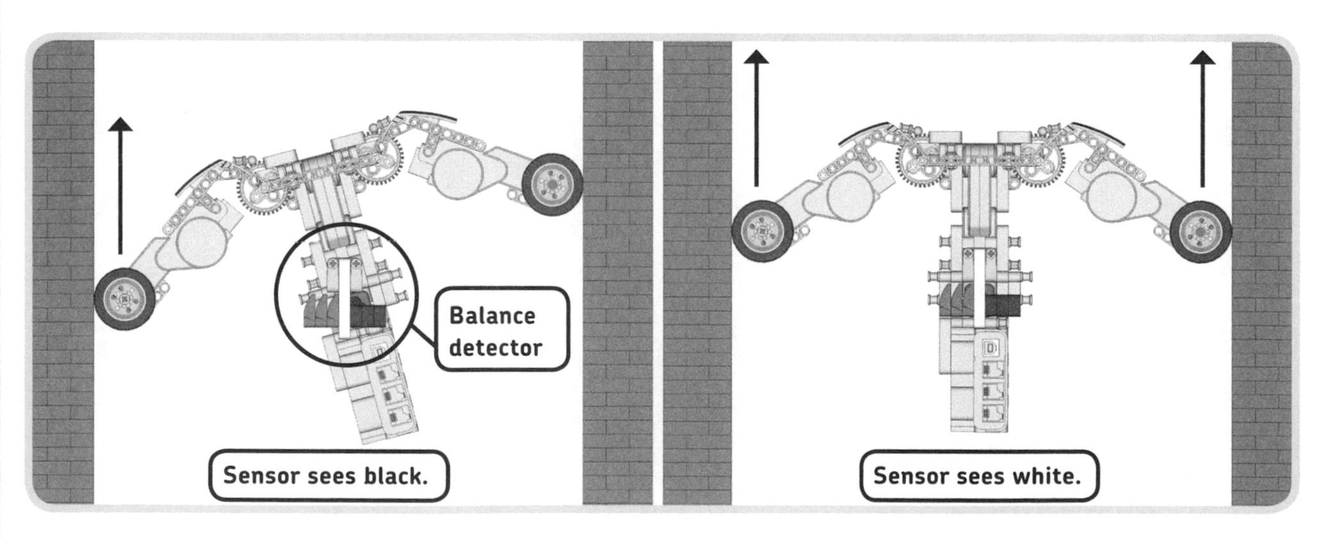

Figure 15-4: Detecting and correcting balance errors on the y-axis

building the compact chimney climber

Now build this robot by following the instructions on the next pages. First select the required pieces for the robot, as shown in Figure 15-5.

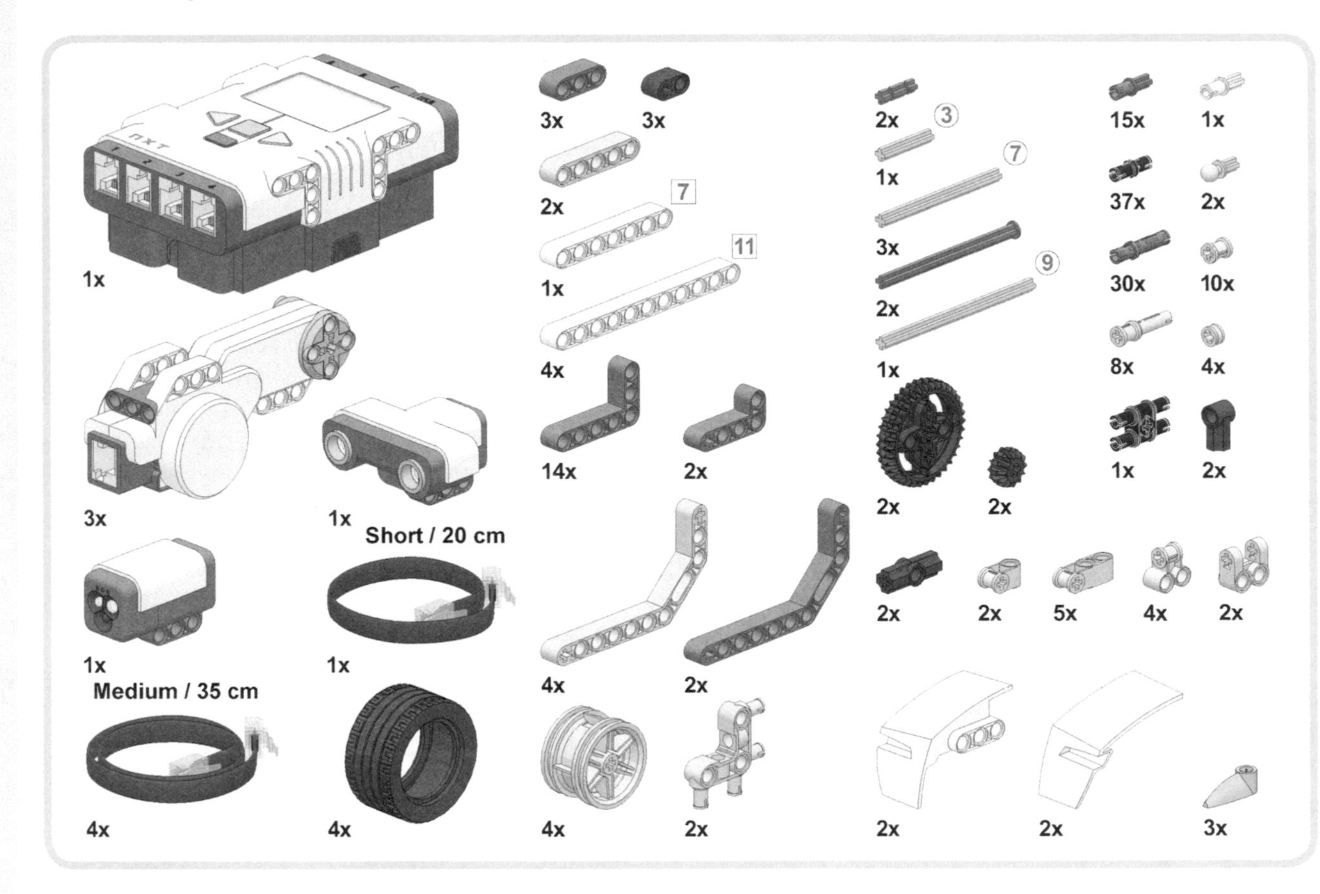

Figure 15-5: Required pieces for the Compact Chimney Climber

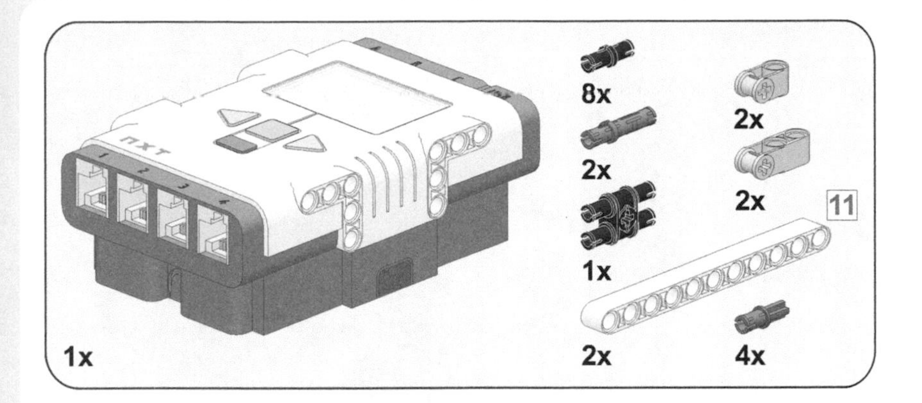

8x
2x
2x
2x
2x
11
1x
1x
2x
4x

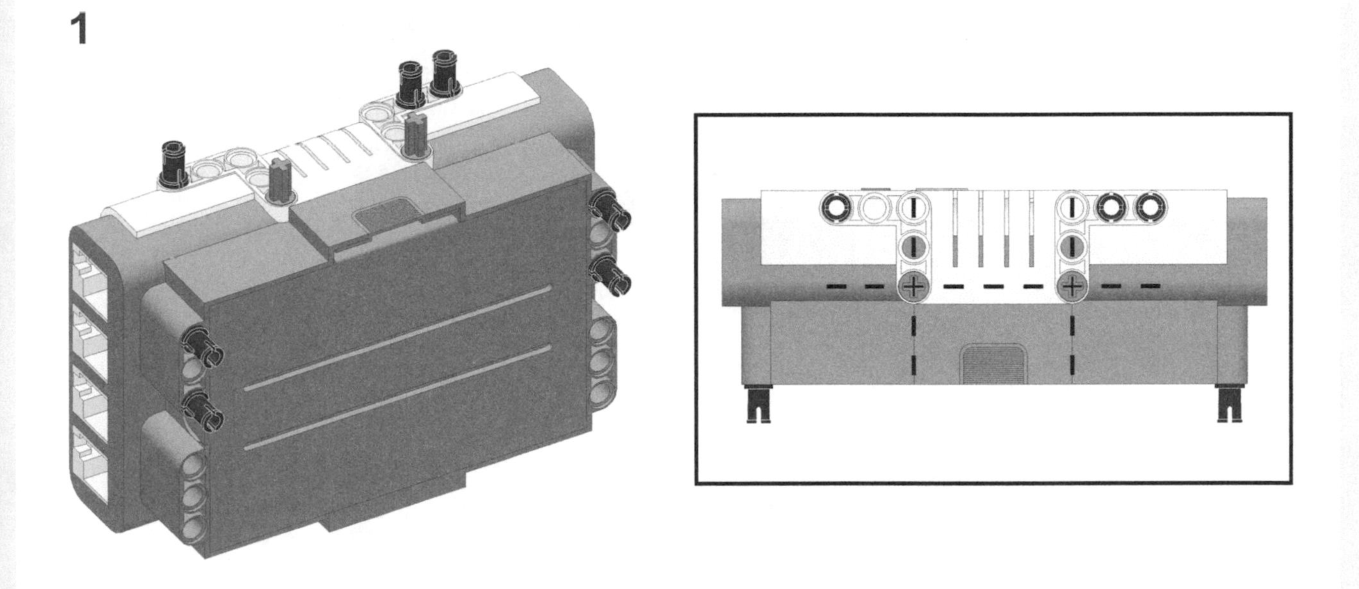

1

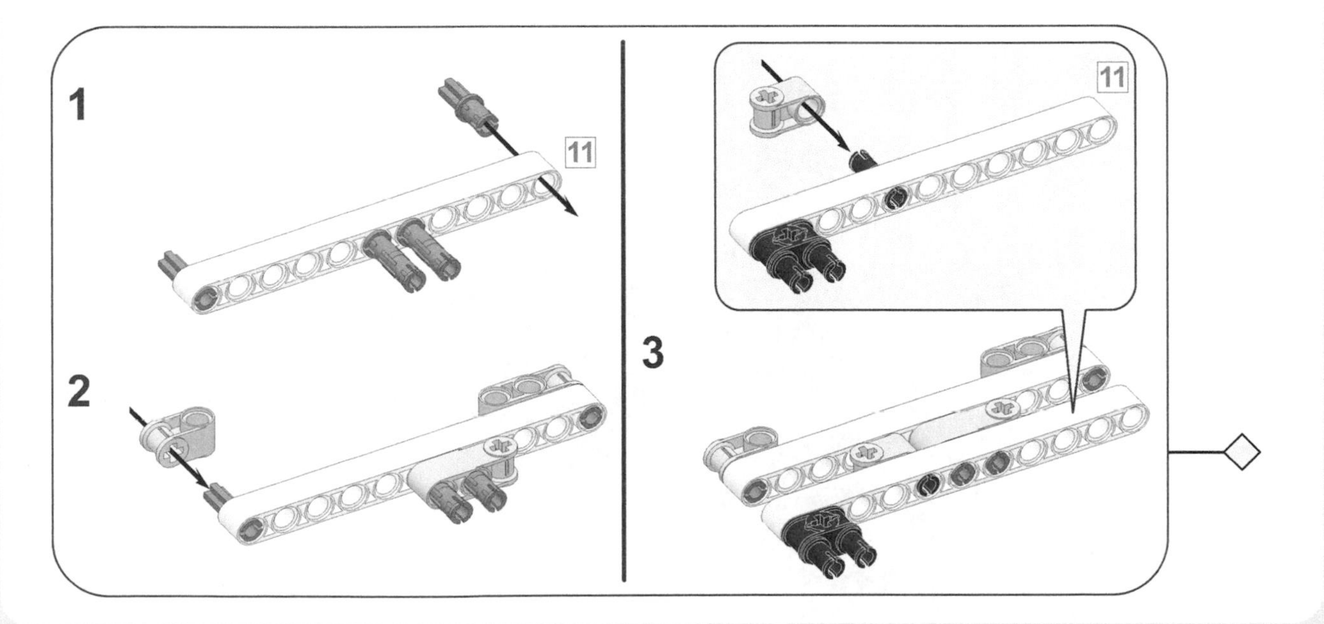

1
11
2
3
11

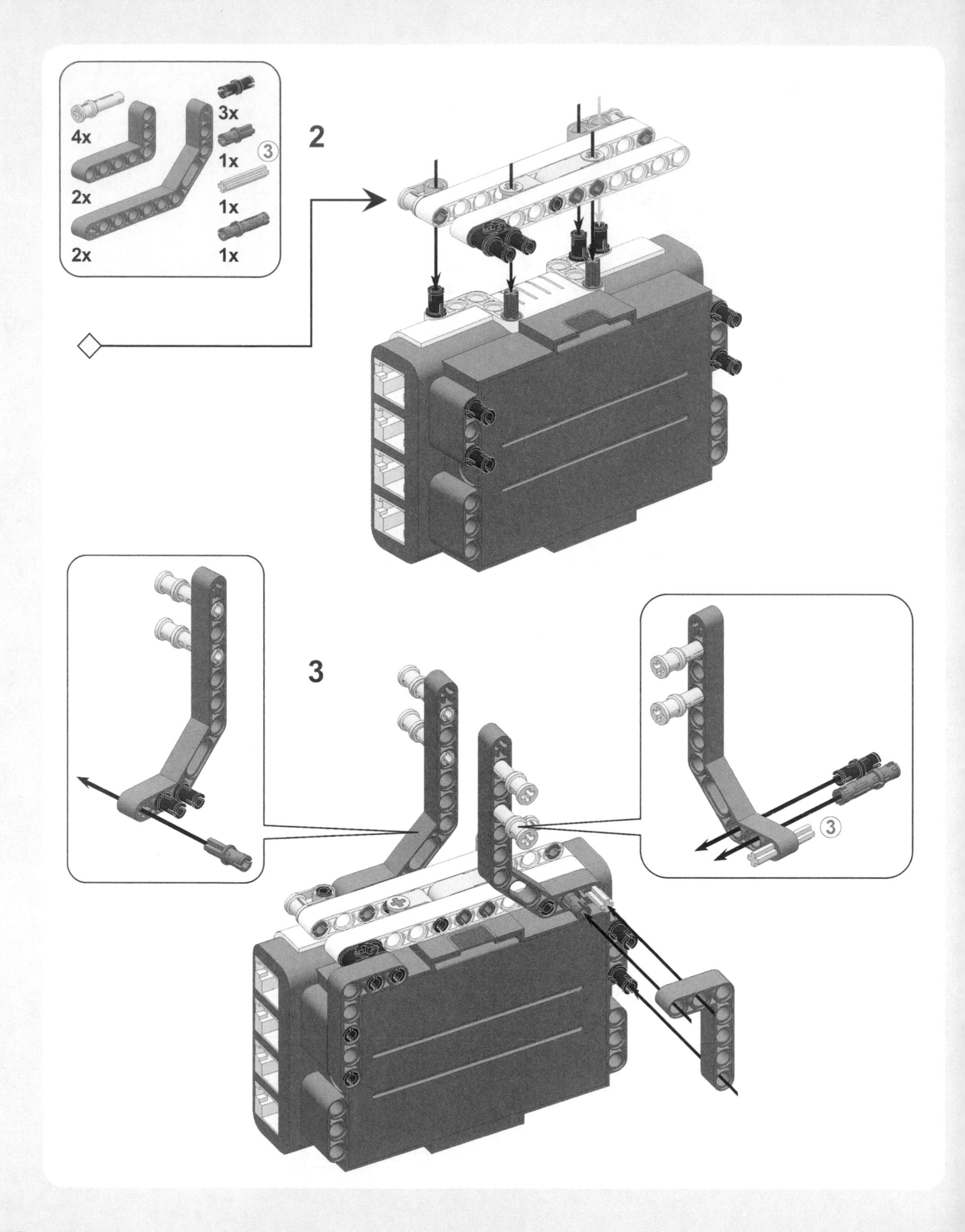
4x
2x
2x
3x
1x
3
1x
1x
2
3
3

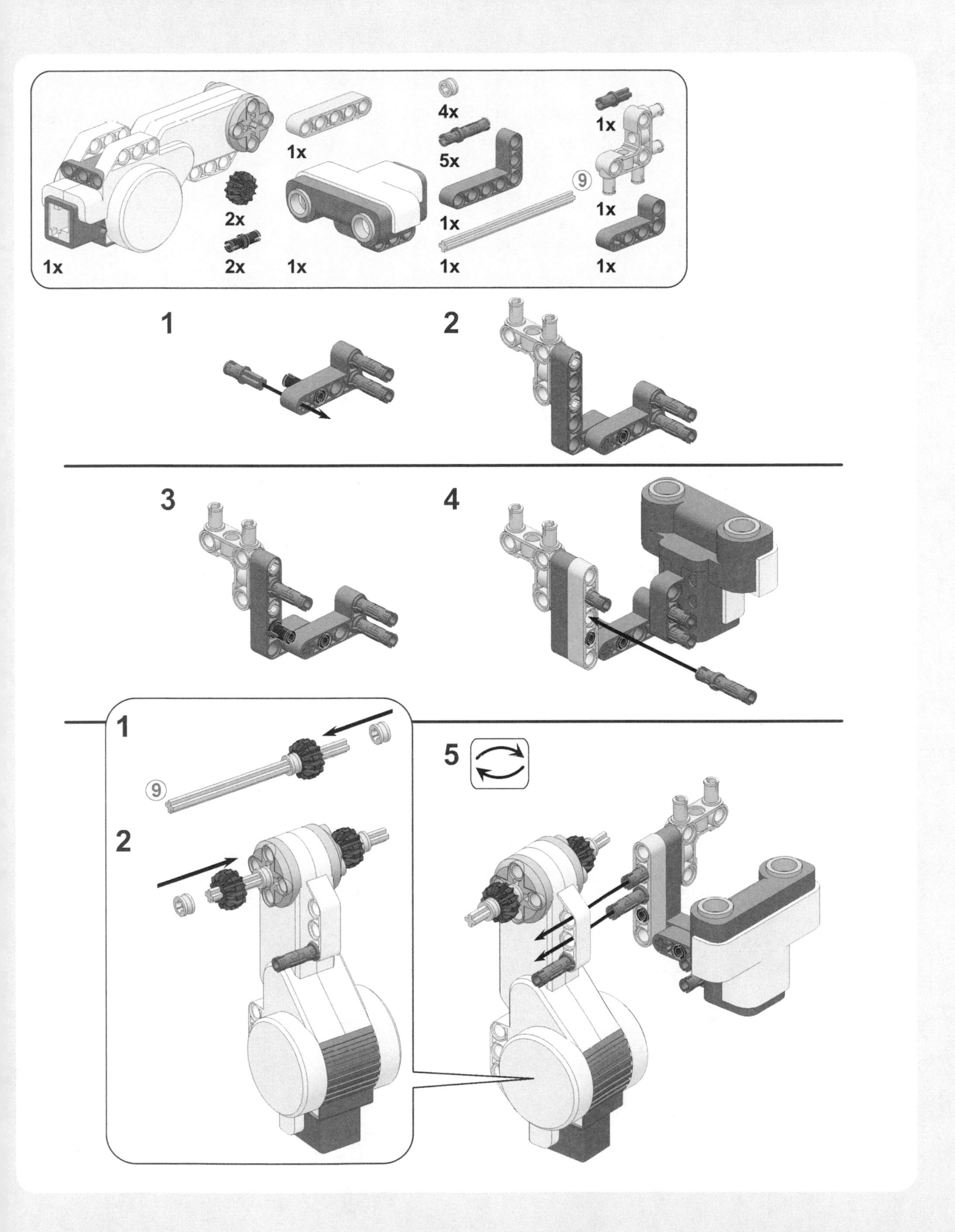
1x
1x
2x
2x
1x
1x
4x
5x
1x
1x
9
1x
1x
1x
1x
1
2
3
4
1
9
2
5

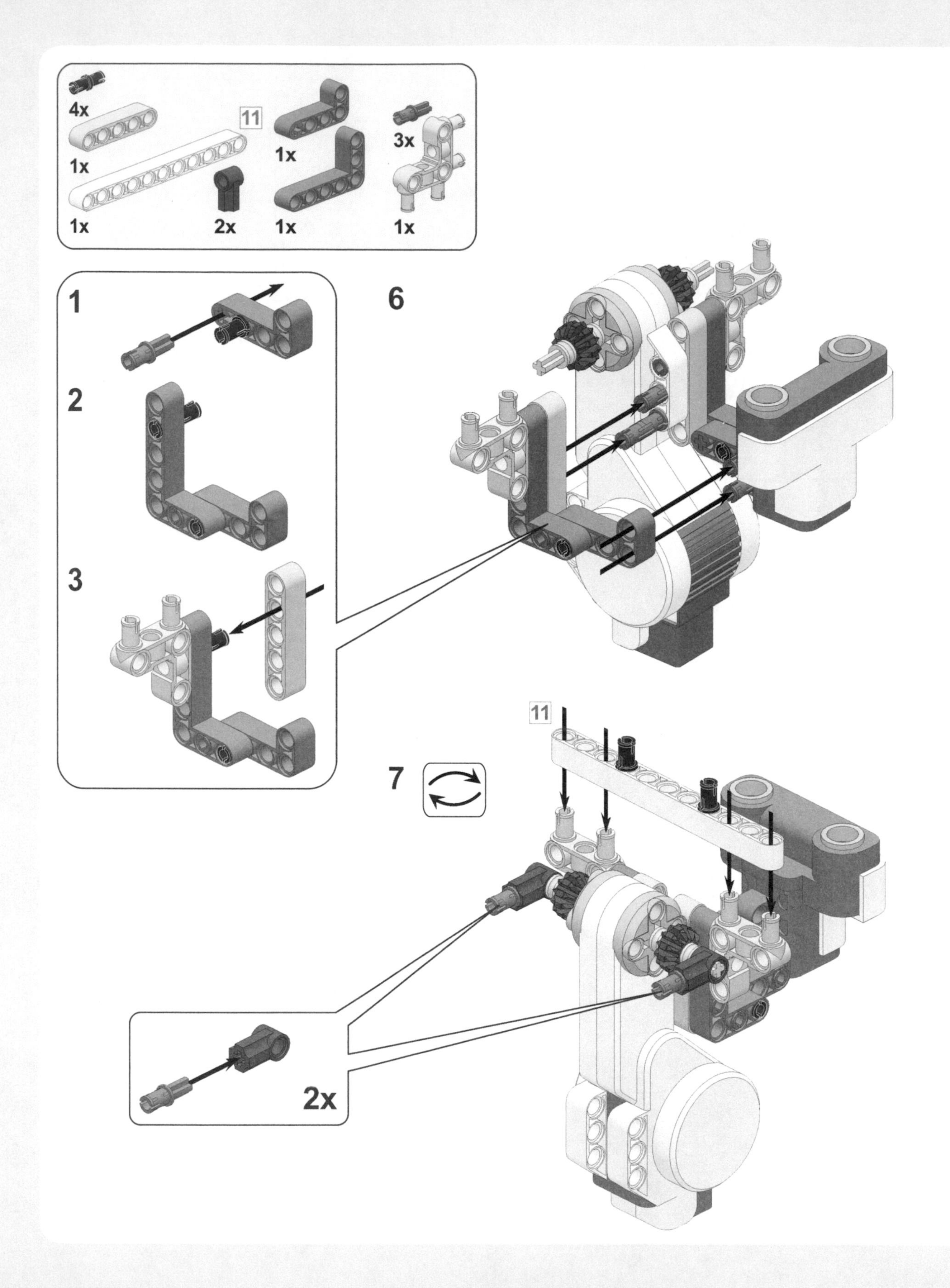
4x
11
1x
1x
1x
1x
2x
1x
3x
1x
1
2
3
6
11
7
2x

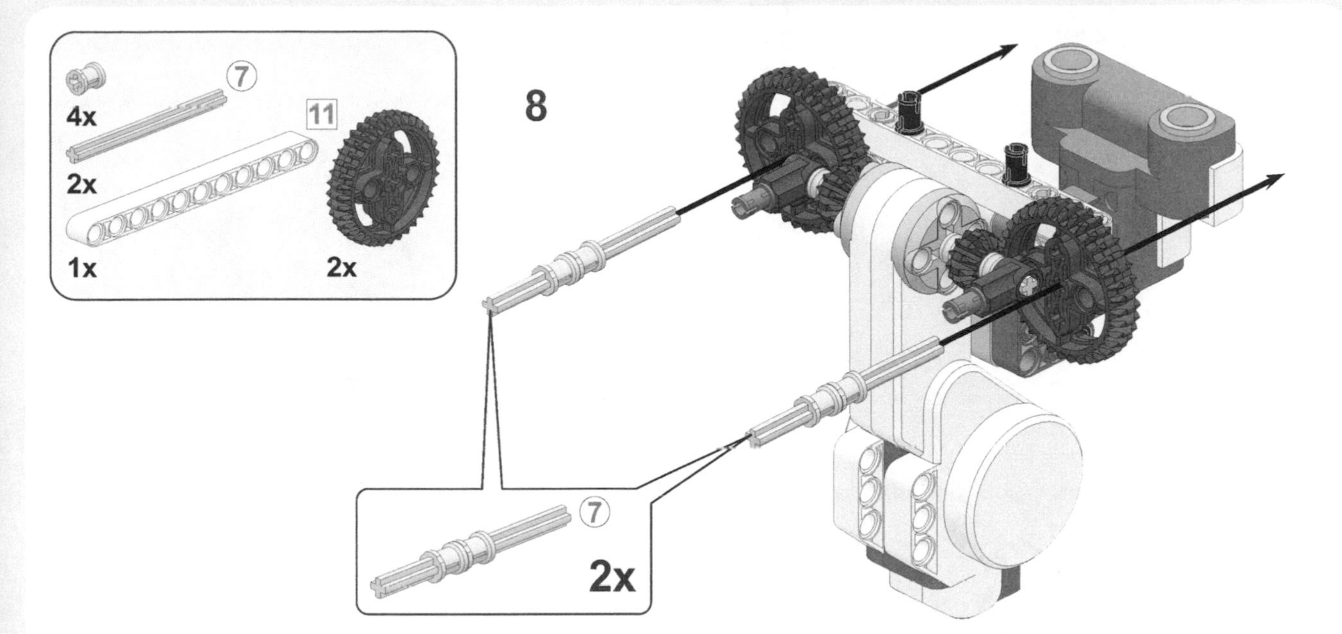
4x
7
11
2x
1x
2x
8
7
2x

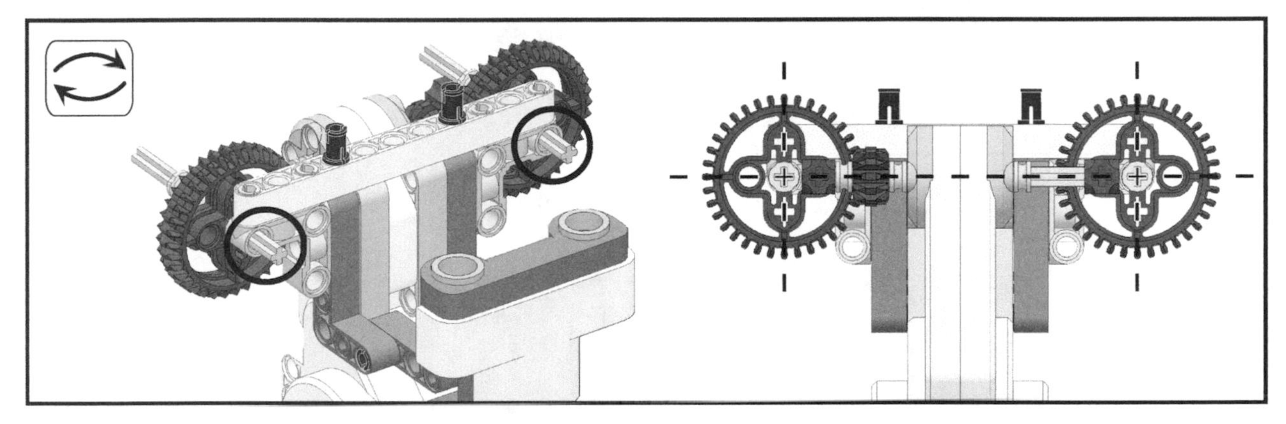

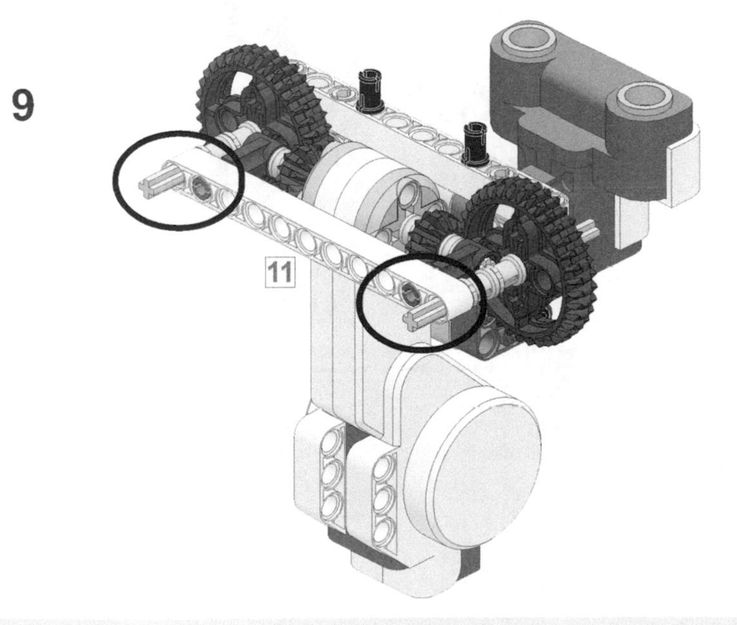
9
11

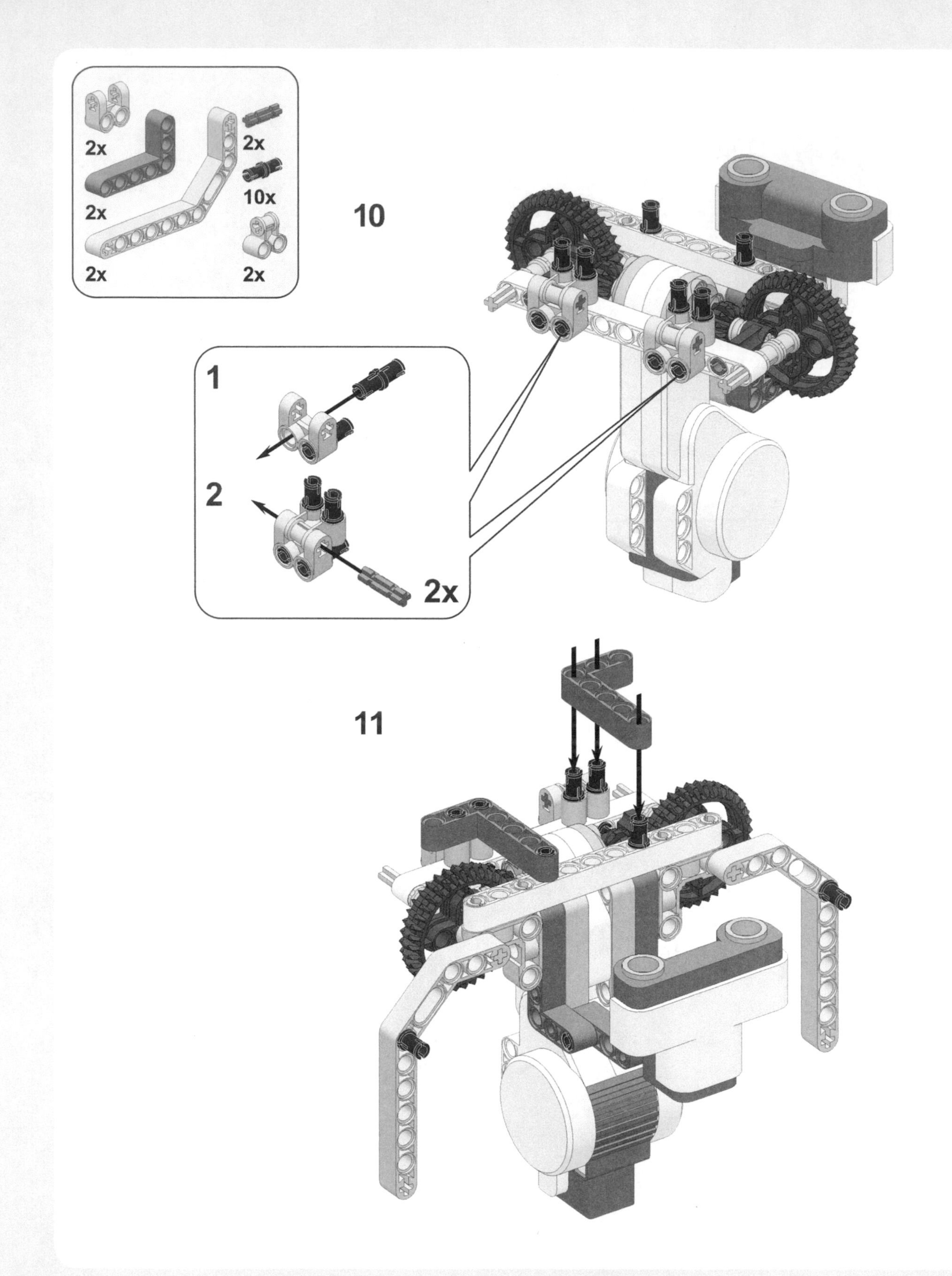
2x
2x
10x
2x
2x
2x
2x
10
1
2
2x
11

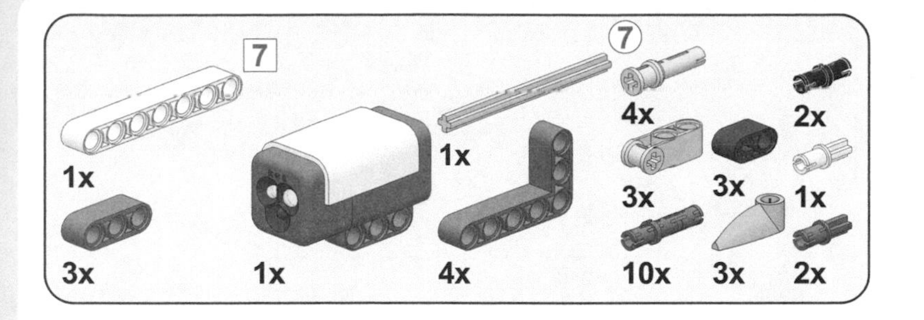
7
7
1x
1x
4x
2x
3x
1x
3x
3x
1x
3x
1x
4x
10x
3x
2x

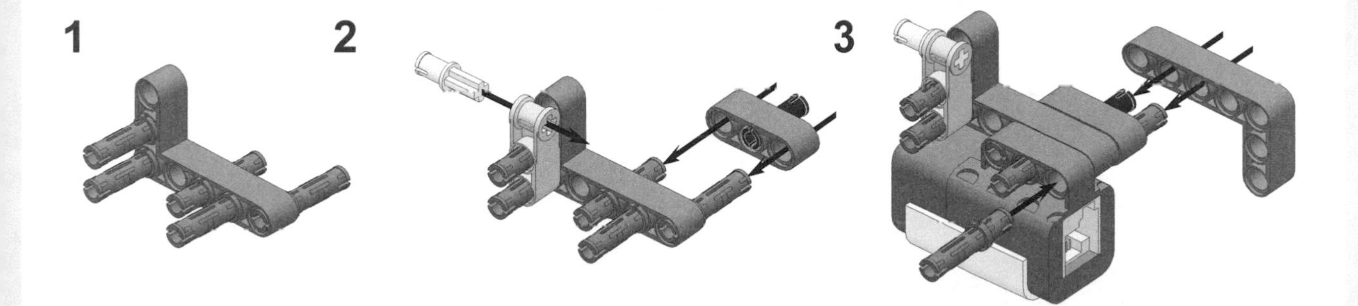
1
2
3

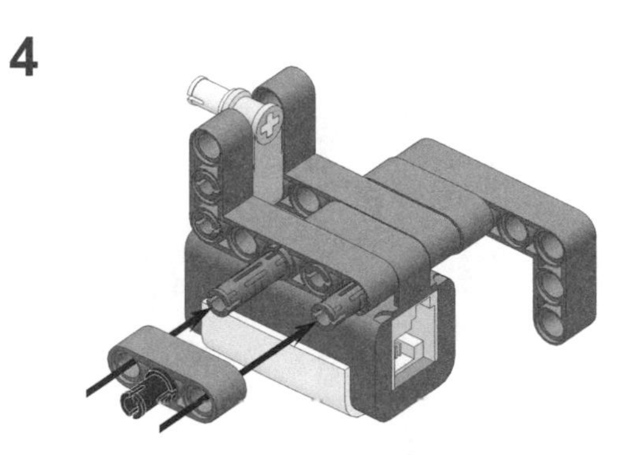
4

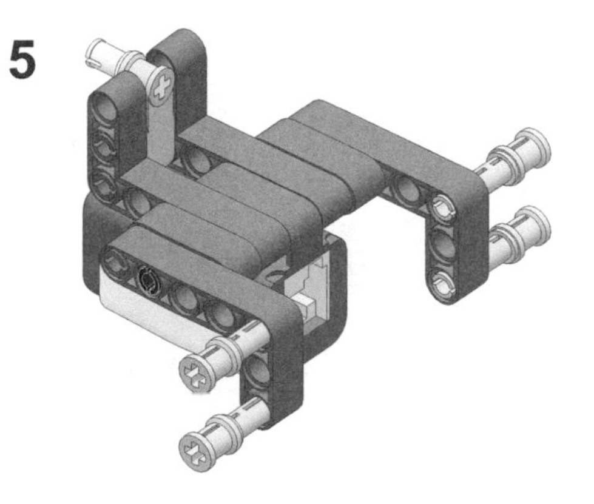
5

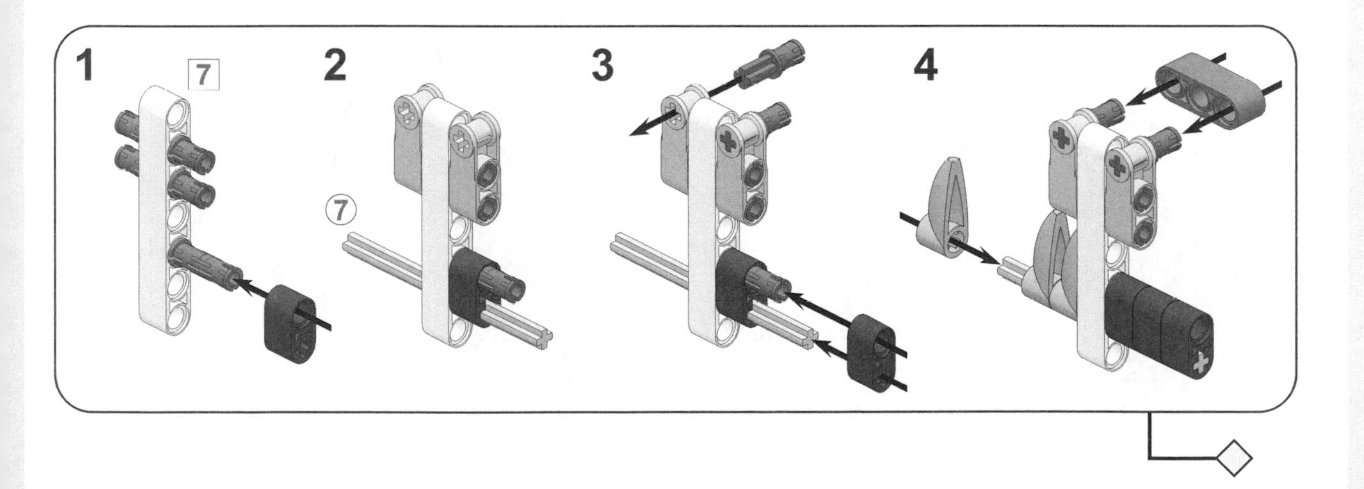
1
7
2
7
3
4

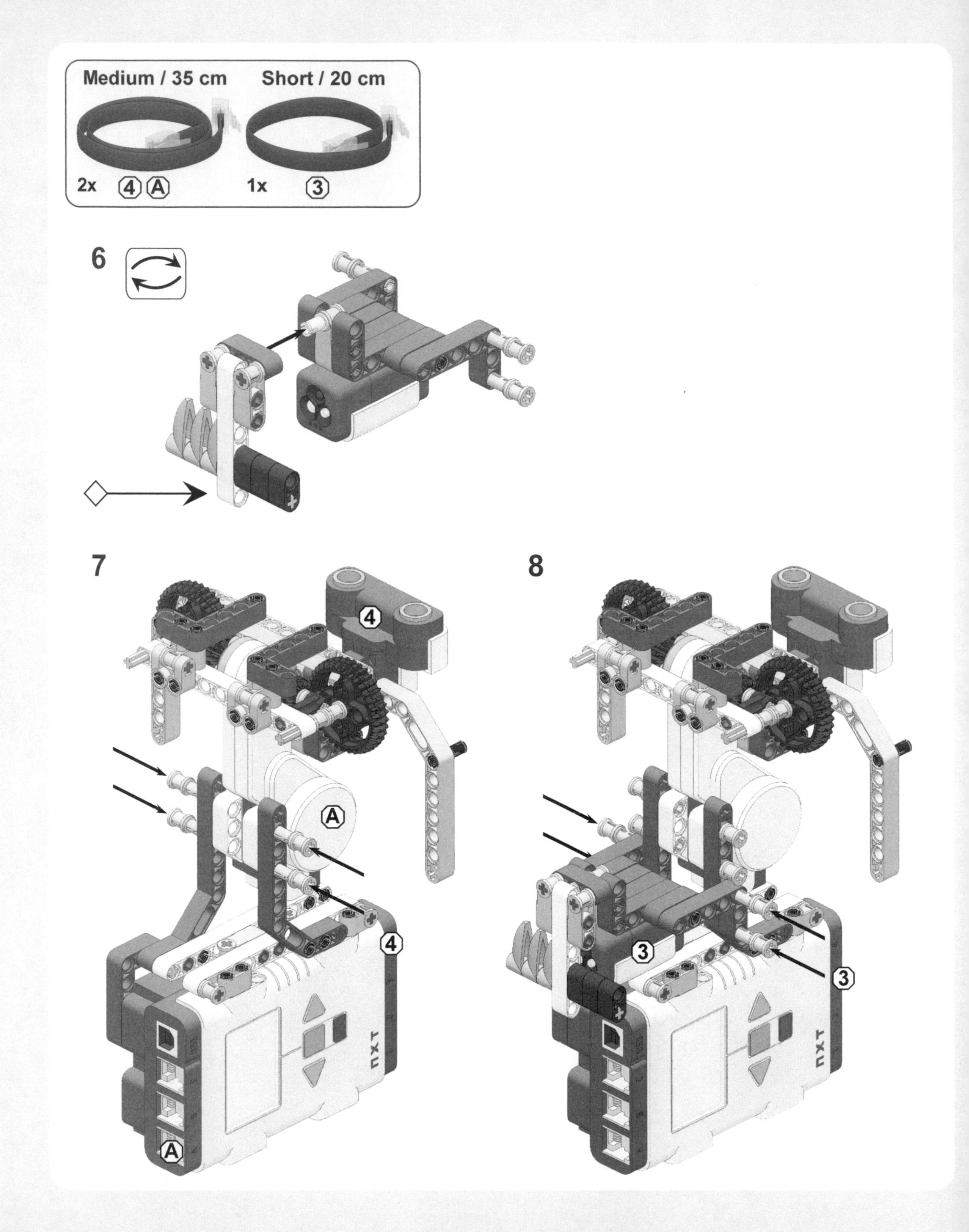
Medium / 35 cm
Short / 20 cm
2x 4 A
1x 3
6
7
4
A
4
A
8
3
3
NXT
NXT

4x
2x
4x
2x
2x
4x
6x
2x
12x
4x
1
2
3
2x

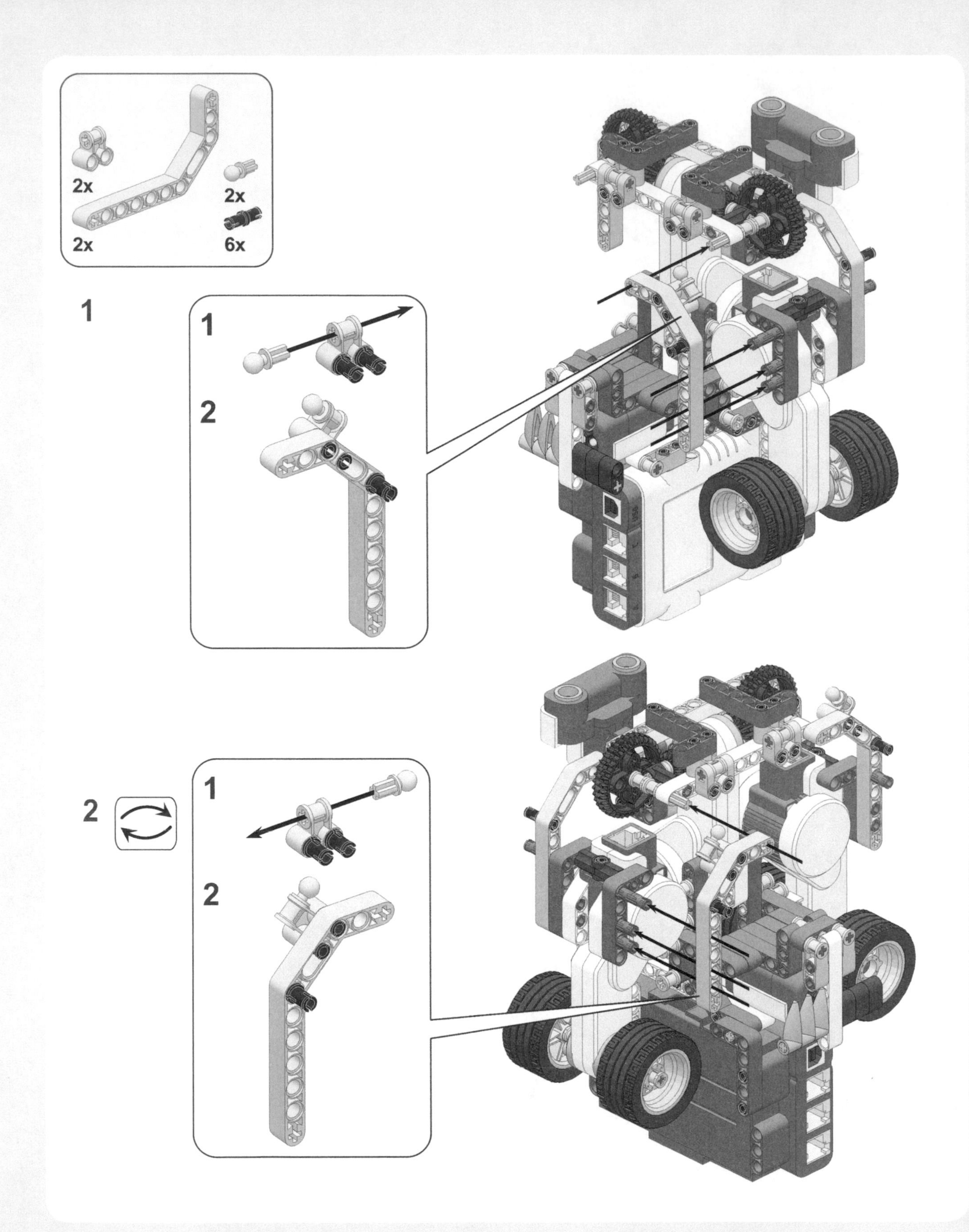
2x
2x
2x
6x
1
1
2
2
1
2

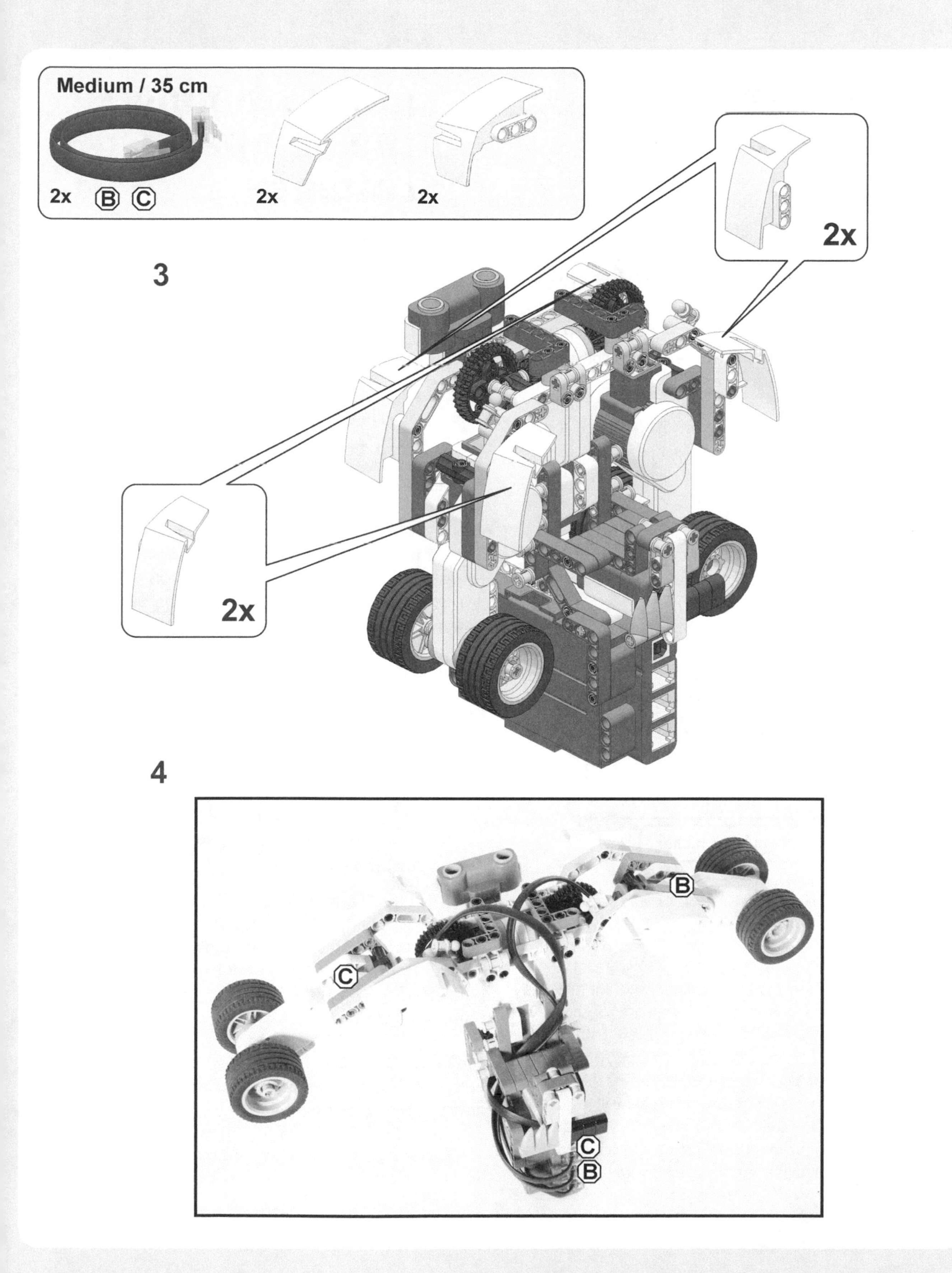
Medium / 35 cm
2x B C
2x
2x
3
2x
2x
4
B
C
C
B

preparing a chimney

Before programming this robot, you'll need to find an appropriate "chimney" for the robot to climb, such as the crevasse made between a wall and the side of a bookcase, as shown in Figure 15-6. Whatever your choice, your chimney's walls should be as follows:

* Solid and unable to move while the robot climbs. Perhaps try the edge of a desk or strong, heavy moving boxes.
* Perfectly parallel to each other.
* About 32 cm (12.5 inches) apart. A slightly smaller (30 cm) or wider (35 cm) gap is fine too. Adjust the width of the gap by moving one of the two walls if possible.

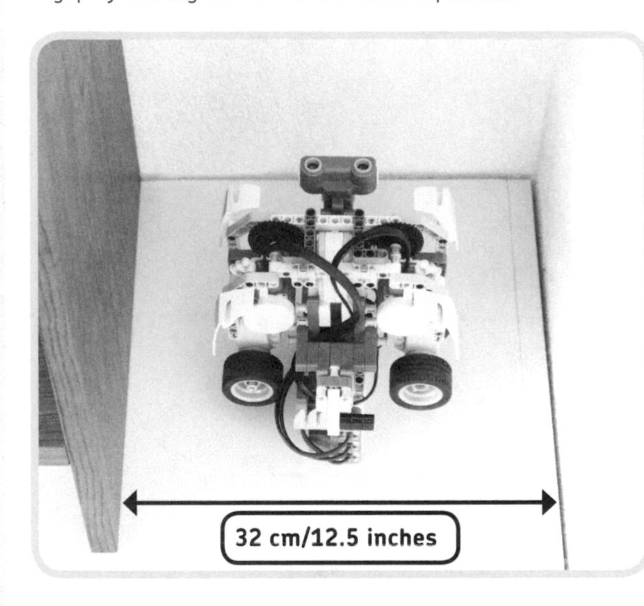

Figure 15-6: An example of a chimney suitable for your CCC

Even if your chimney meets these requirements, be very careful when using the CCC robot because it may still fall. When first testing your robot, make sure to guide it with your hands so that you can instantly grab it if it starts to become unstable. For extra security, put a pillow at the bottom of the chimney just after the robot lifts off the floor.

The program you'll create for this robot will make CCC move up until it spots the ceiling. If one or both walls do not reach the ceiling, your robot won't be able to see it. In that case, keep your hand in front of the sensor to tell the robot to go down again. And if you damage any LEGO pieces, well, you've been warned!

programming the compact chimney climber

Now that you've built the CCC and found an appropriate chimney for it to climb, you're ready to program it. Your CCC program will allow the robot to climb the chimney toward the ceiling and return to its starting position. "Further Exploration" on page 283 will give you some ideas to expand the program.

step 1: extending the arms

First, the program switches on the Wheel motors B and C. Next, the arms are extended to touch the walls of the chimney by rotating the motor on port A inside the climber backward. As the robot climbs, the NXT keeps applying power to this motor so that the robot's wheels are pressed against the walls, giving them more grip. Configure the blocks that perform these actions as shown in Figure 15-7.

NOTE Unless specified otherwise in the figures, all the Motor blocks in the CCC program have the following configurations: Motor Power: *Checked*; Duration: *Unlimited*. Also, the blocks in steps 1 and 2 are set to make the motor turn backward, and those in step 3 make a motor run forward.

step 2: climbing and staying balanced

Recall that the CCC robot uses the Color Sensor to detect balance errors on the y-axis (see Figure 15-4). While climbing, the robot repeatedly polls this sensor and controls the Wheel motors accordingly to maintain balance until the Ultrasonic Sensor detects the top of your chimney, as shown in Figure 15-8.

NOTE When the Color Sensor sees something red, it's actually detecting the orange parts of the swing that balances in front of the sensor.

Follow Figure 15-9 to turn this programming structure into real programming blocks.

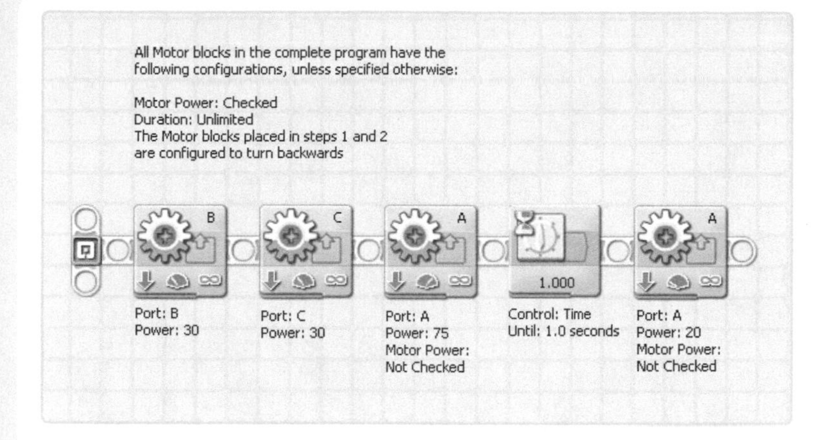

Figure 15-7: The configuration of the blocks that switch on the Wheel motors and extend the arms

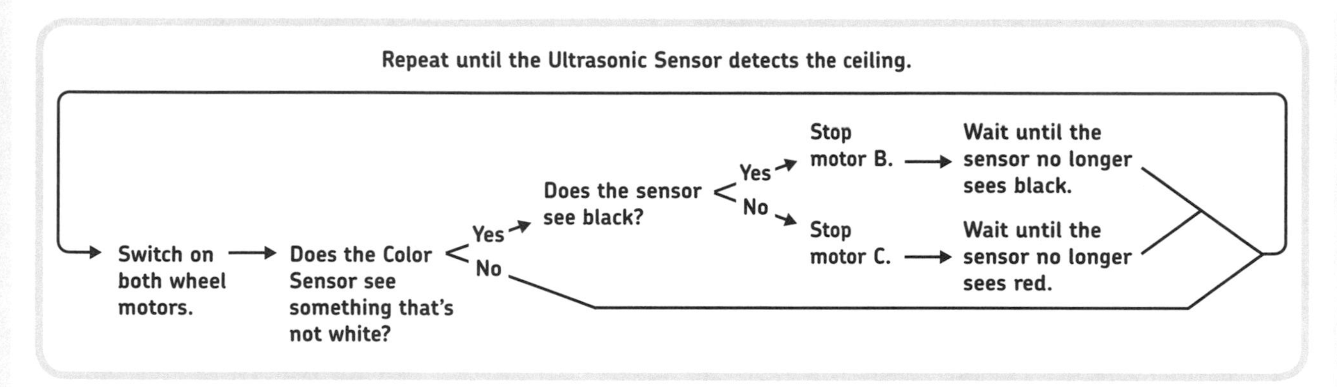

Figure 15-8: An overview of the control technique used to keep the robot balanced. If the sensor reads white, the robot is correctly balanced, so none of the Wheel motors should be stopped.

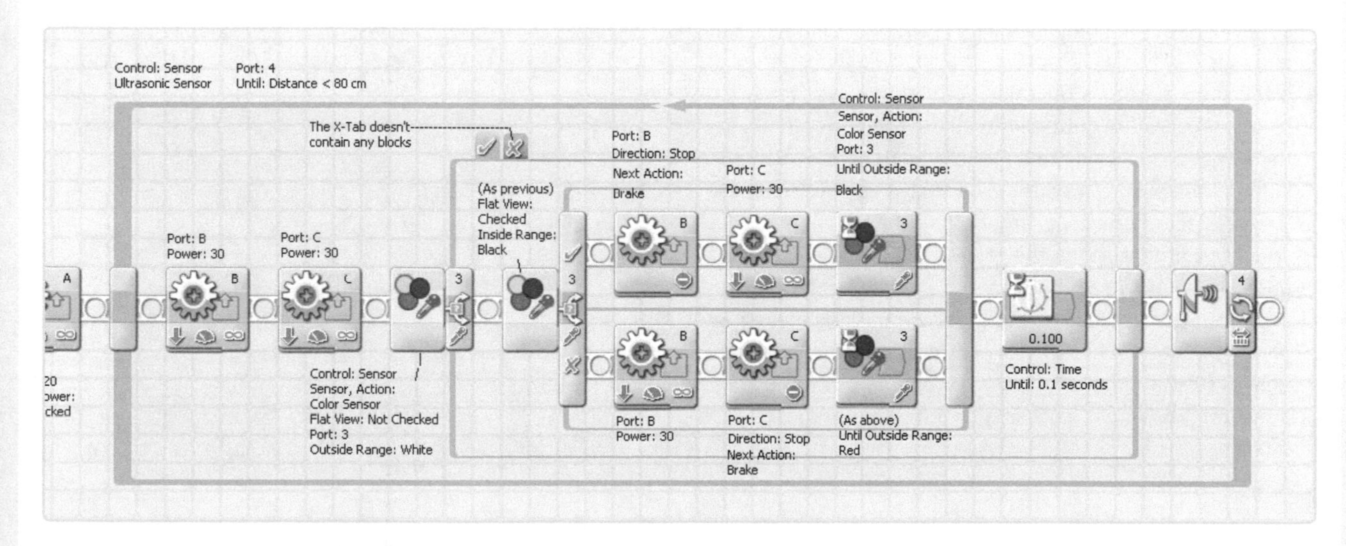

Figure 15-9: The configuration of the blocks that keep the robot balanced as it climbs

step 3: going down, staying balanced, and stopping

Once at the top, the robot stops moving, plays a sound, and starts going down. The blocks that control the robot to go down are similar to the ones you used to climb the chimney, except that the Motor blocks now make the robot move in the opposite direction. Therefore, you can simply copy the previous Loop block and its contents, place it as shown in Figure 15-10, and adjust the settings of each block as indicated.

When you run the program, you'll notice that when going down, the motors spin more slowly and the opposite motor is stopped when a balance error is detected (in other words, black now stops motor C). The Loop block repeats the blocks in it until the Rotation Sensor in motor C reports value 0, indicating that the robot has reached its starting position, at which point the program ends, and the motors stop moving. Because the motor on port A is no longer keeping the arms outstretched, they are released from the wall.

troubleshooting the CCC

Running the program for this robot requires a bit more attention than the other programs you've made so far. Before you run it, make sure that the robot is positioned in the middle of the chimney as shown in Figure 15-6. When starting the program for the first time, carefully follow the robot on its way up with your hands to make sure it won't fall! If you're having problems making the robot work, see the following troubleshooting tips, or visit the companion website for more help:

* There may be something wrong with your chimney. Be sure to check the guidelines in "Preparing a Chimney" on page 280.
* If the robot becomes unstable because it moves too fast or too slow, you might need to adjust the power levels in the Motor blocks that control the Wheel motors. Also, try running the program with new or fully charged batteries.
* Before starting the program, both of the robot's arms should point straight to the ground (Figure 15-6). If only one arm points straight down while, for example, the other sticks out to the right, carefully review building steps 8 and 9 on page 273 to fix the problem. You might also be able to fix this simply by applying some force on the arm in question (you'll hear a clicking sound), but don't do this too often.

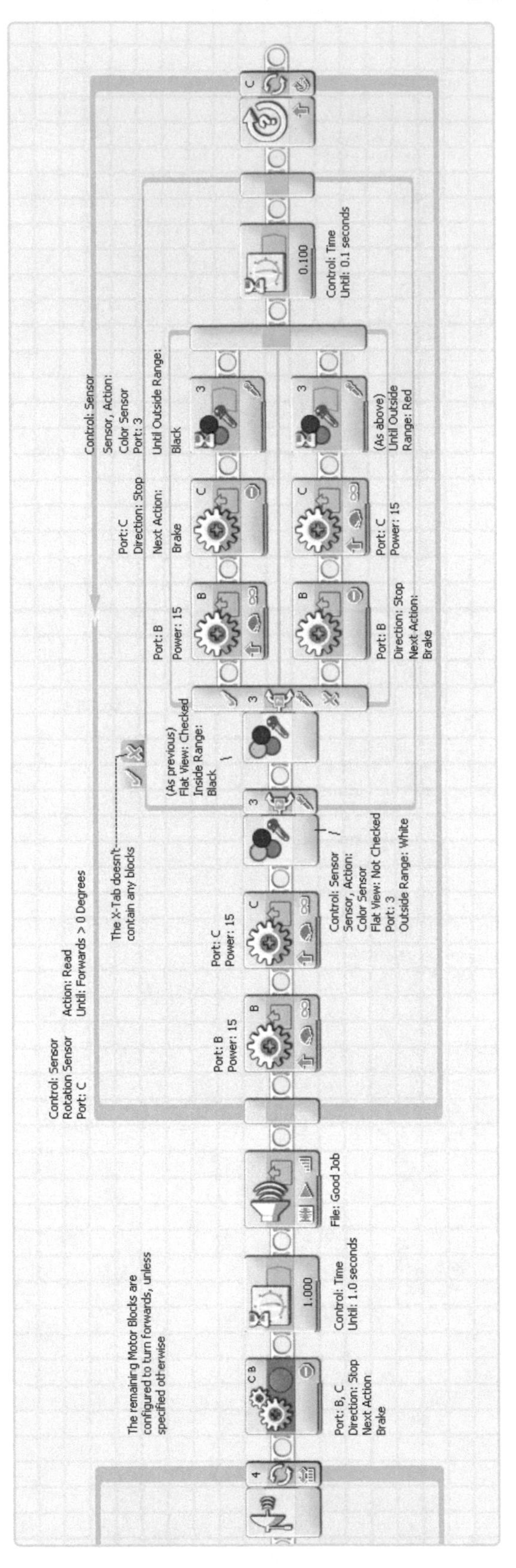

Figure 15-10: The configuration of the blocks that are run when the robot spots the ceiling. The robot pauses for a second, plays a sound, and returns to the ground.

further exploration

You've reached the end of this book. Congratulations! I hope you enjoyed learning the ins and outs of the LEGO MINDSTORMS NXT 2.0 robotics kit, as well as building and programming the robots in this book. But the fun doesn't end here. You're now ready to start creating robots on your own and share your ideas with the world. Whether your robots drive, shoot, walk, grab, sort, or climb, the possibilities are endless with LEGO MINDSTORMS NXT!

But before you close this book, try solving the following two discoveries that will let you do more with the Compact Chimney Climber and the knowledge you gained in this chapter.

DISCOVERY #72: ALTIMETER!

Difficulty: Hard
Use the NXT screen to display the altitude of the climber as it moves. Each rotation of the motor wheels should make the robot increase its altitude by a certain number of centimeters. But, how many exactly, and how do you calculate the altitude (the traveled distance)?

HINT Use a separate Sequence Beam to calculate and display the altitude every second.

BUILDING DISCOVERY #15: AERIAL TRAM!

The CCC you built in this chapter was an uncommon vehicle, and in this discovery you are challenged to build a robot that doesn't touch the ground. Wire a rope strong enough to carry the weight of your NXT with some motors across your room. Can you design a robot that travels through the air, hanging below this rope?

HINT Start by making a frame to carry the NXT, and allow it to move freely along the rope without falling. For example, you could create a frame with holes in the top for the rope to move through and then connect a motor to this frame that makes a wheel move over the rope. As you turn this motor, the whole robot should slide along the rope through the air.

troubleshooting and solving connection problems

When building and programming the robots in this book, it's usually not very difficult to make your robot work by following the instructions. However, sometimes you may run into problems transferring your programs to the NXT. When you do, this appendix will help you find solutions to your problems so that you'll be able to make your robot work successfully!

NOTE If this appendix doesn't answer you questions, post them to the book's companion forums (*http://discovery.laurensvalk.com/*), where you'll be able to ask about your specific problem. By the same token, if you can download a program to the NXT but your robot doesn't work the way you want it to, there may be a mistake in your program. Visit the book's companion forums for help in these cases, too.

using the NXT controller to download programs to the NXT

Once you've finished creating a program, you transfer it to the NXT by pressing **Download** or by pressing **Download and Run** on the NXT controller in your NXT-G software, as shown in Figure A-1. For the transfer to work, your robot has to be connected to the computer with the USB cable. (For transfers via Bluetooth, see "Connecting to the NXT with Bluetooth" on page 289.)

In addition to these two options, you can also click **Download and Run Selected** to transfer only selected blocks to your NXT. For example, if your program consists of two Sound blocks and two Move blocks and you select only the Move blocks, pressing **Download and Run Selected** will transfer only the selected Move blocks to the NXT.

To abort a running program, press the **Stop** button on the NXT controller.

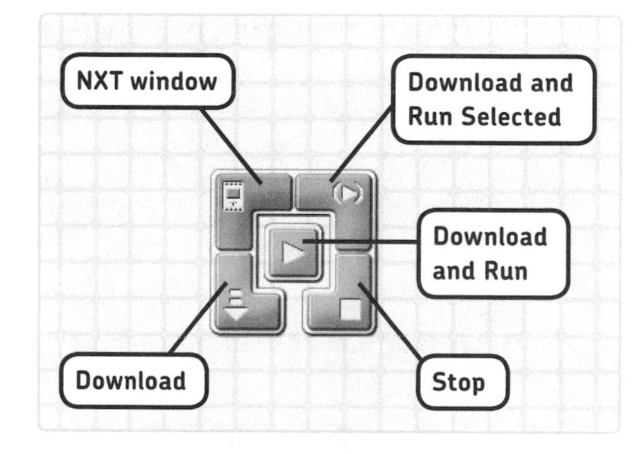

Figure A-1: Buttons on the NXT controller at the bottom right of the NXT-G programming software allow you to transfer programs to an NXT robot that is connected to your computer.

using the NXT window

When you use one of the NXT controller buttons as discussed in the previous section, your computer should attempt to connect to your robot before downloading the program automatically. If the computer fails to connect, you can try setting up the connection manually using the NXT window (Figure A-2). To open this window, click the **NXT Window** button on the NXT controller (Figure A-1).

To connect to an NXT with USB, select it from the list (the Status column should read *Available*), and click **Connect**. To refresh the list of available NXTs, click **Scan**. Once connected to an NXT, the NXT window should show the power left in the NXT's batteries, and the Status column should read *Connected*.

changing the name of an NXT

Once connected to the computer, you can personalize your NXT by changing its name (its name is "NXT" by default) using the NXT window (Figure A-2). To do this, enter a name in the Name box at the top right of the window, and click the arrow to the right of it to confirm. The name you enter should appear at the top of the NXT's screen. The next time you use the NXT window to connect to an NXT, the list of NXT devices should show your NXT's personalized name.

problems connecting to an NXT with USB

If you're unable to connect an NXT to a computer using the method discussed in "Using the NXT Window," follow these directions and then try connecting again:

1. Make sure the NXT is turned on.
2. Make sure the NXT is connected to the computer with the USB cable (be sure to check both ends of the cable). The connection is made properly if the NXT displays the letters *USB* on the top left of its screen.

If this doesn't help, follow these steps:

1. Turn the NXT off and back on.
2. Close and restart the NXT-G software.

If you still can't solve your problem, post your issues with a clear description of the symptoms to the book companion forums (*http://www.discovery.laurensvalk.com/*).

NOTE When you try to connect an NXT to your computer for the first time, you may need administrative rights on your computer in order to succeed. If you encounter this problem, ask your system administrator to log in, launch the NXT-G software, and make the connection to the robot. When you're done, you should be able to connect to the NXT even using your own account. This should be true for both USB and Bluetooth connections.

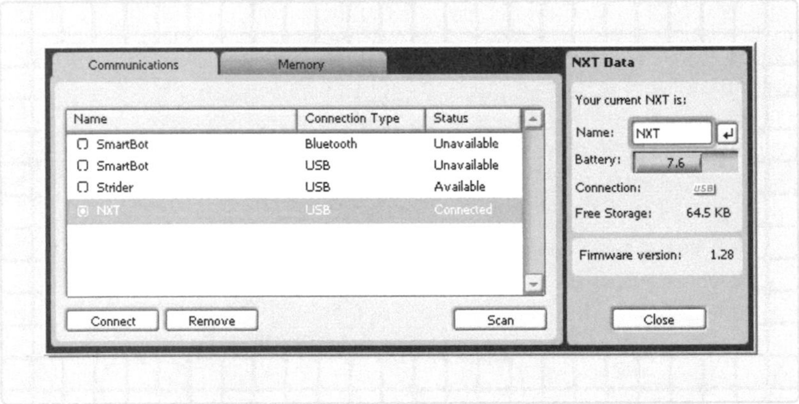

Figure A-2: The NXT window lists the NXT devices that the computer can connect to and those it has been connected to previously. The Connection Type column tells you whether the found NXT can be connected to USB or Bluetooth.

problems downloading programs to the NXT

If you can't transfer a program to the NXT, there may be several causes. Before you proceed to possible solutions, make sure that your robot is connected to the computer by making sure that the Status column in the NXT window says *Connected* (Figure A-2). If not, your problem is most likely a connection issue. To resolve it, read "Problems Connecting to an NXT with USB" on page 286.

If your robot is physically connected to your computer and you still receive an error message when you try to send a program to the robot (such as the one shown in Figure A-3), carefully read the error message and follow the appropriate instructions in the following sections.

The NXT device is no longer connected.

Although your NXT was connected to the computer, it automatically disconnected. This error message appears when you try to download a program after the USB cable is removed or after the NXT switched itself off after not being used for a long period.

Solution: Turn the NXT on, connect the USB cable, and retry downloading the program.

The NXT device is out of memory.

The programs you download to your NXT are stored in its memory. When you create a program that plays sound files, these files are transferred to the NXT as well. Your NXT has only limited memory, so after making several programs, especially ones using sound files, the NXT will eventually run out of memory.

Solution: Remove files that you no longer need from the NXT's memory, as shown in Figure A-4. When you're done, you should be able to download new programs to the robot.

For more on managing files in memory, see the NXT-G's help files, and click **Files and Memory on the NXT** in the menu on the left of the help pages.

data wire error messages

When you create a program with incorrectly wired data wires, your program won't transfer to the NXT. Depending on the type of mistake in your program, you'll see one of the following error messages.

- There is a data wire that is connected to a plug of the wrong type (see "The Broken Data Wire" on page 166).
- There is more than one data wire connected to the same input plug (see "Connecting Multiple Wires to One Data Plug" on page 167).

Figure A-3: A sample error message that may appear when downloading a program to a robot

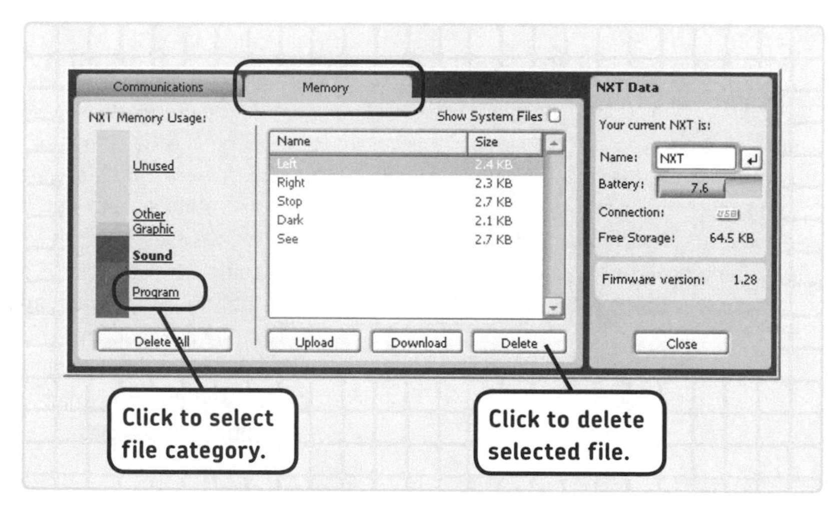

Figure A-4: To remove files from the NXT's memory, go to the NXT window and open the Memory tab. This will show a list of files that are currently on the brick, sorted by category. (Here you see only the sound files.) To view files in a particular category, click the category on the left (such as Program Files). Once you've located the file you want to remove, select it, and click the ***Delete*** *button.*

* There is a data wire that does not have a data source (see "Using Settings with Both Input and Output Plugs" on page 167).

Solution: Reread the appropriate paragraph in this book to fix incorrectly connected wires. If fixing just one wire doesn't help, delete some more or all data wires in your program, and reconnect them appropriately.

Cannot download needed file for the program.

This error may occur in programs that use sound or image files that you made yourself. If the software is unable to locate them (most likely because you removed them), you'll see this error.

Solution: Re-create the files using the sound and image editor in the software.

The program is broken. It may be missing required files.

You see this error when you try to send a program to the NXT that uses a My Block that the programming software can't find because you deleted or moved it.

Solution: Re-create the My Block, or move it back to *My Documents\LEGO Creations\MINDSTORMS Projects\Profiles\Default\Blocks\My Blocks*. Next, remove the My Block from the program that uses it, and choose a new My Block from the Custom Palette. If this doesn't help, create a new My Block with a different name.

The file is currently in use on the NXT device.

This error pops up when you try to download a program to your robot with the **Download** button on the NXT while that same program is currently running.

Solution: Use the **Download and Run** button instead, or abort the running program by clicking the **Stop** button before you download it to the NXT.

using Bluetooth to download programs to the NXT

Instead of using the USB cable to transfer programs to your robots, you can plug a Bluetooth dongle (shown in Figure A-5) into a USB port of your computer, which sends the programs to the NXT wirelessly. Once you've set up the connection between your computer and the NXT, you transfer programs to the NXT as you would with the USB cable.

Figure A-5: Just one example of a compatible Bluetooth dongle

Bluetooth transfers makes programming a lot easier, because when using Bluetooth, you don't have to repeatedly connect and disconnect the USB cable each time you download a program.

finding a Bluetooth dongle

There are many compatible Bluetooth dongles, some of which cost less than $10. Generally, it's not the dongle hardware but its drivers in combination with your computer's operating system that cause the dongle to work or not. In many cases, you'll be able to simply plug the dongle into your computer, wait for the drivers to install automatically, launch the NXT-G software, and follow the connection procedure in the next section. The drivers you'll need will depend on your operating system and your Bluetooth dongle. Be sure to visit the companion website for links to recommended Bluetooth dongles.

connecting to the NXT with Bluetooth

Follow the next steps to set up your first Bluetooth connection between the computer and the NXT:

1. Plug a compatible Bluetooth dongle into a free USB port on your computer. Depending on your operating system, some drivers are automatically located and installed. Usually, it is *not* necessary to install the additional drivers that come with your dongle.
2. Activate Bluetooth on the NXT by turning the NXT on and navigating to **Bluetooth ▸ On/Off ▸ On**.
3. In NXT-G, go to the NXT window, and click **Scan** to search for your NXT via Bluetooth. This search process may take about 30 seconds. When ready, the list of NXT devices is updated with the NXTs that are available for a Bluetooth connection. If your NXT isn't listed, click **Scan** again.
4. Select the NXT you want to connect to (the Connection Type column should read *Bluetooth*), and click **Connect**.
5. When you connect an NXT for the first time, the software may ask you to enter a password to protect your connection, but I recommend that you stick with the default password (1234) because this is easier, especially when errors occur, forcing you to run these steps more than once. Click **OK** to confirm this password.
6. If you had to enter a password in step 5, the NXT should now make a sound and prompt you to enter the same password. Since you're using the default password, just select the *V* sign by pressing the orange **Enter** button on the NXT. Your computer should now automatically install some extra drivers. If no error messages appear, your NXT should now be connected to the computer via Bluetooth, and you should be ready to start downloading programs to it.

You can tell whether an NXT has made a working Bluetooth connection by looking at the top left of the NXT screen, which shows <> when connected and < when not connected to a computer. If you have just one NXT, you can skip all of these steps once you've gone through them once. If you try to transfer a program to your NXT the next time you launch the software, the software should automatically try connecting to the NXT it was last connected to, which is, in this case, your NXT via Bluetooth.

problems connecting to an NXT with Bluetooth

Once you've set up the Bluetooth connection, you usually won't have a problem downloading programs to the NXT. If downloading fails but the Bluetooth connection appears to be working correctly, follow the directions in "Problems Downloading Programs to the NXT" on page 287.

However, in some cases, setting up the Bluetooth connection itself is problematic. There is no easy way to solve Bluetooth problems because they vary depending on, for example, which Bluetooth dongle and which operating system you use. Still, the following solutions may help in some cases.

* Make sure you have administration rights on your computer.
* Restart the NXT-G software, and restart your NXT; then retry connecting.
* In the NXT window, remove all the NXT devices that the computer previously connected to by selecting them and clicking **Remove**. Then click **Scan**, and when it's finished searching for NXTs, retry the connection.

If you continue to have problems using Bluetooth, post details of your problems to the book's companion forums.

conclusion

I hope that this short appendix has helped you find a solution to your problem. Of course, only a few problems and solutions are listed here, and you may have many other questions relating to one of the building or programming instructions in this book. Feel free to post your question to the companion website (*http://www.discovery.laurensvalk.com/*).

index

D

E

F

G

H

I

L

R

S

T

U

V

W

X

Y

More no-nonsense books from **no starch press**

The LEGO® Technic Idea Books

Simple Machines | Wheeled Wonders | Fantastic Contraptions

by YOSHIHITO ISOGAWA

The LEGO Technic Idea Books offer hundreds of working examples of simple yet fascinating Technic models that you can build based on their pictures alone. Colors distinguish each part, showing you how the models are assembled. Each photo illustrates a different principle, concept, or mechanism that will inspire your own original creations. The Technic models in *Simple Machines* demonstrate basic configurations of gears, shafts, pulleys, turntables, connectors, and the like, while the models in *Wheeled Wonders* spin or move things, drag race, haul heavy gear, bump off walls, wind up and go, and much more. *Fantastic Contraptions* includes working catapults, crawling spiders, and bipedal walkers, as well as gadgets powered by fans, propellers, springs, magnets, and vibration. These visual guides are the brainchild of master builder Yoshihito Isogawa of Tokyo, Japan.

OCTOBER 2010, 168 PP., 144 PP., AND 176 PP., *full color*, $19.95 EACH
ISBNS 978-1-59327-277-7, 978-1-59327-278-4, 978-1-59327-279-1

The Unofficial LEGO® MINDSTORMS® NXT 2.0 Inventor's Guide

by DAVID J. PERDUE *with* LAURENS VALK

The Unofficial LEGO MINDSTORMS NXT 2.0 Inventor's Guide helps you to harness the capabilities of the NXT 2.0 set and effectively plan, build, and program your own exciting NXT robots. After examining the pieces in the NXT 2.0 set and the roles they play in construction, you'll learn practical building techniques, like how to build sturdy structures and work with gears. Next, discover how to program with the latest version of the official NXT-G programming language, and how to troubleshoot if something doesn't work right the first time. Finally, follow step-by-step instructions for building, programming, and testing six all-new robots, including Sentry-Bot, a robot guard that shoots balls at intruders; the Jeep, a four-wheeled steering vehicle that avoids obstacles and follows lines; and the Lizard, a large walking robot with a swinging tail that uses the color sensor to detect different colored balls and respond. Each robot can be built using just one NXT 2.0 set, so there's no hunting for obscure parts.

DECEMBER 2010, 336 PP., $29.95
ISBN 978-1-59327-215-9

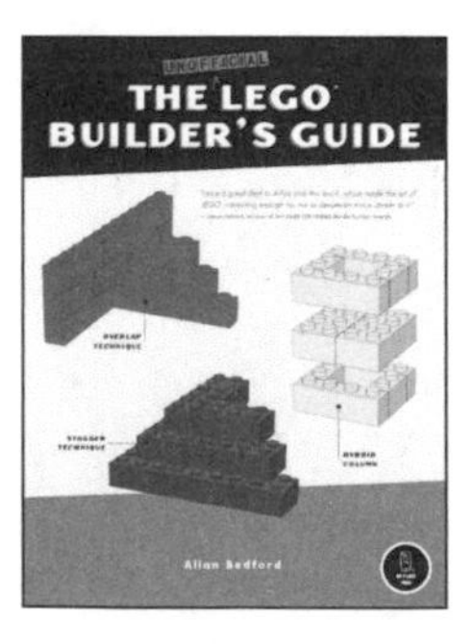

The Unofficial LEGO® Builder's Guide

by ALLAN BEDFORD

The Unofficial LEGO Builder's Guide combines techniques, principles, and reference information for building with LEGO bricks that go far beyond LEGO's official product instructions. You discover how to build everything from sturdy walls to a basic sphere, as well as projects including a mini space shuttle and a train station. The book also delves into advanced concepts such as scale and design. Includes essential terminology and the Brickopedia, a comprehensive guide to the different types of LEGO pieces.

SEPTEMBER 2005, 344 PP., $24.95
ISBN 978-1-59327-054-4

Forbidden LEGO®

Build the Models Your Parents Warned You Against!

by ULRIK PILEGAARD *and* MIKE DOOLEY

Written by a former master LEGO designer and a former LEGO project manager, this full-color book showcases projects that break the LEGO Group's rules for building with LEGO bricks—rules against building projects that fire projectiles, require cutting or gluing bricks, or use nonstandard parts. Many of these are backroom projects that LEGO's master designers build under the LEGO radar, just to have fun. Learn how to build a catapult that shoots M&Ms, a gun that fires LEGO beams, a continuous-fire ping-pong ball launcher, and more! Tips and tricks will give you ideas for inventing your own creative model designs.

AUGUST 2007, 192 PP., *full color*, $24.95
ISBN 978-1-59327-137-4

Badass LEGO Guns

Build 5 Working Guns with LEGO Components!

by MARTIN HÜDEPOHL

Badass LEGO Guns is packed with building instructions for six impressive-looking "weapons" built entirely from LEGO Technic parts. In this heavily illustrated two-color book, you learn how to use LEGO Technic pieces to build working model guns like the Warbeast, a sophisticated, fully automatic submachine gun; the Parabella, a semi-automatic pistol; and the Thriller, a slide-action crossbow pistol with smooth cocking and chambering mechanisms. With the help of a bit of sanding, some oil, and Krazy Glue, each gun actually shoots LEGO bricks or rubber bands at high speed, with surprising accuracy. The building instructions for each model are easy to follow and include detailed parts lists.

DECEMBER 2010, 240 PP., *two color*, $29.95
ISBN 978-1-59327-284-5

Visit *http://nostarch.com/catalog/lego* for a full list of titles.

PHONE:
800.420.7240 OR
415.863.9900
MONDAY THROUGH FRIDAY,
9 AM TO 5 PM (PST)

FAX:
415.863.9950
24 HOURS A DAY,
7 DAYS A WEEK

EMAIL:
SALES@NOSTARCH.COM

WEB:
WWW.NOSTARCH.COM

companion website

Visit *http://www.discovery.laurensvalk.com/* for updates, errata, downloadable programs, hints for solving the discoveries, contests, and more. Be sure to visit the website's forum if you have questions about the book and to share your discoveries with the world!

The LEGO MINDSTORMS NXT 2.0 Discovery Book is set in Chevin. The book was printed and bound by Transcontinental, Inc. at Transcontinental Gagné in Louiseville, Quebec, Canada. The paper is Domtar Husky 60# Smooth, which is certified by the Forest Stewardship Council (FSC). The book has an Otabind binding, which allows it to lie flat when open.